The Chemistry of Soil Constituents

The Chemistry of Soil Constituents

Edited by

D. J. Greenland
Department of Soil Science
University of Reading

and

M. H. B. Hayes
Department of Chemistry
University of Birmingham

nipa

NEW INDIA PUBLISHING AGENCY

New Delhi – 110 034

NEW INDIA PUBLISHING AGENCY
101, Vikas Surya Plaza, CU Block, LSC Market
Pitam Pura, New Delhi 110 034, India
Phone: + 91 (11)27 34 17 17 Fax: + 91(11) 27 34 16 16
Email: info@nipabooks.com
Web: www.nipabooks.com

Feedback at feedbacks@nipabooks.com

First Indian Reprint, 2016
Reprinted with arrangements with the authors.

ISBN: 978-93-85516-21-4

Contributors

P. W. ARNOLD	*Department of Soil Science, University of Newcastle upon Tyne, School of Agriculture, Newcastle upon Tyne,* NE1 7RU
G. BROWN	*Department of Soils and Plant Nutrition, Rothamsted Experimental Station, Harpenden, Herts.*
V. C. FARMER	*Department of Spectrochemistry, The Macaulay Institute for Soil Research, Craigiebuckler, Aberdeen,* AB9 2QJ
D. J. GREENLAND	*Department of Soil Science, University of Reading, London Road, Reading,* RG1 5AQ
M. H. B. HAYES	*Department of Chemistry, University of Birmingham, P.O. Box 363, Birmingham,* B15 2TT
C. J. B. MOTT	*Department of Soil Science, University of Reading, London Road, Reading,* RG1 5AQ
A. C. D. NEWMAN	*Department of Soils and Plant Nutrition, Rothamsted Experimental Station, Harpenden, Herts.*
J. H. RAYNER	*Department of Soils and Plant Nutrition, Rothamsted Experimental Station, Harpenden, Herts.*
R. S. SWIFT	*Soil Science Department, Edinburgh School of Agriculture, West Mains Road, Edinburgh,* EH9 3JG
A. H. WEIR	*Department of Soils and Plant Nutrition, Rothamsted Experimental Station, Harpenden, Herts.*

Contents

Foreword

Prof. E. W. Russell, CMG, MA, PhD, DSc, FIBiol, FInstP, FIAgr.E.

The application of research technology to crop production—whether of food crops, industrial crops or forest products—is becoming of increasing importance in a world with a rapidly rising expectation of higher standards of nutrition and living. The fulfilment of these expectations can only be brought about by raising the levels of crop production and yield, and since effectively all crop production in the world takes place in the soil, there has been a very great increase in the resources devoted to soil research and the interrelation between soil properties and the demands the crop makes on the soil.

The earliest application of science to the soil and to soil fertility was chemistry, and understanding the chemical processes going on in the soil, and their relevance to crop production is still of major concern to the soil scientist. In the early decades of this century, a great deal of soil chemistry had a very inadequate scientific base, and one very important task of the soil chemist has been to construct a much more secure scientific base on which the applied soil chemist could develop improved field practices. Also it is only if the scientific basis of new techniques developed for a particular soil and a particular environment is fully understood, that results and techniques developed in one region for a particular group of soils can be easily transferred to other regions or other soils.

During the last few decades great advances have been made in the rigorous application of well established chemical and physico-chemical theories to the elucidation of the more important chemical processes operative in soils; and this has resulted in a much clearer understanding of the complexities of many of these processes and of the numerous interactions between processes taking place as a result of weathering, plant growth, and biological activity in the soil. But this has had the inevitable consequence that as newer techniques are developed to study particular processes in more depth, the whole subject becomes split up into special sections, each developing its own language and techniques, with the inevitable consequence that it is becoming difficult for any one scientist to be fully familiar with the techniques and theories being developed at important growing points of the subject. Further, the subjects are developing so rapidly, thanks to the considerable resources being allocated to this work, that individual soil scientists, and particularly scientists just entering this field, can only hope to keep abreast of the developments if he has available up-to-date critical reviews of the present state of knowledge in these various sections. Further, the more the soil scientist is concerned with problems of plant growth, whether as an ecologist, agriculturalist or forester, the more he recognizes that many of the processes which control plant growth straddle the division into which soil chemistry is being broken down in the basic soil science research laboratories.

It is thus an essential duty of at least a few soil scientists and scientists in associated disciplines, to ensure that, from time to time, there are authoritative and critical accounts of the present state of knowledge, covering a wide field of expertise. But such accounts are dependent on the willing co-operation of specialists who are not only fully familiar with the research problems and limitations in their own special field but are also aware of their relevance to the complex topic of soil chemistry.

The chapters in this volume could only be written because the authors were willing to give up time they would otherwise spend on their own important research projects so that they could produce a book dealing with the present state of knowledge in the particular fields they have covered.

Preface

Since the beginning of this century production of chemical fertilizers has increased many fold, and the production of pesticides for addition to the soil has become a new major industry. These are not the only chemicals added to the soil, because by accident or design many other chemicals which enter the environment end up in the soil. The world population, increasing at a conservative estimate of 2 per cent per annum, doubles in less than thirty-five years. The soil has to produce the bulk of the food and fibre to sustain this growing population.

If the soil is to continue to be both the source of our foodstuffs and the sink for many of our wastes, it is obviously desirable that we understand its complex chemistry.

Much scientific endeavour has been directed to this end in the past few decades, resulting in a large volume of published literature, including several books concerned with the chemistry of the mineral or organic components of soils, or with certain chemical processes in soils. No recent book has however attempted to produce a coherent account of soil chemistry as a whole, suitable for both students of environmental chemistry and soil science, and for the very wide range of research and other scientists who are interested in the chemistry of soils. The original intention of the editors of the present book was to attempt to fill this gap. Because of the very diverse topics involved it was felt that it would have to be written by several authors if it was to be appropriately authoritative.

The chemistry of soils falls naturally into two parts, the chemistry of the various soil components, and the chemistry of soil processes. It was at first intended that these, the static and dynamic aspects of soils, should form two parts of a single volume. However to deal with the topics involved at the intended level would have produced an unwieldy volume, and so two companion volumes are being issued, 'The Chemistry of Soil Constituents' and 'The Chemistry of Soil Processes'.

In this first volume a short historical outline of the development of soil science is given, touching briefly on soil formation, soil physics, and soil biology, as it can be misleading to regard soils simply as chemical entities. It is also important to have some appreciation of the several processes which have produced the particular soil found at any one place. The major soil types of the world differ according to their origin. Several systems exist for classifying them, and different names are in use for the same major soil types. The most common names of these are therefore introduced. The following two chapters deal with the inorganic and organic components of soils respectively. The chemical structures of the major inorganic components are now reasonably well known, but this is not true of the organic (or humic) materials in soils. The relevant chapter presents an account of what has been experimentally established regarding the constitution of the peculiarly intractable complex of organic compounds found in soils.

Chemical processes in soils are largely determined by reactions at the surfaces of the soil colloids. The final three chapters are therefore concerned with the nature and extent of the surfaces of soil colloids, their electrical characteristics, and the ways in which ions and water are held and arranged at the surfaces.

It is hoped that a sufficient understanding will have been provided of the chemistry of the major constituents of soils to enable the processes described in the companion volume to be fully comprehended.

The editors are grateful to those who have contributed for their collaboration and patience with our requests, to their publishers for their encouragement and tolerance of delays, and to many authors and publishers who have given permission for material to be reproduced. Specific acknowledgements of the latter are made in the text.

We have endeavoured to adhere to SI units throughout. The Ångstrom has been reluctantly abandoned in favour of the nanometre, but we have found it necessary to retain 'equivalent' and 'milliequivalent', as there appears no suitable alternative for expressing the ions retained per unit weight or unit area of a soil colloid.

D. J. GREENLAND
M. H. B. HAYES

CHAPTER 1

Soils and soil chemistry

D. J. Greenland
Department of Soil Science, University of Reading
M. H. B. Hayes
Department of Chemistry, University of Birmingham

1.1 THE IMPORTANCE OF SOILS

Soils are the medium in which crops grow, and so they are essential to the provision of the food and fibre to sustain mankind. But besides the immense importance of this role they are an essential component to life as we know it because of their action as a buffer to control the flow of water between sky, land and sea. They are also a sink, and usually a safe sink, for many of the chemicals that might otherwise make life difficult or unpleasant.

As with other natural resources, it is important that soils are used conservatively, so that the advantages we enjoy from them now can continue to be enjoyed in the future. To do this we need to understand them, to understand the flow of nutrients from soil to plants, and back to the soil or the sea in food chains and water flow. We need to understand the processes by which rocks alter to form soils, and how fast or how slowly the process occurs in different conditions. We need to understand the decay of plant materials in the soil, and how the products of decay help to make the soil a better place for plants to grow.

The chemistry of the constituents of soils, and of the processes by which changes occur, is not simple. Nevertheless we have learnt much about it, and can apply that knowledge to ensure that soils continue their important functions, and are not exploited so that they become infertile, nor used in such a way that the fertile topsoil is lost by erosion by wind or water.

1.2 THE SCIENTIFIC STUDY OF SOILS

1.2.1 The Origins of Soil Chemistry and Pedology

Historically there have been two different approaches to the scientific study of the soil. One has concentrated on the study of the soil as the medium for plant growth, and the other on the study of the soil as a natural body. In England, the application of the subject to crop production has long been dominant, because of the success of Lawes at Rothamsted Manor in Hertfordshire, in first recognizing the fact that shortage of phosphate was limiting plant growth, then demonstrating that the shortage was rectified by using bones as manure, and showing that if the bones were first made more soluble by treating them with sulphuric acid the response of plants to addition of the phosphate could be obtained with much smaller additions of 'manure'. Lawes' success was based on his initial chemical work confirming the importance of phosphate as a plant nutrient. In association with Gilbert he worked at Rothamsted for over fifty years, establishing by careful chemical analysis of soils, plants and manures the basis of proper fertilizer practice, and so laid the foundations of the fertilizer industry of today.

Boussingault in France and Liebig in Germany, amongst others, were conducting similar chemical studies in relation to crop production at the start of the nineteenth century, but it was Lawes and Gilbert who made the widest impact. Not only did their field and laboratory experiments demonstrate plant needs for different nutrient elements, but their fertilizer plant was a considerable commercial success. Thus 1843, the year Lawes and Gilbert came together, and Lawes first sold his 'super phosphate of bone' may be considered the first important milestone marking the beginnings of soil chemistry (Russell, 1942).

In the course of the nineteenth century the study of the soil as a component of the natural environment was also developing. Here the most important milestone was the publication by Dokuchaiev in Russia of the first report which clearly related soil features to the different climates and vegetation as well as parent rocks from which the soils were formed (Dokuchaiev, 1883). This was followed shortly afterwards by his presentation at the International Exhibition in Paris of 1889 of the first map of the 'zonal soils' of the northern hemisphere.

Dokuchaiev described soils in terms of their 'profile morphology' (Figure 1.1) which he observed to differ between the major climatic zones of European Russia. The differences in readily observable profile characteristics are reflected in chemical, mineralogical, physical and biological differences in the soils.

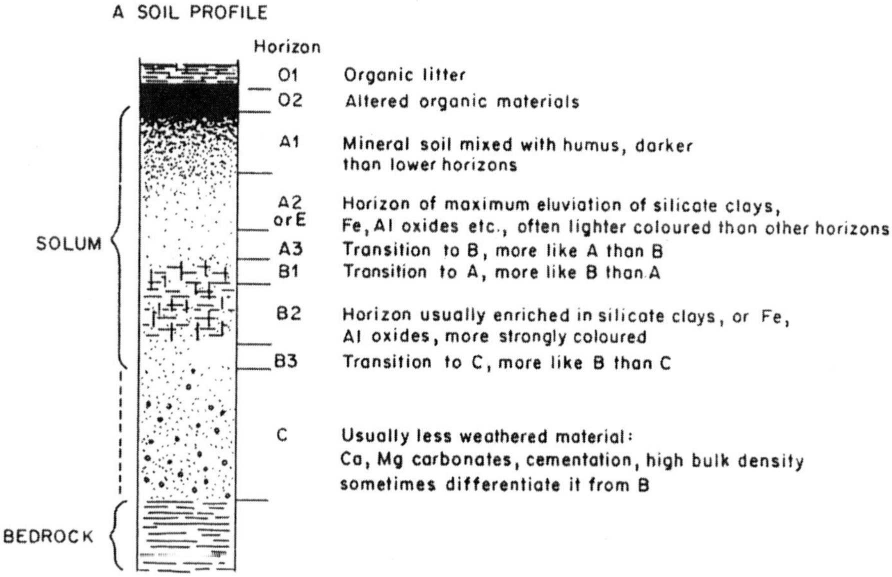

Figure 1.1 A soil profile, showing the O, A, B, C horizon nomenclature commonly used. Further subdivisions and subscripts are often employed, e.g. Ap, to refer to the plough layer

Following Dokuchaiev's work many studies have been made establishing the 'Great Soil Groups' of the world, the factors which determine the development of different soils (Jenny, 1941) and of the principles on which their classification into coherent groupings can be made (Buol, Hole and McCracken, 1973). In recent years the first attempt at a fully comprehensive system of soil classification has been published (Soil Survey Staff, 1975) and the first authoritative 'Soil Map of the World' has been prepared by the Food and Agricultural Organization of the United Nations (FAO, 1974).

Other branches of soil science—soil physics, soil biology, soil mineralogy—have had a lesser impact on the development of the subject than soil chemistry and pedology, the term usually used to refer to the study of soil characteristics in relation to soil genesis and classification. Nevertheless they have also been the subjects of scientific study for many years.

1.2.2 Soil Physics

The physical properties of soils are of course important to agriculture, because the pore space of soils governs gaseous transfer with the atmosphere, as well as water movement and storage, root growth, and the ease or difficulty of tillage. Early studies of soil physics (Warington, 1900) were concerned with all these aspects, and such studies still continue (Baver, Gardner and Gardner, 1972). However the largest volume of work has been concentrated on the development of the physical theory of water movement in soils (Childs, 1969; Kirkham and Powers, 1972). The problems of describing mathematically such movement are considerable. Soils are heterogeneous, and the distribution of pore sizes as well as the continuity of the pores often change as water content changes. A comprehensive mathematical treatment has not been developed except for relatively simple systems. However a considerable body of empirical knowledge exists on which practical methods for controlling water movement and storage in soils is based.

Good drainage is essential to crop production, and much of the world's food production depends on irrigation. Paradoxically, areas of water deficit which require irrigation are also amongst those which most frequently require drainage. The reason for this is that irrigation water almost invariably contains some dissolved salts. If the soil is not naturally well drained, and sufficient water applied to wash these salts out of the soil, they will accumulate and the soil will become too saline for plants to grow. Many of the early civilizations of the Middle East, based on irrigation agriculture, failed because of their inability to deal with the problem. Modern studies of soil physics have therefore tended to develop in areas of irrigation agriculture (Arnon, 1972; Hagan, Haise and Edminster, 1967). The success of many modern irrigation developments is closely associated with the appreciation of the salinity problem, and of methods to control it.

Studies of the physics of soil tillage have had a somewhat chequered history. In the eighteenth century Jethro Tull, a pioneer in designing tillage machinery, advocated vigorous cultivation of the soil to pulverize it into fragments sufficiently small to be absorbed by plant roots. We now know that this was a totally false concept, plant nutrients being absorbed as soluble ions.

Another false idea about tillage developed at the end of the nineteenth century. This was that in dry areas tillage was necessary to break the capillaries which conduct water to the soil surface and allow it to evaporate. 'Bare fallowing' was advocated, in which the soil was kept clean weeded for a year. Vigorous tillage was practised to destroy weeds and break the capillaries, and so preserve water falling on the fallow for crops to be planted in the following year. These crops could then use the water stored in the soil as well as that falling on the crop. This system of 'bare fallowing' was used in the western United States and in Australia. The soils were worked to a fine powder by cultivations which destroyed weeds, but left the bare soils sadly exposed to wind and rain. When it rained the soils were washed away, and when it was dry the fine soil was blown into the air. By the 1920s the problem was so acute that the area around Oklahoma became famous as the 'Dust Bowl'.

The need for tillage to preserve water was again based on a fallacy. Although water does move upward in 'capillary size' pores (2–50 μm diameter approximately) and the proportion of those present can be reduced by tillage, the upward movement is confined to ten to twenty centimetres from a free water surface, or water table, in the soil. In semi-arid areas during dry periods water tables are normally much deeper than this. Thus the tillage operation could have had little effect in conserving water, and it was highly disadvantageous in exposing soils to erosion. Subsequently practices have been greatly modified to ensure erosion is controlled (Bennett, 1939).

The real need for tillage has always been to reduce competition by weeds, and to establish a suitable medium for seed germination. With the advent of cheap and effective herbicides in the 1950s tillage became no longer the only method available to control weeds. Many farms around the world have adopted no-tillage or minimum-tillage systems of crop production, based on weed control using herbicides. The rapid increase in the use of herbicides and other pesticides has led to much detailed study of their sorption and movement in the soil (Goring and Hamaker, 1972).

1.2.3 Soil Biology

Many chemical changes occurring in the soil are brought about by the action of bacteria. Early attention was focused on the transformations of nitrogen. The nitrogen in organic compounds falling on the soil is usually rapidly converted into ammonium ions, which are in turn converted to nitrite and nitrate. The conversion of the phytotoxic nitrite ion normally occurs much more rapidly than that of ammonium, so that nitrite does not accumulate in the soil. This process of nitrification was first described in sewage purification by Schloesing and Muntz in 1877. The recognition that it also occurs readily in soils was largely due to Warington, again at Rothamsted. His work settled an old controversy about whether ammonium or nitrate was taken up by plants. We now know of course that both are utilized, but not organic nitrogen, or only to a very small extent. More controversy centred around the question of whether plants or other organisms could utilize nitrogen directly from the air. Liebig had argued vigorously that plants acquired the nitrogen they need by absorption of ammonia from the air. Lawes and Gilbert had argued on the basis of their field experiments that plants have to derive their nitrogen from the soil, or from 'manures' applied to the soil. Again Lawes and Gilbert were right, but Liebig might plead that he had correctly observed that the atmosphere could contribute to the nitrogen in some plants.

Several workers in the latter half of the nineteenth century were involved in the discovery that bacteria living in nodules on the roots of legume plants could absorb nitrogen from the air and convert it to a form which plants can utilize. Major credit for this discovery is usually given to Hellriegel and Wilfarth (1888) in Germany. Only in recent years have careful biochemical studies enabled the processes involved to be described in detail (Bergersen, 1971; Dilworth, 1974). There are of course several organisms living freely in the soil capable of 'fixing' atmospheric nitrogen, in addition to the symbiotic *Rhizobia* which infect leguminous plants. The study of the nitrogen fixing organisms and their ecology is an important part of soil biology. Selection of appropriate strains of *Rhizobia* for inoculation of legume seeds has established a minor industry.

Many other important biological processes occur in soils. Soil animals are responsible for much of the comminution and initial alteration of plant materials falling on the soil (Burges and Raw, 1967). These are further altered by bacterial and fungal attack, to produce the 'humus' of the soil (Chapter 3).

The persistence of pesticides in the soil can be a potential hazard, and so factors controlling the length of time they exist in the soil have received much attention. An understanding of the processes involved requires an appreciation of the physics and chemistry of soils, but because degradation is largely due to the action of bacteria and fungi, particularly of their biology.

Clearly it is impossible to develop a complete understanding of soil behaviour if biological activity is ignored. Thus in subsequent chapters detailed consideration is given to the constitution of the organic components of soils (Chapter 3) and to many biochemical processes.

1.2.4 Soil Mineralogy

Many soils have features which relate them to the parent rock from which they were formed. Sometimes where the rock is a sediment, itself formed from the weathering and deposition of material arising from alteration of primary or other igneous rocks, the soil may resemble the 'grandparent' from which the sedimentary rock was formed rather than the immediate parent. The resemblance is most often expressed by the persistence of certain minerals from the parent rock in the soil. Naturally some are more resistant than others, notably quartz, muscovite mica, and some feldspars, the chemically more acid minerals, and so these tend to become predominant in soils. The more basic minerals, such as pyroxenes and amphiboles, are more readily altered by chemical weathering and so are less commonly found (Figure 1.2).

The great successes of the mineralogists of the nineteenth century were achieved by using the optical microscope to describe the geometry of individual crystals. The identification of crystals of specific minerals was supported by the finding that each consisted of a limited number of chemical elements present in simple ratios of atomic numbers. Their success paralleled that of the organic chemists in establishing the purity of organic compounds by measurements of boiling and freezing points, and the establishment of the simple ratios of atoms in the molecules of such compounds. When mineralogists attempted to study soils, real problems arose. The sand-sized grains (larger than 0.02 mm diameter) could be relatively easily recognized as minerals or rock fragments, the silt-sized material (0.002 to 0.02 mm) was difficult to study by optical microscopy, and the clay fraction impossible.

In addition to the problems they posed to the optical microscopist, the fine fractions of soils were also not amenable to the methods of the early organic chemists, and in terms of chemical composition they were complex and highly variable. Nevertheless their importance as the store of most plant nutrients was recognized, as well as their major influence on the physical properties of soils.

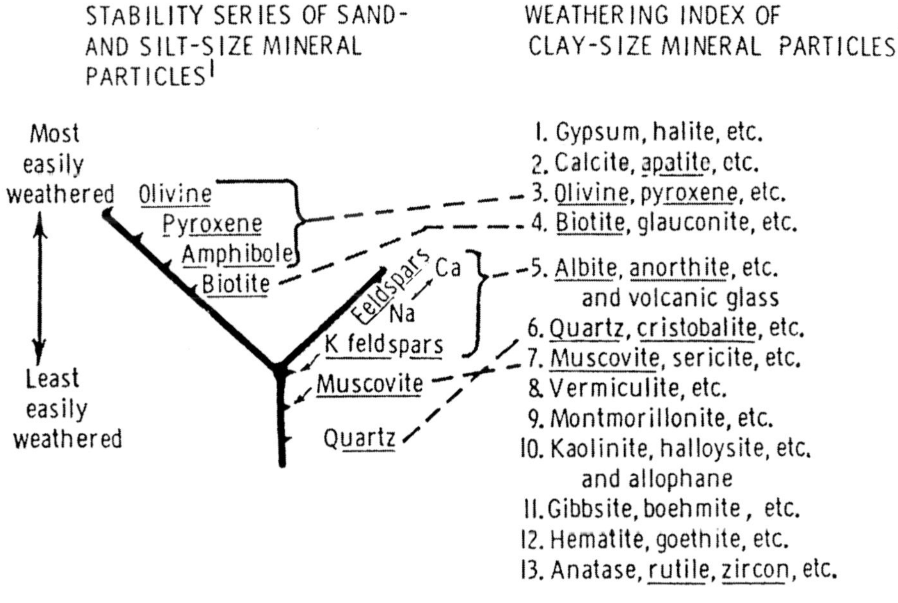

[1] Goldich 1938. Primary minerals are underlined in this figure.
[2] Jackson 1968.

Figure 1.2 The relative stability of coarse and fine grained minerals in soils. The first series consists of primary minerals arranged, from top to bottom, in the order of their crystallization from molten material, and also in the order of decreasing ease of weathering. The second series consists of a condensed version of the first in which the positions of muscovite and quartz have been interchanged because of the greater stability in soils of clay-sized mica. At the top and in most of the lower part of this series are secondary minerals. Reprinted by permission from *Soil Genesis and Classification* by S. W. Buol, F. D. Hole and R. J. McCracken, © 1972 by The Iowa State University Press, Ames, Iowa 50010

Speculation about the constitution of clays was imaginative: most views tended to the concept that they were a mixture of hydrous gels of silicon, aluminium and iron, with no crystalline structure. Some tended to think that kaolin, or china clay, could be pure (it was white) and have a definite atomic structure, although in soils it was 'contaminated' by other amorphous materials.

A revolution in concepts of the structure of clays came in 1923 when Rinne in Sweden and Hadding in Germany first succeeded in obtaining X-ray diffraction patterns from soil clays. Their work was rapidly followed by other studies of X-ray diffraction by clays, which established that the great majority had a well organized crystal structure. The details of the crystal structures were not elucidated until the 1930s. The milestone was the publication by Pauling (1930) of a series of X-ray diffraction studies of the silicate minerals, in which he showed that the micas, chlorites, pyrophyllite, talc, kaolinite, and gibbsite, all had a unit cell with dimensions in the plane of cleavage of approximately 0.50×0.88 nm.

This he suggested arose because all were based on layer structures in which sheets of octahedrally coordinated aluminium atoms were stacked sequentially, as in gibbsite ($Al(OH)_3$), or were fused with tetrahedrally coordinated silicon atoms to give layered structures (Figures 2.14 and 2.40).

It is now well established that soil clays consist of mixtures of aluminosilicate minerals. The chemical composition of most members can vary quite considerably because of the wide possibilities of isomorphous substitutions in their crystal structures. Amorphous materials also occur in soils, but these are the dominant constituents of the clay fraction only in soils which have been developed relatively recently (by which is meant younger than 5000 years) from volcanic ash. The rapid cooling of lava ejected from volcanoes leads to deposition of much of the ash as amorphous glasses. The decomposition of these glasses in the soil proceeds slowly, with the release of amorphous gels of silica and alumina, and the simultaneous or subsequent formation of the amorphous silica-alumina polymer called allophane. These amorphous materials alter in time to clay minerals. The behaviour of the great majority of soils is dominated by the layer lattice silicates which compose the clay minerals, and it is the surface chemistry of these minerals, modified to a greater or lesser extent by the amorphous hydrous oxide material and organic compounds which may be present, that determines the course of many of the chemical processes occurring in soils.

1.3 SOIL FORMATION AND SOIL CONSTITUENTS

To understand the complex chemical constitution of soils, and the behaviour of chemicals in the soil, it is essential to have some understanding of the processes involved in soil formation. Ultimately the origin of most soils can be traced back to the alteration of igneous rocks, although many soils originate more immediately from the already weathered products which have been deposited to become sedimentary rocks. Some were subjected to intense heat and pressure to form the gneisses and schists of the metamorphic rocks. Also precipitated limestones or chalks can form the parent materials of soils, although these are rather special cases, as much of the calcium carbonate is dissolved during soil formation, and the soil is largely formed in the 'impurities' present in the limestone or chalk.

As a first step towards understanding soil constitution we can consider in more detail the alteration of rock minerals by weathering processes. In the following section (1.4) the other processes in soil development will be discussed.

1.3.1 Chemical Constitution of Rocks

Igneous rocks were formed when the original magma, molten at the ambient very high temperatures, was cooled. Two types of magma are recognized on the basis of their contents of silica. One, described as acid and granitic, is composed of greater than 60 % SiO_2, has relatively high contents of sodium and potassium, but is relatively poor in iron, manganese, and calcium. The second is basic and basaltic, contains less than 50 % SiO_2, and is relatively rich in iron, magnesium, and calcium. A further broad classification is based on the silica contents of the rocks. Those which contain more than 66 % SiO_2 are acid, those with 66 to 52 % SiO_2 are intermediate, while rocks having 52–45 % are basic, and ultrabasic rocks have less than 45 % SiO_2.

It cannot be ascertained that all of the same species of igneous rocks were derived from the same type of parent magma, nor indeed that any particular rock was derived from any specific parent magma. Much depends on the cooling environments of the lavas. Studies of flows from modern volcanoes, for instance, have shown that vapours and gases, such as water, carbon dioxide, sulphur dioxide, and lesser amounts of hydrogen, carbon monoxide, sulphur, chlorine, nitrogen, argon and others are lost during the extrusion of the lavas. The escape of these gases leads to increased viscosities in the residual lavas and modifies their subsequent cooling and the crystallization of minerals (Bowen, 1928). Compositions and structures of the rocks formed will be influenced by the amounts of the various elements which escaped from the magma and by the rates of solidification.

Table 1.1, compiled from data by Turekian and Wedepohl (1961), gives average compositions of the most abundant elements in various types of igneous rocks. Data for oxygen, which is the most abundant in all of the rock species, are not included. Earlier data of Clarke and Washington (1924) for the average composition of the elements in igneous rocks give values of 46.41 % for oxygen, 27.58 % for Si, 8.08 % for Al, 5.08 % for Fe, 3.61 % for Ca, 2.83 % for Na, 2.58 % for K and 2.09 % for Mg. On average these eight elements compose 98.26 % of the igneous rocks, while the minor elements in Table 1.1 (Ti, P, Mn) generally are less than 1.2 % of the composition.

Table 1.1 Average elemental compositions of some igneous rocks (from Turekian and Wedepohl, 1961)

Element	Concentration in parts per thousand			
	Ultrabasic rocks	Basaltic rocks	High Ca granitic rocks	Low Ca granitic rocks
Si	205	230	314	347
Al	20	78	82	72
Fe	94	86	30	14
Ca	25	76	25	5
Mg	204	46	9	2
Na	4	20	28	26
K	0.04	8	25	42
Ti	0.30	14	3.4	1.2
Mn	1.6	1.5	0.54	0.4
P	0.22	1.1	0.92	0.6

SiO_2 is the most abundant component in all of the igneous rocks, but the proportions of oxides of Al, Fe, Mg, Ca, Na, and K will vary depending on the mineral composition. For instance Al is an important component of the feldspars, of some amphiboles, and of the mica group of minerals. Fe(II), Fe(III), and Mg are present in the ferromagnesian mineral group which includes olivine and biotite, and K and Na contribute significantly to the composition of the alkali feldspars, orthoclase, microcline, and plagioclase, found in the acid igneous rocks. Ca is a major component of anorthite, a plagioclase feldspar which is present in greatest abundance in basic and in some ultrabasic igneous rocks.

1.3.2 Chemical and Physical Weathering of Rocks

Two principal weathering processes are recognized: (1) disintegration, involving physical or mechanical breakdown of rocks and minerals, and (2) decomposition of the parent rocks and minerals in a manner which causes their chemical as well as physical characteristics to be altered and new minerals to be formed.

Disintegration, or physical weathering, results from processes such as differential thermal expansion of different minerals or along different crystallographic axes when rocks are heated, from the expansion of water in rock fissures where it forms ice, and from the abrasion and grinding action of moving ice and water, particularly when transporting rock debris. Earth movements contribute to a lesser extent to disintegration on a global scale, although earthquakes, volcanic explosions, and landslides can achieve striking results over limited areas.

Decomposition is brought about by chemical processes such as hydrolysis, hydration, oxidation, reduction, carbonation and solution effects. Curtis (1976) has presented a 'cycle' diagram in which the components of the lithosphere (or solid earth), atmosphere, hydrosphere, and biosphere interact in a weathering reaction partially powered by solar energy. In this reaction the chemical weathering products are classed as: (a) dissolved materials composed of ionic and neutral organic and inorganic components; (b) solid organic residues; and (c) newly formed silicate, oxide, hydroxide, carbonate, sulphide, and other solids. The physical weathering products not chemically altered are classed as 'resistates' or fragmented components of the original lithosphere. Other authors give emphasis to the importance of the effects of organic acids produced by lichens and other biological colonizers of rock surfaces.

Considerable information about the course of chemical weathering can be obtained from calculations of the energy changes involved when minerals alter to more stable weathering products. Loughnan (1969, p. 60) has surveyed the work which has tried to establish relationships between crystal structure and resistance or susceptibility to weathering, and concluded that although the role played by crystal structure is still far from understood, some trends are dis-

Table 1.2 The structural classification of the silicates

Classification	Structural arrangement	Silicon–oxygen ratio	Examples
Nesosilicates	Independent tetrahedra	1:4	Olivines $(Mg, Fe)_2SiO_4$ Forsterite, Mg_2SiO_4
Sorosilicates	Two tetrahedra sharing one oxygen	2:7	Akermanite, $Ca_2MgSi_2O_7$
Cyclosilicates	Closed rings of tetrahedra each sharing two oxygens	1:3	Benitoite, $BaTiSi_3O_9$ Beryl, $Al_2Be_3Si_6O_{18}$
Inosilicates	Continuous single chains of tetrahedra each sharing two oxygens	1:3	Pyroxenes, e.g. enstatite, $MgSiO_3$
	Continuous double chains of tetrahedra sharing alternately two and three oxygens	4:11	Amphiboles, e.g. anthophyllite $Mg_7(Si_4O_{11})_2(OH)_2$
Phyllosilicates	Continuous sheets of tetrahedra each sharing three oxygens	2:5	Talc, $Mg_3Si_4O_{10}(OH)_2$ Phlogopite, $KMg_3(AlSi_3)O_{10}(OH)_2$
Tectosilicates	Continuous framework of tetrahedra each sharing all four oxygens	1:2	Quartz, SiO_2 Nepheline, $Na_3K(AlSi)_4O_{16}$ Feldspars

cernible. For instance, silicate minerals are formed by condensation of SiO_4 tetrahedra, and as the degree of linkage of SiO_4 tetrahedra increases through the formation of —Si—O—Si— bonds, resistance to decomposition also increases. Thus the pyroxenes, classed as inosilicates (Table 1.2) and composed of continuous single chains of tetrahedra, in which two oxygens are shared by one silicon, decompose more readily than the amphiboles (continuous double chains formed by 'cross-linking' of two single chains) and the phyllosilicates which include the micas and clay mineral structures. Quartz (SiO_2), a tectosilicate, is composed of a continuous framework of tetrahedra in three dimensions and all of the four oxygens are shared. Predictably it has a high degree of resistance to weathering, although it does dissolve to some extent in natural waters to form silicic acid (H_4SiO_4). Though the dissolution is slow it predictably follows the Arrhenius equation (1.1)

$$k = A \exp(-E_a/RT) \tag{1.1}$$

where k is the rate of the dissolution, A is the frequency factor, representing the total frequency of encounters between two particles, E_a is the activation energy, or the energy required to overcome the energy barrier which impedes the reaction, R is the gas constant and T is the temperature (K). Because the reaction is slow under weathering conditions the activation energy for the formation of silicic acid from quartz must be high. However the rate of the reaction is temperature dependent, and like all chemical reactions its rate will be approximately doubled, within certain limits, for every rise in temperature of 10 K. Thus, where the same chemical environment prevails, the weathering of rocks will be faster in tropical than in temperate climates.

Garrels and Christ (1965) have reviewed the thermodynamics of rock weathering to more stable end products. Only brief mention will be made here of the principles involved. Chemical weathering can be regarded in the same manner as any other chemical reaction and the formation or otherwise of particular products from given minerals can be predicted from free energy considerations. Thus for a reaction to proceed spontaneously the change in the free energy for the reaction (ΔG_r) is governed by the free energies of formation (ΔG_f) of the products and reactants as indicated in equation (1.2).

$$\Delta G_r = \Delta G_f\,(\text{products}) - \Delta G_f\,(\text{reactants}) \qquad (1.2)$$

and the free energy of the products will be less than that for the reactants. Energy is liberated for such spontaneous reactions and the products are more stable than the reactants. The standard free energy change for the reaction (ΔG_r°) can be obtained from equation (1.3).

$$\Delta G_r^\circ = \Delta H_r^\circ - T\Delta S_r^\circ \qquad (1.3)$$

when the changes in enthalpy or heat content (ΔH_r°) and entropy (ΔS_r°) are known for standard conditions at unit fugacity and temperature T.

Data are available for the standard free energies of formation (ΔG_f°), the standard enthalpies of formation (ΔH_f°) and the standard entropies (S°) at 25 °C for vast numbers of compounds of different elements. The data for hydrous aluminium and iron oxides, for some silicate minerals and for clay minerals are of especial interest, as they enable the course of probable mineral alteration in the soil to be predicted. The available data are still subject to revision, and they fail to give information about the rates at which equilibria are attained, which are often extremely slow. An introduction to the information available is given by Garrels and Christ (1965, particularly pages 321 and 403–429), and Hem (1972). The relative stabilities obtained are in line with those shown in Figure 1.2.

1.3.3 Formation and Stability of Clay Minerals

The resistance of different minerals to chemical breakdown has been extensively studied. Goldich (1938) showed that minerals, such as those in the olivine series $(Mg, Fe)_2\,SiO_4$ and calcium plagioclase ($CaAl_2Si_2O_8$), formed at high temperatures and pressures, weathered more rapidly than muscovite and quartz minerals which are formed at relatively low temperatures. This trend suggests that minerals formed in environments resembling those where weathering is taking place will be highly resistant, an observation which can be rationalized on the basis of thermodynamic considerations.

When hydrous oxide species of aluminium such as gibbsite, $Al(OH)_3$, and silicic acid, H_4SiO_4, are present in appropriate concentrations in solution, or are simultaneously released during rock weathering, the formation of silico-aluminium copolymers is thermodynamically possible. For this to take place a critical solution concentration of 3×10^{-5} moles dm^{-3} of silica, or the equivalent of 2 ppm of dissolved SiO_2 is required (Hem, 1972). In the pH range 5–9 the solubilities of hydrous oxides of aluminium are low and that of silica is relatively high. Thus, where leaching conditions predominate, as in some tropical soils of perhumid areas, aluminium hydroxides, particularly gibbsite, with some admixture of iron and manganese oxides will, if equilibrium is attained, dominate the composition of the residual soils (Patterson, 1967). Where drainage is less free and silica concentrations are commonly greater than 2 p.p.m., kaolinite becomes stable relative to gibbsite (Curtis and Spears, 1971). Thus most tropical soils in equilibrium with their environment contain kaolinite as the principal component of the clay fraction. Whenever conditions in the soil are such that the more basic cations, such as calcium and potassium, as well as silicon, are not rapidly removed 2:1 type clay minerals, such as the smectites and micas predominate. Typically soils in poorly drained conditions, soils in arid environments, and soils in the early stages of weathering where insufficient time has elapsed for removal of the basic cations, contain clays dominated by 2:1 type layer silicates.

The pH of the soil or weathering environment is governed by its initial base status, and by the extent to which cations such as Ca, Mg, and K are removed by leaching or by plants. The biological activity and the resultant release of CO_2 and the formation of carbonic acid (H_2CO_3) is also important, as well as processes such as the biologically catalysed oxidation of metal sulphides (e.g. pyrite, FeS_2) to form sulphuric acid, and the release of organic acids from plant materials or as products of microbial metabolism and oxidation of humic substances. Highly acid (pH 2–4) soil environments and drainage waters arise in the upper horizons of certain soils (termed podsols or spodosols) and some upland peats formed from mosses and heather (*Calluna vulgaris*). At such low pH values the relative stabilities of the oxides of silicon and of the hydrated oxides of aluminium are reversed and quartz-rich soils result. Iron and aluminium are translocated in the form of organic complexes, usually to be deposited partially or completely in a lower soil horizon.

Bicarbonate is the most abundant anion in weathering systems, as judged by its concentration relative to other anions in river waters (Livingstone, 1963), and therefore considerable attention has been given to studies of its weathering effects. It arises from the dissociation of carbonic acid (H_2CO_3) formed partly from dissolution of CO_2 from plant and microbial respiration processes, and from atmospheric CO_2 dissolved in rainwater:

$$H_2O + CO_2 \rightleftharpoons H_2CO_3 \rightleftharpoons H^+ + HCO_3^- \qquad (1.4)$$

The weak acidity of carbonic acid will always tend to produce soils which have been leached of basic cations, and in which few 'primary' minerals formed during crystallization of rock magma survive.

The alteration of calcium, sodium and potassium feldspars, to the phyllosilicates kaolinite and muscovite can be formulated:

$$\underset{\text{plagioclase}}{CaAl_2Si_2O_8} + 3H_2O + 2CO_2 \rightarrow \underset{\text{kaolinite}}{Al_2Si_2O_5(OH)_4} + Ca^{2+} + 2HCO_3^- \quad (1.5)$$

$$\underset{\text{albite}}{2NaAlSi_3O_8} + 11H_2O + 2CO_2 \rightarrow \underset{\text{kaolinite}}{Al_2Si_2O_5(OH)_4} + 2Na^+ + 2HCO_3^- + 4H_4SiO_4 \quad (1.6)$$

$$\underset{\text{orthoclase}}{6KAlSi_3O_8} + 4H_2O + 4CO_2 \rightarrow \underset{\text{muscovite}}{K_2Al_4(Si_6Al_2)O_{20}(OH)_4} + \underset{\text{quartz}}{12SiO_2} + 4K^+ + 4HCO_3^- \quad (1.7)$$

The process of alteration involves dissolution of the primary mineral (the feldspar) followed by reaction between the products. The product formed depends on the intensity of the leaching process, as well as the mode of weathering of the parent rock, and the length of time for which alteration has been proceeding. In general the more intense the weathering environment, the more does silicic acid tend to be removed. Thus quartz may be formed as a secondary mineral by alteration of orthoclase under mild weathering, whereas in tropical conditions silicic acid tends to be leached, and kaolinite to be the only secondary mineral formed.

The course of alteration will also depend on whether free drainage occurs at the point of weathering. If the products persist at the site of weathering, the cations such as Ca^{2+}, Mg^{2+} and K^+ will participate in the formation of secondary minerals. As noted above, when Ca^{2+} (and Mg^{2+}) are present, minerals of the smectite–vermiculite type are formed, and when K^+ is present, mica-type minerals are formed.

Obviously climate can have a very significant effect on the clay composition of a soil. Montmorillonite, vermiculite and hydrous mica or illite, 2:1 type clays, predominate over the 1:1 layer clays, such as kaolinite and halloysite, when solution silicate concentrations and pH values are relatively high. The 2:1 clays will also predominate under mild alteration conditions, such as those implied in reaction (1.7) for the transformation of feldspars. The preponderance of kaolinite usually indicates advanced weathering or the presence of synthesis conditions which preclude the formation of 2:1 type minerals. Although kaolinite is commonly the predominant clay mineral in tropical soils it is also a major component of many soils in temperate areas, and although 2:1 clays are abundant and widely distributed in the more temperate and cooler regions they are also widely found in warmer, wet conditions.

1.3.4 Properties of Inorganic and Organic Soil Colloids

Inorganic colloids

The clay component of soils is sometimes referred to as the 'active fraction'. This is because many soil properties are related to adsorption reactions on clay surfaces. The specific surface areas of clay particles range from 1 $m^2\,g^{-1}$ for particles with equivalent spherical diameters close to the upper limit for clays of 2.0 μm to over 760 $m^2\,g^{-1}$ for smectites and vermiculites, which have internal surfaces accessible to water and electrolytes and some organic compounds. Some poorly crystalline hydrous oxides such as the amorphous silica-alumina copolymer called allophane also have surface areas in aqueous suspension of the order of several hundreds of square metres per gram. The surfaces of soil clay particles are usually highly charged. Most layer silicates have negatively charged surfaces, the surface charge being balanced by an excess of cations in the soil solution adjacent to the clay surface. These cations are readily exchangeable against others. The 'exchange capacity' of the clay will depend on the specific surface area and the surface charge density.

Plant growth generally depends on the exchange of hydrogen ions exuded by roots for calcium, potassium, magnesium and other cations held by the clay. Thus the ability to exchange cations is an essential property of the soil if it is to sustain plant growth.

In the layer silicates the negative charge arises from isomorphous substitutions within the clay lattice, of an ion of higher positive charge by one of lower—mostly Al^{3+} for Si^{4+} and Mg^{2+} for Al^{3+}. However a second mechanism contributes to the cation exchange capacity of most soils. This is the charge produced by dissociation of hydroxyl groups, for instance Si—OH, on silica surfaces. The extent of dissociation depends on the pH, the electrolyte concentration of the medium and the ion species involved. Thus layer silicate surfaces tend to have a 'constant charge' due to isomorphous substitutions, whereas hydrous oxide surfaces have a pH dependent charge due to dissociation of surface hydroxyls, and are of 'constant potential' relative to the electrolyte solution in which they are immersed. These properties are discussed fully in Chapters 4 and 5.

Silanol, SiOH, groups dissociate to $Si—O^-$, but do not accept a proton above pH 2 and so are not amphoteric in the pH range (3 to 10) found in soils. Aluminol (Al—OH) and ferrol (Fe—OH) groups however are amphoteric. Below pH 8, aluminol groups may accept a proton and become positively charged $(Al—OH_2)^+$. Ferrol groups may be positively charged below about pH 7. At pHs above 8 and 7, respectively, they both become negatively charged. The pH at which they are uncharged is referred to as their 'point of zero charge' (p.z.c.).

Clearly, in soils where hydrous oxide type surfaces are dominant, the soil can have no fixed cation exchange capacity. Most soils in fact have only a minor contribution to their charge characteristics from hydrous oxides, so that the common practice of ascribing a definite cation exchange capacity value to a soil is usually valid.

Hydrous oxide surfaces, particularly aluminol and ferrol surfaces, also participate readily in ligand exchange reactions with anions such as phosphate:

$$\rangle Al{-}OH + H_2PO_4^- \rightleftharpoons \left[\rangle Al{-}O{-}\overset{O}{\overset{\|}{P}}{-}OH \atop \quad\quad\quad\quad \searrow O\right]^- + H_2O \qquad (1.8)$$

Phyllosilicates, particularly micas, smectites and vermiculities, normally repel anions. The phosphate sorption reaction is one of the important processes giving rise to the so-called 'phosphate fixation' of soils. It often determines the fate of phosphate added to soils. Organic anions, silicate and others can react similarly, and as the reactions are not readily reversible clean aluminol and ferrol surfaces cannot be expected to be found in many soils. Because iron and aluminium hydroxides can be positively charged, it is sometimes suggested that they are involved in phosphate fixation because of the association between the negatively charged anion and the positive charge. This is quite incorrect, as has been shown by many studies of the phosphate sorption reaction in the presence of large excesses of other weak anionic ligands such as chloride. Anion sorption reactions in soils are considered further by Mott in a chapter in the companion volume to this book.

Although hydrous oxides only form the major component of the clay fraction of soils in some intensively weathered, and extremely old tropical soils, they are sometimes considered to be more widely important because they form coatings on the surfaces of the phyllosilicates. The question of how widely such coatings occur is still not settled, but they are certainly the exception rather than the rule for most soils of the temperate zone. A special example of soils whose properties are dominated by hydrous oxide behaviour are young soils on volcanic ash, whose clay fraction consists almost entirely of allophane.

In addition to the difference in the ion sorption reactions, phyllosilicate and hydrous oxide surfaces differ markedly in their interaction with water (Chapter 6). The surfaces of low charge 2:1 phyllosilicates, such as those of talc and pyrophyllite, are relatively hydrophobic. However when the surfaces are strongly charged, as in the micas, vermiculites and smectites, the exchangeable cations held at their surfaces interact strongly with water, adsorbing initially a hydration shell, and subsequently further water molecules which hydrogen bond to the water in the hydration shell and perhaps weakly to oxygens of the clay surface: further water may be adsorbed by what is essentially an osmotic process, leading to dilution of the cations as they form a double layer at the clay surface (Chapter 5).

On hydrous oxide surfaces water may hydrate the counterions which are present, but also becomes directly hydrogen bonded to hydroxyl groups of the surface.

Whether the water is associated with cations at the surface, or directly with the surface constituents, it is very firmly held. Even when several molecular layers of water are present on clay surfaces, the adsorption energy exceeds the forces which plants can exert to extract water from soil. This force is equivalent to about 15 atmospheres, and can be defined in terms of the radius of a capillary from which the water can be extracted, i.e. the size of soil pore corresponding to the point at which plants wilt. This is a capillary of approximately 200 nm diameter, or 400 molecular layers of water on adjacent surfaces. Water at clay surfaces is important in relation to ion exchange reactions, and its involvement in adsorption processes. It also has a major effect on forces between soil particles, but it does not directly affect the availability of water to plants, or the flow of water in soils. The water involved in these processes is well removed from particle surfaces.

Organic colloids

So far only the properties of the inorganic colloids have been discussed. In surface soils, and sometimes in subsoils also, an important part of the colloidal fraction consists of organic polymers. The constitution and chemistry of these polymers is discussed fully in Chapter 3. They originate in plant and animal débris falling on the soil. This débris is altered physically and chemically by soil animals, fungi and bacteria, so that the soil usually contains some organic materials in process of alteration, as well as the relatively stable end-product or 'humus'. Because fresh and partially altered organic material of biological origin can usually be separated from the remainder of the soil by virtue of its lower specific gravity, it is sometimes referred to as the 'light fraction'. The more humified material is 'weighted' by its intimate association with inorganic soil colloids, in the so-called 'clay–organic complex'. Thus the incompletely decomposed 'light fraction' (Plate 1.1) can be floated off in liquids of specific gravity around 1.8 to 2.0. The interaction between the inorganic and organic colloids in soils is a complex process. The mechanisms involved are discussed in the Chapter on adsorption in the companion volume, and have been reviewed by Greenland (1965, 1971a), Mortland (1970) and Theng (1974). The most important mode of interaction probably involves association of polyvalent metal ions at clay surfaces, or in the surfaces of hydroxy polymers, with ligand groups such as carboxyl of the organic polymers.

In spite of their involvement with the clay fraction, the carboxyl groups of the organic polymers are also a source of negative charge in the soil. Like the hydroxyl groups on hydrous oxides, their extent of dissociation depends on pH, electrolyte concentration and cation species. Depending on conditions and the nature of the polymer, their negative charge is in the range 2.0 to 4.0 meq g^{-1}, which may be compared with the charge on phyllosilicates, which lies between 0.01 and 2.0 meq g^{-1}. Thus even though the organic matter content of most surface soils is less than 10 per cent by weight, it can make a substantial contribution to the cation exchange capacity of a soil.

The organic fraction of soils is also important because it carries virtually all the nitrogen and sulphur reserves of the soil, and some of the phosphorus and trace element reserves. The processes by which the organic compounds are mineralized so that the nutrients are released in inorganic forms available to plants are again the subject of a chapter (by Jenkinson) in the companion volume.

The ability of some of the soil organic compounds to form stable complexes of heavy metals is important, in relation to the translocation of iron and aluminium in the soil, to the retention of trace elements important to plant nutrition and for the immobilization of potentially toxic heavy metals.

One of the most important properties of humic materials is their influence on the stability of soil pores. As mentioned above, water in fine pores is not available to plants, as it is held too strongly. It is the very coarse pores (>50 μm e.c.d.) that are essential to air and water movement in the soil, and pores which are still relatively large (approximately 0.5 to 50 μm e.c.d.) that hold the water on which plants depend for growth (Table 1.3). Such pores can exist between silt and sand grains, but clay particles when present in sufficient amount may fill them. This in fact can happen in a 'structurally unstable soil'. However in most soils the particles tend to be molded into aggregates or peds, either physically by shrink–swell processes associated with wetting and drying or freezing and thawing, or by biological molding processes due to soil animals and possibly the action of

Table 1.3 A functional classification of soil pores (Greenland, 1977)

Name	Function	Equivalent cylindrical diameter, μm
Transmission pores	Air movement and drainage of excess water	>50
Storage pores	Retention of water against gravity, and release to plant roots	0.5–50
Residual pores	Retention and diffusion ions in solution	<0.5
Bonding spaces	Support major forces between soil particles	<0.005

Table 1.4 A simple classification of soil aggregates by size (Greenland, 1971b)

Name	Description	Size range
Clod	Clusters of aggregates	>5 mm
Aggregate	Clusters of microaggregates and sand particles	0.5–5 mm
Microaggregate	Domains, silt and fine sand particles bonded by organic polymers	5–500 μm
Domain	Oriented clusters of clay crystals	<5 μm

plant roots. If the aggregates so formed are stable, transmission and storage pores are formed in the interstices between them. The soil water can then move freely, and exchange of gases between soil and the atmosphere can proceed. Further, the forces between soil particles, which are dependent on their mean separation, are less, so that the soil is easier to cultivate and more amenable to root growth.

Clay crystals are normally clustered into domains (Table 1.4). The domains however will only interact weakly with each other, and organic materials are often required to act as bonding agents to retain them in microaggregates, and to act as bonding agents between microaggregates. The production of these aggregates in surface soils is closely associated with the action of the soil fauna. Worms in particular appear to create aggregates in their castings, and to add polysaccharide and humic materials to the molded aggregate to stabilize it (Plate 1.2).

Emerson (1959) and others have attempted to depict the manner in which soil particles cluster to form soil crumbs. Figure 1.3 is an attempt to illustrate the structure of an aggregate. Although the importance of organic matter as a stabilizing agent has been emphasized here, inorganic materials precipitated in the soil can also act as interparticle cements. These include iron and aluminium oxides and hydroxides, silica and calcium carbonate. If the precipitates only extend between or around a few particles they influence only the stability of aggregates, but if conditions are such that the crystals are initiated throughout the soil, and can grow sufficiently to link up with each other, then the whole soil mass can be hardened. Where ferrous iron enters a soil in solution, and is

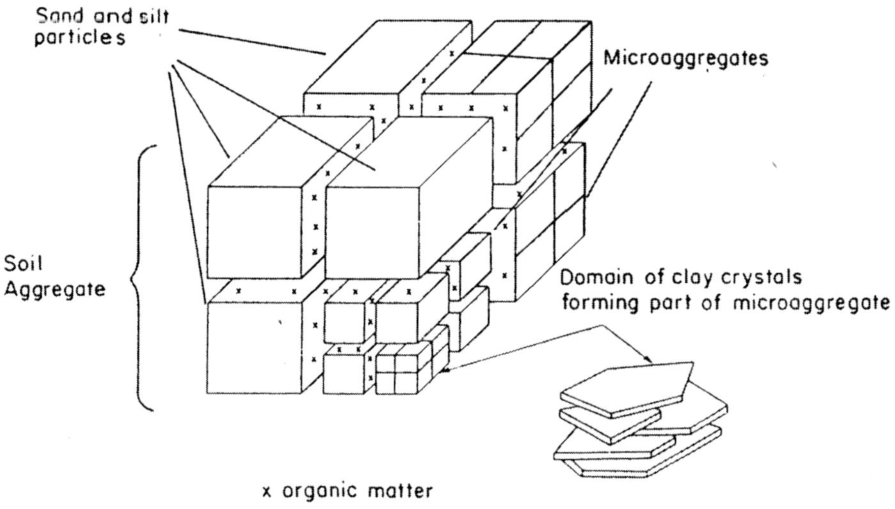

Figure 1.3 A hypothetical model of a soil aggregate, illustrating the clustering of clay crystals to form domains, of domains to form microaggregates, and of microaggregates to form aggregates. Molecules of soil organic matter acts as bonding agents between domains and microaggregates, and sand and silt particles (after Williams, Greenland and Quirk, 1967)

oxidized to form goethite (α-FeOOH) the goethite crystals may form a honeycomb through the whole soil mass. Usually other iron hydroxides or oxides are crystallized at the same time, and aluminium may also be involved. The process is called laterization, and the material laterite or plinthite. In the wet soil, the crystallization of the iron is incomplete, and the soil may be soft. On exposure to the air, it dries, the crystallization of the hydrous oxides proceeds and the soil hardens (Sivarajasingham *et al.*, 1962).

Continued precipitation of calcium carbonate leads to the appearance of material known as calcite, and of silicon to silcrete.

While these processes may be important in certain classes of soil, the effects of organic materials are of much more general significance. Polysaccharide gums are an important constituent of soil organic matter (Chapter 3) and are certainly involved in the bonding of soil particles in aggregates (Greenland, Lindstrom, and Quirk, 1962; Finch *et al.*, 1971). Humic materials also contribute, at least in soils of moderate or large contents of organic matter (greater than about 4 per cent by weight of soil) (Stefanson, 1971; Hamblin and Greenland, 1977).

Humus synthesis and degradation in the soil is a dynamic process. Although humic materials have a degree of resistance to microbial breakdown they must be continually renewed in the soil environment if a constant level is to be maintained. This renewal cannot take place unless fresh organic debris is continuously made available to provide a substrate for the microbial metabolism and transformation processes, which lead eventually to humus formation. An equilibrium is established in aerobic soils which have supported a natural vegetation cover, such as grass or forest, for many years. However, the soil organic matter content will be changed when this equilibrium is disturbed, as will happen, for instance, when the water-table level, or the plant nutrient status of the soil is altered, or when the natural vegetation is removed and the soil cultivated. Jenkinson, in a chapter in the companion volume to this book, shows that it is necessary to add large amounts of organic matter to cultivated soils in order to compensate for the losses which result from the increased biological oxidation brought about when the soil is disturbed. Such additions are seldom made in modern agricultural practices and thus continuous cultivation usually leads to the depletion of the humus. As a result the cation retention and natural nutrient reserves fall, the stability of the structure decreases and soil fertility declines. For these reasons an essential part of good soil management involves maintenance of an adequate level of organic matter in the soil. Nevertheless soils can be successfully farmed even when they have very low organic matter reserves, by judicious use of inorganic fertilizers and skilled cultivation techniques, but it is much easier to maintain high levels of productivity in soils of better organic matter status.

1.4 THE DIVERSITY OF SOILS

Soils are very diverse in composition and behaviour. Some are very acid, others alkaline, some reduced, some strongly oxidized. Thus if we wish to know

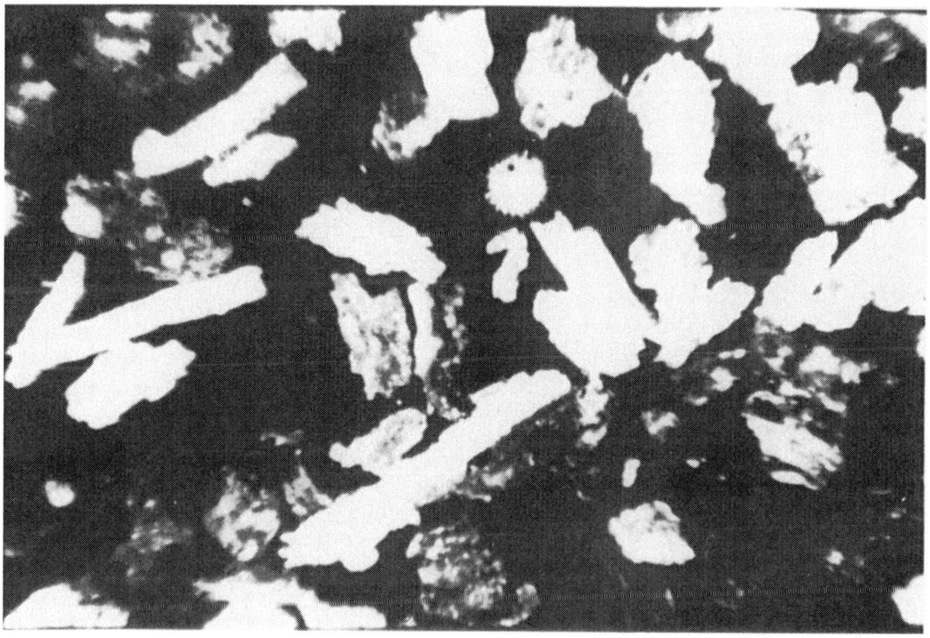

Plate 1.1 Partially decomposed plant remains (the 'light fraction') separated from an arable soil. This material still shows the cellular organization of higher organisms, unlike the bulk of the 'humified' soil organic matter. It is rapidly decomposed, and so in a highly dynamic equilibrium, but often represents 5 to 20 per cent of the total organic matter in surface soils. × 500

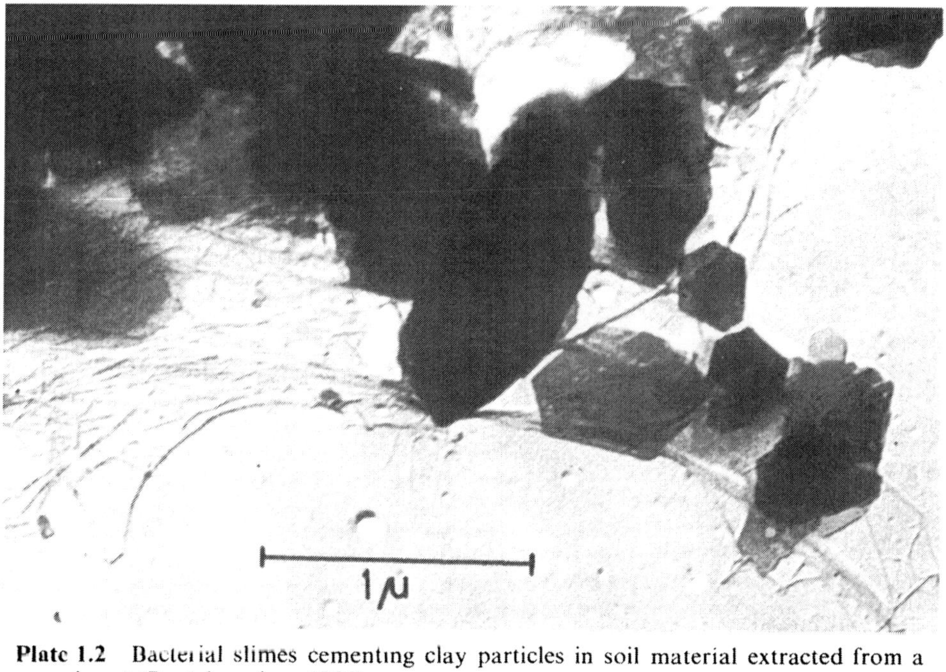

Plate 1.2 Bacterial slimes cementing clay particles in soil material extracted from a worm's gut (Beutelspacher, 1955)

what happens to a chemical when it is added to 'the soil', we should specify the soil very carefully if any answer is to be meaningful. There has in the past been a strong tendency to ignore this simple essential. One of the reasons has been the inadequate and sometimes confusing systems, which have been developed for classifying and naming soils. However soil chemistry is of much greater value within a context of soil classification, and an attempt is therefore made here to outline the basis of systems for classifying soils.

1.4.1 Factors of Soil Formation

The soil at any one place is seldom simply the product of weathering of the underlying rock. It arises from a complex of factors, involving additions of material by deposition from air and water, or by surface creep, or in solution, and from vegetation and man's activities, and is also affected by removals by erosion and leaching. In many areas there have been periods of denudation when soils were almost completely stripped from the landscape, for instance by the action of ice sheets moving as glaciers and pushing the soil before them. The material ground and mixed by the action of ice was then redeposited in other areas. When the ice retreated new soils developed. As the last ice age in the northern hemisphere was less than 20,000 years ago, many soils in the areas affected by the ice are relatively young, in contrast to soils of most areas of the tropics and Australia which are mostly very old. Soils in areas of recent volcanic activity, and those where the landscape has been rejuvenated by other action, may also be relatively young.

Following Dokuchaiev's early work, many studies of the genesis of soils have been made. It is now appreciated that soils are the result of the interaction of five major factors of soil formation (Jenny, 1941), viz. parent material, climate, time, relief and natural organisms. This can be expressed as:

$$s = f(p, \mathrm{cl}, t, r, o) \tag{1.9}$$

The importance of the climatic factor to soil formation has lead to the 'zonal' concept of soil distribution, certain major differences between soils developed in different climatic zones being apparent, e.g. between soils of temperate and tropical regions, and between soils of humid regions and soils of arid regions. However some soils are not in equilibrium with the climate in which they occur and do not conform to zonal concepts. The influence of age or parent material may then be dominant.

1.4.2 Definition of Soil Classes

Any classification of soils which groups them according to their mode of genesis suffers from the disadvantage that a correct interpretation of their formation is necessary before they can be classified. The complexities of soil genesis should already be apparent. Many soils are in fact now known to be

Table 1.5 Differentiating characteristics of the categories of the USDA soil classification system 'Soil Taxonomy'. Reprinted by permission from *Soil Genesis and Classification* by S. W. Buol, F. D. Hole and R. J. McCracken, © 1972 by The Iowa State University Press, Ames, Iowa 50010

Category	Number of of taxa	Nature of differentiating characteristics
Order	10	Soil-forming processes as indicated by presence or absence of major diagnostic horizons.
Suborder	47	Genetic homogeneity. Subdivision of orders according to presence or absence of properties associated with wetness, soil moisture regimes, major parent material, and vegetational effects as indicated by key properties; organic fiber decomposition stage in Histosols.
Great group	206 (approximate)	Subdivision of suborders according to similar kind, arrangement, and degree of expression of horizons, with emphasis on upper sequum; base status; soil temperature and moisture regimes; presence or absence of diagnostic layers (plinthite, fragipan, duripan).
Subgroup		Central concept taxa for great group and properties indicating intergradations to other great groups, suborders, and orders; extragradation to "not soil".
Family		Properties important for plant root growth; broad soil textural classes averaged over control section or solum; mineralogical classes for dominant mineralogy of solum; soil temperature classes [based on mean annual soil temperature at 50 cm (20 in.) depth].
Series	10,000 (approximate) in United States	Kind and arrangement of horizons; colour, texture, structure, consistence, and reaction of horizons; chemical and mineralogical properties of the horizons.

polygenetic, i.e. to have been involved in several sequential periods of soil formation when different factors were dominant. The proper assignment of soils in terms of genesis is then particularly difficult.

The alternative to a genetically based classification is one based on the morphology and properties of the soil profile. Such systems are less subjective and do not require extensive investigation of each soil before its assignment to a particular category. They suffer from the disadvantage that there is no coherent basis for their groupings. It is, however, possible to develop a system in which generalized genetic principles are used to link like soil categories, but the details of the assignments to categories are made on the basis of profile morphology and soil properties. Most modern systems of soil classification have been developed in this way. Of these systems the most comprehensive is that developed by the Soil Survey Staff (1975) of the United States Department of Agriculture. In this system, published as 'Soil Taxonomy', soils are divided into ten orders, 47

Table 1.6 Terminology used in some soil classification systems, and names used for soil mapping units

Older names		USDA Soil Taxonomy	FAO/ UNESCO Soil Map of the World	Soil map of Africa
Azonal soils	Lithosols Regosols Alluvial soils	Entisols	Lithosols Regosols Fluvisols Arenosols	Raw mineral soils Weakly developed soils
Volcanic soils	Andosols Basisols	Inceptisols	Andosols Cambisols	Eutrophic brown soils of tropical regions
Desert soils	Sierozems Solonchak Solonetz	Aridisols	Xerosols Yermosols Solonchak Solonetz	Brown and reddish-brown soils of arid and semi-arid regions Halomorphic soils
Brown and black soils	Grumosols Black cotton soils	Vertisols	Vertisols Chernozems	Calcimorphic soils including vertisols
	Terra rossa Rendzinas	Mollisols	Phaeozems Rendzinas Gleysols	Red and brown Mediterranean soils
Podsols		Spodosols	Podsols	
Red loams	Krasnozems	Alfisols	Luvisols Nitosols	Ferruginous tropical soils
Latosols	Red and yellow podsolics	Ultisols	Acrisols	Podsolic soils Ferrisols
	Red earths Lateritic soils	Oxisols	Ferralsols	Ferrallitic soils
Intrazonal soils	Peats Mucks Bog soils	Histosols	Histosols	Hydromorphic soils

The terms used in different systems are seldom exactly equivalent. Those given here are arranged so that there is a very rough equivalence between terms at the same level on the page. The names all refer to very broad groups of soils, defined at the 'order' level. Recent accounts of different systems are given by Buol, Hole and McCracken (1973) and Young (1976). The American system has been described in a handbook of the U.S. Department of Agriculture 'Soil Taxonomy' (Soil Survey Staff, 1975). The units used for the Soil Map of Africa are described in the 'Explanatory Monograph to the Soil Map of Africa, Soils 1 to 500,000' (d'Hoore, 1964). The units used by FAO for the Soil Map of the World are described in the 'Legend to soil units for the soil map of the world' (FAO, 1974).

suborders, and 206 great groups which may be further subdivided into subgroups, families and series (Table 1.5). The soil series is the most widely used mapping unit, and the recognition of soil series is the usual starting point for mapping work in the field.

Table 1.7 Mapping units used in the FAO/UNESCO 'Soil Map of the World'

Name	Short description
Fluvisols	Water deposited soils with little alteration
Regosols	Thin soils with little or no development over unconsolidated rock material
Arenosols	Soils formed from sand (e.g. on sand dunes)
Gleysols	Soils with mottled or reduced horizons due to wetness
Rendzinas	Shallow soils over limestones
Rankers	Thin soils over siliceous materials
Andosols	Soils formed from volcanic ash with dark surface horizons
Vertisols	Clay soils, rich in smectitic clay minerals
Yermosols	Desert soils
Xerosols	Dry soils of semi-arid regions
Solonchaks	Soils with salt accumulation
Solonetz	Soils of high exchangeable sodium content
Planosols	Soils with an abrupt change of texture and other properties at the junction of surface (A) and subsoil (B) horizons, usually in poorly drained areas
Castanozems	Soils developed under steppe vegetation with chestnut-coloured surface horizons
Chernozems	Soils developed under prairie vegetation with black surface horizons of high humus content
Phaeozems	Soils with dark surface horizons, but more leached than Castanozem or Chernozem
Greyzems	Grey forest soils of cool temperate latitudes
Cambisols	Soils of light colour, with structure and/or consistence change in the profile due to weathering
Luvisols	Soils of medium to high base status, with an argillic[a] horizon
Podzoluvisols	Soils with a leached horizon tonguing into an argillic B horizon
Podzols	Soils with a strongly bleached eluvial subsurface (A_2 or E) horizon and subsoil accumulation of iron, aluminium and humus
Acrisols	Highly weathered soils with an argillic horizon, and of low base saturation
Nitosols	Soils with clay fractions of low cation exchange capacity in an argillic horizon, usually developed from basic rocks.
Ferralsols	Soils of tropical regions with sesquioxide rich clay fractions
Histosols	Organic soils
Lithosols	Shallow soils over hard rock

[a] Argillic horizon = A soil horizon enriched in clay relative to the horizon above, and with clay skins around aggregates, indicative of the inward translation of clay.

1.4.3 The Major Classifications of the Soils of the World

The Orders of Soil Taxonomy (Table 1.6) are defined according to morphological properties, and subdivided according to features associated with certain genetic factors (Table 1.5). The suborder names are made by using a formative element, such as 'fluv' from fluvial, to denote soils influenced by fluvial deposition, with the formative element for the soil order. Thus a 'Fluvent' is an Entisol, which has developed in fluvial deposits.

Approximately equivalent names for the soil orders are indicated in Table 1.6, but because of differences in manner of definition these are only rough guides. The correct classification of a soil requires careful characterization of the soil and determination of its properties.

To prepare the FAO/UNESCO Soil Map of the World it was necessary to obtain international agreement on the mapping units to be used. A mapping unit is not a category in a classification system, although it may often be similar to one. The names of the units (Table 1.7) are often terms which have previously been used in classification systems. Very brief descriptions are given here. Full definitions are given in the FAO report (FAO, 1974) and a general description by Dudal (1968).

Many other regional systems are in use, each emphasizing the more important factors of soil formation where they originated. Several of these are discussed by Buol, Hole and McCracken (1973) and Young (1976). Hopefully in time soil descriptions will become standardized, using the comprehensive system of 'Soil Taxonomy' or perhaps by further development of the FAO Units for the Soil Map of the World, into a classification system with a greater range of subordinate classes.

REFERENCES

Arnon, I. (1972), *Crop Production in Dry Regions*, Leonard Hill, London.

Baver, L. D., Gardner, W. H. and Gardner, W. R. (1972), *Soil Physics*, 4th edition, Wiley, New York.

Bennett, H. H. (1939), *Soil Conservation*, McGraw Hill, New York.

Bergersen, F. J. (1971), Biochemistry of symbiotic nitrogen fixation by legumes, *Ann. Rev. Plant Physiology*, **22**, 121–140.

Beutelspacher, H. (1955), Interaction between the inorganic and organic colloids in soil, *Z. Pflernahr. Dung. Bodenk.*, **69**, 108–115.

Bowen, N. L. (1928), *Evolution of Igneous Rocks*, Princeton University Press.

Brady, N. C. (1974), *The Nature and Properties of Soils*, 8th edition, Macmillan, New York.

Buol, S. W., Hole, F. D. and McCracken, R. J. (1973), *Soil Genesis and Classification*, Iowa State University Press, Ames.

Burges, A. and Raw, F. (1967), *Soil Biology*, Academic Press, New York.

Childs, E. C. (1969), *An Introduction to the Physical Basis of Soil Water Phenomena*, Wiley–Interscience, London.

Clarke, F. W. and Washington, H. S. (1924), Composition of the earth's crust. US Geological Survey Professional Paper 127, Government Office, Washington.

Curtis, C. D. (1976), Chemistry of rock weathering: fundamental reactions and controls, in E. Derbyshire (ed.), *Geomorphology and Climate*, pp. 25–57, Wiley, New York and Chichester.

Curtis, C. D. and Spears, D. A. (1971), Diagenetic development of kaolinite, *Clays and Clay Minerals*, **19**, 219–227.

d'Hoore, J. (1964), Soil map of Africa scale 1:500,000. Explanatory monograph, CCTA Publication 93, Lagos.

Dilworth, M. J. (1974), Dinitrogen fixation, *Ann. Rev. Plant Physiology*, **25**, 81–114.

Dokuchaiev, V. V. (1883), *Russian chernozem*, Monograph, Sankt–Peterburg. (Also in Dokuchaiev, Sochineniya (collected works), Vol. 3, Izdatel' stvo AN SSSR, Moskva–Leningrad, 1949.)

Dudal, R. (1968), Definitions of soil units for the soil map of the world. FAO World Soil Resources, Rept. 33.

Emerson, W. W. (1959), The structure of soil crumbs, *J. Soil Sci.*, **10**, 235–244.

Finch, P., Hayes, M. H. B. and Stacey, M. (1971), The biochemistry of soil polysaccharides, in A. D. McLaren and J. Skujins (eds.), *Soil Biochemistry*, Vol. 2, pp. 257–319, Marcel Dekker, New York.

FAO (1974), Legend to soil units for the FAO/UNESCO soil map of the world, UNESCO, Paris.

Garrels, R. M. and Christ, C. L. (1965), *Solutions, Minerals and Equilibria*, Harper and Row, New York.

Goldich, S. S. (1938), A study in rock weathering, *J. Geol.*, **46**, 17–58.

Goring, C. A. J. and Hamaker, J. W. (1972), *Organic Chemicals in the Soil Environment*, Vols. 1 and 2, Marcel Dekker, New York.

Greenland, D. J. (1965), Interaction between clays and organic compounds in soils. Part I. Mechanisms of interaction between clays and defined organic compounds, *Soils and Fertilizers*, **28**, 415–425. Part II. Adsorption of soil organic compounds and its effect on soil properties, *Soils and Fertilizers*, **28**, 521–532.

Greenland, D. J. (1971a), Interactions between humic and fulvic acids and clays, *Soil Sci.*, **111**, 34–41.

Greenland, D. J. (1971b), Changes in the nitrogen status and physical condition of soils under pastures, *Soils and Fertilizers*, **34**, 237–251.

Greenland, D. J. (1977), Soil damage by intensive arable cultivation: temporary or permanent? *Phil. Trans. Roy. Soc. (London), B*, **281**, 193–208.

Greenland, D. J., Lindstrom, G. R., and Quirk, J. P. (1962), Organic materials which stabilise natural soil aggregates, *Soil Sci. Soc. Amer., Proc.*, **26**, 366–371.

Hagan, R. M., Haise, H. R. and Edminster, T. W. (eds.) (1967), *Irrigation of Agricultural Lands*, American Society of Agronomy, Madison, Wisconsin.

Hamblin, A. P. and Greenland, D. J. (1977), Effect of organic constituents and complexed metal ions on aggregate stability of some East Anglian soils, *J. Soil Sci.*, **28**, 410–416.

Hellriegel, H. and Wilfarth, H. (1888), *Untersuchungen über die Stickstoffnahrung der Gramineen und Leguminosen.* Beilagehelft zu der Ztschr. Ver. Rübenzucker-Industrie Deutschen Reichs, 234.

Hem, J. D. (1972), Aluminium; behaviour during weathering and alteration of rocks, 13-G; Solubilities of mineral species which may control aluminium concentrations in natural water; Adsorption processes, 13-H, in K. H. Wedepohl (ed.), *Handbook of Geochemistry*, Vol. 2–1, Springer Verlag, Berlin.

Jackson, M. L. (1968), Weathering of primary and secondary minerals in soils, *Trans. 9th Int. Congr. Soil Sci.*, Adelaide, 4, 281–292.

Jenny, H. (1941), *Factors of Soil Formation*, McGraw Hill, London.

Kirkham, D. and Powers, W. L. (1972), *Advanced Soil Physics*, Wiley–Interscience, New York.

Livingstone, D. A. (1963), Chemical composition of rivers and lakes. U.S. Geological Survey Professional Paper 440-G. Government Printing Office, Washington.

Loughnan, F. C. (1969), *Chemical Weathering of the Silicate Minerals*, Elsevier, Amsterdam.

Mortland, M. M. (1970), Clay-organic complexes and interactions, *Advances in Agronomy*, **22**, 75–117.

Olness, A. and Clapp, C. E. (1975), Influence of polysaccharide structure on dextran adsorption by montmorillonite, *Soil Biol. Biochem.*, **7**, 113–118.

Patterson, S. H. (1967), *Bauxite reserves and potential aluminium resources of the world*, U.S. Geol. Survey Bull. 1228, 176 pp.

Pauling, L. (1930), The structure of micas and related minerals, *Proc. Nat. Acad. Sci., Washington*, **16**, 123–129 (see also pp. 578–582 of same volume).

Russell, E. J. (1942), *British Agricultural Research: Rothamsted*, Longmans Green, London.

Sivarajasingham, S., Alexander, L. T., Cady, J. G. and Cline, M. (1962), Laterite, *Advan. Agron.*, **14**, 1–60.

Soil Survey Staff (1975), Soil taxonomy: a basic system of soil classification for making and interpreting soil surveys. USDA Agriculture Handbook No. 436, U.S. Government Printing Office, Washington.

Stefanson, R. C. (1971), Effect of periodate and pyrophosphate on the seasonal changes in aggregate stabilisation, *Aust. J. Soil Res.*, **9**, 33–42.

Theng, B. K. G. (1974), *The Chemistry of Clay–Organic Reactions*, Adam Hilger, Ltd., London.

Turekian, K. K. and Wedepohl, K. H. (1961), Distribution of the elements in some major units of the earth's crust, *Geol. Soc. America Bull.*, **72**, 175–191.

Warington, R. (1900), *Physical Properties of Soil*, Oxford University Press, London.

Williams, B. G., Greenland, D. J. and Quirk, J. P. (1967), The effect of polyvinyl alcohol on the nitrogen surface area and pore structure of soils, *Aust. J. Soil Res.*, **5**, 77–83.

Young, A. (1976), *Tropical Soils and Soil Survey*, Cambridge University Press, London.

CHAPTER 2

The structures and chemistry of soil clay minerals

G. Brown, A. C. D. Newman, J. H. Rayner and A. H. Weir

all of Department of Soils and Plant Nutrition, Rothamsted Experimental Station

2.1 INTRODUCTION

Textural assessment has long been used to indicate quality in agricultural soil and although it contains an element of subjective judgement, a skilled surveyor can differentiate many grades of texture. Soil texture correlates with the distribution of particle size in the soil, though this is not the only component in texture. Although particle size distribution is an objective attribute of soils and can, in principle, be measured with accuracy, the determination is in practice quite difficult and certainly laborious, so that it is customary in soil mechanical analysis to divide the soil into several broad fractions: sand, silt and clay are the principal categories. The clay fraction, which comprises the smallest particles in the soil, is generally defined as the fraction smaller than a nominal diameter of 2 μm. The practical division is, however, based on the velocity of fall of soil particles through a fluid, calculating the diameter of the particles from the equation of Stokes,

$$v = \frac{2}{9}\frac{(p_s - p_e)gr^2}{\eta}$$

which is valid for spheres large enough to be unaffected by Brownian motion. In this equation, v is the velocity of fall, p_s and p_e are the densities of the soil particles and the fluid, g is the acceleration due to gravity, r is the sphere radius and η the coefficient of viscosity. Much of the clay fraction is disc-like rather than spherical, and shape places a limitation on the application of Stokes' equation for calculating particle size from the velocity of fall. Consequently, the clay fraction so separated is said to have an 'equivalent sphere diameter', abbreviated e.s.d., of less than 2 μm, that is, it is composed of particles that settle with a velocity equal to or smaller than 2 μm diameter spheres of the same density.

In most soils and sediments the clay fraction has properties of water retention and ion exchange that it would not have if it consisted of fine particles of the minerals found in the larger size fractions, which contain detrital primary minerals, mainly quartz and other forms of silica, with some feldspar, mica and heavy minerals. This change to material with active surface properties occurs somewhere in the particle size range 5 to 0.5 μm.

The clay fraction of soils contains both organic and inorganic material; this chapter is concerned with the structural chemistry of the minerals that form the inorganic part of the clay fraction. Most soil clays are a mixture of one or more aluminosilicate clay minerals with lesser amounts of iron oxides and oxyhydroxides; quartz and feldspars, oxides and hydroxides of aluminium and manganese and oxides of titanium are frequently present in small amounts, and carbonates, principally calcite, are also widespread. In special situations certain zeolites, iron sulphides, and sulphates such as gypsum, jarosite and alunite, phosphate minerals of the plumbogummite group and soluble salts such as halite may be present. Clays in soils formed from deposits of volcanic origin often consist largely of the poorly ordered or structurally disordered materials imogolite and allophane, but small amounts of so-called amorphous oxide gels containing silica, alumina and iron may occur in many soils; the latter are thought to occur both as discrete particles and as coatings on the surfaces of the crystalline particles. A small proportion of organic matter (1 % of the fraction or less) is commonly associated with the clay fraction, but the proportion may be much larger in the surface layers of soils.

2.1.1 Crystal Structure Concepts

Solids range from materials in which the internal atomic arrangement repeats periodically with perfect regularity in three dimensions over many thousands of units of structure, to glasses in which there is no periodic repetition and the only regularity is that imposed by chemical bonding of the constituent atoms. Reference has already been made to the crystalline nature of most of the components of clays, and it is necessary to define the term 'crystalline' as it is used with respect to clays.

The external regularity in the shape of crystals has long been attributed to the regular repeated arrangement in space of small identical structural units. X-ray diffraction has shown that most solids are characterized by atomic arrangements that are regular and periodic in three dimensions over distances large compared with the size of atoms. The arrangement of atoms in space that produces a particular crystalline substance can be specified in terms of the size and shape of a three-dimensional structural building unit, termed the unit cell, and the pattern of atoms within it. The shape and size of the unit cell, which must be a parallelepiped, is specified by the length of its edges, a, b and c

and the angles between them α, β and γ (Figure 2.1). The unit cell contains the complete unit of pattern, and is repeated by regular translations of a, b and c in three non-planar directions in space, the x, y and z axes respectively, to build up the crystal. When unit cells of a particular substance are placed together in space their corners define a space lattice, consisting of a regular three-dimensional arrangement of points. The space lattice, which is solely a geometrical concept, can be thought of as a reference framework or scaffolding on which the unit cells are hung. The origin of the unit cell can be taken at any point within the structure without changing its size or shape, and each different choice of origin produces a different but exactly parallel space lattice.

The position of each of the various atoms within the unit cell is given by its coordinates, x, y and z, which are usually presented as decimal fractions of the cell edges a, b and c respectively. The size of the unit cell of crystalline materials varies greatly from substance to substance, but for minerals the length of the cell edges is almost always in the range 0.3 to 3.0 nm. For example the common iron oxide mineral, goethite, has a unit cell in which the angles are $\alpha = \beta = \gamma = 90°$ and the sides are $a = 0.4605$, $b = 0.9952$ and $c = 0.3021$ nm.

The mathematically ideal crystalline state is one of perfect regularity, in which an identical pattern of atoms is infinitely repeated by exactly regular translations in three directions in space. No real material conforms to this ideal; all real crystals are limited in extent and depart from perfection to some degree. The majority consist of a three-dimensional mosaic of small blocks of almost perfect structure that are in slightly less than exact parallel alignment; these blocks are about 1 μm^3 in volume.

Many kinds of departures from perfect regularity are found in real crystals; some that are of frequent occurrence in clay minerals are outlined below.

Substitution disorder occurs when the structural pattern is repeated correctly in every unit cell but the atomic composition varies from cell to cell. For example, suppose Si is found at position (x, y, z) in one cell but in the next cell there is Al at (x, y, z), and perhaps Si in the following cell, and so on through the structure. If Si and Al alternate regularly on passing from cell to cell, the substitution is ordered and a larger unit cell would encompass an exactly repeating structural unit. A random, or at least a non-regular pattern of Al:Si replacement cannot, however, be usefully expressed by expanding the size of the unit cell and this type of non-regular replacement is a common form of substitutional disorder in clay minerals.

Many clay minerals have layer structures (defined and described on page 46) in which the atoms within a layer are strongly bonded to each other but the bonding between the layers is very much weaker, so that each layer is, to a good approximation, an independent structural unit. There are often several ways in which one layer can stack on top of another to give almost but not quite identical atom-to-atom contacts at the interface between the layers. These alternative arrangements, specified by directional displacements and rotations do not differ much in free energy because this depends mainly on the interactions

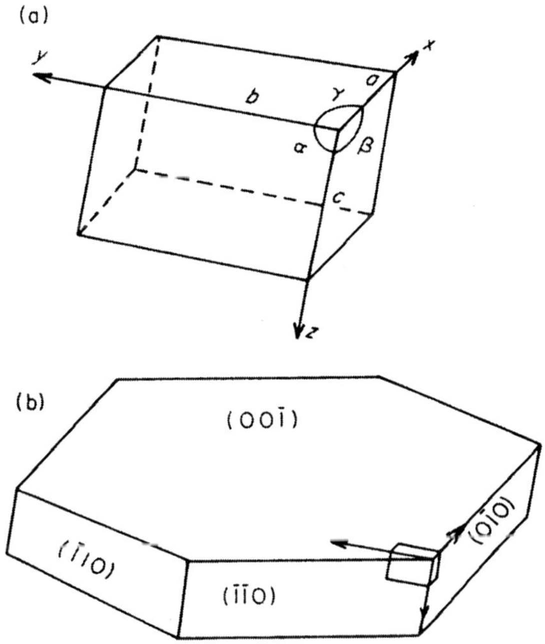

Figure 2.1 The unit cell of a crystal. It is a parallelepiped with angles α, β and γ and edges *a*, *b* and *c*. Positions of atoms in the crystal are usually given as *x*, *y*, *z* coordinates scaled to be fractions of the corresponding cell edges *a*, *b* and *c*. (a) Cell for kaolinite. (b) The outline of a crystal of kaolinite showing the orientation of one of its unit cells. The unit cell is $a = 0.515$ nm, $b = 0.895$ nm, $c = 0.715$ nm, $\alpha = 91.8°$, $\beta = 104.8°$, and $\gamma = 90.0°$. The crystal is about 9.0 nm wide, 10.3 nm long and 2.1 nm thick, about 10 cells by 20 cells by 3 cells, say 450 cells in all. This crystal is at the smaller end of the range of kaolinite crystals found in soils. The Miller indices of the planes forming crystal faces are shown (see Figure 2.5)

between nearest neighbour atoms. Thus, it is common to find layer silicate structures composed of structurally similar layers but differing in the ways that the layers are stacked. This is a special form of polymorphism called polytypism. When a given chemical compound exists in two or more forms with different crystal structures, these are called polymorphs. For example calcite and aragonite, both $CaCO_3$, are polymorphs. Polytypism is a special kind of polymorphism in which identical planes or layers can have different stacking arrangements and hence build up different regular structures. If the displacements are not the same at each layer-to-layer junction less regular structures are found. Structures with irregular layer displacements are common among clay minerals.

Because the surfaces of the layers of different kinds of clay minerals are structurally similar, layers of *different kinds* can fit together to form relatively stable structures. This kind of mixed structure is so common among clays that Section 2.3.8 is devoted to these mixed-layer or interstratified minerals.

The irregularities described above lessen the extent of structurally ordered regions. In so-called non-crystalline or amorphous solids, regions of regularly repeating structure are very small, extending over distances of no more than a few atom diameters. Usually such solids are called amorphous when the investigator has not found evidence of a periodically repeating structure. For example, failure to obtain an X-ray diffraction pattern is frequently taken as evidence that the material is non-crystalline. The distinction between crystalline and non-crystalline material is, however, imprecise, depending, as it does, on the experimental techniques used and the skill of the investigator.

When the term crystalline is used with reference to clays it usually means that there are regions of regularly repeating structure at least 10 to 20 atom diameters in size (about 3–6 nm). Materials in which the constituent atoms exist in a random arrangement or in which regularity exists over distances of less than 1.0 nm can probably be considered non-crystalline. In terms of these distinctions, most soil clays consist largely of crystalline materials, but the range of crystalline perfection is very great.

2.1.2 Crystal Chemistry

Full wave-mechanical treatments of complex solids such as the aluminosilicate minerals are impossible, but many features of the crystal structures of minerals can be visualized and understood on the basis of the *ionic model* which in its simplest form supposes the atoms to be fully ionized solid spheres of invariable radius; thus silicon appears as the cation Si^{4+}, aluminium as Al^{3+} and oxygen as the anion O^{2-} and so on. When a crystalline solid is formed the ions take up the arrangement in space that minimizes the electrostatic potential subject to the condition that the spheres come no closer than their contact distance.

Early crystal structure determinations on simple compounds such as halides and oxides showed that the distance between atoms could be represented as the sum of fixed ionic radii for each ion, so establishing the ionic model as a good approximation to interionic distances. Corrections were required to improve the fit when more structures were considered. Appropriate factors were deduced that allowed the tabulated radii to be adjusted when the coordination differed from the standard value of six. These corrections were relatively small for anions but appreciable for cations. Over the years several tables of ionic radii have been published (e.g. Shannon and Prewitt, 1969). The recent tabulation of Whittaker and Munkus (1970) is specially suitable for minerals and has been used in this chapter. For each ionic species radii are given for each coordination state found.

In all these tables, anions stand out as being generally larger than cations; the anions common in minerals, O, OH and F, have radii near 0.13 nm, whereas Si^{4+}, Al^{3+}, Mg^{2+} and Fe^{3+} are smaller, and of the common cations, only Na^{+}, K^{+} and Ca^{2+} are comparable in size with the anions (Figure 2.2). In a 'spheres in contact' model, the packing together of the anions dominates the structural framework and the smaller cations fit into the interstices of the anion matrix, so that anions form coordination groups around the cations. The commonest regular groupings that recur throughout mineral structures are four anions arranged tetrahedrally (Figure 2.3) and six anions arranged octahedrally (Figure 2.4) around the central cation.

The number of anions around a cation is controlled by the relative sizes of the anions and cations, and the simple ionic model imposes the conditions that (i) every anion in a coordinated group must be in contact with the central cation, and (ii) the maximum number of anions group themselves around the central cation, subject to the limitation imposed by (i). With these restrictions, simple geometry shows that the minimum radius ratio r_c/r_a(r_c = cation radius, r_a = anion radius) is 0.225 for tetrahedral coordination and 0.414 for octahedral coordination. More than six anions can be accommodated around the cation if $r_c/r_a > 0.645$.

The simple ionic model is only approximate; atoms in minerals are only partly ionized and some electrons are shared to form partially covalent bonds. Distances between bonded cations and anions vary depending on coordination number and other structural factors, so that the concepts of the simple ionic model and tables of ionic radii should not be interpreted too rigidly. Nevertheless, in practical structural terms it is found that for radius ratios between 0.2 and 0.4 the coordination polyhedra are usually fairly regular tetrahedra; if the

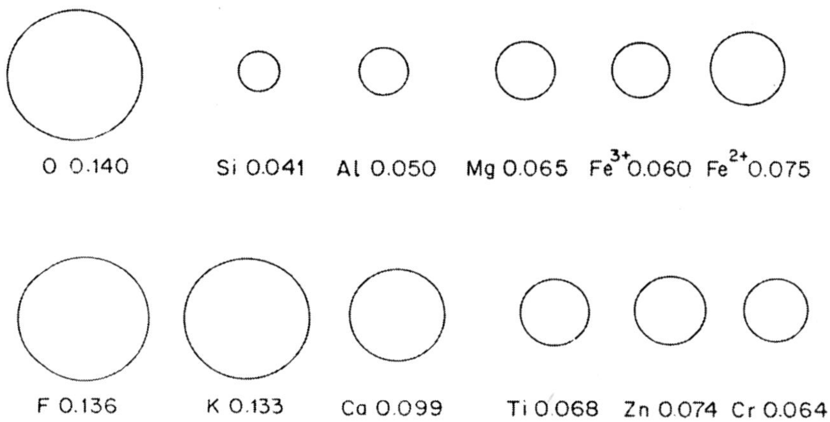

Figure 2.2 Ionic radii and relative sizes of ions commonly occurring in phyllosilicates (radii from Whittaker and Muntus, 1970, in nm)

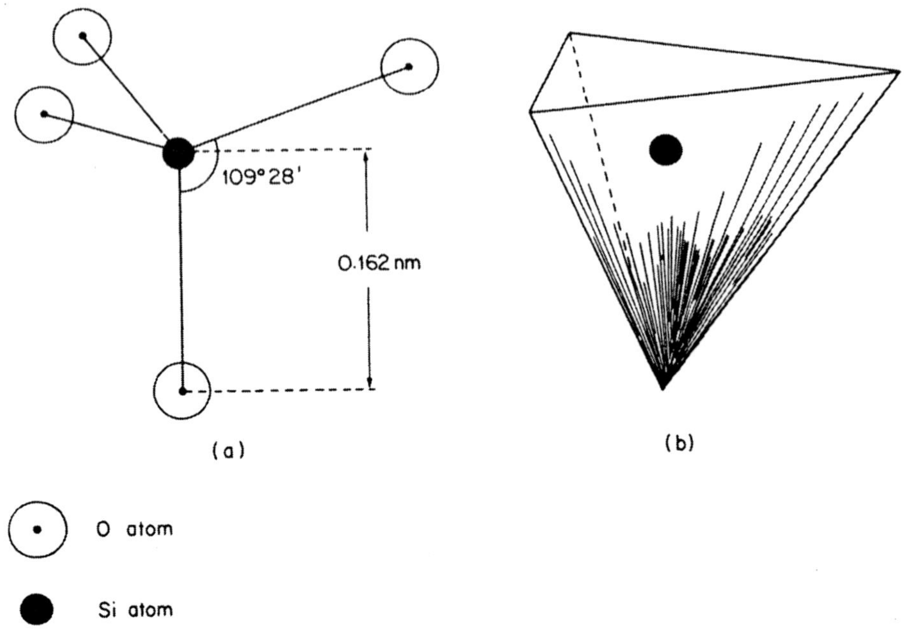

Figure 2.3 The tetrahedron formed by coordination of Si by four oxygens: (a) bond lengths and angle, (b) isometric representation

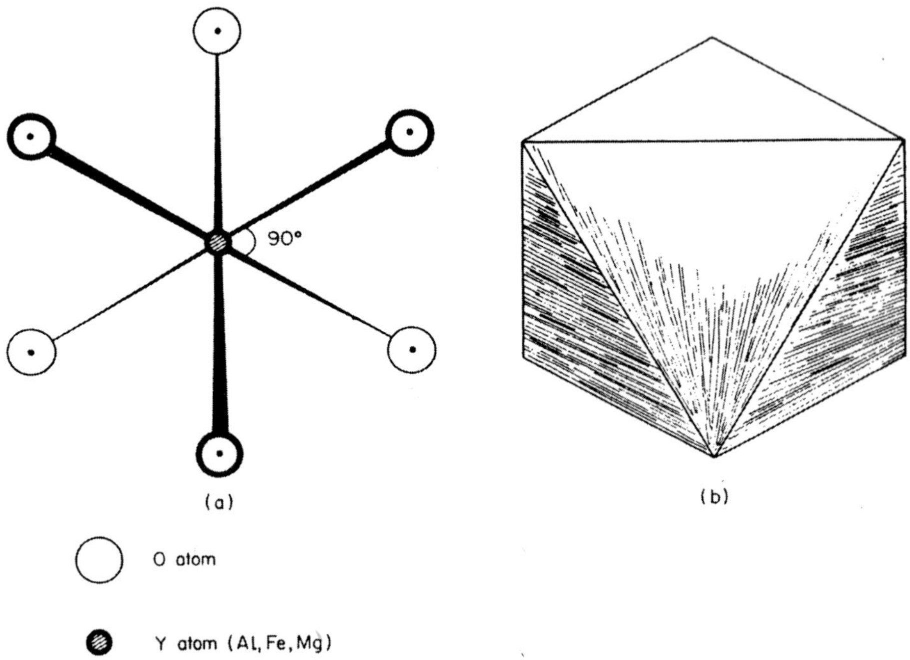

Figure 2.4 The octahedron formed by coordination of a cation (such as Al) by six oxygens: (a) as a 'ball and spoke' model and (b) an isometric representation

radius lies between 0.4 and 0.6 to 0.7, octahedral coordination is usually adopted, but the octahedra are frequently less regular than the tetrahedra. For ratios greater than 0.7, more than six anions surround each cation. In the minerals found in soils Si^{4+} is always found in tetrahedral coordination, Fe^{2+} and Mg^{2+} are almost invariably in octahedral coordination and the large alkali and alkaline earth cations, principally K^+, Na^+ and Ca^{2+}, are surrounded by more than six anions. The common cations Al^{3+} and Fe^{3+} are intermediate in size between Si^{4+} and Mg^{2+}, which are tetrahedrally and octahedrally coordinated respectively. It is therefore not surprising that Al^{3+} and Fe^{3+} may be found in both tetrahedral and octahedral sites.

Tetrahedral and octahedral groupings of O^{2-} (and OH^-) are the basic structural units of minerals. They occur as separate tetrahedra or octahedra but more commonly they are linked together by sharing corners, edges and, less frequently, faces. Pauling (1929) deduced a set of principles, now commonly known as Pauling's Rules, that describe the way polyhedra are likely to be linked in ionic crystals such as minerals. These rules are based on minimization of electrostatic potential in the crystal by means of balance of positive and negative charge locally on the atomic scale. The more important rules are:

1. Each cation is surrounded by anions, and the number of anions around a cation is governed by their relative sizes.
2. The formal charge on each anion is balanced within a small fraction of the charge of a single electron by the cations with which it is in contact. The charge of the cation is considered to be equally divided between the anions in its own polyhedron.
3. Corners are readily shared between adjacent polyhedra. Edges are less frequently shared and the sharing of faces is avoided. These restrictions are more important for small cations with larger charge than for large cations with smaller charge.

The formal charge on an ion is the charge the ion would have if fully ionized. For example the formal charge on an oxygen anion is -2. Thus Rule 2 is a statement of the fact that charges are balanced locally. Sharing of edges and faces is not favoured (Rule 3), because it tends to bring highly charged cations close together leading to strong electrostatic repulsion. When edges or faces are shared, the polyhedra are often distorted in a way that increases the separation of cations while maintaining the coordination.

In the fifty years since their formulation, the general validity of these rules for mineral structures has been amply verified by many structure determinations. The rules, particularly the second, impose quite stringent restrictions on structural arrangements of anions and cations, and can be used to forecast from chemical composition the most likely of several possible structures. This is particularly useful for clay minerals, for which detailed structure determinations are rarely possible, but chemical composition is more readily established.

2.2 EXPERIMENTAL METHODS

Many different experimental methods have contributed to present day knowledge of the crystal chemistry of soil minerals. In addition to chemical analysis probably the most important are diffraction methods using X-rays, electrons and neutrons. From the geometry of diffraction patterns of single crystals the size and shape of the unit cell and information on the symmetry of the atomic arrangement can be found; from the intensities of the diffracted beams the positions of the atoms, averaged over many unit cells, can be calculated. Supplementary to the diffraction methods various resonance techniques provide information about the local environment of specific types of atoms. The most widely used of these methods is infrared absorption spectroscopy which depends on the frequencies of vibrations or rotations of neighbouring atoms in the crystal. Mössbauer (recoil-less gamma-ray) resonance spectroscopy is now widely applied to study of the location of iron in mineral structures and recently electron paramagnetic resonance techniques have been applied to the study of clays (Angell *et al.*, 1974; Meads and Malden, 1975). Imaging techniques, which provide a recognizable 'picture' of the object, also provide structural information.

It is not the purpose of this Chapter to provide a text on experimental methods; an outline only will be given of the more important techniques indicating how they have been of value in soil mineral studies, and references to other publications where the methods are fully described.

2.2.1 Diffraction Methods

When a monochromatic beam of X-rays strikes a crystal each atom scatters a small fraction of the incident radiation. If the scattered X-rays are to produce diffracted beams of measurable intensity, the scattering from different atoms must reinforce each other. Bragg and Bragg (1913) showed how this complex problem could conveniently be regarded as one of reflection from sets of parallel planes in the space lattice of the crystal. The condition for reinforcement and therefore for production of a significant diffracted beam is

$$\lambda = 2d \sin \theta$$

where λ is the wavelength of the X-rays and θ is the angle the incident beam makes with the set of planes with interplanar spacing of d. Diffracted beams of measurable intensity are obtained only when these conditions, often known as Bragg's Law, are fulfilled. For X-rays, λ is of similar magnitude to the separation d, of parallel atomic planes in crystals, so that a convenient range of values is found for θ. There are many different sets of parallel planes in a particular space lattice; they are specified by Miller indices, (hkl), which are related to the intercepts made by the planes on the crystal axes, a, b and c (Figure 2.5). Different sets of planes usually have different interplanar spacings and it has become customary to refer to the diffracted maximum from the set of planes (hkl) as the hkl reflection.

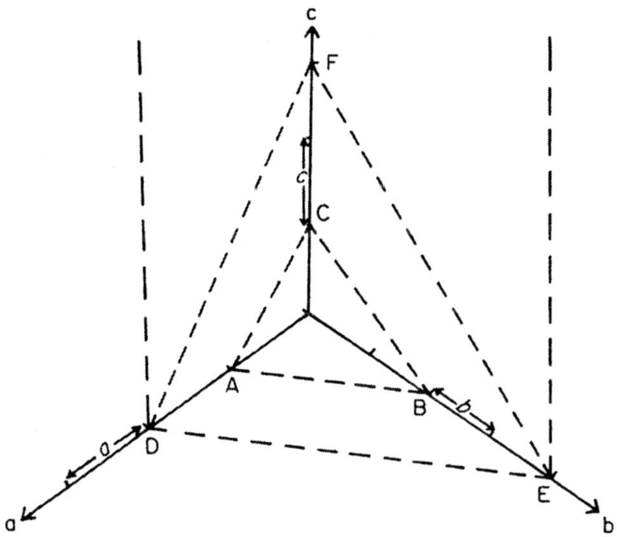

Figure 2.5 Derivation of Miller indices for planes intersecting a set of 3 axes (after Klug and Alexander, 1974). Reproduced by permission of John Wiley & Sons, Inc., New York. Planes cut the axes at integral multiples of lengths *a*, *b* and *c*; broken lines show where they cut the axial planes. The plane through D and E that runs parallel to the c-axis never cuts it, but can be thought of as cutting it at infinity, so that the reciprocal of the intercept is zero.

Plane	Intercepts	Reciprocals	Cleared of fractions	Miller indices of planes
ABC	*a*, *2b*, *c*	$\frac{1}{1}, \frac{1}{2}, \frac{1}{1}$	2, 1, 2	(212)
DEF	*2a*, *4b*, *3c*	$\frac{1}{2}, \frac{1}{4}, \frac{1}{3}$	6, 3, 4	(634)
DE	*2a*, *4b*, *c*	$\frac{1}{2} \; \frac{1}{4} \; \frac{1}{\infty}$	2, 1, 0	(210)

If the planes represent crystal faces, then *a*, *b* and *c* are not absolute lengths; but a ratio *a*:*b*:*c* can be found for which all the faces of a crystal have rational Miller indices. If *a*, *b* and *c* are true distances between identical repeating points in the crystal, then they are the edges of a unit cell (not necessarily the smallest). X-ray diffraction from the crystal can then be treated as reflection from parallel planes all with the same Miller index and this index, written without brackets, is used to name the diffraction maximum, e.g. the 003 reflection. It is sometimes convenient to represent the higher order maxima from a plane by Miller indices containing a common factor, e.g. 222 for the second order of diffraction from planes (111). Once an origin has been chosen it is possible for planes to cut the crystal axes along the negative direction of the axes. A bar is then placed over the appropriate index (cf. Figure 2.1)

Determination of the crystal structure by diffraction methods depends on two steps. First the size and shape of the unit cell is determined by the positions of the diffracted beams. This information taken together with the chemical formula and the density of the crystal allows the number of each of the different kinds of atoms in the unit cell to be found. The arrangement of the atoms in the unit cell is obtained by what is essentially a trial and error method. Taking account of likely atomic arrangements and interatomic distances possible structures are postulated consistent with the symmetry of the crystal. The intensities of many different reflections that would be given by the trial structures are calculated and compared with the measured intensities. When the intensities calculated for a feasible structure are in reasonable agreement with those observed, the structural model is 'refined' by altering the coordinates of the atoms until good agreement is obtained (Lipson and Cochran, 1953).

Determination of the structure of unknown material requires small single crystals (0.1 mm^3) unless the structure is very simple. Powders (which consist of many very small crystals oriented more or less randomly) also give diffraction patterns but in these, instead of beams, the reflections from the many small crystals are spread out into cones obeying Bragg's Law around the direct beam direction. The d spacings are derived from the wavelength λ, and the semi-angle of the cone, which equals 2θ. A clear description of X-ray diffraction methods is given by Cullity (1956) and Azaroff and Buerger (1958).

The random orientation of the crystallites in a powder specimen greatly decreases the proportion of the specimen that can contribute to any reflection. This, combined with the spreading out of diffracted beams into circles or ellipses, decreases the intensities greatly, so that very many fewer reflections can be observed from powders than from single crystals. In addition, unless the unit cell is small and symmetry is high, reflections from different planes often overlap, especially at larger values of 2θ, and so intensities cannot be unequivocally attributed to particular (hkl) planes. This makes structure determination of unknown material impossible using powder diffraction from any but the simplest materials. Nevertheless, X-ray powder diffraction from crystalline materials in finely divided form is of great value in soil mineralogy as a method of identification and semi-quantitative assessment (see page 41), as the d-values of the principal atomic planes are characteristic of each clay mineral.

Single crystal structure determinations by X-ray diffraction provide the basis of much of our current knowledge of the crystal chemistry of minerals. Positions of individual atoms can be found with an error of less than 0.001 nm. X-ray diffraction can locate all the atomic species common in minerals except hydrogen, but there is considerable difficulty in distinguishing atoms that are adjacent in the periodic table. This occurs because the main factor governing scattering power is the number of electrons in the atom and when fully ionized atoms such as Si, Al and Mg have the same number of electrons. Only the most refined methods can distinguish between Si and Al, both of which can occupy tetrahedral sites, on the basis of scattering power; however, their distribution is often inferred from the distance between the tetrahedrally coordinated cation (T) and the surrounding oxygens. The Si—O distance is about 0.162 nm; the Al—O distance for tetrahedral coordination is about 0.177 nm. If in structure determination of an alumino-silicate the average T—O distance for a particular site is found to be intermediate, it is inferred that these T sites may be occupied by either Si^{4+} or Al^{3+} and the T—O distance is determined by the proportion of each. For example, a T—O distance of 0.165 nm would signify an occupancy of $(Si_{0.75}Al_{0.25})$ averaged over many unit cells, i.e. that there is a 75 per cent probability that any of this set of T sites would be occupied by an Si^{4+} ion. By these methods X-ray single crystal diffraction can locate with great accuracy the atoms in crystals and give good estimates of the proportions of the different atomic species in structurally similar sites.

Procedures similar to those for X-rays are used when diffraction involves electrons or neutrons. Neutron single crystal diffraction gives structure determinations of accuracy similar to X-ray diffraction. The relative scattering factors of different atomic species for neutrons are different from those for X-rays. Hydrogen scatters more strongly and can be located accurately, and also because the difference in scattering between Si^{4+} and Al^{3+} is greater for neutrons, Si and Al can be more readily distinguished. However, a nuclear reactor is required to supply the neutrons for neutron diffraction and a single crystal not smaller than 1 mm^3 is required, much larger than the 0.1 mm^3 favoured for X-ray single crystal diffraction.

Electron diffraction (McConnell, 1967) is a poorer method than X-ray diffraction for accurate structure determination. Because the absorption and scattering of electrons is much greater, multiple scattering can occur even in crystals much thinner than 1 μm, so that the theory relating intensities to crystal structures is much more complex. Structure determinations made with electrons, therefore, are likely to be much less accurate than those made with X-rays or neutron diffraction.

Diffraction patterns are useful not only for determining crystal structures but for identifying crystalline materials; X-ray diffraction is probably the most widely used technique for identifying fine-ground crystalline materials. Each crystalline substance has a different crystal structure which produces a characteristic diffraction pattern that can be used for identification. An X-ray powder diffraction pattern is characterized by the positions and relative intensities of the reflections. The positions of the reflections are related by Bragg's equation to the *d* spacings of the reflections. The most convenient method of cataloguing the information is by means of lists of the *d* spacings of the reflections and their relative intensities; the latter are usually scaled so that the strongest reflection is given the value of 100. The Powder Diffraction File (Joint Committee on Powder Diffraction Standards, 1975) lists powder diffraction patterns of more than 25,000 substances and the Mineral Sub-file contains patterns of about 1900 minerals accompanied by an index. If the pattern of an unidentified mineral is in the File systematic methods are available to locate the pattern and so identify the unknown. The method can also be applied to identify the several components of a mixture of crystalline materials.

The powder diffraction pattern of a mixture is the superimposed patterns of the components, the contribution of each being proportional to its amount. This provides the basis for quantitative analysis of mixtures provided adequate standards are available and specimens can be made in which the crystallites are randomly oriented. These requirements can be met fairly well for some of the minerals occurring in soils but not for all. The phyllosilicate minerals are prone to preferred orientation because of their platey habit. For this reason and because of their chemical complexity and wide range of structural perfection, it is difficult if not impossible to find standards for general application. Feldspars are subject to similar difficulties. Reasonably accurate quantitative analyses can be made only for quartz, calcite and similar relatively simple substances.

The platey phyllosilicates tend to produce specimens in which the arrangement of the crystallites is not random. For example, when a suspension of clay is dried on to a flat surface such as a glass slide, the particles tend to lie flat, parallel to the glass surface. If such a specimen is examined by the normal reflection method in an X-ray powder diffractometer, reflections from the basal planes are greatly enhanced in intensity relative to other reflections, and so much so that frequently only the basal reflections can be seen. While this adds to the difficulty of making quantitative analyses, it is valuable for identification of the clay minerals, *sensu stricto*, which pose special problems when the conventional powder diffraction methods are used. As they are based on similar structural schemes, the majority of the reflections from most of the minerals are very similar in spacing. The main structural difference between the different clay mineral species is the repeat distance perpendicular to the aluminosilicate layers, often loosely referred to as the 'layer thickness'. The reflections from planes parallel to the layers (basal or $00l$ reflections), which give a measure of the 'layer thickness', are increased in relative intensity by preferred orientation. This forms the basis of the most commonly used technique for identification of clay minerals in materials such as soil clays. Different clay mineral groups have different layer thicknesses or can be made to have different layer thicknesses by a range of treatments. For example, in the normal air-dry condition some clays contain water molecules between the aluminosilicate layers; the amount of water absorbed and the layer thickness depends on the nature of the interlayer cation (*vide infra*). Heating drives off the interlayer water and decreases the 'layer thickness'. Certain organic liquids such as ethylene glycol or glycerol can replace the water between the layers and change the 'layer thickness' (Section 2.3.6). Examining specimens that have been subjected to a series of treatments usually allows recognition of the main types of clay minerals present in an unknown clay.

Electron diffraction can also be used for identification. Polycrystalline specimens give ring patterns analogous to X-ray powder diffraction patterns (Plate 2.7c). The measured spacings are considerably less accurate than those obtained by X-ray diffraction and because the specimen takes the form of a thin film problems caused by preferred orientation are more severe. The technique known as selected area electron diffraction in which the diffraction pattern from a small (< 1 μm) single crystal is recorded, is more valuable. The pattern consists of a series of spots each of which, if the specimen is of the correct thickness, represents a single *hkl* reflection. These spot patterns give information about the shape and size of the unit cell and about the symmetry of the atomic arrangement which is often sufficient for the identification of sub-micron particles.

2.2.2 Resonance Methods

Infrared absorption is the resonance method most widely used for structural studies of soil minerals (Farmer, 1974). Mössbauer spectroscopy has more recently been found to be valuable for studying the distribution of Fe^{2+} and Fe^{3+} ions in mineral structures. Both depend on the absorption of radiation of the appropriate frequency by processes involving quantized changes in energy levels in atoms or by groups of atoms. Interpretation of the spectra in both techniques is largely empirical in the sense that assignments of absorption bands to specific structural features have been built up by comparison of spectra of simple compounds and minerals for which the interpretation is unequivocal.

Infrared spectroscopy

When infrared radiation (4000–200 cm^{-1}) of a given frequency strikes a specimen containing a bond between two atoms with a corresponding vibration frequency, energy is absorbed; radiation of other frequencies is not absorbed. Most infrared absorption bands result largely from motions involving adjacent atoms and so they are sensitive to local (nearest neighbour) arrangements of atoms. This contrasts with X-ray diffraction which relies mainly on the extended periodic repetition of a pattern of atoms and gives structural information averaged over many units of structure. Infrared radiation is equally applicable to gases, liquids and glasses, as well as crystalline solids; it is therefore valuable in soil studies for investigating poorly crystalline and amorphous materials that are not so readily amenable to study by diffraction methods.

A major use of infrared absorption spectroscopy in mineralogy has been to provide information on the bonding of hydrogen, e.g. whether it is present as OH or H_2O (Chapter 6). In addition, analysis of the OH and (H_2O) vibrations gives information not only about the hydrogen but also about the cations in the immediate environment to which the OH or (H_2O) groups are bonded. Replacement of hydrogen by deuterium changes the frequency of the vibrations in which they are involved. This is often useful in confirming assignments and also can often provide a method for recognizing hydrogens in structurally distinct positions. For example, water sorbed on surfaces or in accessible interlayer regions of expanding clays can be readily exchanged by treatment with D_2O at room temperature whereas deuteration of structural hydroxyl groups usually requires treatment at 300 °C or so to effect exchange of D for H.

Information about the way sorbed molecular species are linked to surfaces can be obtained by studying changes in infrared spectra brought about by sorption (Sheppard, 1959).

Infrared absorption spectra can also be used for qualitative and quantitative analysis of minerals. Although its application is more limited it offers some advantages and has some disadvantages compared with the more widely applicable X-ray powder diffraction method. The infrared spectrum is sensitive mainly to chemical rather than structural features, and so it allows ready recognition of broad classes of compounds such as nitrates, carbonates, and hydrates, in which the particular chemical grouping gives rise to a characteristic absorption band. It requires only small amounts of material and for certain materials, for example quartz and gibbsite, it is extremely sensitive providing an excellent rapid method for quantitative analysis.

However, for complex mixtures such as soil clays that are broadly similar chemically, it is difficult if not impossible to disentangle the overlapping absorption bands and so although some components can be identified with certainty even when present in very small amounts, others present as major components cannot. Its extreme sensitivity to chemical changes means that quite small variations in composition within a single mineral group can make large changes in the infrared spectrum and so identification of the components of a multi-component mixture is frequently impossible from the infrared spectrum alone.

Mössbauer spectroscopy

In 1957 R. L. Mössbauer discovered the phenomenon of recoil-less emission and resonant absorption of nuclear gamma rays which occurs when the emitting and absorbing nuclei are restrained from recoil by chemical bonds in solids. The narrow line width of these absorption bands allows nearby energy levels in the nucleus to be observed. The energy of the emitted gamma rays obtained from a suitable radioactive source (for example ^{57}Co in copper) is varied using the Doppler effect, by moving the source relative to the absorber. The Mössbauer spectrum is displayed as a graph that plots absorbance against energy, which is usually given in terms of the relative velocity of source and absorber.

The Mössbauer effect can be observed with a limited number of nuclei of which only ^{57}Fe is sufficiently abundant in soil minerals to make the technique useful. The iron atoms in a mineral can be in the ferrous or ferric form and each of these species may be situated in several distinct positions in the crystal structure. For ^{57}Fe in minerals, there are two absorption bands for each different type of iron atom. These doublets are characterized by two parameters, quadrupole splitting, given by the energy separation of the doublet peaks, and chemical isomer shift, which is the displacement of the midpoint of the doublet relative to some standard substance such as iron foil. The quadrupole splitting is determined by the electric field gradient at the iron nucleus and the chemical isomer shift is proportional to the electron density at the site of the nucleus. Mössbauer spectra therefore give information about the electronic and electrostatic environment of iron atoms in crystals.

Absorption bands are broad and from most minerals the spectrum consists of several overlapping bands that have to be resolved into their component doublets. This is done by computer fitting of the observed spectrum to one or more pairs of doublets on the basis of certain constraints relating to peak shape. As in infrared spectroscopy, assignment of bands is mostly based on knowledge derived from comparisons of the spectra of simple iron-bearing compounds.

Mössbauer spectra can generally indicate the oxidation state and the nature of the coordination and bonding of the different types of iron atoms in a mineral structure. Quantitative estimates of the proportions of the different kinds of iron atoms can be made from the area under the resolved absorption peaks. When properly calibrated Mössbauer spectroscopy provides an accurate non-destructive method for determining Fe^{3+}/Fe^{2+} ratios that requires only a small sample (50 to 100 mg for a mineral with about 10 % Fe). In structures such as biotite micas in which Fe^{2+} and Fe^{3+} may occur in either or both the crystallographically distinct octahedral cation sites, the proportions of Fe^{2+} and Fe^{3+} in each of the sites can be determined more readily than by X-ray structure determinations.

2.2.3 Imaging Techniques

The most familiar imaging technique is the microscope which, using visible light, produces enlarged images of the object. The petrological microscope was for long the main tool of mineralogists and geologists for identifying minerals and studying their arrangement in rocks and it is still widely used. Its resolution however, limits its use to objects larger than 5 to 10 μm.

Electrons, however, have a much shorter wavelength than light and can also be focused to form enlarged images. Scanning electron microscopy (SEM) which makes use of back-scattered electrons allows the three-dimensional arrangement of submicron particles exposed on surfaces to be examined directly. Transmission electron microscopy (TEM) readily provides information on the shape and size of separate sub-micron particles at magnifications up to 100,000 or more (Plate 2.7). The magnification provided by SEM is less than in TEM but depth of focus is greater (Plate 2.6). In addition, textural features such as dislocations and twinning on a submicron scale can be observed by suitable TEM techniques on specimens that are sufficiently thin. Recently it has been shown that the specialized technique referred to as 'high resolution electron microscopy' can be used to provide two-dimensional images of crystal structures that have a resolution of 0.3 to 0.4 nm (Plate 2.3(c)). This so-called 'lattice imaging' or 'structure imaging' can in favourable circumstances show the relative positions of the heavier atoms, i.e. provide a much-magnified image of a projection of the crystal structure in which the positions of atoms can be seen.

The morphological features shown by electron microscopy are rarely sufficiently characteristic to allow unequivocal identification of minerals in soil clays. However, TEM combined with selected area diffraction is a powerful tool for identifying very small isolated particles. It is also invaluable in providing information about relative size, shape and surface morphology of soil particles (Section 2.7).

In this brief account it is not possible to deal fully with any of the techniques discussed above. In addition to the references in the text, more extensive accounts of the various techniques are covered in the books and articles mentioned below. The principles of crystallography are dealt with by Phillips (1955) and the application of X-rays to the study of crystals by Bragg (1933). The fundamental principles of X-ray diffraction are explained in James (1954) and Warren (1969) and the determination of crystal structures by X-ray diffraction is the subject of a book by Lipson and Cochran (1953). A general text suitable for beginners on the interpretation of X-ray diffraction patterns is provided by Henry, Lipson and Wooster (1951), whereas the books by Cullity (1956) and Klug and Alexander (1974) are particularly concerned with X-ray diffraction by polycrystalline (powdered) material. Bragg and Claringbull (1965) provide an excellent brief account of the analysis of crystal structures and the general principles governing the structures of minerals as well as a comprehensive survey of the atomic structures of a wide range of minerals. Brown (1961) deals with the application of X-ray diffraction to the study of clays. The study of minerals by electron microscopy and electron diffraction is covered by McConnell (1967) and Wenk (1976), while Gard (1971) is particularly concerned with the application of these techniques to clays. Allpress and Sanders (1973) give a clear account of 'lattice imaging'. Information on infrared absorption spectroscopy and its application to clay studies is given by Lyon (1967) and Farmer (1974). Bancroft's (1973) book gives a detailed account of Mössbauer spectroscopy and the papers by Bancroft and Maddock (1967), Bancroft and Stone (1968) and Bancroft and Brown (1975) exemplify the application of this technique to the study of silicates.

2.3 LAYER SILICATE MINERALS

2.3.1 Structural Principles

Layer silicates, also called phyllosilicates, are defined (Bailey *et al.*, 1971b) as containing 'continuous two-dimensional tetrahedral sheets of composition Z_2O_5' (where Z is the tetrahedrally coordinated cation, usually Si^{4+} or Al^{3+}), 'in which individual tetrahedra are linked with neighbouring tetrahedra by sharing three corners each. The fourth tetrahedral corner may point in any direction' (but in most layer silicates the fourth corners all point in the same direction). 'Tetrahedral sheets are linked in the unit structure to octahedral sheets, groups of coordinated cations or individual cations.' This definition was made complex to include all known layer silicate structures, but as it is difficult for the newcomer to clay minerals to visualize these structures without further explanation, the first part of this section illustrates and elaborates this definition.

Layer silicates commonly contain, besides oxygen and hydrogen, the elements Si, Al, Fe, Mg, Ca, Na and K; small amounts of Ti and Mn may also be present, and uncommon, rare or synthetic layer silicates are known that contain substantial quantities of Li, Be, N, Ga, Ge, V, Cr, Ni, Cu and Zn. The structural classes of layer silicates are most conveniently described by the ways that the tetrahedral ZO_4 (Z = Si, Al) and octahedral YO_6 (Y = Al, Fe, Mg, etc.) coordination groups that they contain are linked together in three dimensions. The geometry of the SiO_4 group varies little in silicate minerals; the O—Si—O bond angles are all very close to the tetrahedral angle 109°28′ and the Si—O bond length is 0.162 nm (Figure 2.3a). It is therefore a useful simplification in any mineral structure to describe first the arrangement of the ZO_4 groups before considering the larger coordination groups. Sometimes the ZO_4 group is represented as bond length and angles, as in Figure 2.3a, at other times it may be clearer to represent the group as a solid tetrahedron formed by joining the centres of the 4 oxygen atoms (Figure 2.3b), the silicon atom at the centre being optionally omitted. Both representations of ZO_4 and linked ZO_4 groups will be used in this chapter; likewise, octahedral YO_6 groups may be represented either in the bond length and angle form (Figure 2.4a) or as the solid form (Figure 2.4b). It is now necessary to define *plane*, sheet, and layer. We will follow Bailey *et al.* (1971a) who state: 'Recommended usage is as a single *plane* of atoms, a tetrahedral or octahedral *sheet*, and a 1:1 or 2:1 *layer*. Thus plane, sheet and layer refer to increasingly thicker arrangements. A sheet is a combination of planes, and a layer is a combination of sheets. In addition, layers may be separated from one another by various *interlayer* materials, including cations, hydrated cations, organic molecules and hydroxide octahedral groups and sheets'.

The architecture of the layer silicates is in most instances comparatively simple. Each ZO_4 coordination group shares oxygen atoms with three neighbouring ZO_4 groups to form rings containing 6Si and 6O, each ring being joined to a neighbouring ring through shared oxygen atoms (Figures 2.6a and b). The oxygen atoms shared between every ZO_4 group are approximately coplanar forming a basal plane of oxygens that extend through the structure (Figure 2.7) and are called the *basal oxygen atoms*, coded O_B. The fourth oxygen atom of each ZO_4 is, except in sepiolite and palygorskite (*q.v.*), oriented in approximately the same direction normal to the basal plane. These are called the *apical oxygen atoms*, coded O_A. The apical oxygen atoms are also coplanar, and the part of the structure between this plane and the plane of the basal oxygen atoms is the *tetrahedral sheet*. This structural element is present in all layer silicates.

The other main structural element in layer silicates is an *octahedral sheet* that contains cations in YO_6 coordination between two planes of oxygen atoms (Figure 2.8). The octahedral sheet can be regarded as built up from octahedra sharing edges, that is, each Y cation shares two oxygen atoms with the Y cation in an adjacent octahedron. There are two ways that this sharing occurs. If all the

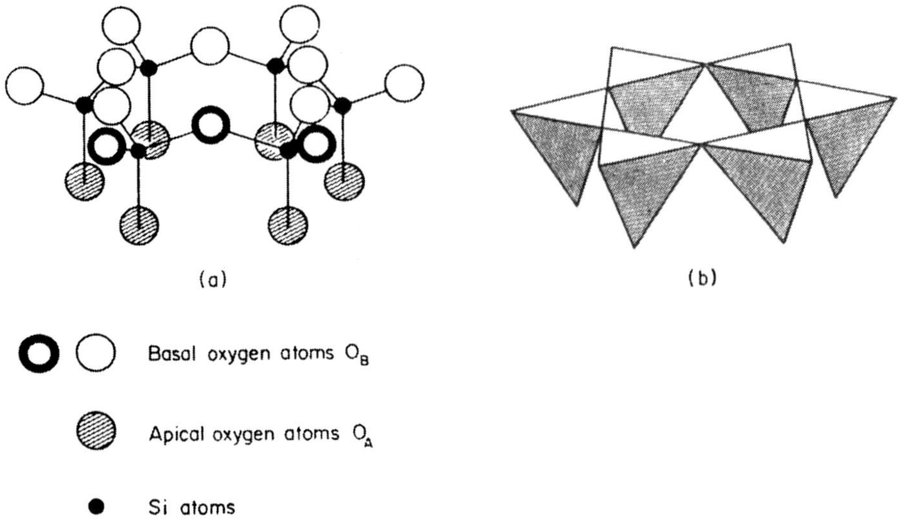

Figure 2.6 Linked SiO_4 groups in one Si_6O_6 ring (a) as a 'ball and spoke' model and (b) as linked tetrahedra

octahedral sites contain cations, each shares two oxygen atoms with each of six neighbouring cations (Figure 2.9a), whereas if only two thirds of the sites contain cations, each Y shares two oxygens with only 3Y neighbours (Figure 2.9b). When all or nearly all the interstices are occupied, the sheet is said to be *trioctahedral* whereas a sheet with only two thirds of the octahedral interstices filled is *dioctahedral.*

The anion groups coordinated to the Y cations are $(OH)_6$, $(OH)_4O_2$ or $(OH)_2O_4$, depending on the structural class of the layer silicate, and as the cation can occupy either four or six of the octahedral interstices, there are six possible types of octahedral sheet, with the following compositions:

$$\text{Dioctahedral: } Y_2(OH)_6; \quad Y_2(OH)_4O_2; \quad Y_2(OH)_2O_4$$

$$\text{Trioctahedral: } Y_3(OH)_6; \quad Y_3(OH)_4O_2; \quad Y_3(OH)_2O_4$$

(a) If all the anion groups are OH, the sheet is complete without further coordination of the oxygens; such sheets occur singly, alternating with silicate layers, in the chlorites, and comprise the only structural units in hydroxide minerals such as brucite, $Mg(OH)_2$ and gibbsite, $Al(OH)_3$; they are known as *hydroxide sheets* (Section 2.4).

(b) Octahedral sheets of composition $(OH)_4O_2$ are present in the kaolinite and serpentine group minerals. One plane of groups is entirely (OH), the other has the composition $(OH)O_2$ and the oxygen atoms are shared with tetrahedral Z atoms, forming the apical oxygen atoms of a tetrahedral sheet superimposed on the octahedral sheet; this junction is illustrated in Figure 2.10, which represents the ideal structure of the clay mineral kaolinite. Each layer in this group of

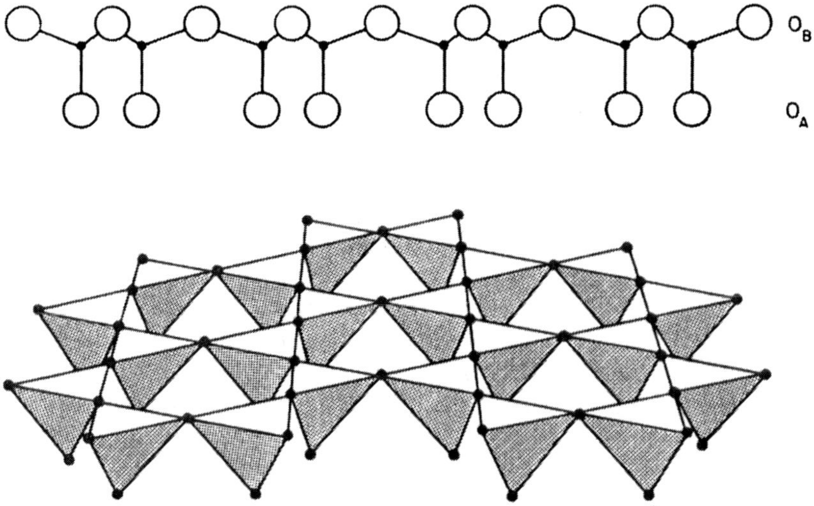

Figure 2.7 Linked Si_6O_6 rings in 1:1 and 2:1 layer silicates

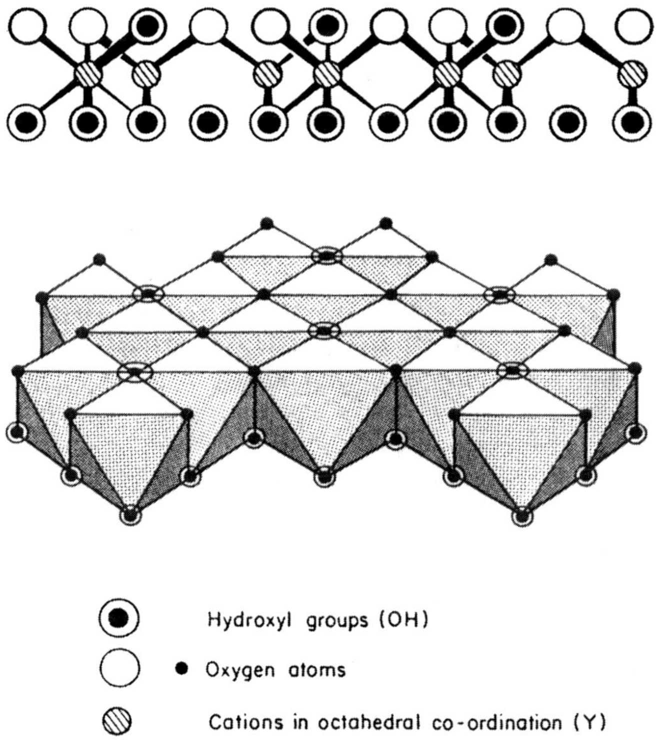

Figure 2.8 An ideal structure for a dioctahedral sheet in kaolinitc

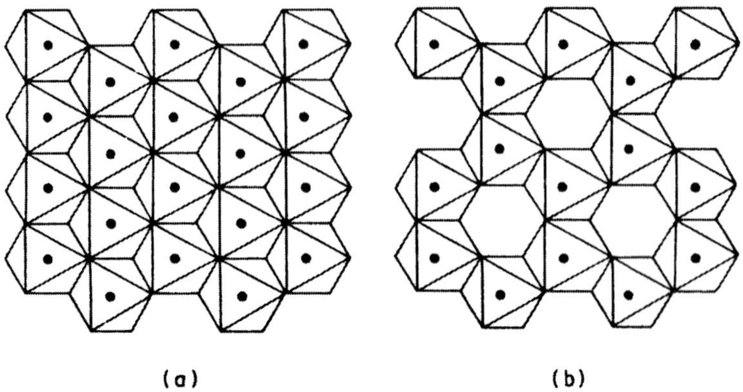

Figure 2.9 Types of octahedral sheet: (a) trioctahedral, and (b) dioctahedral

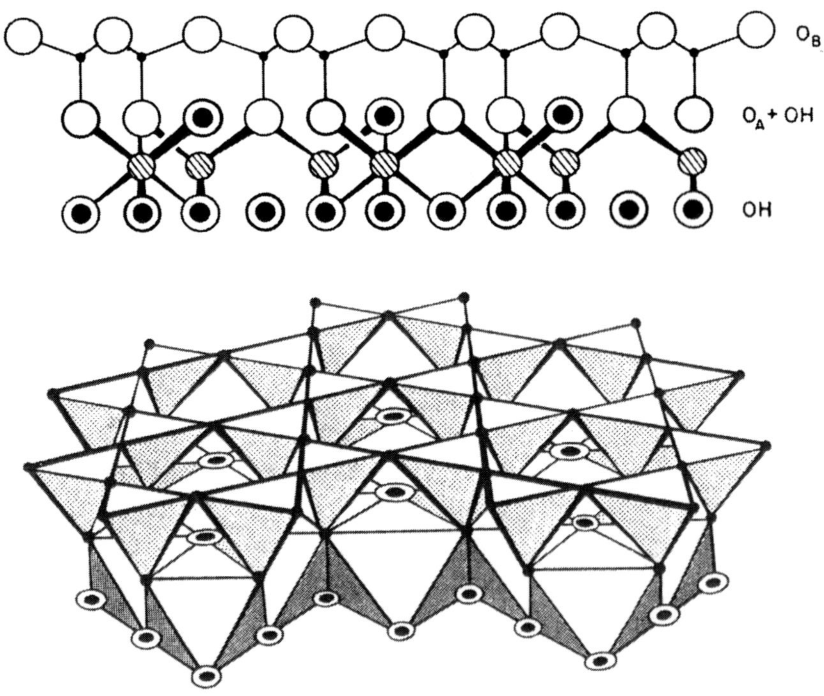

Figure 2.10 An ideal structure for a 1:1 dioctahedral layer silicate

minerals contains one tetrahedral sheet and one octahedral sheet and the minerals are termed 1 : 1 layer silicates.

(c) Octahedral sheets of composition $(OH)_2O_4$ occur in the mica, vermiculite, smectite, pyrophyllite and talc minerals. In these, both planes have the composition $(OH)O_2$ and tetrahedral sheets are superimposed on both sides of the

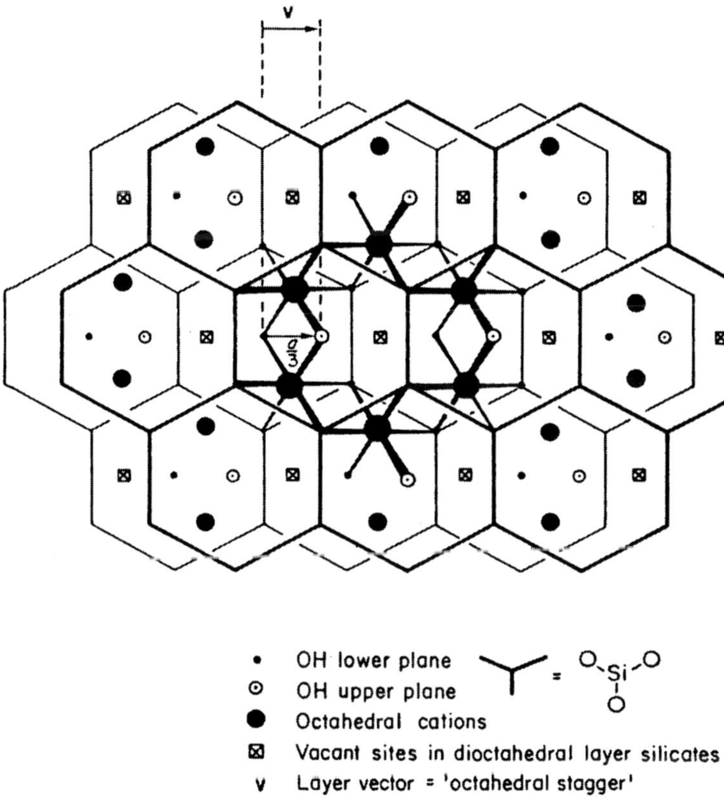

Figure 2.11 An ideal structure for a 2:1 dioctahedral unit layer

octahedral sheet, one tetrahedral sheet being inverted with respect to the other. The layer unit consists, therefore, of two outer tetrahedral sheets and an inner octahedral sheet, and minerals containing such layers are called 2:1 layer silicates. Because the oxygen atoms in the upper plane of the octahedral sheet are displaced relative to those in the lower plane (by the 'octahedral stagger'), the Si_6O_6 rings of the upper tetrahedral sheet are also displaced with respect to those of the lower tetrahedral sheet (Figure 2.11).

(d) The chlorites contain two types of layer, an hydroxide sheet alternating with a 2:1 silicate layer, and they have sometimes been referred to as 2:1:1 or 2:2 minerals; it is now recommended that they are called 2:1 minerals with an hydroxide interlayer (Bailey *et al.*, 1971b).

From these structural considerations, it can be seen that the layer silicates can be represented by ions in tetrahedral and octahedral coordination and that the numbers of such ions bear a relatively simple relationship to the oxygen and hydroxyl construction of the mineral. In general, the chemical constitutions of the layer silicates are complex and are described in the sections on individual mineral groups; however, a few have relatively simple compositions and exemplify the 1 : 1 and 2 : 1 groups. Thus in kaolinite, in which silicon occupies all the tetrahedral sites and aluminium two thirds of the octahedral sites (the remaining third of the sites being vacant), the composition is $Si_4^{iv}Al_4^{vi}O_{10}(OH)_8$ where ^{iv} is used to indicate tetrahedrally coordinated ions and ^{vi} those octahedrally coordinated. This may be represented in terms of the number of ions on each plane in the unit cell in a single alumino silicate layer as:

		Charges −	Charges +
6O		12	
	4Si		16
4O+2OH		10	
	4Al		12
6OH		6	
		28	28

Clearly the charge on the unit cell is balanced.

In pyrophyllite, a dioctahedral 2:1 mineral, the corresponding formula is $Si_8^{iv}Al_4^{vi}O_{20}(OH)_4$ and the number of ions in each plane represented by

		Charges −	Charges +
6O		12	
	4Si		16
4O+2OH		10	
	4Al		12
4O+2OH		10	
	4Si		16
6O		12	
		44	44

Both minerals have almost invariant compositions but this is exceptional in layer silicates; in most of these minerals some Al replaces Si in tetrahedral sites, some Mg and Fe replace Al in dioctahedral sheets, and an even wider range of elements can replace Mg in trioctahedral sheets. The convention for expressing these replacements in a general way is to write the formulae as

$$(Si, Al)_8^{iv}(Al, Fe, Mg)_4^{vi}O_{20}(OH)_4 \text{ for the dioctahedral}$$

and

$(Si, Al)^{iv}_8(Al, Fe, Mg, Ti)^{vi}_6O_{20}(OH)_4$ for the trioctahedral 2:1 minerals.

2.3.2 Chemical Constitution of the Major Clay Mineral Groups

Minerals are defined and classified not only by their structure but also by chemical composition, determined by analysis and initially expressed as an elemental composition per unit weight of mineral. Before the structures of minerals were known, these chemical compositions were recalculated as relative proportions of oxides, for instance the chemical composition of kaolinite was written $Al_2O_3 \cdot 2SiO_2 \cdot 2H_2O$. This only gave satisfactory formulae for minerals with a simple composition. Many in fact have complex chemical constitutions. For example the compositions of two micas (biotites), calculated in this way and normalized in terms of their alumina contents, might be written

$$35SiO_2 \cdot 20Al_2O_3 \cdot 20MgO \cdot 5.6FeO \cdot 5K_2O \cdot 12H_2O$$

and

$$37SiO_2 \cdot 20Al_2O_3 \cdot 26.4MgO \cdot 0.9FeO \cdot 5.7K_2O \cdot 13.3H_2O.$$

Although the relative proportions of oxides in these formulae are fixed by the chemical analyses, the actual numbers are arbitrary, and do not adequately reflect their mineralogical similarity.

Rational structural formulae

The application of X-ray diffraction to minerals showed that their composition was related to structure and could be recalculated to give the numbers of atoms in the unit that repeats throughout the crystal, the unit cell. For crystals of uniform and fixed composition, and with dimensions of at least 0.1 mm, the method is usually straightforward: the unit cell volume is calculated using lattice parameters determined by X-ray diffraction and the density of the crystal is determined by one of the standard methods, such as liquid displacement or flotation. The total mass of the unit cell in atomic mass units (1 a.m.u. = atomic weight of carbon/12) is

$$U = NDV \times 10^{-21} \text{ a.m.u.}$$

where D is the density in g cm^{-3}, V is the volume in nm^3 and $N = 6.022 \times 10^{23}$, Avogadro's number. In many methods of analysis, the chemical composition is determined as the element concentration in the solid and then recalculated as percent oxide for each constituent, say q percent Q_xO_y, that is, every 100 a.m.u. contains q a.m.u. of Q_xO_y, or xq/M atoms of Q and yq/M atoms of O, M being the formula weight of Q_xO_y. Therefore by simple proportion the number of Q atoms in the unit cell mass U is $602.2DVxq/100M$, and the number of O atoms is $602.2DVyq/100M$.

For example, the density of alpha quartz is 2.65 g cm^{-3} and the unit cell volume is 0.113 nm^3 so that the unit cell mass $U = 602.2 \times 0.113 \times 2.65 = 180.3$ a.m.u. The determined composition might be 46.6 % Si, with 0.2 % other cations; oxygen, by difference, is therefore 53.2 %. The number of Si atoms per unit cell is $(180.3 \times 46.6)/(100 \times 28.086) = 2.992$, and the number of O atoms is $(180.3 \times 46.6)/(100 \times 15.9994) = 5.996$, so that the rational formula would be Si_3O_6. The convention for expressing unit cell contents when these differ from the normal chemical formula is to use the symbol 'Z' for the integer that multiplies the chemical formula to give the unit cell contents; in the present example, the formula is SiO_2 with $Z = 3$. Although in this simple example, there is little advantage of the rational formula over the simple formula SiO_2, the derivation of unit cell contents is necessary for a chemically more complex mineral like the mica muscovite, and the calculation for an example of this mineral is given in Table 2.1.

This calculation shows that the *total* number of cations in the unit cell is 36.1 and the *total* number of oxygen anions is 48.01, even though the numbers of each individual cation are, in most instances, not close to a whole number. A survey of the structural formulae of many minerals shows that while the total number of *cations* in a unit cell need not be integral, *the total number of anions invariably is a whole number*, and further, that for minerals in the same structural group, *the number of anions* is the same in each mineral, *or is a small integral multiple of the number in the smallest unit cell in the group.* This latter qualification is necessary because in some layer silicates (muscovite is one such example) successive layers differ in orientation and the crystallographic repeat unit spans more than one layer unit, but the chemical composition is conventionally referred to the anion content of a single layer, using Z to specify the number of formula units in a unit cell.

The non-integral numbers of individual cations show that these must, in many cases, be distributed statistically between many unit cells. Detailed crystal structure determinations of site occupancy or bond lengths is needed to reveal whether there is any tendency to ordering of cations into particular positions in the structure.

For the layer silicates, the anion contents per unit layer of the main structural classes are:

1:1 minerals: O_{18}, of which 8 are usually (OH),
2:1 minerals: O_{24}, of which 20 are oxygen and 4 are (OH) though sometimes F replaces (OH),
2:1 + hydroxide: O_{36}, of which 20 are oxygen and 16 are (OH).

Structural formulae for clays

A rigorous recalculation of chemical composition to unit cell contents is often impossible for clays. Firstly, clay minerals, and soil clays in particular frequently have disordered structures so that unit cell and lattice parameters can no longer be determined. An assumed structural unit has to be used to

Table 2.1 Calculation of the structural formula for muscovite, a mica mineral
The unit cells is monoclinic:

$$a = 0.518 \text{ nm}, \quad b = 0.902, \quad c = 2.004 \text{ nm}, \quad \beta = 95°35'$$

Unit cell volume $= abc \sin\beta = 0.932 \text{ nm}^3 = 932 \times 10^{-24} \text{ cm}^3$
Density—$D = 2.834 \text{ g cm}^{-3}$
$N = 6.022 \times 10^{23}$ molecules per mole
Unit cell mass $U = DVN = 2.834 \times 932 \times 10^{-24} \times 6.022 \times 10^{23} = 1591$

(1) Oxide	(2) Oxide formula weight	(3) Composition per cent	(4) Formula per cent	(5) Formula per unit cell mass	(6) Cations per unit cell	(7) Anions per unit cell
SiO_2	60.09	45.55	0.7580	12.06	12.06	24.12
Al_2O_3	101.94	36.89	0.3619	5.76	11.52	17.27
TiO_2	79.9	0.26	0.0033	0.05	0.05	0.10
Fe_2O_3	159.7	0.39	0.0024	0.04	0.08	0.12
FeO	71.85	0.86	0.0120	0.19	0.19	0.19
MgO	40.32	0.58	0.0144	0.23	0.23	0.23
Na_2O	61.98	0.80	0.0129	0.21	0.42	0.21
K_2O	94.2	10.17	0.1080	1.72	3.44	1.72
$H_2O(+)$	18.016	4.59	0.2548	4.05	8.11	4.05
Totals		100.09			36.10	48.01

Column (4) = Column (3) ÷ Column (2)
Column (5) = Column (4) × 1591 ÷ 100
Column (6) = Column (5) × number of cations in oxide formula
Column (7) = Column(5) × number of oxygen atoms in formula

Each unit cell spans two aluminosilicate layers so that the layer or basal spacing for muscovite is $\frac{1}{2}c \sin\beta = 0.997$ nm. The layer unit contains 18.05 cations and 24 anions, and the unit formula for one layer unit is

$$Si_{6.03}Al_{5.76}Ti_{0.03}Fe^{3+}_{0.04}Fe^{2+}_{0.10}Mg_{0.12}Na_{0.21}K_{1.72}O_{19.95}(OH)_{4.05}$$

with Z = 2, that is, there are two unit formulae per unit cell. This may be written

$$[K_{1.72}Na_{0.21}]^{xii}[Si_{6.03}Al_{1.97}]^{iv}[Al_{3.79}Ti_{0.03}Fe^{3+}_{0.04}Fe^{2+}_{0.10}Mg_{0.12}]^{vi}O_{19.9}{}^{5}(OH)_{4.05}$$

interpret the chemical analysis. Secondly, colloidal-sized mineral particles have a large specific surface that is often charged and hydrated, and the true particle density is difficult to define and determine. Furthermore, the expanding-layer clays, vermiculites and smectites, contain variable amounts of water between individual layers and, for any particular clay, the amount of water present can only be defined under carefully controlled relative humidities. Consequently, the most satisfactory way to express the chemical composition of clays on a comparable basis is by reference to the weight of clay after ignition to 950 °C. This temperature not only expels hydration water but decomposes the mineral, hydroxyl groups forming oxygen atoms and water (dehydroxylation)

$$4(OH^-) \rightarrow 2(O)^{2-} + 2H_2O \uparrow$$

so that only anhydrous oxides remain. This enables a check to be made that no constituents have been overlooked in the analysis, for the sum of the individual oxide components should total 100 per cent.

Structural formulae for clays can be calculated from the chemical composition by assuming the anion composition of the original clay, using X-ray diffraction to identify the type of clay mineral present. For example, a clay mineral identified as a smectite has a 2:1 layer structure with an assumed anion composition of $O_{20}(OH)_4$ for a unit layer, and the negative charge on the anion framework must be exactly balanced by the total positive charge from the interstitial cations. The procedure is the following:

1. The weight of each oxide present, in g per 100 g is divided by the molecular weight of the oxide, to give 'moles' oxide per 100 g;
2. The calculated 'moles' per 100 g are multiplied by the number of atoms of cation in the oxide formula and by its valency to give the relative cation charges per 100 g;
3. The sum of the relative positive charges of all the cations is divided into the charge from the anion framework, in this instance 44, to give a scaling factor, f:

$$f = 44/\sum \text{relative cation charges};$$

4. The relative charges for each cation are multiplied by f and divided by valency to give numbers of cations per $O_{20}(OH)_4$.

An example of this calculation is given for a smectite mineral, a montmorillonite from Woburn, that has been treated with a solution of a sodium salt to replace all the exchangeable cations by Na (Table 2.2).

The original chemical analysis contains two determinations of water content, $H_2O(-)$ and $H_2O(+)$, that is water evolved between room temperature and 105 °C, and between 105 and 950 °C, respectively. For the reason given above, however, these analyses are disregarded and the oxide composition is expressed on an anhydrous (950 °C) basis in column 4 in Table 2.2.

Assignment of cations to structural sites

The calculation of rational formulae is completed by assigning cations to tetrahedral, octahedral and larger coordination groups in the anion framework. Considerations based on ionic radii and packing of spheres indicate that Si and Al can be tetrahedrally coordinated to O, that Al, Fe, Ti, Mg and Mn can be octahedrally coordinated to O, and that Ca, Na and K have coordination numbers greater than 6, usually in the range 8–12.

Using these assumptions, cations are assigned to sites in the following manner:

(1) Si atoms are placed in tetrahedral sites; if the number of Si atoms is insufficient to fill the number of sites available, this is completed with Al, and sometimes Fe, though this is seldom necessary.

(2) Remaining Al, and Fe, Ti, Mg and Mn are placed in octahedral sites.

(3) Ca, Na and K are assigned to sites between the aluminosilicate layers.

With expanding layer silicates containing exchangeable cations, this assignment is not possible if the exchangeable cations include cations normally present in the octahedral sites, such as Mg or Al. The best solution in such instances is to replace all the interlayer cations by another cation that is too large to substitute in the octahedral sites, as was done for the montmorillonite quoted in Table 2.2.

For the muscovite example (Table 2.1) there are 8 tetrahedral sites, 6 octahedral sites and 2 interlayer sites. The cations are assigned as follows:

Tetrahedral:	$Si_{6.03}Al_{1.97}$;	total = 8
Octahedral:	$Al_{3.79}Ti_{0.03}$ $Fe^{3+}_{0.04}Fe^{2+}_{0.10}$ $Mg_{0.12}$;	total = 4.05
Interlayer:	$Na_{0.21}K_{1.72}$	total = 1.93

As 4.05 out of six octahedral sites are filled the mineral is dioctahedral.

Montmorillonite (Table 2.2), like muscovite, has 8 tetrahedral, 6 octahedral sites and up to 2 interlayer sites. The sites are filled as follows:

Tetrahedral:	$Si_{7.72}Al_{0.28}$	total = 8
Octahedral:	$Al_{2.18}Fe^{3+}_{1.14}Ti_{0.04}$ $Mg_{0.48}Fe^{2+}_{0.11}$	total = 3.95
Interlayer:	$Na_{0.91}K_{0.03}Ca_{0.02}$	total = 0.96

As 3.95 out of six possible octahedral sites are filled, the mineral is dioctahedral. The cations assigned to the interlayer positions, Na, K and Ca, should be exchangeable. As the clay was sodium saturated before analysis, the K and Ca must be residual exchangeable cations not replaced by sodium, or be present in small amounts of impurity in the clay, possibly mica and feldspar. This indicates a possible source of error in the structural formula derived by this calculation.

If the exchange capacity is taken to be represented by the sodium content, 3.2 g Na_2O per 100 g air-dry clay or 3.8 g Na_2O per 100 g ignited clay, the 'cation exchange capacity' is

$$1000 \times 3.2/30.99 = 103 \text{ meq Na per 100 g hydrated clay}$$

or

$$1000 \times 3.8/30.99 = 123 \text{ meq Na per 100 g ignited clay.}$$

The exchange capacity can also be calculated from a structural formula by expressing the exchangeable cation content as a proportion of the molecular

Table 2.2 Calculation of the structural formula for montmorillonite, a smectite mineral

The clay mineral, from Woburn Sands, Bedfordshire, was dispersed by sodium saturation and purified by separating the size fraction <0.25 μm (Schultz, 1969)

(1) Oxide	(2) Formula weight	(3) Composition per cent hydrous mineral	(4) Composition per cent ignited mineral	(5) Formula units per 100 a.m.u.	(6) Cations per 100 a.m.u.	(7) Cations charges per 100 a.m.u.	(8) Cations per $O_{20}(OH)_4$	(9) Cation charges per $O_{20}(OH)_4$
SiO_2	60.09	52.7	62.6	1.0418	1.0418	4.1671	7.72	30.88
Al_2O_3	101.94	14.2	16.9	0.1658	0.3316	0.9947	2.46	7.35
TiO_2	79.9	0.39	0.46	0.0058	0.0058	0.0230	0.04	0.16
Fe_2O_3	159.7	10.4	12.3	0.0770	0.1540	0.4621	1.14	3.45
FeO	71.85	0.88	1.04	0.0145	0.0145	0.0290	0.11	0.22
MgO	40.32	2.2	2.6	0.0645	0.0645	0.1290	0.48	0.96
CaO	56.08	0.10	0.12	0.0021	0.0021	0.0043	0.02	0.04
Na_2O	61.98	3.2	3.8	0.0613	0.1226	0.1226	0.91	0.91
K_2O	94.2	0.16	0.19	0.0020	0.0040	0.0040	0.03	0.03
$H_2O(-)$	18.016	10.4						
$H_2O(+)$	18.016	5.4						
Totals	100.03	100.01				5.9358	12.91	44.00

Column (5) = Column (4) ÷ Column (2)
Column (6) = Column(5) × number of cations in oxide formula
Column (7) = Column (6) × cation valency
Column (8) = Column (6) × 44 ÷ Σ Column (7) = Column (6) × 7.413
Column (9) = Column (8) × cation valency, to check that the sum = 44

The structural formula for one layer units is therefore

$$Si_{7.72}Al_{2.46}Ti_{0.04}Fe^{3+}_{1.14}Fe^{2+}_{0.11}Mg_{0.48}Ca_{0.02}Na_{0.91}K_{0.03}O_{20}(OH)_4$$

which may be written

$$Na_{0.91}K_{0.03}Ca_{0.02}[Si_{7.72}Al_{0.28}]^{iv}[Al_{2.18}Ti_{0.04}Fe^{3+}_{1.14}Fe^{2+}_{0.11}Mg_{0.48}]^{vi}O_{20}(OH)_4$$

where the Na, K and Ca are 'exchangeable'.

weight of the unit layer. For example, the molecular weight of the Woburn montmorillonite, containing 0.91 gram atoms Na per unit molecular weight, is 741.8, and so its exchangeable cation content is 1000 × 0.91/7.418 meq Na per 100 g ignited clay, that is 123 meq Na per 100 g as before.

2.3.3 Real Structures

The general description of the structural classes of layer silicate minerals given above (2.3.1) is based on the geometrical simplification that the highly symmetric tetrahedral and octahedral sheets described superimpose exactly at the apical oxygen atoms. This is rarely true, and for most minerals in which the positions of atoms have been determined accurately by X-ray analyses, the oxygen atoms are displaced from the regular packing arrangements described in

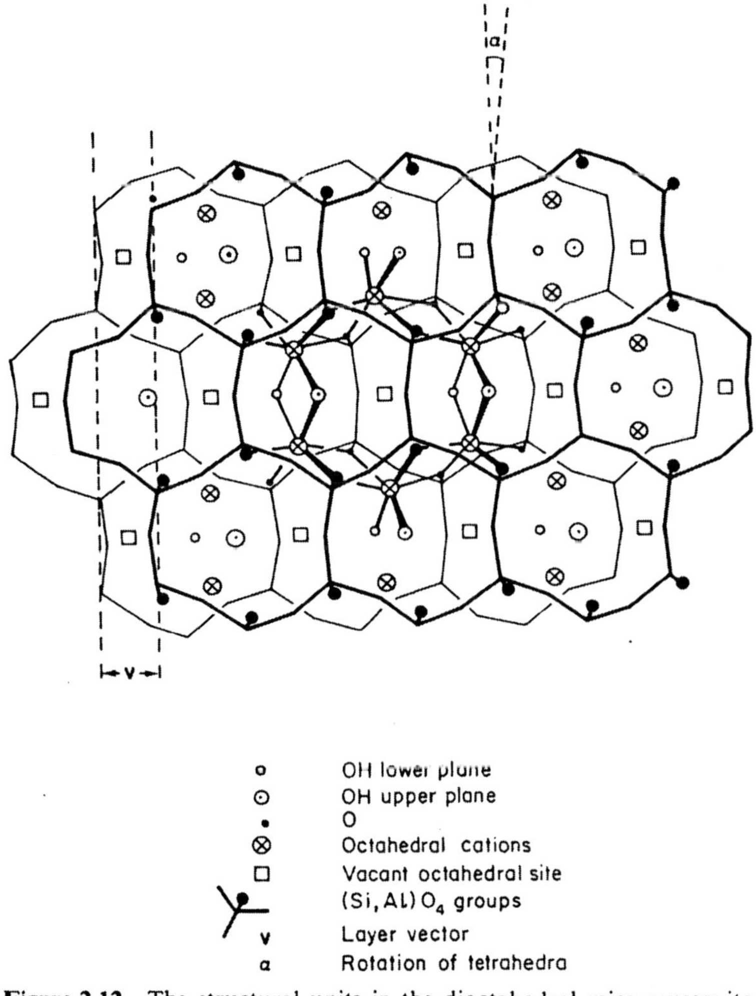

Figure 2.12 The structural units in the dioctahedral mica muscovite, showing twist of the (Si, Al)O_4 groups and distortion of the octahedral sheet

2.3.1. An example of this is given in Figure 2.12, which should be compared with Figure 2.11. The reason for these displacements seems to be that unconstrained tetrahedral and octahedral sheets will only superimpose exactly for unusual combinations of cationic compositions of the sheets, and that for most of the common layer silicates, the cationic composition is such that the tetrahedral and octahedral sheets must accommodate to each other by distortion from the highly symmetrical structure of the unconstrained sheets. These distortions can be simply expressed by comparing a lattice parameter that defines the lateral dimension of the repeat unit in the basal plane with theoretical values calculated from the chemical composition. Usually the *b*-parameter is used because this can be readily calculated from the position of the 060 reflection or 06 band in the X-ray diffraction pattern of a layer silicate.

A tetrahedral sheet with hexagonal symmetry and containing only Si in the tetrahedra has a calculated repeat unit length of 0.916 nm, based on an inter-atomic distance of 0.162 nm for the Si—O bond; as the Al—O bond length in tetrahedra is longer (0.177 nm), substitution of Al for Si increases the lateral dimension for an unconstrained sheet.

For a regular system of joined hexagonal rings (Figure 2.7) simple geometry shows that

$$b_{tetr} = 2\sqrt{3} \times \bar{D}_B$$

where b_{tetr} is the calculated size of the repeat unit along the *y*-axis, and $\bar{D}_B$ is the mean distance between adjacent basal oxygen atoms. It has seldom been possible to distinguish tetrahedra containing Al from those containing Si (for an exception, see Bailey, 1975b), so that a mean O_B—O_B distance is obtained from most refined crystal structure determinations. When b_{tetr} was calculated from the mean O_B—O_B distance in a wide range of layer silicates, it was found to vary from 0.905 nm where no Al replaced Si to 0.95 nm when about half the Si atoms were replaced by Al (Bailey, 1966). This calculation showed that Al-for-Si substitution significantly increased the size of an unconstrained regular tetra-hedral sheet.

Turning now to the octahedral sheet, a regular structure containing only Mg has a calculated *b* parameter of 0.891 nm based on a bond length of 0.210 nm for Mg—O; the bond length for Al—O in octahedra is 0.191 nm and b_{calc} = 0.810 nm. The measured parameter in brucite, $Mg_3(OH)_6$, is 0.936 nm and in gibbsite, $Al_2(OH)_6$, is 0.865 nm; the octahedra in both structures are distorted by cation repulsion.

These calculations show that, in general, regular unconstrained tetrahedral and octahedral sheets have different lateral dimensions and that some deviation from these ideal configurations must occur in structures where tetrahedral and octahedral sheets are joined at the apical oxygen atoms, as in the layer silicates. The adjustments to the structures that are found depend on whether the calcu-lated unconstrained dimension of the tetrahedral sheet is larger or smaller than that of the octahedral sheet.

Commonly the tetrahedral sheet is larger than the octahedral and it accommo-dates to the octahedral sheet by rotation of alternate tetrahedra clockwise and anticlockwise about axes normal to the basal plane, so that the $(Si, Al)_6O_6$ rings (Figure 2.13, upper) no longer have hexagonal but trigonal symmetry (Figure 2.13, lower). If α is the angle through which the tetrahedra are twisted from hexagonal symmetry.

$$\cos \alpha = b_{obs}/b_{tetr}$$

The maximum possible value of α is 30°, and for this angle of rotation the

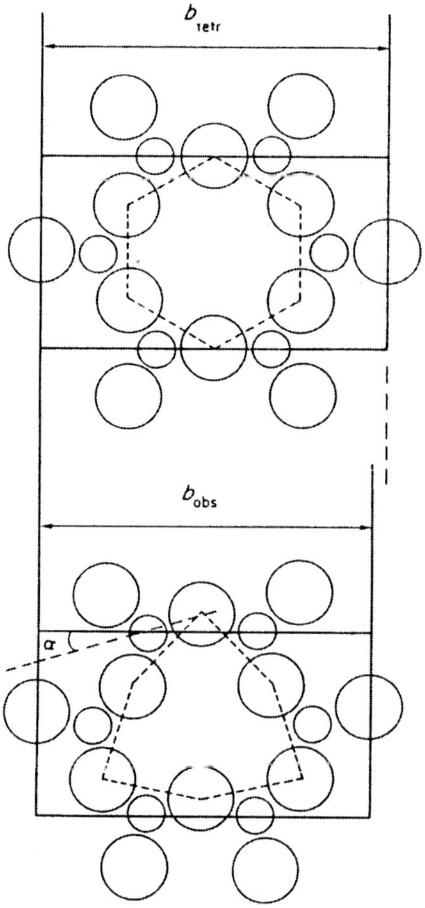

Figure 2.13 The effect of a tetrahedral twist α on the lateral dimension of a tetrahedral sheet (b_{obs})

lateral contraction is 13.4 per cent. Thus there is potentially a considerable range of adjustment in the tetrahedral sheet.

By contrast, the octahedral sheets tend to increase their lateral dimensions. In dioctahedral sheets where two thirds of the octahedral sites are occupied, the trivalent cations repel each other and move further apart, so that the two oxygen atoms shared between adjacent cations approach each other more closely, shortening the shared edges of the octahedra and lengthening the unshared edges, so that there are two types of octahedra, one larger and unoccupied, and two smaller containing the cations. This has the effect of making the sheet thinner and slightly larger in lateral dimensions. In trioctahedral minerals, cation repulsion is approximately equal in three directions at 120 ° and the oxygen atoms cannot move directly along shared edges but only inward along a normal to the basal plane. In consequence the octahedral flattening is much smaller in trioctahedral minerals, only about 2% in contrast to 10% in dioctahedral, but the lateral extension is larger. Neither of these displacements from the ideal symmetry in the octahedral sheet permits as much variation in lateral dimension as rotation of tetrahedra and so the *b* lattice parameter of layer silicates is much more influenced by the composition and the average cation–oxygen bond length of the octahedral layer than by the composition of the tetrahedral layer. The structure of muscovite (Figure 2.12) illustrates most of these points.

In Al-rich dioctahedral minerals, however, in which substitution of Al for Si in the tetrahedral sheet approaches 50%, the disparity in the sizes of the unconstrained sheets cannot be entirely accommodated by tetrahedral rotation and the tetrahedra not only twist but also tilt so that the basal oxygen atoms are no longer coplanar. The basal oxygen surface takes on a small corrugation in which two basal oxygens are raised slightly from the basal plane and the third oxygen is depressed towards the octahedral sheet. This type of structure is exemplified by the rare brittle mica margarite.

In the trioctahedral 1:1 layer silicates, particularly where there is no substitution for Si in the tetrahedral sheet, the octahedral sheet is larger in all lateral directions and so the structural adjustments described so far cannot compensate for the misfit of the two sheets. In these minerals, the tetrahedra tend to tilt away from the octahedral sheet and layer curvature results, the structure being no longer planar. In chrysotile, this results in a tubular morphology and high-resolution electron micrographs clearly demonstrate the curved form of the layers. By contrast, in antigorite, the curling structure is reversed periodically every 6–8 unit cells in the *a* lattice direction.

Interlayer cations

Many layer silicates, most importantly the micas, contain essential interlayer cations, usually K, Na or Ca, that are situated between adjacent layers in the twelve-coordinate sites that result when the Si_6O_6 rings in one layer superimpose on the Si_6O_6 rings in the adjacent layer. For the layer silicates in which these rings have trigonal rather than hexagonal symmetry, three oxygen atoms in each Si_6O_6 ring are much closer to the interlayer cation than the remaining three (see Figure 2.13), and this close approach exercises some restraint on tetrahedral rotation and the *b* lattice parameter, so that if K, a large cation, is replaced by a smaller cation, some change in b_{obs} occurs. This aspect is discussed later in the section on micas.

2.3.4 Principal Structural Groups

There are six main structural groups of layer silicate minerals. One of these is based on 1:1 layers as already discussed, and four are based on 2:1 layers, and

Table 2.3 Classification and generalized structural formulae of phyllosilicates

Layer type	Group	Octahedral occupancy	Negative charge per unit[a]	Unit formula: Cations[b] Oct.	Unit formula: Cations[b] Tet.	Unit formula: Anions	Layer thickness nm	Interlayer composition: Cations[c]	Interlayer composition: Hydroxide sheet[d]	Interlayer composition: Water	Occurrence in soils
1:1	Kaolinite	Di	0	Y_4	Z_4	$O_{10}(OH)_8$	0.7	None			Common
	(Halloysite)	Di	0	Y_4	Z_4	$O_{10}(OH)_8$	1.0~0.7	None		0.6 to $4H_2O$	Common
	Serpentine	Tri	0	Y_6	Z_4	$O_{10}(OH)_8$	0.7	None			Rare
2:1	Pyrophyllite	Di	0	Y_4	Z_8	$O_{20}(OH)_4$	0.92	None			Rare
	Talc	Tri	0	Y_6	Z_8	$O_{20}(OH)_4$	0.9	None			Rare
	Micas	Di		Y_4	Z_8	$O_{20}(OH)_4$	1.00	X'_2			
		Tri	2	Y_6	Z_8	$O_{20}(OH)_4$	1.00	X'_2			Common
	Brittle Micas	Di		Y_4	Z_8	$O_{20}(OH)_4$	1.00	X''_2			
		Tri	4	Y_6	Z_8	$O_{20}(OH)_4$	1.00	X''_2			Rare
	Chlorites	Di		Y_4	Z_8	$O_{20}(OH)_4$	1.40		$A'''_4(OH)_{12}$		
		Di, tri	Variable	Y_4	Z_8	$O_{20}(OH)_4$	1.40		$A''_6(OH)_{12}$		Common
		Tri		Y_6	Z_8	$O_{20}(OH)_4$	1.40		$A''_6(OH)_{12}$		
	(Swelling chlorite)		Variable				$\geqslant 1.40$		$\sim A'''_4(OH)_{12}$	nH_2O	Common
	Smectites	Di	0.5–1.2	Y_4	Z_8	$O_{20}(OH)_4$	$\geqslant 0.96$	$X'_{0.5-1.2}$		nH_2O	Common
		Tri	0.5–1.2	Y_6	Z_8	$O_{20}(OH)_4$	$\geqslant 0.96$	$X'_{0.5-1.2}$			
	Vermiculites	Di	1.2–1.9	Y_4	Z_8	$O_{20}(OH)_4$	$\geqslant 0.94$	$X'_{1.2-1.9}$		nH_2O	Common
	Palygorskite	Tri	1.2–1.9	Y_6	Z_8	$O_{20}(OH)_4$	$\geqslant 0.94$	$X'_{1.2-1.9}$			
			?	Y_4	Z_8	$O_{20}(OH)_2$ $(OH_2)_4$[e]		X ?		$4H_2O$	Uncommon except in arid areas
	Sepiolite		?	Y_8	Z_{12}	$O_{30}(OH)_4$ $(OH_2)_4$		X ?		$8H_2O$	Uncommon

[a] Negative charge per formula unit is twice that given by Bailey *et al.* (1971a) because the formula unit used here applies to the contents of a volume defined by the a × b unit cell base area that is one layer thick. For example for chlorites this volume is approximately $0.5 \times 0.9 \times 1.4\ nm^3$, and for palygorskite it is $1.8 \times 0.5 \times 0.65\ nm^3$.

[b] Y represents cations in octahedral coordination, most commonly Al but many others of similar dimensions can substitute for it isomorphously. Z represents cations in tetrahedral coordination, most commonly Si, but Al frequently replaces some Si in a regular or random manner, e.g. replacement of one in four occurs in the micas and gives rise to the negative charge of two per formula unit; in the brittle micas 2Al replace two of each four Si.

[c] X′ represents a monovalent cation, and X″ a divalent cation.

[d] A‴ is a trivalent cation (usually Al) and A″ is a divalent cation (usually Mg) but extensive isomorphous substitution may occur.

[e] Following Bradley (1940); according to Gard and Follett (1968) the anions are $O_{20}(OH)_3(OH_2)_3$.

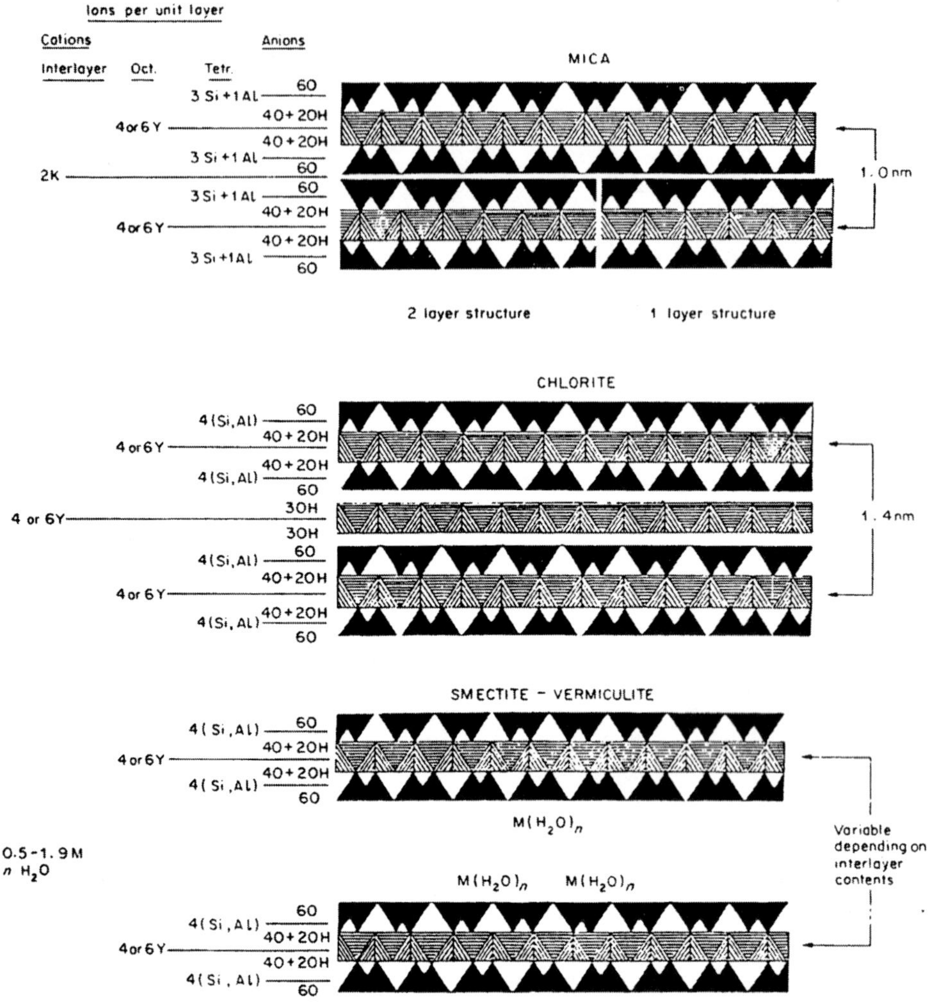

differing according to interlayer occupancy. These five groups are:

the kaolinite–serpentine group with uncharged 1:1 layers repeating at 0.7 nm when unhydrated but at approximately 1.0 nm in hydrated halloysite which contains a sorbed interlayer of water;
the pyrophyllite–talc group with uncharged 2:1 layers;
the mica group with charged 2:1 layers and interlayer potassium, repeating at 1.0 nm;
the smectite–vermiculite group with charged 2:1 layers and interlayer cations whose hydration state varies with humidity, and which therefore have interlayer repeat distances dependent on the interlayer occupancy;

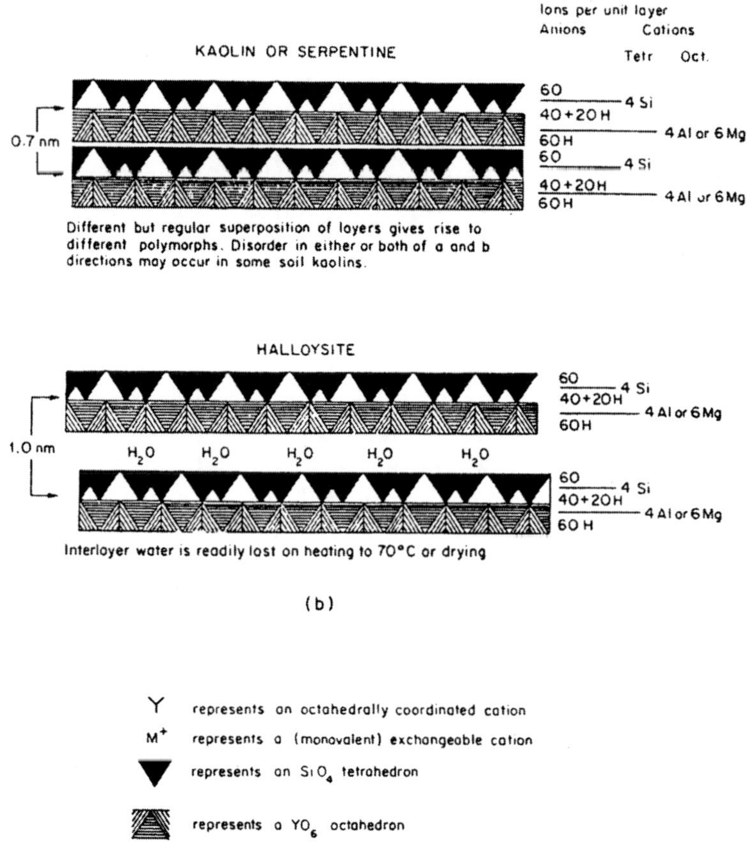

Figure 2.14 Schematic representation of structures of principal phyllosilicate minerals: (a) 2:1 minerals, and (b) 1:1 minerals. Pyrophyllite and talc are similar to the micas, except that the tetrahedral occupancy is 4Si, so that the layer change is zero and there are no interlayer cations.

the chlorite group with charged 2:1 layers and positively charged hydroxide interlayers repeating at 1.4 nm.

The sixth is

the palygorskite–sepiolite group, with fibrous morphology and narrow ribbons of 2:1 layers linked by their edges to form a network of alternating channels occupied by water molecules.

This classification and the structural formulae of the mineral types is summarized in Table 2.3 and Figure 2.14; the following sections (2.3.5 to 2.3.8) describe each mineral group in more detail.

2.3.5 1:1 Layer Silicates $(Si_4)(Al_4)O_{10}(OH)_8$ and $(Si_4)(Mg_6)O_{10}(OH)_8$

This group of minerals is divided into two series, dioctahedral and trioctahedral, that in many ways provide a complete contrast.

The dioctahedral minerals kaolinite, dickite and nacrite are true clay minerals that vary little in chemical composition but differ in the way that successive 1:1 layers superimpose; the halloysites have a similar layer composition but additionally contain interlayer water and often a distinctive crystal morphology. Kaolinite in particular is commonly present in soils and is the dominant clay mineral in many highly weathered tropical soils (Oxisols). It is formed by weathering of primary minerals; its formation from alkali feldspars (Helgeson, 1974) and micas (Eberl and Hower, 1975) is probable and it is readily synthesized in the temperature range 150–400 °C. Conditions for its formation are moderate acidity ($\log(a_{alk}/a_H) < 4$, where a_{alk} = alkali metal activity) and Si/Al ratios in the range 2 to 0.6.

By contrast, the trioctahedral series of the 1:1 layer silicates shows a considerable range of chemical composition. The magnesian end members, the serpentines, are analogous to the kaolinite minerals with 3Mg in place of 2Al in the octahedral sheet, and there are several structurally distinct minerals with this composition; they are formed principally by hydrothermal alteration of ultrabasic rocks such as dunites, pyroxenites and peridotites, although metamorphosed dolomites may also contain serpentine minerals. The ferroan end member greenalite contains Fe(II) in place of Mg. In the other members of the 1:1 trioctahedral series, Al substitutes for divalent cations in the octahedral sheet, and also for Si in the tetrahedral sheet in equal proportions so that the net charge of the layer is zero, excess positive charge in the octahedral sheet balancing the charge deficit in the tetrahedral sheet. Many of these minerals are restricted in distribution or rare, and are seldom found in soils; only berthierine (0.7 nm chamosite) is properly a clay mineral, sometimes found in lateritic clay deposits and in sedimentary ironstones.

Dioctahedral 1:1 minerals

Kaolinite, *dickite* and *nacrite* have essentially the same layer structure with one octahedral and one tetrahedral sheet joined through the apical oxygens of the tetrahedra. The outer plane of the octahedral sheet is entirely hydroxyl anions, whereas the inner plane contains O_{apical} and OH in the proportion 2:1; the tetrahedral sites contain Si, and Al occupies two thirds of the octahedral sites. The ideal structural formula is therefore $Si_4Al_4O_{10}(OH)_8$. Most natural samples closely approach this composition and very little isomorphous substitution occurs. Jepson and Rowse (1975) by electron microscope microprobe analysis (EMMA) on individual clay crystals of kaolinite showed that mean Al:Si ratios for kaolinite from St Austell sized <1 µm was 0.967 ± 0.018, whereas Georgia kaolinite gave 0.997 ± 0.018 showing that the English kaolinite departed significantly from the ideal ratio of 1, whereas the American kaolinite did not. Ratios of Fe, Ti, Mg and K to Si are given in Table 2.4; small amounts of each of these elements are thought to be present in the structure of both kaolinites.

Electron spin resonance and Mössbauer spectral studies have shown that

Table 2.4 Mean Mg:Si, K:Si, Ti:Si and Fe:Si atom ratios for St Austell and Georgia kaolinites (0.9–1.0 μm fractions)

	St Austell	Georgia
Mg:Si	0.00438 ± 0.0155	0.00246 ± 0.00200
K:Si	0.00185 ± 0.00139	0.00186 ± 0.00044
Ti:Si	—	0.00221 ± 0.00140
Fe:Si	0.00137 ± 0.00100	0.00248 ± 0.00053

Fe^{3+} substitutes for Al^{3+} in the octahedral sheet (Meads and Malden, 1975) but on the evidence of the EMMA results in Table 2.4 the extent of the substitution is very limited. Many chemical analyses of kaolinites give larger impurity concentrations but these are almost certainly due to the presence of separate mineral phases.

Aluminium occupies the octahedral sites in an ordered arrangement in which each vacant site is surrounded by 6 occupied sites (Figure 2.9b). The octahedra containing Al are distorted by contrarotation of the upper and lower anion triads so that the edges shared between adjacent Al cations are shortened and the unshared edges lengthened (Figure 2.15). Misfit to the octahedral sheet requires that the lateral dimensions of the tetrahedral sheet are decreased by rotation of alternate tetrahedra clockwise and anticlockwise (Figures 2.13 and 2.15 lower part), a displacement that may be emphasized by interlayer hydrogen bonds between the outer hydroxyl plane of one layer and the basal oxygen anions of the next (Figure 2.14).

Kaolinite, dickite and nacrite differ mainly in the manner that successive layers superimpose (Bailey, 1963), although the individual layers in the three minerals also differ slightly in tetrahedral twist and lateral repeat distance. Each layer has a plane of symmetry that intersects the inner OH groups (at the centre of the Si_6O_6 rings in the projection shown in Figure 2.15) and also passes through the vacant octahedral sites. The position of the vacant octahedral site relative to the Si_6O_6 ring defines the type, orientation and position of each layer, and the superposition of successive layers ('stacking') can also be shown by representing each layer as a ditrigonal ring and a vacant site (Figure 2.16a).

In kaolinite, each layer has the vacancy in the same site and is displaced by $-a/3$ with respect to the preceding layer (Figure 2.16b). The interaction between each outer OH of the lower layer and the basal O of the next layer is equalized by this displacement, so that each hydrogen bond is approximately the same length (Brindley, 1961). There are two possible structures of this kind depending on whether the left hand or right hand site (referred to the a-axis direction) is chosen for the vacancy; these structures are related by object and mirror image symmetry (enantiomorphs) and cannot at present be distinguished in clay sized material by X-ray methods, and so this choice is arbitrary.

The stacking of dickite (Newnham, 1961) is similar to that in kaolinite but the

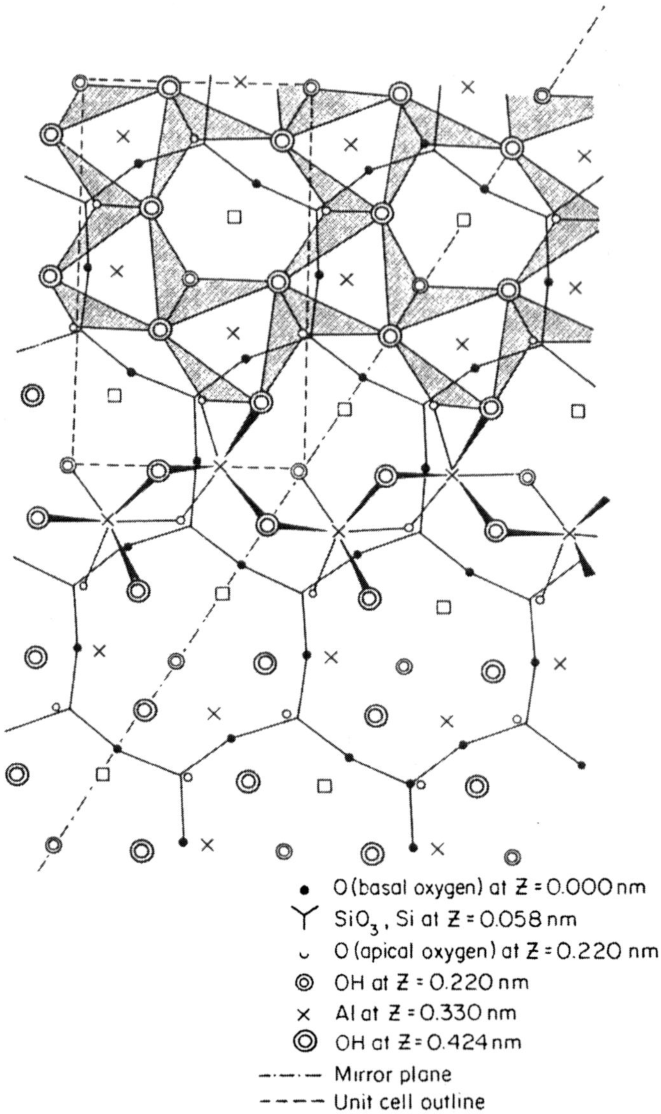

Figure 2.15 Projection on the basal plane of one 1:1 layer in kaolinite

vacant site alternates between left and right sites in successive layers (Figure 2.16c) so that a two-layer repeat unit results. The hydrogen bonding between the layers is similar to that in kaolinite, alternating between left and right hand forms.

Nacrite has a different stacking sequence (Blount, Threadgold and Bailey, 1969) in which alternate layers are rotated through 180° and displaced by $-\frac{1}{3}$ of

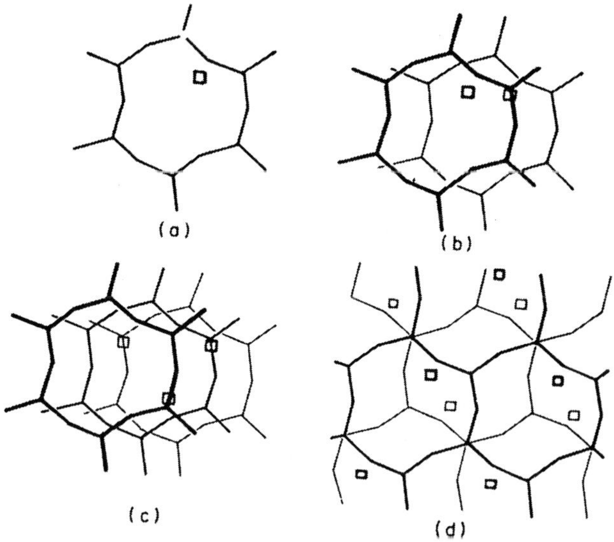

Figure 2.16 Layer stacking in the kaolinite minerals (schematic): (a) a single 1 : 1 layer, showing the position of the vacancy (□) relative to the Si_6O_6 ring, (b) two successive layers in kaolinite, (c) three successive layers in dickite, and (d) two successive layers in nacrite

the 0.89 nm lateral repeat unit, as opposed to $-\frac{1}{3}$ of the 0.51 nm repeat in kaolinite and dickite. In consequence the pattern of successive Si_6O_6 rings and vacancies in nacrite and the hydrogen bonding between the layers is quite different from that in kaolinite and dickite (Figure 2.16d).

Of the three polymorphs, kaolinite is the commonest and most likely to be found in soils.

Disordered structures. Kaolinites from different sources giving a normal sequence of basal reflections (00l) from the 0.715 nm layer repeat unit may nevertheless vary considerably in the non-basal reflections given. The differences are most evident for reflections in which the k index is not a multiple of 3 ($k \neq 3n$). The diffraction effects indicate that the structures of the minerals giving these weak or diffuse reflections are disordered to varying extents. The hydroxyl anions in the outer plane of the octahedral sheet lie nearly on lines parallel with the b-axis and occur every $b/3$ distance in this direction. Consequently it is possible to displace a layer by $b/3$ or $2b/3$ relative to the layer below or above without greatly altering the interlayer bonding; a displacement of $3b/3$ or b is equivalent to zero displacement. Random displacements by $n_b/3$ ($n \neq 3$) break up regular stacking sequences of Si_6O_6 rings and vacancies described for kaolinite and alter the sharpness and intensities of the reflections for which $k \neq 3n$; kaolinites showing such diffraction effects are described as *b*-axis disordered. The powder patterns for a *b*-axis disordered kaolinite, a triclinic kaolinite and a soil kaolinite are contrasted in Figure 2.17.

In order to test how disorder changes X-ray diffraction patterns of kaolinites over the region $d = 0.45$ to 0.36 nm, Noble (1971) supposed that random lateral displacements occurred every M layers in a crystal 70 layers thick and calculated diffraction profiles for values of M between 1 and 70. This model matched calculated and observed diffraction profiles of kaolinites with $M < 6$ but was less satisfactory for the kaolinites with most disorder, for instance Pugu kaolinite (Figure 2.17). It seems possible that these poorly crystallized kaolinites, which are quite widely distributed in soils, may be disordered not only by stacking faults, but also in other ways, perhaps by within-layer disorder.

Anauxite is a name given to kaolinite-like materials that contain a substantial excess of SiO_2, sometimes as high as $3SiO_2:Al_2O_3$. Recently it has been established by selective dissolution and X-ray diffraction studies that the excess SiO_2 is amorphous and interspersed between well-crystallized stacks of kaolinite (Langston and Pask, 1969; Bailey and Langston, 1969) and it is now recommended that anauxite should not be recognized as a separate mineral species.

Halloysites have a very similar $SiO_2:Al_2O_3$ ratio to kaolinites but contain additional water that is largely driven out by heating to 110 °C; unlike most smectites and vermiculites (q.v.), halloysites do not regain their hydration water on cooling. Structurally they contain 1:1 aluminosilicate layers similar to those in kaolinite, but fully hydrated halloysites have a layer repeat distance of 1.01 nm, and a strong *hk* band at about 0.44 nm indicating a highly disordered structure. On dehydration the layer spacing decreases to 0.72 to 0.74 nm, an appreciably larger spacing than that in kaolinite, 0.716 nm. The composition of the fully dehydrated phase, referred to as metahalloysite, is close to $2SiO_2 \cdot Al_2O_3 \cdot 2H_2O$ and the fully hydrated mineral has the composition $2SiO_2 \cdot Al_2O_3 \cdot 4H_2O$. The additional $2H_2O$ is interspersed between the alumino-silicate layers as a single layer of water molecules 0.29 nm thick. Many electron micrographs of minerals with these properties show that they have a fibrous morphology and it has been inferred that the fibres consist of rolled up layers, the curvature resulting from the relief of strain arising from the misfit of the tetrahedral and octahedral sheets. More recently however, halloysites with spheroidal (Askenasy, Dixon and McKee, 1973; Sudo and Yotsumoto, 1977; Plate 2.1) and tabular (Souza Santos *et al.*, 1966) morphologies have been reported. The fibrous/tubular morphology sometimes results from scrolling of single layer sheets during the dispersion and air drying. In addition, it is possible to introduce into the interlayer regions of many kaolinites salts such as potassium acetate and some other compounds (Chapter 7 and Olejnik, Posner and Quirk, 1970) so that the original distinction between kaolinite and halloysite has to some extent become blurred; Churchman and Carr (1975) review the literature on the definitions of halloysite. Hydrated halloysite forms complexes with many organic compounds (Carr and Chih, 1971); the complexes with diols, particularly ethylene glycol, are stable to heat and to vacuum and are useful for identification.

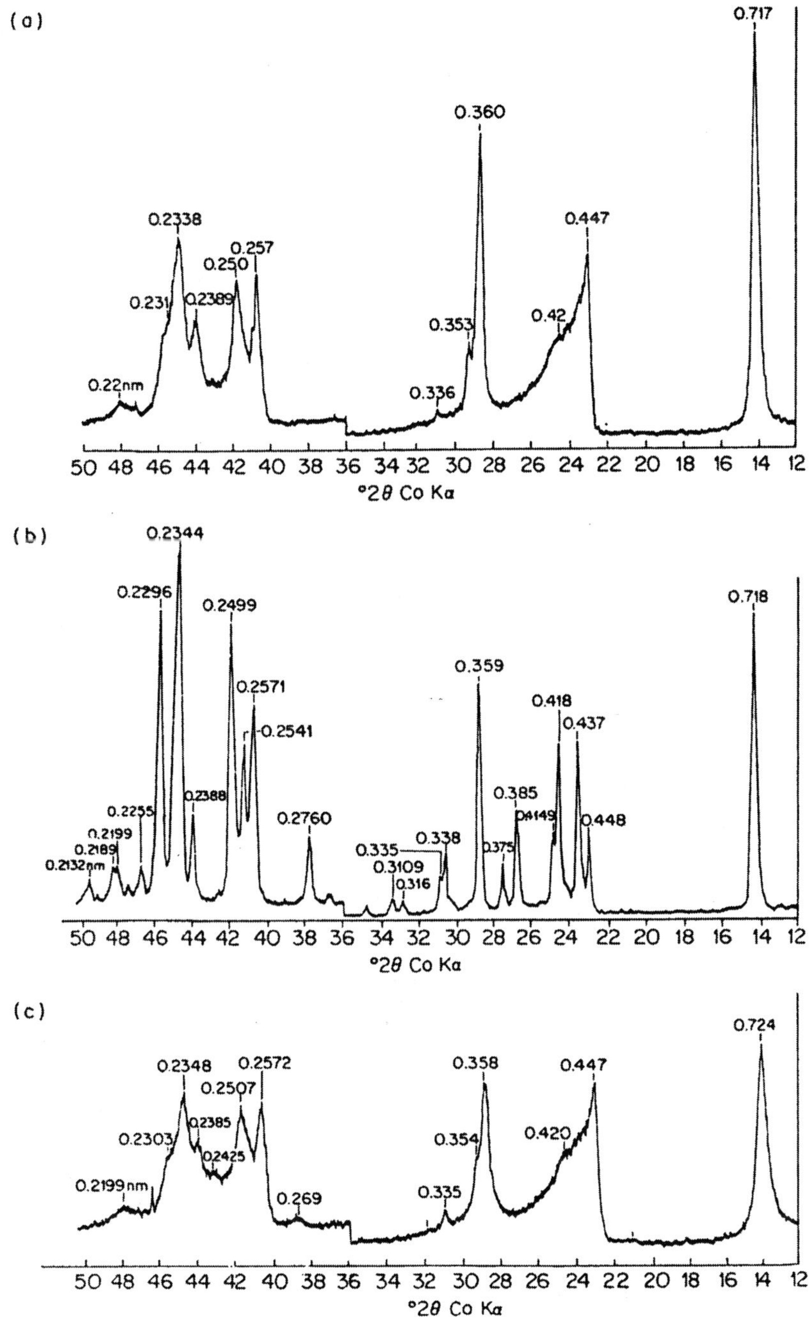

Figure 2.17 X-ray diffractometer traces obtained from: (a) 'b-axis disordered' kaolinite, Pugu D, Tanzania; (b) Triclinic kaolinite, Mexico; and (c) Soil kaolinite, Ikenne, Nigeria

Trioctahedral 1:1 minerals (Brindley, 1961)

The ideal chemical compositions of the end-member minerals are:

Serpentine:	$Mg_6Si_4O_{10}(OH)_8$
Greenalite:	$Fe(II)_6Si_4O_{10}(OH)_8$
Amesite:	$(Mg_4Al_2)(Si_2Al_2)O_{10}(OH)_8$
Berthierine:	$(Fe(II)_4Al_2)(Si_2Al_2)O_{10}(OH)_8$
Cronstedtite:	$(Fe(II)_4Fe(III)_2)(Si_2Fe(III)_2)O_{10}(OH)_8$

They all contain one tetrahedral and one octahedral sheet joined through apical oxygen atoms; all the octahedral sites are occupied in the forms containing only divalent cations in the octahedral sheet, but vacancies occur in some aluminian varieties, up to about 0.5 per 6 sites.

The three main structural forms of serpentine are the fibrous asbestos mineral chrysotile, lizardite and antigorite. As the octahedral sheet is larger laterally than the tetrahedral sheet, the misfit at the apical oxygen atoms cannot be relieved by tetrahedral rotation and curvature of the layers occurs, the basal oxygen surface being concave (Plate 2.2). In chrysotile, the layers are rolled about the fibre axis, either as concentric tubes or as several independent spirals, with the hydroxyl surface outside, in contrast to halloysite in which the basal oxygen surface should be external. Lizardite is fine grained and less is known about its structure. Antigorites have an interesting wave-like structure in which the layer curvature reverses after approximately 8 repeats of the normal a parameter of 0.54 nm. This occurs at an inversion of the tetrahedral sheet where the octahedral sheet is formed on the opposite side of the basal oxygens. In the minerals amesite, berthierine and cronstedtite the octahedral–tetrahedral misfit is less but the layers have complex stacking arrangements (Bailey, 1969). Nickel serpentines occur in the ore, garnierite (Brindley and Maksimovic, 1974) and cobalt serpentines have been synthesized.

2.3.6 2:1 Layer Silicates

Pyrophyllite and Talc $(Al_4)(Si_8)O_{20}(OH)_4$ and $(Mg_6)(Si_8)O_{20}(OH)_4$

Pyrophyllite and talc are regarded as the prototypes for the 2:1 minerals in that their chemical compositions are the simplest in the group and their structures are analogues for the remaining minerals. Pyrophyllite is a relatively uncommon mineral formed by hydrothermal alteration of feldspars, and talc occurs as a high temperature alteration product of ultrabasic rocks and is also formed during low-grade thermal metamorphism of siliceous dolomites. They are rarely found in soils, and then only in small or trace amounts.

Their structure consists of a central octahedral sheet joined through oxygen atoms to tetrahedral sheets above and below, these oxygen atoms of the octahedral sheet forming the apical oxygens of the tetrahedra (Figure 2.14). In their ideal mineral composition, the tetrahedra contain only Si, and the octahedral sites are filled by Mg in talc, and two thirds are filled by Al in pyrophyllite; they are therefore respectively tri- and dioctahedral end members of the 2:1 group, their structural formulae being talc $Mg_6Si_8O_{20}(OH)_4$, and pyrophyllite $Al_4Si_8O_{20}(OH)_4$. Until recently, it was thought that little variation occurred in the composition of pyrophyllite, but it is now known that some natural samples contain more H and Al and less Si than the ideal composition; it is suggested that the excess Al replaces Si in the tetrahedral sheet and that some basal oxygen atoms are protonated to become OH (Rosenberg, 1974).

Natural talcs show a wider range of composition and may contain several weights per cent of FeO, the ferrous atoms substituting for Mg in the octahedral sheet. In a rare variant, minnesotaite, the octahedral sheet contains only Fe(II), but there does not seem to be a simple replacement series between talc and minnesotaite (Forbes, 1969). Talcs containing octahedral Ni are also known from natural sources (de Waal, 1970) and Co-, Zn-, and Cu-substituted talcs have been synthesized (Wilkins and Ito, 1967).

Pyrophyllite and talc illustrate well the structural distortions that result from misfit of octahedral and tetrahedral sheets. In talc, the Si—O bond length is 0.1622 nm and the calculated *b*-parameter for a hexagonal tetrahedral network is 0.9175 nm; $b_{(measured)}$ is 0.9179 nm, so that there is an almost exact match between a regular tetrahedral sheet and the *b*-parameter, and the rotation of the tetrahedra is very small, only 3.4° (Rayner and Brown, 1973). The measured *b*-parameter for pyrophyllite is 0.8959 nm because the dioctahedral sheet is smaller in lateral dimensions so that the tetrahedral rotation is about 10° (Rayner and Brown, 1965; Wardle and Brindley, 1972) giving a trigonal symmetry to the Si_6O_6 rings.

The anion charge of the $O_{20}(OH)_4$ framework is balanced within the 2:1 layers by the interstitial cations and the net layer charge is zero. There are, in consequence, no interlayer cations in either talc or pyrophyllite, nor is there any tendency for the minerals to swell by intercalation of water or organic liquids between the layers, and the basal spacings are invariant below their decomposition temperatures.

The mica group $(X_2)(Y_{4\text{ or }6})(Z_8)O_{20}(OH)_4$

The micas are a fundamentally important group of minerals because they have a very wide range of possible chemical compositions and are very widely distributed. They can occur in alkali-rich igneous rocks from acid to ultra-basic composition and in metamorphic rocks, phyllites, schists and gneisses; they are almost universally present in sedimentary rocks, where they may be of detrital or authigenic origin.

They have a 2:1 layer structure like pyrophyllite and talc but unlike the latter, the intralayer charge balance between cations and anions is incomplete, there being a net layer charge deficit of 2 units of negative charge per $O_{20}(OH)_4$ anions, or 4 units in the rare brittle micas. This net charge deficit is balanced by alkali or occasionally alkaline earth cations situated between the 2:1 layers and coordinated into the hexagonal/trigonal $(Si, Al)_6O_6$ rings, so that the nominal coordination of these interlayer cations is 12 oxygens, 6 from one 2:1 layer and 6 from the immediately adjacent layer. The interlayer cations thus 'key' one layer to the next throughout the structure.

General chemical composition. There are three main classes of micas, dioctahedral with two thirds of the octahedral sites occupied, trioctahedral with a fully occupied octahedral sheet, and lithium micas that tend to be intermediate in composition; the latter are comparatively rare minerals and have not been reported in soils. In the trioctahedral micas, the deficit of layer positive charge originates from substitution of Al for Si in the tetrahedral sheets, but in dioctahedral minerals, deficit also occurs from replacement of trivalent octahedral cations by divalent.

Ferrous iron replaces Mg, and Fe(III) replaces octahedral Al in all the classes, but the range of replacement is much greater in the trioctahedral minerals. Because of the wide ranging possibilities of isomorphous replacement in the micas, discussion of their composition is simplified by the 'end-member' concept, in which ideal compositions are represented for the extreme limits of substitution in the structure; micas with end-member composition are extremely rare or even unknown except from synthetic preparations and most natural mineral compositions lie between the end-members. Table 2.5 relates these extreme end-member compositions, with the given mineral names; some of these composition fields are still incompletely defined.

In soils, we are mainly concerned with clay micas and these introduce additional complexities which are often difficult to resolve satisfactorily. The term *illite* was introduced by Grim, Bray and Bradley (1937) '. . . not as a specific mineral name but as a general term for the clay mineral constituent of argillaceous sediments belonging to the mica group', the name being derived from the State of Illinois where the authors had frequently encountered these clays. They are now known to occur almost universally in sediments and soils, in one form or another, and are frequently interstratified. The relationship of illite to the interstratified minerals common in soils is described in a later section.

Detailed structure analyses. Many micas occur as macroscopic crystals suitable for refined structure analysis from X-ray diffraction measurements; a number of such determinations have been completed in the last fifteen years and have greatly expanded our understanding of the principles governing the detailed structure of layer silicates. The main departures from ideal structure arise from the ordering of trivalent cations in the octahedral sheet, which decreases the size of the occupied and increases the size of the unoccupied octahedra, and from misfit of the octahedral and tetrahedral sheets, which distorts the Si_6O_6 rings into a ditrigonal network defined by clockwise and anticlockwise rotation of

Table 2.5 Approximate cationic compositions for some micas and brittle micas of general formula $[X_2][Y_{4-6}][Z_8]O_{20}(OH,F)_4$

	Interlayer cations, X	Octahedral cations, Y	Tetrahedral cations, Z
Dioctahedral micas			
Muscovite	K_2	Al_4	Si_6Al_2
Phengite	K_2	$(Al, Fe^{3+})_3 (Mg, Fe^{2+})$	Si_7Al
Celadonite	K_2	$(Al, Fe^{3+})_2 (Mg, Fe^{2+})_2$	Si_8
Paragonite	Na_2	Al_4	Si_6Al_2
Dioctahedral brittle mica			
Margarite	Ca_2	Al_4	Si_4Al_4
Trioctahedral micas			
Phlogopite	K_2	Mg_6	Si_6Al_2
Biotite	K_2	$(Al, Fe^{3+})_y (Mg, Fe^{2+})_{6-(x/2)-(3y/2)}$	$Si_{6-x}Al_{2+x}$ $(x<1, y<2)$
Lepidomelane	K_2	$Fe^{3+}_2\ Fe^{2+}_3$	Si_6Al_2
Siderophyllite	K_2	$Al^{3+}_2\ Fe^{2+}_3$	Si_6Al_2
Trioctahedral brittle mica			
Clintonite	Ca_2	$Al_x(Mg, Fe^{2+})_{6-x}$	$Si_{4-x}Al_{4+x}$
Lithium micas			
Lithian muscovite	K_2	$Li_{0-1.5}Al_{4-3.5}$	Si_6Al_2
Trilithionite	K_2	Li_3Al_3	Si_6Al_2
Polylithionite	K_2	Li_4Al_2	Si_8
Ferroan lepidolite[a]	K_2	$Li_3(Fe^{3+}, Al)_{2.67}$	Si_7Al
Zinnwaldite	K_2	$Li_{0-3}(Al, Fe^{3+})_{1.5-2.7}(Fe^{2+})_{1.0}$	$Si_{5.5-7}Al_{2.5-1}$
Taeniolite	K_2	Li_2Mg_4	Si_8

[a] Sometimes called cryophyllite.

alternate tetrahedra through an angle α (Figures 2.12 and 2.13). In the micas, an additional factor that might control tetrahedral rotation is the size of the interlayer cation, because its detailed coordination changes as the $(Si,Al)_6O_6$ rings depart from hexagonal symmetry.

McCauley and Newnham (1971) applied a series of statistical tests to the structures available at that time to find out if the misfit and interlayer cation factors could be isolated and, if so, to determine which was more important. As the tetrahedra rotate, 3 oxygens in the Si_6O_6 rings approach the interlayer cation and 3 recede, so that in general the interlayer cation has an inner and an outer coordination sphere each of 6 oxygens, 3 from one layer and 3 from the adjacent layer; there are therefore two mean interlayer cation–oxygen bond lengths. McCauley and Newnham found that there is an almost exact linear relation between the difference in these mean bond lengths (Δ) and the tetrahedral rotation α, in the determined structures, so that either parameter could be used in the statistical tests. Using a field strength parameter to characterize the interlayer cation–oxygen interaction, and tetrahedral and octahedral cation—

oxygen bond lengths to define the sizes (and therefore the misfit) of the tetrahedral and octahedral sheets, they showed that most of the variance in α (97 %) was accounted for by the misfit parameter, deducing the following multiple regression equation:

$$\alpha(°) = 218.0(b_t/b_0) - 1.5(\text{field strength}) - 221.5$$

This equation can be used to predict the amount of tetrahedral rotation in micas from the chemical composition data and published bond lengths, and is thus an outstanding vindication of the crystallochemical approach to mineral structures. Hazen and Wones (1972) have clarified the calculation of the configuration of the octahedral sheet from mean cation–oxygen bond length for a much wider range of cation substitutions.

Mica polymorphs (Smith and Yoder, 1956). Individual layers in 2:1 minerals have a plane of symmetry that intersects the octahedral hydroxyl groups, and the upper $(Si,Al)_6O_6$ rings are offset with respect to the lower rings by the displacement of the apical oxygen atoms in the upper and lower planes of the octahedral sheets; this displacement of approximately $a/3 = 0.17$ nm parallel to the mirror plane is represented by a layer vector, $\boldsymbol{v}$ (Figure 2.12).

The interlayer cations in micas key the position of one 2:1 layer to the next through the adjacent $(Si,Al)_6O_6$ rings but do not necessarily define the orientation relationship of successive layers. For ideal structures, it is possible to superimpose one layer on another in 6 possible ways, by rotating the layer vector through 0°, 60°, 120°, 180°, 240° and 300°, while still retaining the same relative positions of the adjacent $(Si,Al)_6O_6$ rings. The following layer can also be oriented in six possible ways, and so on, so that the theoretical number of stacking sequences is infinite. In practice, the number of stacking sequences, or polytypes as they are called, is quite small, and it appears that the ditrigonal configuration of the $(Si,Al)_6O_6$ rings favours rotations of 0° or ±120°, so that the coordination of the interlayer cation to the closest basal oxygens is approximately octahedral rather than prismatic. If the rotation is 0°, the structure repeats after one layer; this polytype is designated 1M. If layer 2 is rotated through +120°, layer 3 may rotate through +120° or −120° and so on. When the rotation is always in the same sense the structure repeats only after 3 layers (3 × (+)120°) and the polytype is called 3T; there are two forms of this polytype related by object and mirror image configurations depending on whether the initial rotation is left handed or right handed. 3T polymorphs are uncommon. When the rotation alternates regularly +120° and −120°, the structure repeats after two layers and the polytype designation is 2M; this polytype is common in dioctahedral micas.

A further type of stacking pattern (or lack of it) occurs frequently in clay micas; the characteristic reflections used to identify 1M or 2M polytypes are absent or represented by broad diffuse maxima, rather analogous to the patterns of *b*-axis disordered kaolinites. It is believed that such micas have disordered

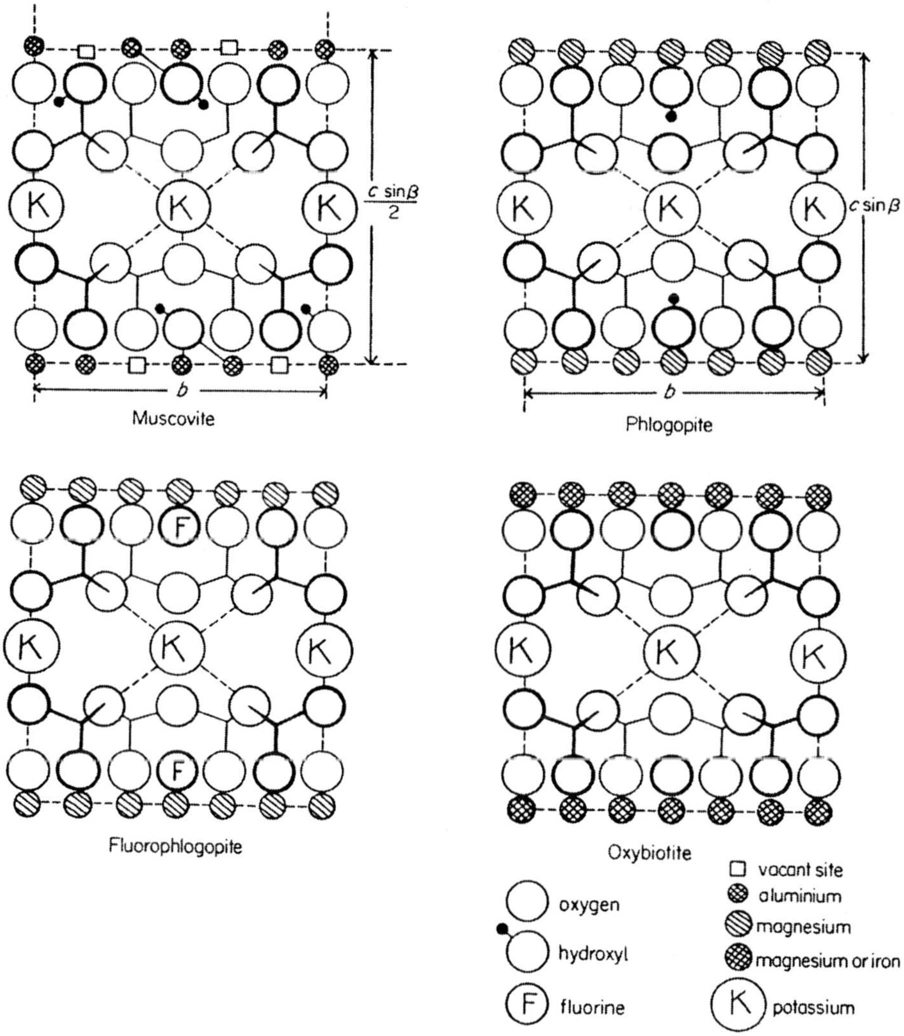

Figure 2.18 Idealized structures of muscovite, phlogopite (biotite), fluorphlogopite and oxybiotite, showing the proximity of potassium to OH or F

stacking sequences and are given a general designation 1Md; they are typically low temperature crystallization forms.

Additional factors in potassium coordination. Most micas contain potassium as the interlayer cation and clays containing mica layers are the common potassium-bearing minerals in soils. It has long been known that micas can weather to vermiculite (q.v) by a process in which relatively little change occurs to the 2:1 silicate layers but the interlayer potassium is replaced by a hydrated cation.

Micas differ greatly in their ability to weather by this mechanism and there has been much interest in establishing the crystallochemical factors that affect potassium bonding in the interlayer sites of micas.

Bassett (1960) was probably the first to realize that the hydroxyl groups in the octahedral layers could influence the strength of potassium bonding, and it has become evident since then that this is probably the predominant factor affecting the rate of mica weathering.

A recent neutron diffraction study of phlogopite (Rayner, 1974) has shown that the potassium–hydrogen distance is 0.31 nm; the K-basal O bond lengths in the same mica are 0.2965 nm and 0.3365 nm, so that the K—O and K—H distances are very similar. The important difference is that hydrogen is a positive ion and exerts an antibonding force on K^+. This short K—H distance arises because phlogopite is a trioctahedral 2:1 silicate and the OH group is coordinated to $3Mg^{2+}$, each of which repels the H^+ equally towards the interlayer cation, so that the OH orientation is normal to the basal plane.

Neutron diffraction analysis has also located the position of H in muscovite (Rothbauer, 1971) and here the configuration is different. The hydroxyl is coordinated to $2Al^{3+}$, but as the mineral is dioctahedral, the third site, which is occupied by Mg^{2+} in phlogopite, is vacant. Consequently H^+ in muscovite is deflected sharply towards the vacant site and the OH orientation is at an angle of 74° to the normal to the basal plane; with this configuration the K—H distance is approximately 0.40 nm and the interaction with K is much smaller than in phlogopite.

The OH orientations in these two minerals are illustrated in Figure 2.18.

Furthermore, two additional factors can influence K bonding in micas. Fluorine can partially or completely replace hydroxyl, particularly in trioctahedral micas; natural phlogopites with an anion composition $O_{20}(OH)_{1.4}F_{2.6}$ are not uncommon (e.g. Hazen and Burnham, 1973) and a synthetic fluorophlogopite with almost the theoretical formula $O_{20}F_4$ is known (McCauley, Newnham and Gibbs, 1973). Fluorine does not exert an antibonding effect on interlayer K, and the rate of alteration of micas correlates inversely with their fluorine content (Rausell Colom *et al.*, 1964; Newman, 1969).

The other factor known to affect K bonding is deficiency of hydroxyl. Silicates containing hydroxyl and ferrous iron can lose hydrogen by a reaction represented by the equation

$$Fe(II)OH + \tfrac{1}{4}O_2 \rightarrow Fe(III)O + \tfrac{1}{2}H_2O$$

and it is known that biotites that have been oxidized by heating in air release K^+ less readily than before oxidation (Robert and Pedro, 1968; Barshad and Kishk, 1968).

Dioctahedral micas. True micas should have a net layer charge approaching -2 per $O_{20}(OH)_4$ and natural dioctahedral micas appear to cluster in two main fields of composition depending on iron content. The muscovite–phengite field lies between the compositions $K_2(Al)_4(Al_2Si_6)O_{20}(OH)_4$ (muscovite) and $K_2(Al,Fe^{3+})_3(Mg,Fe^{2+})(Al_1Si_7)O_{20}(OH)_4$ (phengite); the total octahedral Fe does not exceed one Fe per unit formula. In muscovite, the deficit of layer charge arises entirely from substitution of Al in the tetrahedral sheet, whereas in phengite, the deficit is partly in the tetrahedral sheet and partly from replacement of Al^{3+} by Mg^{2+} or Fe^{2+} in the octahedral sheet.

Minerals in the glauconite–celadonite field contain more than one Fe atom per unit formula and a smaller substitution of Al for Si in the tetrahedral sheet, and the net negative charge originates predominantly from substitution of divalent Fe and Mg for trivalent cations in the octahedral sheet; the celadonite end-member composition appears to be $K_2(Fe_2^{3+})(Mg,Fe^{2+})_2Si_8O_{20}(OH)_4$ with Mg as the dominant divalent cation.

Many glauconites are not true micas in the sense that they contain only one half to three quarters of the theoretical amount of potassium, and are interstratified mixtures of mica and smectite layers. Glauconites are of marine biological origin and often contain impurities like iron oxides, apatite and organic matter.

Trioctahedral micas. The trioctahedral micas are common constituents of igneous and metamorphic rocks and many hundreds of analyses of them have been published.

The end member of the trioctahedral series is phlogopite, $K_2(Mg_6)(Si_6Al_2)O_{20}(OH,F)_4$; natural phlogopites with total Fe less than 1% FeO are rare, and micas containing more than about 6% FeO are called biotites. Simple replacement of Mg by Fe^{2+} is never found and natural micas always contain octahedral Al^{3+}, Fe^{3+} and Ti (grouped together as R^{3+}); the additional positive charge from these polyvalent cations in the octahedral sheet is balanced in two ways: (1) Al substitutes for Si in the tetrahedral sheet up to about $Si_{5.3}Al_{2.7}$ and balances the equivalent excess charge in the octahedral sheet; (2) R^{3+} replaces R^{2+}, not atom for atom, but in the proportion $2R^{3+}:3R^{2+}$ so that some octahedral sites remained unfilled. In many formulae, both types of charge compensation occur.

These two modes of substitution imply a much wider composition range than is actually found in natural micas: Hazen and Wones (1972) point out that the octahedral Al content increases as Fe^{2+} replaces Mg and summarize the structural limitations on the substitutional pattern. A plot of natural composition on a triangular diagram bounded by the proportions of Mg, $(Fe^{2+} + Mn^{2+})$ and R^{3+} in the octahedral sheet (Foster, 1960) shows that the proportion of Fe^{2+} to R^{3+} is mostly between 4:1 and 3:2 with a central value of 7:3.

Formulae in which the number of octahedral R^{3+} exceeds 1.8 per 6 sites are rare, and the number of unfilled octahedral sites is seldom greater than 0.8, which is evidence against the existence of a complete series between the dioctahedral and trioctahedral micas except when lithium is also present.

Foster (1960) found that most of the analyses of the trioctahedral micas formed three natural groups and suggested the following terminology. Micas in which more than 70 per cent of the occupied octahedral sites contained Mg^{2+} she termed *phlogopites* whereas in *biotites* 20–60 per cent of the occupied sites contained Mg; the biotites were further subdivided into *magnesian biotites* with between 40 and 60 per cent Mg and *ferroan biotites* with 20–40 per cent Mg. Finally, the micas with Mg in less than 10 per cent of the occupied sites were divided into *siderophyllites* in which the dominant trivalent cation is Al, and *lepidomelanes* in which Fe^{3+} is the main trivalent cation.

Trioctahedral micas are readily weathered so that trioctahedral analogues of the hydrous dioctahedral micas and illites are rarely found; the only reported example of a trioctahedral illite (Walker, 1950) was too impure for its chemical composition to be established. Vermiculite deposits often contain hydrobiotite which has a mixed structure composed of vermiculite and potassium-containing layers, sometimes in random sequence but often with a tendency for alternation of vermiculite and potassium-containing layers; these minerals are intermediate in composition between biotites and vermiculites, with most of the iron in the ferric state (Boettcher, 1966).

Clay micas. Micas with a particle size smaller than 2 µm equivalent spherical diameter often contain less potassium and more water than the coarser grained micas described in the previous section; this trend is even more evident if a finer clay fraction, say less than 0.2 µm, is taken.

Sometimes this is because the particles are not structurally homogeneous so that although some or even the majority of the interlayers in a mineral grain contain potassium, others contain hydrated cations such as Ca, Mg, or Al, and these hydrated interlayers break up the otherwise regular succession of mica layers repeating at 1.0 nm spacing. Such clays are described as mixed-layer or interstratified minerals and give X-ray diffraction patterns that differ from those of true micas (see 2.3.8 below).

Another reason for the deficit of K is that as the particle size becomes smaller, the surface area of the mica increases. Layer silicate minerals have a platy morphology and much of the area consists of the basal surfaces. The cations on these surfaces are exposed and much more readily hydrated and exchanged than the interlayer K in the interior of the crystal. The aspect ratio (the ratio of plate diameter to plate thickness) of micas is often about 10:1, and a particle of 1 µm diameter could have a thickness of about 0.1 µm and contain about 100 2:1 layers 1.0 nm thick, with a ratio of internal to external charge of 99:1. Exchange of surface K by a hydrated cation decreases the K content of such a particle by 1%, a change not readily detected by chemical analysis. A particle of 0.1 µm diameter, however, would by the same reasoning have a ratio of internal to external charge of 9:1, corresponding with a decrease in K content of 10% and an exchange capacity of 25 meq 100 g^{-1} (cf. Chapter 4 and Table 4.1).

Hower and Mowatt (1966) examined the relationships between K content, exchange capacity and hydrated interlayers within the crystal grains for many micaceous and interstratified clays, and their data extrapolated to 0 % expanding interlayers suggest that clay micas tend towards an exchange capacity of 10 meq 100 g^{-1} and a 25 % deficit of K compared with macroscopic micas.

These comparisons show that neither interstratification nor surface effects entirely account for the K deficit and water excess of clay micas. Other suggestions put forward are that clay micas contain additional hydroxyl groups formed by protonation of basal or apical oxygen atoms, or that the interlayers contain H_3O^+, but neither explanation has yet been sufficiently corroborated.

One consequence of these features is that the definition of 'clay mica' is difficult. 'Illite' is the most widely used group name, and current usage seems to have established the term for hydrous clay micas present in soils and sediments that do not expand from a 1.0 nm basal spacing (Grim *et al.*, 1937) but nevertheless contain less K and more structural hydroxyl than macroscopic micas. This combination of structural and chemical definition seems adequate at present to distinguish clay micas from interstratified minerals but may need to be refined as our understanding improves.

Vermiculites

Material familiar as exfoliated vermiculite is derived from a platy mineral that in its natural state appears to resemble mica but is much softer; it also differs from mica in its ability to expand normal to the basal cleavage when heated rapidly to form a light product that is a good thermal insulator. The natural mineral is formed by the action of supergene solutions on mica present in the host rock: typically it occurs as an alteration product of basic or ultrabasic rocks, in gneisses and schists, in carbonate rocks and in granites (Bassett, 1963). It is readily formed by alteration of biotite mica. Macroscopic vermiculite is a 2:1 layer silicate, invariably trioctahedral in composition, containing much molecular water that is mostly expelled on heating to 300 °C, causing the characteristic exfoliation, and has a large exchange capacity ranging between 120 and 200 meq 100 g^{-1}.

Clay-sized minerals resembling vermiculite are frequently found in soils. These are usually dioctahedral in composition. Some differ from true vermiculite in having interlayer and surface coatings of hydrous oxides that give the clays properties intermediate between vermiculite and chlorite. Because these vermiculite-like materials are usually mixed with other clay minerals in soil clays, their characterization is difficult and nearly always requires a detailed investigation. Consequently much less is known with certainty about their structure and composition. Clay vermiculites may be either dioctahedral or trioctahedral; the latter occur mainly in young soils such as those on glacial tills, and dioctahedral intergrade vermiculites occur in some micaceous soils that do not contain free calcium carbonate.

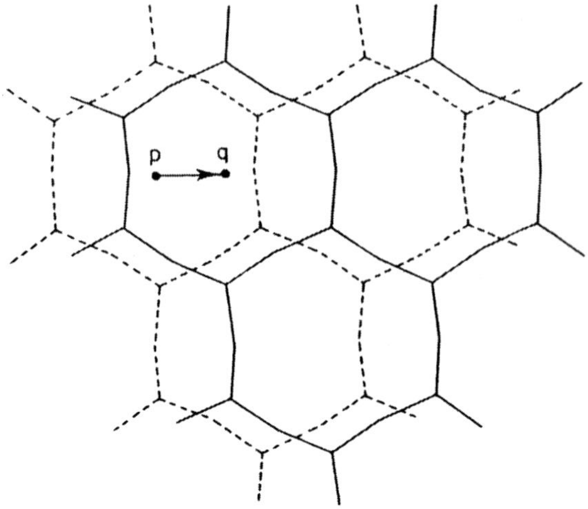

Figure 2.19 Representation of one 2:1 layer in vermiculite, showing the line joining the centre of the basal oxygen atoms in the lowest plane (p) to the centre of the basal oxygen atoms in the uppermost plane (q). This line was used by Mathieson and Walker (1954) to define the relative positions and orientations of successive layers in vermiculite (see Figure 2.21)

Structure. Refined structure determinations done on macroscopic vermiculite have established that the structure of the 2:1 aluminosilicate layers is similar to that of the trioctahedral micas. The central octahedral sheet defines the relative positions of the two outer tetrahedral sheets, and the centre of the upper (Si, $Al)_6O_6$ rings is displaced relative to the lower rings by the octahedral diagonal joining the upper and lower OH groups (Figure 2.19). Misfit of the octahedral and tetrahedral sheets is accommodated by rotation of alternate tetrahedra by 5.5° in Kenya vermiculite (Mathieson and Walker, 1954) and 5.7° in Llano vermiculite (Shirozu and Bailey, 1966), both close to the calculated rotation of 6°.

The 2:1 layers have a net negative charge that arises from isomorphous substitution, mainly Al replacing Si, and is balanced by exchangeable cations situated between the layers. In vermiculites from freshly weathered rock the interlayer cation is usually Mg. Interlayer Mg is not a necessary constituent of vermiculite, however, and it is readily replaced by a different cation if Mg vermiculite is immersed in a solution of the appropriate salt; the essential structure of the 2:1 layers is little changed by exchanging the interlayer cation and Mg-vermiculite is reconstituted by reaction with a solution of a magnesium salt.

The interlayer region of vermiculite contains molecular water, that can be readily removed by desiccation. The amount of interlamellar water is not however continuously variable. Its arrangement at least in macroscopic

Table 2.6 Basal spacings (nm) of powdered macroscopic Kenya vermiculite saturated with various cations at 25 °C (Walker, 1961)

p/p_0	0.3	0.5	0.7	1
Li	1.23	1.50	1.50	[a]
Na	1.23	1.48	1.48	1.48
K	1.05[b]		1.06[b]	1.3[b]
NH_4	1.08[b]		1.08[b]	1.09[b]
Mg[c]	1.44	1.44	1.44	1.47
Ca	1.50	1.50	1.51	1.53
Sr		1.49		1.54
Ba	1.23	1.50	1.52	1.57

[a] High spacings up to a hundred nm.
[b] Very diffuse basal reflections.
[c] Olphen (1969) showed that the one sheet (1.2 nm) phase of Mg-Llano vermiculite is stable in the range p/p_0=0.0005 to 0.015.

Table 2.7 Hydration states of Na- and Mg-Llano vermiculites

Interlayer cation	Stability range (p/p_0) at 25 °C	Basal spacing, nm	$H_2O/O_{20}(OH)_4$	H_2O/cation
Na	<0.07	0.98	0.4	<0.3
	0.07–0.23	1.18	3.3	2.1
	0.45–0.7	1.48	10.4	6.5
Mg	<0.0005	0.93	?	?
	0.0005–0.015	1.16	2.8	3.5
	0.03 –0.20	1.43	8	10
	0.3 –0.45	1.43	8.8	11

vermiculites has much structural order, and the positions of the molecules have been located by X-ray diffraction, although they are more variable than the position of the other atoms in the aluminosilicate framework. The interstitial water molecules increase the separation of the aluminosilicate layers so that the repeat unit of the structure normal to the basal plane is larger than in the micas. The series of basal 00*l* reflections are simple diagnostic evidence of the hydration state of vermiculite, and the basal separation of the layers depends upon which cation is present in the interlayer and upon the relative humidity of the ambient atmosphere (p/p_0) (Tables 2.6 and 2.7 and Figure 6.6, p. 420). Three principal hydration states are recognized: dehydrated vermiculite has a basal spacing of about 1.0 nm, partially hydrated vermiculite a spacing of about 1.2 nm, and when fully hydrated, the spacing is 1.4 to 1.5 nm. These spacings correspond to zero, one and two molecular layers of interlamellar water respectively. For a particular vermiculite containing a specific interlayer cation, intermediate spacings are not observed and a sharp transition from one state to another occurs over a narrow range of relative humidity (Table 2.7). K-, NH_4-, Rb- and Cs-vermiculites do not conform exactly to this general pattern, apparently because of irregularity in the layer separation sequence, and the weak hydration of these cations. Li-vermiculite has the normal zero, one and two water layer states, and also in very dilute solution or water, swells macroscopically to form a gel-like material in which the average layer separation is greater than 3.0 nm, a behaviour analogous to Na-smectite (*q.v.*). A full description of the hydration states of every vermiculite requires that the water content and basal spacings are measured as a function of relative vapour pressures from $p/p_0 = 0$ to 1 at several different temperatures for each cation form; this amount of experimental work has rarely been attempted, but was done by van Olphen (1965, 1969) for Mg- and Na-Llano vermiculite and confirms the stepwise hydration mechanism indicated by the change in basal spacings. The factors controlling the water structure in the interlayer region are discussed more fully in Chapter 6.

Macroscopic Mg-vermiculite has a two-layer monoclinic structure similar to the type of chlorite designated Ia by Bailey and Brown (1962) and discussed below; it differs from it in having an interlayer composition of approximately $Mg_{0.8}(H_2O)_{8.8}$ compared with $M_6(OH)_{12}$ (M = Mg, Fe, Al) in chlorites. The essential features of the Ia structure (Figure 2.20) are;

(i) the 2:1 layer octahedral stagger is in the same direction as the stagger of the octahedrally disposed oxygen atoms in the interlayer;
(ii) tetrahedral atoms in the upper $(Si,Al)_6O_6$ ring of the lower 2:1 layer, the interlayer cation, and tetrahedral atoms in the lower ring of the upper layer are vertically superposed;
(iii) the interlayer oxygen atoms are placed vertically over the octahedral cation in the 2:1 layer.

There are four possible stacking arrangements of successive 2:1 layers in the Ia structure, designated p, q, r, and s by Mathieson and Walker (1954). The relative positions and orientations of successive 2:1 layers may be simply specified in terms of the vector that describes the displacement of the upper $(Si,Al)_6O_6$ ring relative to the lower ring in the same layer (Figure 2.19). In Llano vermiculite, and probably also in the Scotscalder vermiculite formed by weathering of phlogopite mica (Smith Aitken, 1965), these vectors are parallel but displaced by $-a/3$ and alternately $+b/3$ and $-b/3$ in successive layers (Figure 2.21b); this is the 's' stacking arrangement of Mathieson and Walker. In Kenya vermiculite, by contrast, successive layers are also rotated by alternately $+60°$ and $-60°$, an alternation of the 'q' and 'r' stacking sequences (Figure 2.21a).

An interesting feature of both these stacking sequences is that only three out of six tetrahedral sites in each layer are superposed by tetrahedral sites in the succeeding layer, with the interlayer cation placed between these superposed sites (Figure 2.22). In their refined analysis of the Llano-vermiculite structure,

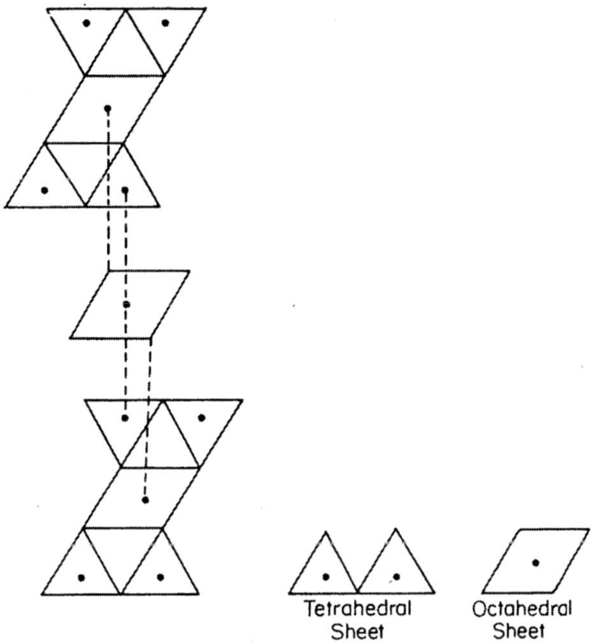

Figure 2.20 The Ia structure of chlorite adopted by vermiculite

Shirozu and Bailey showed that the cation–oxygen bonds were significantly different for these two sets of tetrahedra, being 0.1673 nm in the set that superpose the interlayer cations (T_1 sites) compared with 0.1641 nm for the T_2 set. In layer silicates, the Si—O bond length is 0.162 nm and Al—O is 0.177 nm and from the observed bond lengths it is calculated that the average occupancy of the T_1 sites was $Si_{0.65}Al_{0.35}$, and $Si_{0.86}Al_{0.14}$ for the T_2 sites. The deficit of positive charge that results from Al substituting for Si is therefore concentrated more in the T_1 sites than in the T_2, and the close association of the T_1 sites and the interlayer cation minimizes the separation of positive and negative charges.

Chemistry. As already noted, the net negative charge in macroscopic vermiculite derives from substitution of Al for Si in the tetrahedral sites; this substitution usually exceeds 2Al per $O_{20}(OH)_4$, but is partially balanced by an excess of positive charge in the octahedral sites where trivalent Al and Fe replace divalent cations. The net negative charge of vermiculite layers as a consequence is less than 2 per $O_{20}(OH)_4$, and ranges between 1.2 and 1.9. In all known macroscopic vermiculites, Mg is the dominant cation in the octahedral layer, and in many Mg occupies more than two thirds of the octahedral sites. The composition of vermiculites is therefore more restricted than that of the trioctahedral micas (Foster, 1963). Vermiculites also differ from biotites in containing iron mainly

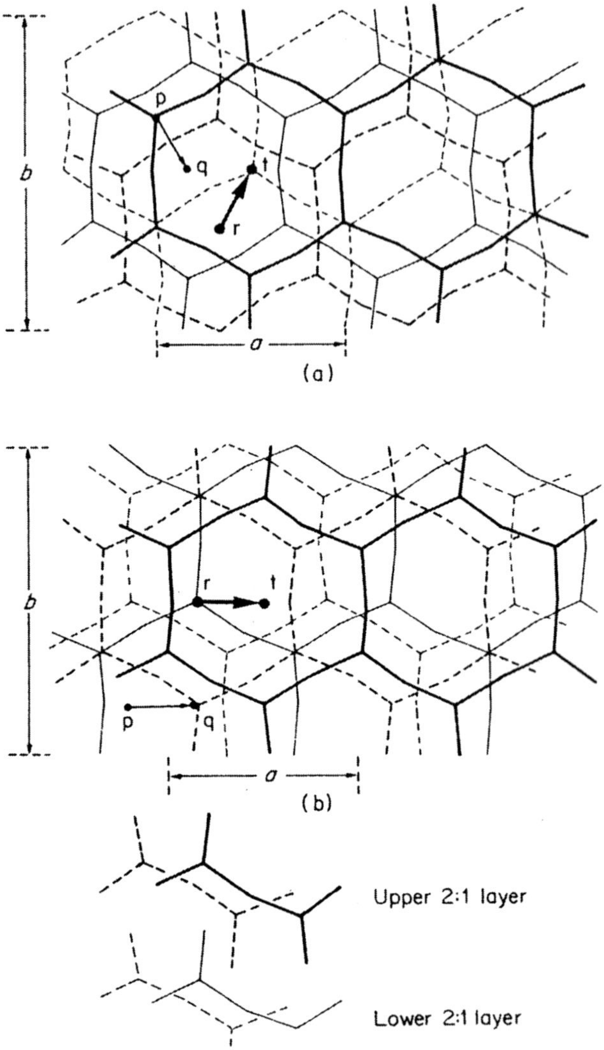

Figure 2.21 Superposition of 2:1 layers in (a) Kenya (b) Llano vermiculites. p and q are the centres of the basal oxygen rings in the lower layer, and r and t in the upper

in the trivalent form and Norrish (1972) has pointed out that the net charge in vermiculites correlates negatively with the Fe(III) content. When mica is altered to vermiculite under natural conditions and interlayer K is replaced by Mg, Fe(II) originally present in the mica is oxidized to Fe(III) and this is the main mechanism decreasing the net negative charge.

Experimental studies have shown that when biotites containing much ferrous

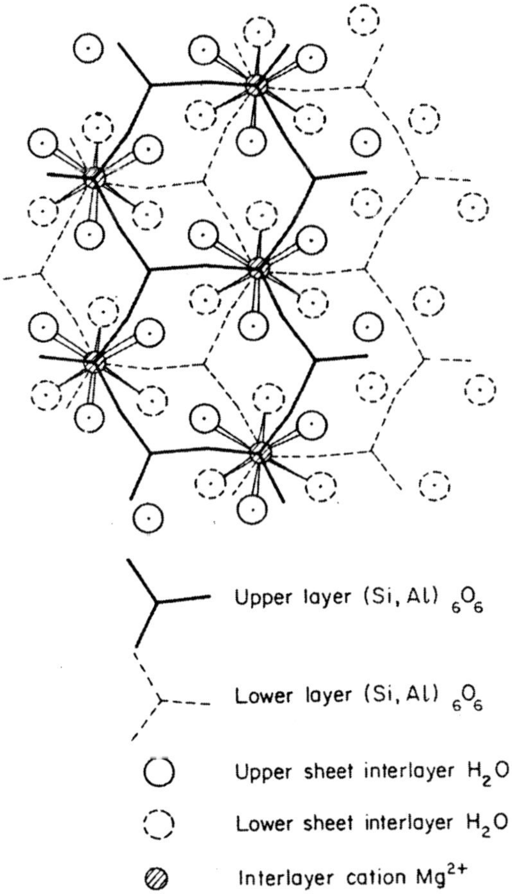

Figure 2.22 Interlayer structure in hydrated (1.43 nm) Mg-vermiculite. Projection on a basal plane

iron are altered and oxidized, the high concentration of trivalent Fe in the octahedral layer caused instability and some Fe is ejected from the structure, which then tends towards a dioctahedral composition (Farmer *et al.*, 1971); it is probable that this sequence of reactions also occurs under natural conditions and accounts in part for the rarity of iron-rich vermiculites. Structural formulae for the two most investigated vermiculites are:

Kenya; $Mg_{4.72}Fe^{3+}_{0.96}Al_{0.32}(Si_{5.44}Al_{2.56})O_{20}(OH)_4 \cdot Mg_{0.64}$

Llano; $Mg_{5.62}Fe^{3+}_{0.13}Al_{0.21}(Si_{5.79}Al_{2.21})O_{20}(OH)_4 \cdot Mg_{0.93}$

Kenya is a medium-to-low-charge vermiculite with a relatively high-iron content, whereas Llano is a high-charge, low-iron vermiculite.

Organic complexes. Like smectites, vermiculites form complexes with many types of organic compounds, although, in general, these are less readily prepared because of the higher charge density on vermiculite layers (Mortland, 1970; Theng, 1974). Alkylammonium and pyridinium cations readily replace inorganic interlayer cations; the bipyridilium herbicides Diquat and Paraquat are strongly sorbed by vermiculite. n-Butylammonium vermiculites swell macroscopically in water to gels analogous to swollen Li-vermiculite. Amines, including NH_3, are sorbed and protonated on vermiculite surfaces, the highly polarized water in the hydration sphere of the interlayer cation acting as the proton donor. Alkylammonium vermiculites have the ability to incorporate additional organic molecules that by themselves do not form complexes with vermiculites containing inorganic interlayer cations; compounds forming association complexes include aldehydes, ketones, carboxylic acids, phenols, esters and even hydrocarbons. Amino acids in the zwitterion form swell vermiculite single crystals, the swelling increasing with the concentration of the amino acid solution (Walker and Garrett, 1961).

The complexes with alcohols and in particular the diol, ethylene glycol and the triol, glycerol are frequently used to distinguish smectites from vermiculites (Brindley, 1966). When Mg is the exchangeable cation, vermiculites generally form one-sheet complexes having basal spacings of 1.29–1.4 nm with ethylene glycol and 1.42–1.48 nm with glycerol, whereas the smectite complexes have spacings of 1.68 nm and 1.78 nm respectively (Walker, 1957, 1958). This is, however, an oversimplification and the complexes formed by vermiculites depend on both the interlayer cation, as well as the charge density; as the vermiculite charge density can range between 1.2 and 1.9 per $O_{20}(OH)_4$, a low-charge vermiculite may form a two molecular layer complex with ethylene glycol.

Clay vermiculites. Mineralogical examination of soil clays by X-ray diffraction sometimes yields the following observations: Mg-saturated clay, air dry, gives a diffraction maximum at 1.4 nm that changes little on treatment with ethylene glycol; K-saturated clay also gives a maximum at 1.4 nm that on heating decreases progressively until at 550 °C the dominant reflection is close to 1.0 nm and no reflection at 1.4 nm remains.

This set of observations cannot be reconciled with the properties of any of the major clay mineral groups: if the 1.4 nm reflection were due to smectite (or smectite-illite mixed-layer mineral) it would not remain unchanged by treatment with K or ethylene glycol; chlorite, if present, would have given a reflection in the region of 1.3 to 1.4 nm on heating to 550 °C, whereas vermiculite on K-saturation normally changes to give a spacing of 1.0 to 1.05 nm.

However, if a soil clay showing this anomalous behaviour is first treated with certain chemical reagent solutions able to bring Al into solution, including KOH+KCl, NaCl+HCl, NH_4F or neutral sodium citrate, it will in many instances expand with ethylene glycol untii the basal spacing is near to 1.7 nm or collapse to 1.0 nm with K-saturation, like normal smectites or vermiculites. These chemical treatments often increase the cation exchange capacity of the clay, and although the reaction solutions contain the elements Si, Al, Fe, Mg and K, showing that some clay has been dissolved, Al is often in excess of the others; it is now generally accepted that clays showing the behaviour described above contain interlayer aluminium, as some form of hydroxy polymer.

In free solution, aluminium is hydrolysed first to $AlOH^{2+}$ or more probably the dimer $(AlOH)_2^{4+}$, and then to polynuclear hydroxyl complexes of the type $Al_6(OH)_{15}^{3+}$ or similar zones of higher molecular weight, with mean ratio of OH:Al close to 2.5. These polynuclear complexes are soluble in water and are only very slowly depolymerized by dilute mineral acid. If solutions containing these hydroxyl complexes of aluminium are added to vermiculite or montmorillonite, the complexes are rapidly and strongly sorbed by the clay, displacing the interlayer cation quantitatively; the sorbed Al is not readily displaced by cation exchange and the more vigorous chemical treatments mentioned above are needed to remove it, though present evidence suggests that the montmorillonite–hydroxyaluminium complex is less stable and is decomposed by potassium salts.

Vermiculite with introduced hydroxyaluminium interlayers has a basal spacing of 1.4 nm unaffected by ethylene glycol or K-saturation (Nagasawa *et al.*, 1974) the reflection of 1.4 nm decreases progressively on heating to 550 °C. Hydroxyaluminium vermiculite has a very much smaller cation exchange capacity than untreated vermiculite but this increases when aluminium is extracted with sodium citrate, showing that the positively charged aluminium polymers partially balance the net negative charge of the aluminosilicate layers.

The behaviour of these laboratory-prepared complexes closely parallels that of the soil clays described earlier, and it is reasonably certain that such soil clays contain vermiculite with interlayers of hydroxyaluminium. It is claimed by some that ferric iron also forms hydroxyl complexes in soils and clays, but the evidence for this is much less convincing.

All degrees of hydroxyaluminium interlayering are possible, from normal vermiculite to a clay in which all the negative charge is countered by aluminium. Some soils show this progression, from a true vermiculite that collapses to 1.0 nm on K-saturation occurring at the base of the profile, to a 1.4 nm mineral in the surface horizon that is unaffected by K. Unfortunately as yet no systematic test has been devised that will estimate the amount of aluminium interlayering in a mixture of unknown clay minerals.

Smectites

In the eighteenth century the name smectite was applied to 'fuller's earth', which is mainly a hydrous aluminium silicate clay containing magnesium and calcium. The name montmorillonite was given to a clay from Montmorillon, France, in the mid nineteenth century, and when the similarity of smectite and montmorillonite was established, the name smectite was discarded. When further work revealed the chemical complexity of these minerals (Ross and Hendricks, 1945; MacEwan, 1961) it became necessary to call them the 'montmorillonite group'. This proved unsatisfactory as 'montmorillonite' *sensu stricto*, that is a mineral with a defined composition, was confused with 'montmorillonites', meaning minerals with a much wider composition range. By international agreement (Brindley and Pedro, 1975) the old name 'smectite' has now been revived as a group name to embrace the dioctahedral minerals montmorillonite, beidellite and nontronite, the trioctahedral minerals saponite and hectorite, and other similar but uncommon minerals such as trioctahedral sauconite containing Zn, dioctahedral volkhonskoite containing Cr, and trioctahedral medmontite containing Cu.

Smectites resemble vermiculites in being 2:1 layer silicates with variable basal spacings and cation exchange properties, but differ in occurring only as clay- and silt-sized particles and in having a lower exchange capacity, or more correctly, a lower negative charge on the aluminosilicate layers. These two differences are sufficient to make many of the properties of smectites very distinct from those of vermiculites.

Bentonites are the best known source of true smectites and are formed by the alteration of volcanic glass. They consist mainly of montmorillonite or less commonly beidellite, but because the parent glass frequently contained silica in excess of that required for the clay mineral composition, free SiO_2 as quartz, cristobalite or amorphous silica may also be present. Bentonites have considerable commercial value; they are used as decolorizers in oil and sugar refining, as catalyst supports, in pharmaceutical preparations, and in their sodium form, as binding aids, plasticizers, and for forming water-impermeable barriers in civil engineering. The American deposits are well known but bentonites occur in many regions of the world and calcium montmorillonite is the main constituent of English fuller's earth.

Many soils and sediments contain swelling clays that resemble smectites, and in the past these clays have often been so identified. Yet chemical analysis frequently reveals that such clays contain structural potassium, which is not present in true smectite. Sometimes this potassium can be allocated to fine-grained mica in a mixed mineral assemblage, but when no discrete mica can be detected, the potassium is assigned to a specific type of mineral in which mica and smectite layers are mixed in one structural unit. These are called mixed-layer or interstratified minerals, and as they require special methods for their description and identification, they are described in Section 2.3.8.

Structure. Smectites have not been found as crystals of sufficient size to provide single crystal X-ray patterns. Powder patterns show much less detail than, for instance, that of well crystallized kaolinite, but they contain two types of reflection; one type is symmetrical and varying with water content (basal reflections, marked B in Figure 2.23), the other type is invariant with some

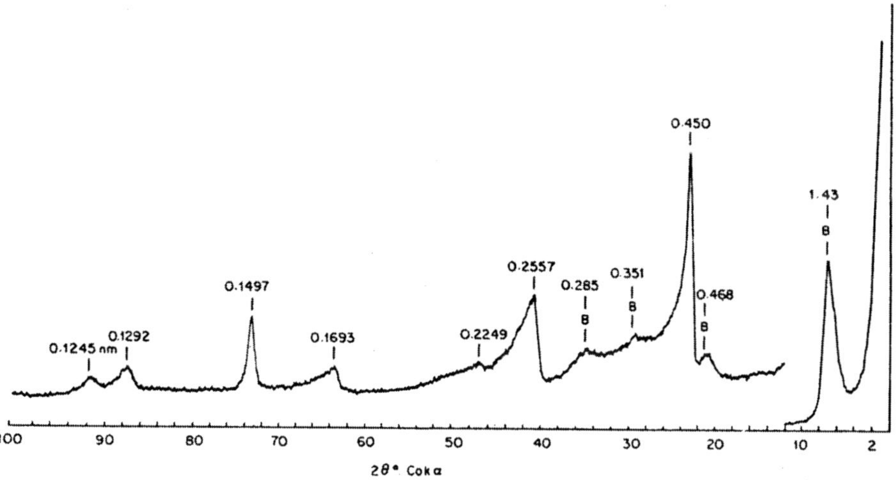

Figure 2.23 Diffractometer trace of Redhill montmorillonite. Basal reflections marked B, are the first, third, fourth and fifth orders (trace from randomly oriented powder).

reflections markedly asymmetrical (non-basal or general reflections). By analogy with the structures of pyrophyllite and micas, the smectite patterns were considered to derive from a 2:1 layer silicate containing interlayer water, and a one-dimensional Fourier synthesis using the basal 00*l* reflections showed that the z-coordinates for the planes of atoms in the structure are consistent with this model (Pézerat and Méring, 1954). The layer structure so calculated is very similar to that in pyrophyllite and mica.

The asymmetry of the *hk*0 diffraction bands of smectites (Figure 2.23) indicate that they arise from irregularly superposed 2:1 layers, each successive layer remaining parallel but randomly orientated and positioned relative to the preceding layer. This is termed a 'turbostratic' structure. Because of this random superposition, little can be learnt from X-ray powder diagrams about the *x*- and *y*-coordinates of the atoms within each layer and many details of the structure are not known for certain. For example, the minerals in the montmorillonite–beidellite series are known from chemical analysis and from the position of the 06 reflection to be dioctahedral in composition, with two thirds of the octahedral sites occupied. By analogy with pyrophyllite, muscovite, and other dioctahedral minerals whose structures are accurately known, the cations are probably ordered, with the vacant site lying on the layer plane of symmetry, but other views have been expressed (Méring and Oberlin, 1971; Güven, 1974). In nontronite, however, the structure is non-centrosymmetric (Goodman *et al.*, 1976).

Cation exchange and hydration. Layer separation in smectites depends both on the interlayer cation and the amount of water associated with it and is readily measured from the 00*l* basal reflections. The interlayer cations are much more accessible than other cations in the structure and are replaced when the clay is wetted with a suitable salt solution. As negative charge on the 2:1 layers arises from replacement of tetrahedral Si by a trivalent cation (usually Al) and from replacement of a trivalent cation in the octahedral sheet by a divalent, the charge is unchanged by exchanging the balancing interlayer cation. Hydration of the cation can however affect the weight basis on which the results are expressed.

The basal spacings of different cation forms of smectites show that only a small number of discrete hydration states are structurally stable (Table 2.8). When dehydrated, smectites have basal spacings of 0.95 to 1.0 nm, but on exposure to water vapour (e.g. atmospheric humidity).they quickly rehydrate and the spacings increase through 1.25 to 1.5 and 1.9 nm at saturation, and to 2.2 nm in the case of lithium-smectite.

Intermediate spacings are apparently observed over small ranges of vapour pressure, but the 00*l* reflections are not rational, that is the *d*-spacings of the higher order reflections are not integral submultiples of the first order. This indicates that there are two or more layer spacings in the structural unit, mixed more or less randomly. For example, Na-Wyoming montmorillonite containing 70 mg H_2O/g clay gives reflections at 1.12, 0.56 and 0.32 nm. These clearly do not represent a true layer separation of 1.12 nm because neither 3×0.32 nor 4×0.32 equal 1.12 nm, and the pattern is interpreted as coming from a structure with approximately 70 per cent of 1.0 nm and 30 per cent 1.24 nm layers randomly mixed (cf. Section 2.3.8).

By combining measurements of basal spacings with adsorption and desorption isotherms for water on various cation forms of smectites, Mooney *et al.* (1952) showed that the 1.24–1.25 nm spacing corresponded to the presence of a

Table 2.8 Basal spacings, nm, of Wyoming montmorillonite–water complexes saturated with various cations at 25 °C and equilibrated at different relative humidities (p/p_0) after Norrish (1954)

Exchange cation	p/p_0			
	0	0.5	0.7	1.0
Li	0.95	1.24	d	M
Na	0.95	1.24	1.51	M
NH_4	1.0	d	d	1.50
K	1.0	1.24	d	1.50
Cs	1.2	1.28	1.28	1.38
Mg	0.95	1.43	d	1.92
Ca	0.95	1.50	1.50	1.89
Ba	0.98	1.26	1.62	1.89

d, diffuse reflections; M, Macroscopic swelling.

monolayer of water in the interlayer region, and analogous reasoning shows that basal spacings of 1.5, 1.9 and 2.2 nm correspond with 2, 3 and 4 layers of water between each aluminosilicate layer. Table 2.8 gives the basal spacings for some cation forms under several humidity conditions; K- and NH_4-smectites do not normally expand beyond 1.5 nm and Cs-smectite has a narrow range of spacing between 1.2 and 1.38 nm (Norrish, 1954). Most Li- and Na-smectites in dilute solution or in water swell macroscopically into a gel-like state in which the average lay·r separation is greater than 4.0 nm and increases in proportion to $1/\sqrt{c}$, where c is the electrolyte concentration in the liquid phase; the increase is continuous and there is no indication of discrete hydration states. This is known as macroscopic or double-layer swelling, as it is associated with the formation and extension of diffuse electrical double layers on each aluminosilicate surface. This contrasts with the swelling between 1.0 nm and 2.2 nm which is called crystalline swelling because definite hydration structures are formed (Chapter 6).

Chemistry. Because only limited structural information about smectites can be obtained by X-ray diffraction, evidence about their chemical composition is very important. It is, however, often difficult to be sure of the purity of a clay mineral, and many early analyses were made on clays subsequently shown to contain impurities; the existence of end-member beidellite was only certainly established in 1962 when Weir and Greene-Kelly described an exceptionally pure example. A clay mineral usually needs to be size-fractionated to exclude non-layer-silicate minerals (<0.2 μm is often necessary), examined by X-ray diffraction to make sure that other crystalline clays are absent, and preferably given selective dissolution treatments to detect X-ray amorphous Si, Al, and Fe hydrous oxides.

Smectites have a net layer charge smaller than 1.3 per $O_{20}(OH)_4$, and this distinguishes them from vermiculites which have a larger charge. In the *dioctahedral series*, two extreme compositions are recognized in which the charge is entirely tetrahedral or entirely octahedral in origin:

$$\left.\begin{array}{ll}\text{beidellite:} & M_x^+(Al_4)(Si_{8-x}Al_x)O_{20}(OH)_4 \\ \text{montmorillonite:} & M_x^+(Al_{4-x}Mg_x)(Si_8)O_{20}(OH)_4\end{array}\right\} 0.65 < x < 1.3$$

Thus beidellite is essentially Mg-free (Black Jack beidellite contains 0.1 % MgO) whereas montmorillonite is Mg-rich (Otay montmorillonite contains over 7 % MgO). Ferric iron may replace Al in the octahedral sheet and montmorillonite from Redhill fullers earth contains about 8 % Fe_2O_3, equivalent to 0.7 Fe/$O_{20}(OH)_4$, but few montmorillonites have as much Fe(III) as this. Iron-rich smectites with more than 2Fe/$O_{20}(OH)_4$ are called nontronites and several examples are known in which Fe(III) replaces Si in the tetrahedral sheet as well as octahedral Al (Goodman *et al.*, 1976).

Few smectites have the ideal compositions given above and most are partly beidellitic and partly montmorillonitic in composition; for classification the division is drawn at the halfway point, with beidellites having more than 50 per cent of their charge from tetrahedral substitution. Montmorillonites have been further subdivided into Wyoming, Otay and Chambers types, originally on the basis of their differential thermal analysis patterns (Grim and Kulbicki, 1961) and subsequently by their composition (Schultz, 1969):

Wyoming:	total charge <0.85; of which tetrahedral charge deficiency accounts for 15–50 %
Otay:	total charge >0.85; of which tetrahedral charge deficiency accounts for <15 %
Chambers:	total charge >0.85; of which tetrahedral charge deficiency accounts for 15–50 %

Schultz has also redetermined the hydroxyl contents of many smectites, and while many are close to the composition 4(OH) per O_{20}, some, which he called 'non-ideal', contain more or less than the theoretical hydroxyl content. If a mineral does not contain the ideal hydroxyl content, the standard method for calculating the structural formula of a clay given on page 56 is invalidated because the anion charge is no longer 44; Schultz discusses the possible alternative methods that could be used, which depend on a structural model for the hydrogen deficit or excess.

When a Li-saturated montmorillonite is heated to 200 °C, the net charge, which has its origin mainly from substitution of divalent cations in the octahedral sheet, decreases permanently and the clay swells less with glycerol; Li-beidellite with only tetrahedral charge, is unaffected by heating. Lithium, in its unhydrated form a relatively small cation (r = 0.082 nm), migrates into the vacant sites in the octahedral sheet and neutralizes the net octahedral charge. These observations are the basis of a test for determining how the charge is distributed between the tetrahedral and the octahedral sheets (Greene-Kelly, 1955) by estimating the proportion of layers that re-expand when treated with glycerol.

The commonest *trioctahedral* smectite, saponite, contains predominantly Mg in the octahedral sheet, with variable substitution of trivalent cations up to about one per six octahedral sites. This gives the octahedral sheet a positive charge that partially balances negative charge in the tetrahedral sheet from replacement of Si by Al. It is not possible to represent the composition range of saponites exactly by structural formulae, but a fairly close approximation is:

$$\text{saponite:}\quad M_x^+(Mg_{6-z}R_z^{3+})(Si_{8-y}Al_y)O_{20}(OH)_4$$

where R^{3+} = Fe or Al, x, the layer charge = sum of sheet charges = $-(z-y)$, and $0.6 < x < 1.2$. In rare instances, up to 1 Mg in 4 is replaced by Fe^{2+}.

Saponites with Ca, Mg, or Ba interlayer cations form zero, one, two and three layer hydrates like the dioctahedral smectites; some Na-saponites swell only to 1.52 nm but Li-saponite shows macroscopic swelling. In swelling behaviour, therefore, saponites are intermediate between beidellites and vermiculites (Suquet, de la Calle and Pézerat, 1975). Like beidellite, there is some tendency for layer stacking and the non-basal reflections are resolved into *hkl* components.

Hectorite contains essential Li and F, but very little Fe or Al, and the net layer charge is due to replacement of Mg by Li in the octahedral sheet:

$$\text{hectorite:}\quad M_x^+(Mg_{6-x}Li_x)Si_8O_{20}(OH,F)_4$$

The net layer charge of hectorites is apparently smaller than that of other smectites.

Organic complexes. Many natural clays, particularly those from soils, contain organic carbon compounds that are difficult to remove even by drastic chemical treatments and it seems certain that clay markedly protects the organic matter from reaction; this strong association is expressed in the term clay–organic complex. The form of clay–organic complexes in soils has not been established with certainty, partly because knowledge about the composition of soil organic matter is incomplete and also because the clays themselves are seldom precisely identified. Consequently, knowledge about clay–organic complexes has derived mainly from studying the interaction between specific chemical compounds and monomineralic clays; montmorillonite from Wyoming has been used more than any other mineral and there is a vast literature on the organic complexes of montmorillonite and vermiculite. This has been reviewed in depth by Theng (1974) and also by Greenland (1965) and Mortland (1970); here only broad outlines are given.

Positively-charged organic cations readily displace inorganic interlayer cations in smectites, and will also displace the interlamellar potassium ions from micas, although less readily (Mackintosh, Lewis and Greenland, 1971). Many alkylammonium complexes of smectites have been prepared, and the n-alkylammonium derivatives have a particular interest. When the chain length is short, they initially lie parallel with the basal plane of the aluminosilicate layers and form one, two or three such parallel layers. When the number of carbon atoms in the chain is increased to twelve or more the chains would overfill the available space if they remained parallel to the basal plane, and at this chain length, the basal spacing increases substantially, with the adsorbed ions standing 'on end' in the interlayer region (Figure 4.12).

Smectites readily form complexes with substituted pyridinium compounds and the bipyridylium herbicides Diquat (1,1′-ethylene-2,2′-bipyridinium dibromide) and Paraquat (1,1′-dimethyl-4,4′-bipyridinium dichloride). X-ray data for complexes of these herbicides with montmorillonite give basal spacings of 1.26 nm which indicates that they are flattened in the interlamellar spaces and the amounts adsorbed are close to the cation exchange capacity of the clay. Microcalorimetry has shown that the interactions are significantly exothermic and this effect can be attributed to the coulombic attractions between the cations and the lamellae and to the formation of charge transfer complexes between the electron-rich clay surface and the flattened electron-deficient bicyclic aromatic rings (Hayes *et al.*, 1978).

Complexes of 1-n-alkyl pyridinium compounds with montmorillonite take more than one form, and when the chain length is eight carbon atoms or less, maximum sorption is not greater than the exchange capacity of the clay, as with the n-alkyl ammonium complexes. Where longer chains are present among the alkyl groups the cations are sorbed in excess of this, and the anion is sorbed simultaneously to maintain charge balance. Thus, for instance, the maximum sorption of n-cetyl pyridinium bromide is determined by the total surface area of the clay, and not by the density of charge (Chapter 4).

Because hydration water coordinated to interlayer inorganic cations has a potential acidity function, and organic bases tend to coordinate protons, they readily react with smectites, a proton being donated to the uncharged base so that it adopts the cationic form, and may displace exchangeable cations initially present (Chapter 6). Urea and amides can be adsorbed in this way.

As the basal surfaces of smectite layers are negatively charged, anions are normally repelled from the interlamellar regions. However some anionic substances are sorbed and two types of association are thought to be responsible. Polyvalent cations, for example, Al^{3+}, are strongly held by smectites, and in acid conditions organic anions with a complexing action may be sorbed at the cation site. For instance, montmorillonite negatively adsorbs the herbicide 2,4-dichlorophenoxyacetic acid ($Cl_2C_6H_3OCH_2COOH$) at pH values between 4 and 10, where the $—COO^-$ species is present but positively adsorbs the compound if the pH is below 4.0, where the —COOH species is dominant (Frissel and Bolt, 1962).

Many uncharged polar organic compounds, including alcohols, ketones, aldehydes, ethers, esters and saccharides, form complexes with smectites (Table 2.9). Infrared evidence shows that, as with water, hydrogen bonding of the interlamellar organic molecules to the siloxane oxygen surface of the clay is weak, but hydrogen bonds to water molecules in the hydration shell of the exchangeable cation are often important. A feature of this type of complex is that it is usually only stable in the presence of an excess of the organic compound. In the presence of excess water the hydrate may be reformed. It is, however, often difficult to remove the last traces of adsorbed organic compounds from smectites, just as the preparation of truly anhydrous clay–organic complexes is also difficult. Thus alcohol-washed clays may retain a little of the solvent even after several cycles of hydration and dehydration.

The complexes with ethylene glycol, $(CH_2OH)_2$, and glycerol, $(CH_2OH)_2$ CHOH, have practical importance in clay mineral identification (Brindley, 1966). These enable smectites, which when Mg is the exchangeable cation form a complex with ethylene glycol with an interlamellar layer two molecules thick and a basal spacing of 1.68–1.70 nm, to be distinguished from Mg-vermiculites, which adsorb only a monomolecular layer so that the spacing is 1.29–1.40 nm. The basal spacings of the complexes with glycerol are 1.78 nm and 1.42–1.48 nm

Table 2.9 Basal spacings, nm, of some smectite–organic complexes (after Theng, 1974)

Smectite	Exchangeable cation	Organic compound	Approx. no. of molecular layers per interlamellar region	$d(001)$ nm
Montmorillonite	Mg	Ethylene glycol	2	1.70
Montmorillonite	Mg	Glycerol	2	1.78
Montmorillonite	Ca	Triethylene glycol	1	1.33
			2	1.73
Montmorillonite	K	Acetone	1	1.34
Montmorillonite	Na	Acetone	1	1.32
Montmorillonite	Ba	Acetone	2	1.73
Montmorillonite	Ca	Acetone	2	1.73
Montmorillonite	Mg	n-tetradecanol	2	5.17
Nontronite	Mg	n-tetradecanol	2	5.14
Beidellite	Mg	n-tetradecanol	2	5.17
Beidellite	Ca	n-tetradecanol	2	4.83
Montmorillonite	—	Methylammonium	1	1.25
Montmorillonite	—	Ethylammonium	1	1.27
Montmorillonite	—	Cetylpyridinium bromide	2	4.16

for montmorillonite and vermiculite respectively. Both complexes have well-defined structure and are readily formed by immersing the clay in the liquid. They are stable indefinitely in the presence of excess liquid, but decompose on prolonged exposure in the air, reforming the hydrate.

Retention of both liquids has been extensively used to estimate the relative surface areas of clays. This application and other methods for surface area measurements are discussed in Chapter 4.

Chlorites

The chlorite group of minerals, named from the green colour of many of them, was established on the basis of chemical and optical properties before any structural information was available. Some minerals included earlier, such as serpentine and amesite, are now known to be 1:1 group minerals. Structural studies from about 1930 showed that most of the minerals in the group have a structure based on 2:1 units similar to those in mica, but with the interlayer space increased and the interlayer cations joined by sharing hydroxyl groups to form a continuous sheet as in brucite ($Mg(OH)_2$) (Figure 2.14). The interlayer sheet is similar in thickness to the $Mg(H_2O)_2$ sheet in vermiculites and it adds about 0.4 nm to the 1.0 nm spacing of mica. The presence of this 2:1 layer plus hydroxide sheet structure is the defining characteristic of chlorite minerals (Bailey *et al.*, 1971b). Within the group, names are given on the basis of chemical composition and structure. The octahedral sheets can be di- or trioctahedral—

the commonest types from rocks are trioctahedral in both sheets whereas soil chlorites are probably dioctahedral in both sheets. As in micas the 2:1 layers have a negative charge and this is balanced by the positive charge of the inter-layer sheet where some R^{3+} substitutes for R^{2+}, usually Al for Mg in the brucite-like hydroxide sheet. The resulting structure is harder and less slippery than uncharged structures such as pyrophyllite and talc, and softer and less elastic than the micas.

Chemistry. Bailey (1975a) describes all the main classifications of these minerals from 1890 to 1964. Here the only classifications discussed are those made after the structural basis for the group had been found. The results of chemical analysis could then be assigned on the basis of an anion framework containing 20O + 16OH (see page 53). The most important parameters used in the classifications are the number of Si atoms per formula unit or per tetrahedral site, the ratio of iron to other ions at octahedral sites, and the ferric iron content. Figure 2.24 (after Foster, 1962) shows one way in which the names of the species can be allocated; the classification is based on replacement of Mg by Fe^{2+}, and replacement of tetrahedral and octahedral Al by Si and Mg, respectively—these are assessed as the Fe^{2+}/R^{2+} ratio to produce three categories which are further subdivided according to the numbers of Si atoms per four tetrahedral sites. Similar schemes were proposed by Orcel *et al.*, (1950) and Hey (1954). Lapham (1958) adapted this type of scheme to chromium-bearing chlorites; those containing below 2% Cr_2O_3 he gave the name of the corresponding iron variety in Hey's system but with a Cr prefix. The samples with higher chromium content were named kämmererite when the Cr was in octahedral sites and kotschubeite when in tetrahedral sites.

In trioctahedral chlorites the charge produced when trivalent ions (mainly Al^{3+}, but occasionally Fe^{3+} or Cr^{3+}) replace Si in the tetrahedral sheet is balanced by trivalent ions in the octahedral and hydroxide sheets. If the number of trivalent cations in the octahedral sheet is greater than the Al content of the tetrahedral sheet then, in order to balance the charges which would result from this, the octahedral occupancy will be lowered by an amount equal to one-half of the excess, to re-establish the charge balance. Foster (1962) shows that this balance (and it would be an automatic consequence for an oxide analysis that totalled exactly 100%, if it were scaled to an anhydrous formula with 28 oxygen atoms) is well kept for chlorites and the occupancy of the twelve octahedral sites runs from 12.1 down to 10.9. The samples with the larger numbers of vacancies tend to be high in iron. Samples with much Al replacing Mg tend to take the replacement further and to be dioctahedral in either one or both of the sheets; that is to have 6 + 4 or 4 + 4 of the 12 sites occupied.

Some ferric iron chlorites have less water (Foster, 1964) than would be expected for an $O_{20}(OH)_{16}$ ideal formula. In these chlorites the hydroxide deficit (measured by the water lost above 110 °C) should balance the Fe^{3+} content, but in many others the Fe^{3+} exceeds the hydroxide deficit and such chlorites

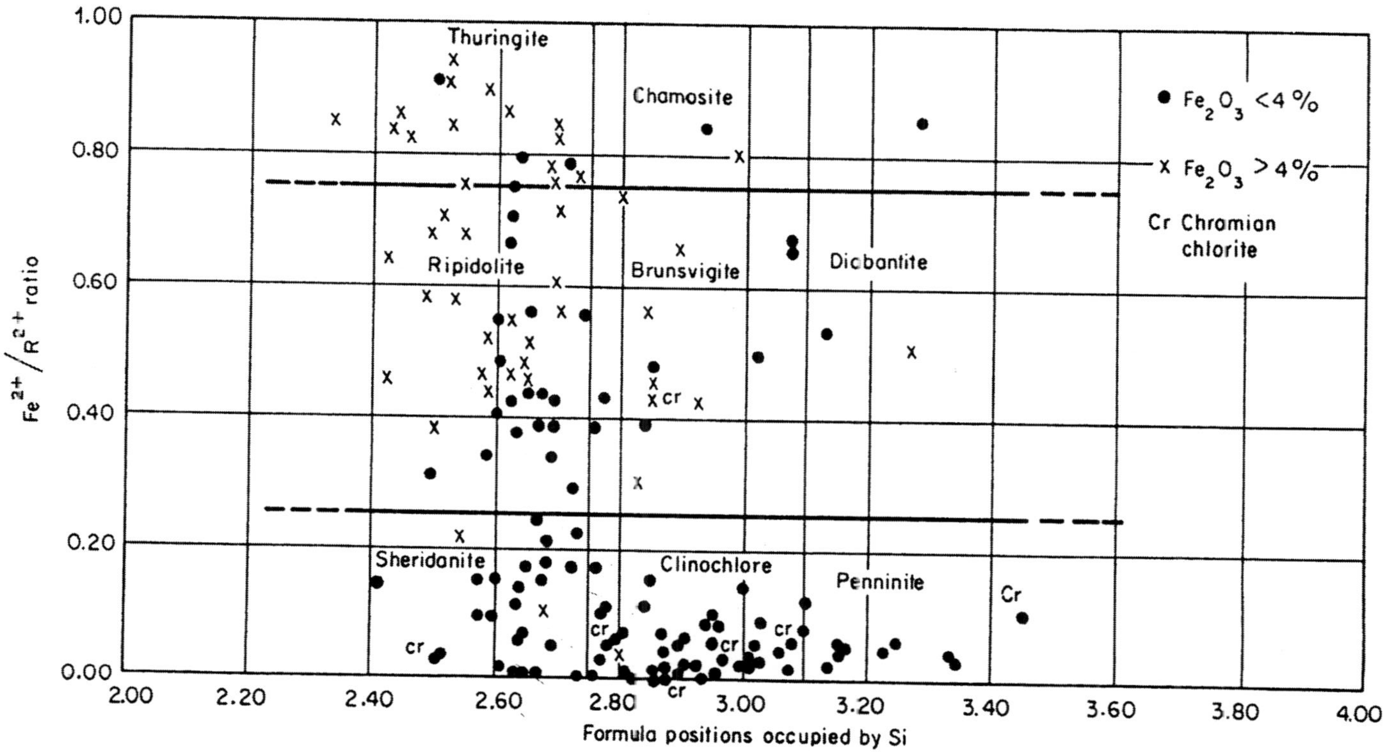

Figure 2.24 Diagram of chemical composition of chlorite species as defined by Foster (1962). Cation contents allocated on the basis of 18(O+OH) atoms. From Bailey (1975a) after Foster

must have crystallized in a more oxidizing environment. It has been postulated, but not proved, that OH vacancies can explain the deficits.

Structure. Two tetrahedral sheets, one on each side, are linked to an octahedral sheet to make the 2:1 layer as in mica, and brucite-like octahedral sheets are interleaved between these layers to give a *c*-spacing of the order of 1.42 nm. In the absence of Al substitution the layers would be uncharged, as in talc and brucite; but the substitution, as well as making the layers the right size to fit together (see page 60), also gives a negative charge to the 2:1 layer and a positive charge to the hydroxide sheet. This increases the attraction between the layers, stabilizes the structure, and explains the greater hardness when compared with talc.

Many stacking arrangements of these alternating layers are possible, but the number is greatly diminished by assuming that the structures are further stabilized by hydrogen bonding between the OH plane of the hydroxide sheet and the outer O plane of the 2:1 layer (Bailey and Brown, 1962). Detailed structural studies of a few single crystal specimens have shown that the structures do indeed belong to some of these theoretical stackings, though with some irregularity of choice among the possible well hydrogen-bonded positions (Steinfink, 1958a and b; Bailey and Brown, 1962; Shirozu and Bailey, 1965). As in other clay minerals these detailed studies have shown adjustment of the various parts to one another. In addition to the changes in sheet dimensions caused by the substitution of tetrahedral and of octahedral ions, as noted by Pauling (1930), rotation of the tetrahedra decreases the area of the tetrahedral sheet and distortion of the octahedra causes the octahedral sheet to become thinner perpendicular to the sheet and thereby increased in area.

As already mentioned (page 46) the individual planes of atoms in Pauling's structure all have sixfold symmetry, and even in structures with rotated tetrahedra they have at least threefold symmetry. The orientation of a layer can be defined uniquely by the direction of 'stagger' from a point in the tetrahedral sheet on one side to a corresponding point on the other side of the layer (Figure 2.19).

Bailey and Brown (1962) described the possible crystal structures by assuming that the surface oxygen planes of the 2:1 layers have sixfold symmetry and the OH planes of the hydroxide sheets must be positioned to give good hydrogen bonding with them. Figure 2.25 shows the two distinct ways the OH plane can be superimposed on the surface oxygen layer of a 2:1 sheet to give this good bonding. The coded circles then give the positions of the hydroxide layer cations in the four possible arrangements. They are coded by assuming that the octahedral stagger in the 2:1 layer is along $-x_1$. If the three hydroxyls in the plane below each cation are found in the $-x$ directions from the cation the code is I, if along $+x$ directions it is II; if the cations lie above tetrahedra (and 2:1 hydroxyls) the code is a, if three cations lie over each hexagon the code is b. The code so far gives the relationship of a hydroxide sheet to a 2:1 layer; the code is completed by a number to show how the next 2:1 layer lies above the

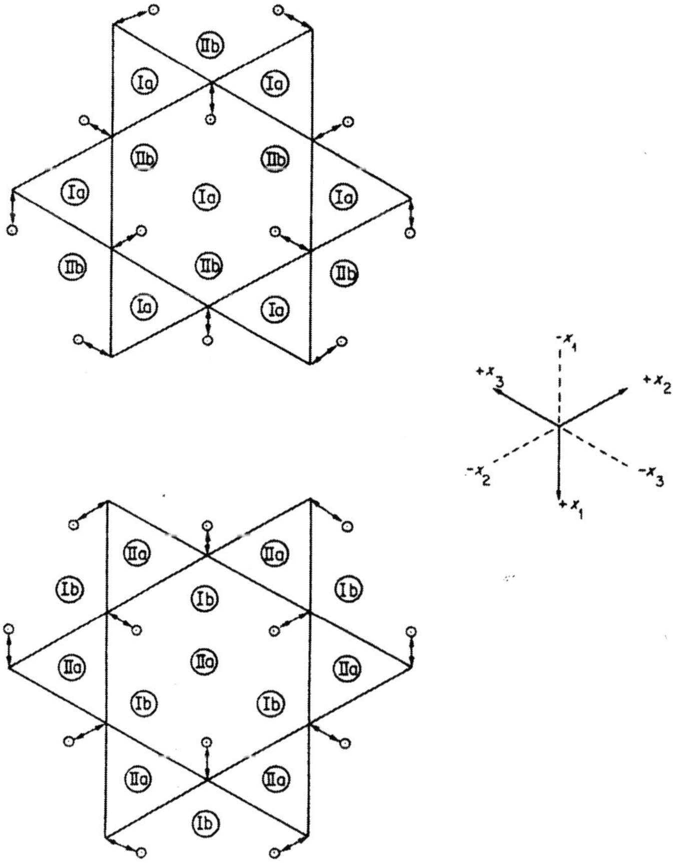

Figure 2.25 Bailey and Brown's (1962) stacking symbols for chlorites. Triangles represent the upper surface oxygen plane of the 2:1 layer. The internal stagger in the layer is along $-x_1$. Small circles represent lower OH plane of the superimposed hydroxide sheet and the double-headed arrows the hydrogen bonds. The large circles with symbols represent the alternative ways of placing the cation plane of the hydroxide sheet

hydroxide sheet (e.g. Ia-4 or IIb-5). It is assumed that this 2:1 layer has the same orientation as the first 2:1 layer, that is that 'one layer' structures are being described in which one layer is related to the next by simple translation. The contact must again be of the well hydrogen bonded type and so could be coded similarly. The code I implies parallel octahedral stagger for the 2:1 layer and hydroxide sheet, and so in one layer structures the code would be the same and need not be repeated. Contacts that would be coded a can occur in three different positions relative to the already stacked layers and these are coded 2, 4 and 6 (even) (Figure 2.26); the three possible b type contacts are coded 1, 3 and 5 (odd).

Soil chlorites are normally less regular and cannot be fitted into this scheme.

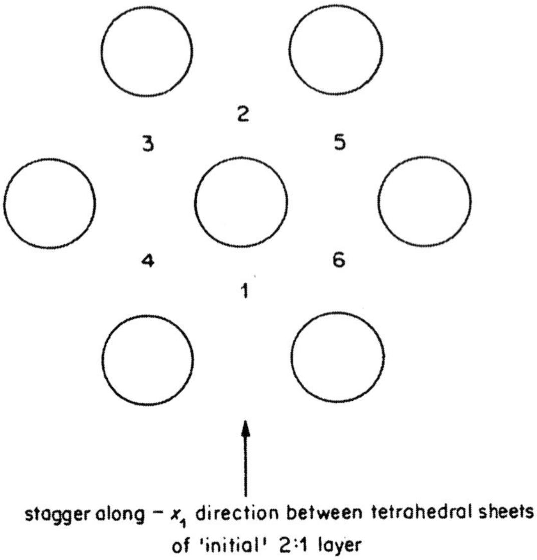

Figure 2.26 This represents the same orientation as Figure 2.25 and shows the symbols for the ways in which a second 2:1 layer can be added. The circles represent the upper OH plane of the hydroxide sheet and the numbers the possible positions for the centre of the hexagon of O atoms on the lower side of the second 2:1 layer

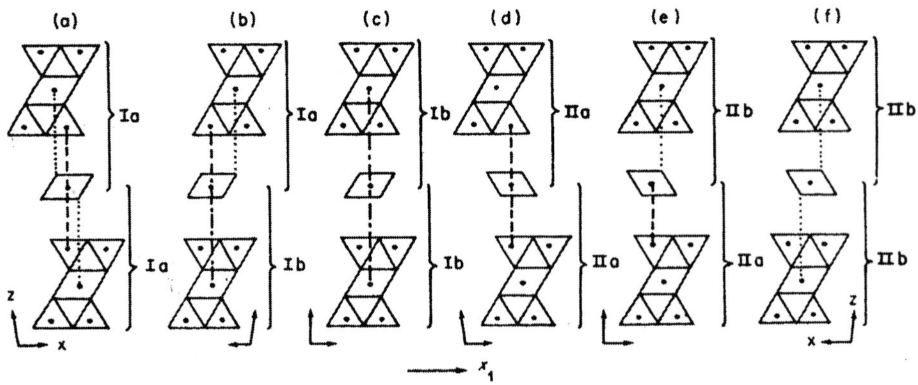

Figure 2.27 b-axis views of 6 stacking arrangements for chlorites, showing the symbols for the two separate 2:1 layer/hydroxide sheet contacts. The symbols for the 6 arrangements are (a) Ia-even, (b) Ib-even, (c) Ib-odd, (d) IIa-odd, (e) IIa-even, and (f) IIb-even

Although the nearest neighbour relations are kept the same, and the contacts are regularly all the same a or b type the choice of 1, 3 or 5 or of 2, 4 or 6 stacking is irregular. Bailey and Brown (1962) call these 'semi-random' layer sequences. Figure 2.27 shows six possible types of semi-random stacking, and were able to allocate 303 chlorites to these types.

The first detailed descriptions of structures of chlorites were by Steinfink (1958a and b, 1961). The first was a prochlorite in which iron, originally largely ferrous (analysed in 1891), had been oxidized until it was largely ferric. The stacking was of the IIb-2 type. The tetrahedra are rotated 9.4° (Figure 2.28) shortening the hydrogen bonds to 0.288, 0.291 and 0.297 nm (Steinfink, 1958a). There are two different tetrahedra and their mean bond lengths can be interpreted (Smith and Bailey, 1963) on the basis of the longer Al—O than Si—O as being occupied by $Si_{0.29}Al_{0.71}$ and $Si_{0.91}Al_{0.09}$. The electron density and bond lengths on the octahedral sites were interpreted (less reliably) as

$$\begin{array}{ll} \text{2:1 layer} & A_1 = \text{Mg},\quad A_2 = 0.75\text{Mg}+0.25\text{Fe}^{2+},\quad A_3 = 0.9\text{Fe}^{3+} \\ \text{interlayer} & A_4 = A_5 = 0.75\text{Al}+0.25\text{Fe}^{3+},\quad A_6 = 0.75\text{Mg}. \end{array}$$

This interpretation of the X-ray results gives a model with 1.5 units of charge per 3 octahedral cations in the hydroxide sheet, and some of this may be balanced by loss of hydroxyl H, but the hydroxide sheet carries a large positive charge.

The other sample studied by Steinfink (1961), a chromium corundophyllite (or perhaps a kämmererite as it contains 2.3% Cr_2O_3) was triclinic in symmetry though monoclinic in cell shape with $a = 0.534$, $b = 0.927$, $c = 1.436$ nm, $\beta = 97°22'$. The stacking is of the IIb-4 type. There is again a rotation of tetrahedra of 7.7° in the direction that improves the hydrogen bonding. The four

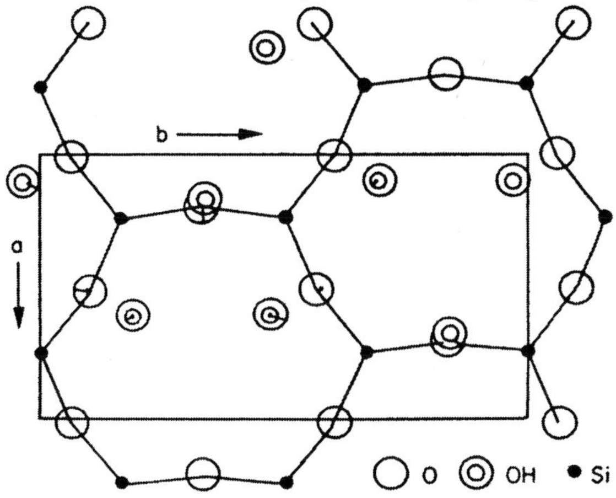

Figure 2.28 Part of the structure of prochlorite (Steinfink, 1958) to show the distortions from the ideal hexagonal structure; arrows show the small displacements. From Deer, Howie and Zussmann (1962) after Steinfink (1958a)

tetrahedra all have the same average bond lengths and so all would be expected to have the same average occupancy of $Si_{2.5}Al_{1.5}$ agreeing with the chemical analysis. There is no evidence of octahedral differences from electron density but a small difference in bond length suggests more Al in the interlayer than in the 2:1 sheet octahedral sites.

Brown and Bailey (1963) refined a kämmererite crystal with much more chromium than the one studied by Steinfink. This is a regular Ia-4 polytype with triclinic symmetry. The tetrahedra are rotated 6.2°. The tetrahedral sites have occupancies of $Si_{0.57}Al_{0.43}$ and $Si_{0.89}Al_{0.11}$, and a one-dimensional electron density projection (Bailey, 1975a) shows that the Cr is concentrated in the interlayer sheet and in a site best placed to give local charge balance with the high Al tetrahedral sites.

Aleksandrova, Drits, and Soklova (1973) determined the structure of an Mg—Al chlorite and found that it is dioctahedral in the 2:1 layers and trioctahedral in the hydroxide sheets. For every formula unit of the structure there are four sites in the octahedral plane of the 2:1 layer with M—O distances averaging 0.2075 nm and two with distances of 0.197 nm. For these they deduce compositions of $Mg_{0.84}Al_{0.16}$ and $Al_{0.73}Mg_{0.27}$. Four octahedral sites in the hydroxide sheet are occupied by Al and the other two are vacant. The total octahedral occupancy is thus $Al_{6.1}Mg_{3.9}$, close to the value in the structural formula derived from chemical analysis: $(Si_{5.72}Al_{2.28})$ $(Al_{5.96}Mg_{3.90}Fe^{3+}_{0.16}Fe^{2+}_{0.28})$ $O_{20}(OH)_{16}$. The repeating pattern of stacking for this structure is twice as high as for the 'one layer' structures. Drits and Karavan (1969) have tabulated the possibilities for such 'two packet' chlorites. For this di-trioctahedral chlorite the stacking is of type II, with the 2:1 layer octahedral stagger opposed to hydroxide layer stagger.

The stabilities of the different stacking arrangements have been compared in a qualitative way in terms of electrostatic interactions firstly within a 2:1 + 1 unit for which the order of stability would be IIb > Ib > Ia > IIa. The forces for the next 1–2:1 interaction can then be given as corresponding to one of these types differently arranged in space. For example, the most stable arrangements of the IIb-even type have two contacts of type IIb whereas IIa-even, or its equivalent IIb-odd, have one type IIb interaction and one type IIa—an arrangement on balance so unstable that no examples are found. In sediments, IIb chlorites are probably detrital, while the less stable Ib type, is probably authigenic.

Identification. Oriented X-ray powder diffraction patterns of chlorites show lines that indicate an interlayer spacing of about 1.42 nm that is not increased by treatment with glycerol and ethylene glycol or much reduced by heating to 700 °C. For some iron-rich chlorites the 001 reflection is weak. In most chlorites this reflection becomes much stronger on heating to 400 °C and other reflections become weaker. The 060 reflection in an unoriented powder pattern can be used to estimate the trioctahedral or dioctahedral nature of a chlorite. Spacings of 0.153–0.156 nm, depending on iron or manganese content, have been found for trioctahedral chlorites and 0.149–0.150 nm (Shirozu, 1958) for dioctahedral ones. If one layer is di- and the other is trioctahedral the values range from 0.149–0.151 nm (Eggleton and Bailey, 1967).

Brydon, Clark and Osborn (1961) have found dioctahedral chlorite in a concretionary brown soil of British Columbia that they suggest was formed in the soil by depositing aluminium hydroxide interlayers into montmorillonite. Many other soils show components in their X-ray patterns that behave partly as chlorite would be expected to do. Sometimes the difference is due to the presence of an interstratified mixture of chlorite with other layer silicates (page 130), in others the hydroxide sheets are not complete and there may be some limited expansion with glycol or glycerol treatment and partial collapse on heating (Rich, 1968, *or* page 133).

2.3.7 Sepiolite and Palygorskite

Sepiolite and palygorskite are fibrous (Plate 2.10 f and g) hydrated magnesium silicate minerals that contain some substitution of aluminium, iron and other ions and balancing exchangeable cations. Early names for white magnesium silicates are 'meerschaum' and 'écume der mer' (sea foam) (Brochant, 1802). Sepiolite (Glocker, 1847) comes from the Greek for cuttle-fish, as the white porous mineral of low density resembles the dried skeleton of the fish. Palygorskites were originally referred to descriptively as mountain cork, leather, paper and fossil skin. They are now named after Palygorsk, a mining district in the Ural Mountains, U.S.S.R. (Ssaftchenkov, 1862). Other names used include pilolite, lassalite and attapulgite, the latter for a compact variety from Attapulgus, U.S.A. (de Lapparent, 1935), but these are all mineralogically indistinguishable from palygorskite and the latter name has priority.

Sepiolite is used for meerschaum pipes, in ceramics and cosmetics where a white mineral with rheological properties somewhat similar to those of smectite are required, and as a zeolite-type absorbent. Palygorskites are used as plasticizers in ceramics, binding agents in moulding sands, zeolite-type absorbers in the treatment of oils and greases, and, because they resist flocculation in sea water, for drilling muds for undersea oil exploration (Ovcharenko, 1964). They occur commonly in soils of arid areas, and sometimes in wetter sites, where they are probably relics of previous arid phases.

Structure. In most discussion of the structures of sepiolite and palygorskite the fibre axis is made the *c*-axis. However, the similarity of these minerals to the other clay mineral groups is more apparent if the *a*- and *c*-axes are interchanged (Zvyagin, Mishchenko and Shitov, 1963; Gard and Follett, 1968 and Martin Vivaldi and Robertson, 1971) so that the a-axis becomes the fibre axis, and this practice will be followed here.

Model structures for sepiolite have been proposed by Nagy and Bradley (1955) and Brauner and Preisinger (1956), but that of the latter is now preferred. In the *bc*-plane perpendicular to the fibre axis, the mineral consists of narrow strips or ribbons of 2:1 layers that are linked in a stepped fashion so that the

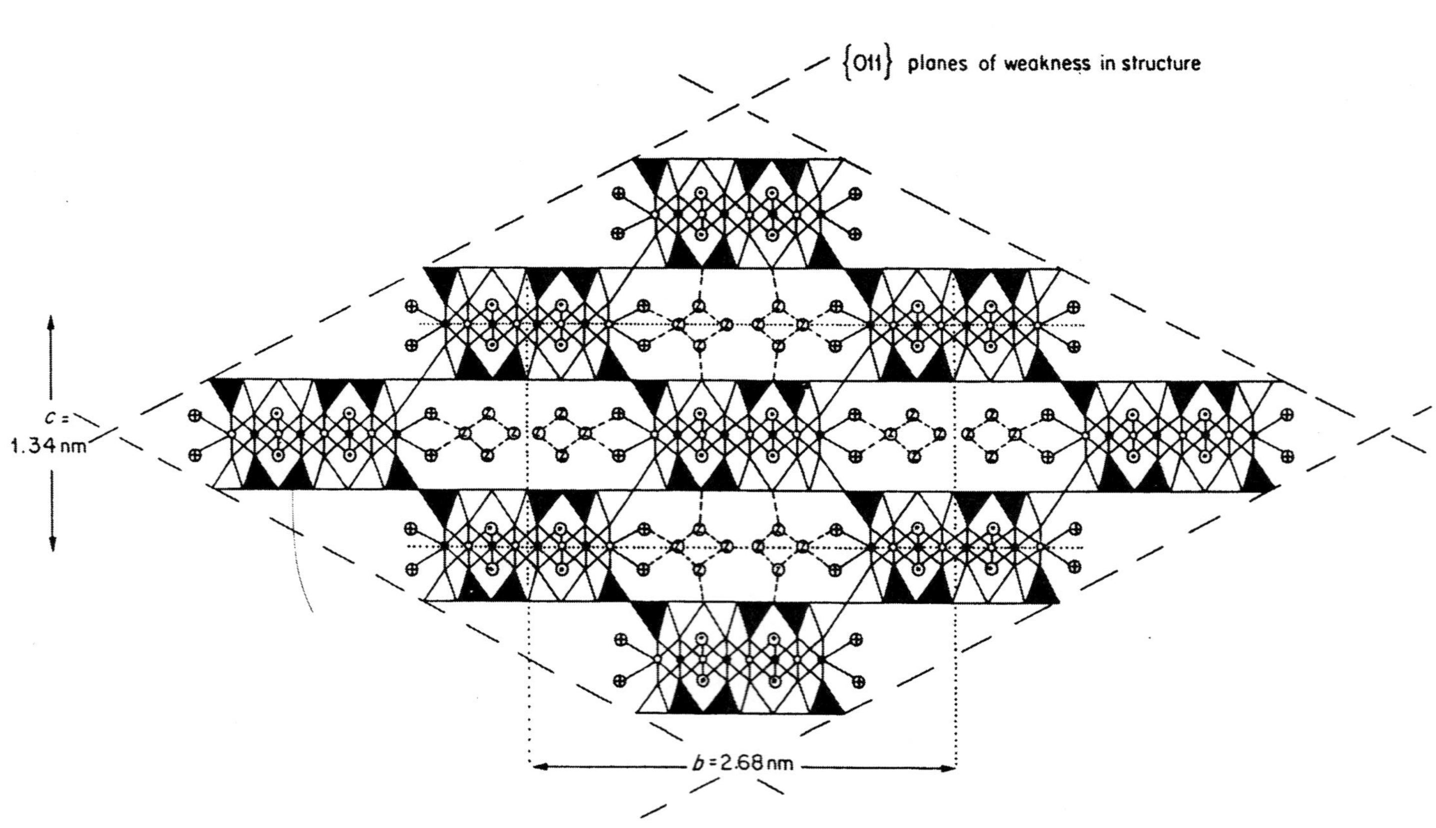

Figure 2.29 The structure of sepiolite (after Brauner and Preisinger, 1956) projected down the a-axis. SiO_4 groups represented by tetrahedra, Mg by small circles, hydroxyl by ⊙, water coordinated to Mg by ⊕ and zeolite water by ⓩ. The {011} planes are found as surface planes in fibres of Plate 2.3. Reproduced by permission of the Mineralogical Society, London

layer of oxygens forming the base of one 2:1 layer strip continues to form the top of the 2:1 layer strips on either side (Figure 2.29). The width of the strips in the *b*-direction equals that of two complete oxygen hexagons and two half hexagons (Figure 2.30a) and is sufficient to contain 12 tetrahedral silicon and 8 octahedral magnesium ions. Thus the basal oxygens form continuous layers, but the rest of the tetrahedra and the octahedra of the 2:1 units are in strips three oxygen hexagons wide. The effect of this is to make a very open structure with channels running the lengths of the fibres that are nearly as large as the 2:1 strips. The channels are filled with exchangeable cations, mainly magnesium and calcium in the natural minerals, water that is bound to the structure and free, or zeolitic, water. The cations may be exchanged for other inorganic or organic cations and the water replaced, at least in part, by many organic compounds.

The structure has lines of weakness along the 011 planes where the 2:1 strips may be divided by the breaking of single Si—O bonds. Sepiolites preferentially form crystal faces of 011 type, as was shown by Rautureau and Tchoubar (1976) by electron microscopy of microtome thin sections cut perpendicular to embedded single fibres (Plate 2.3). The same authors also show that the highly

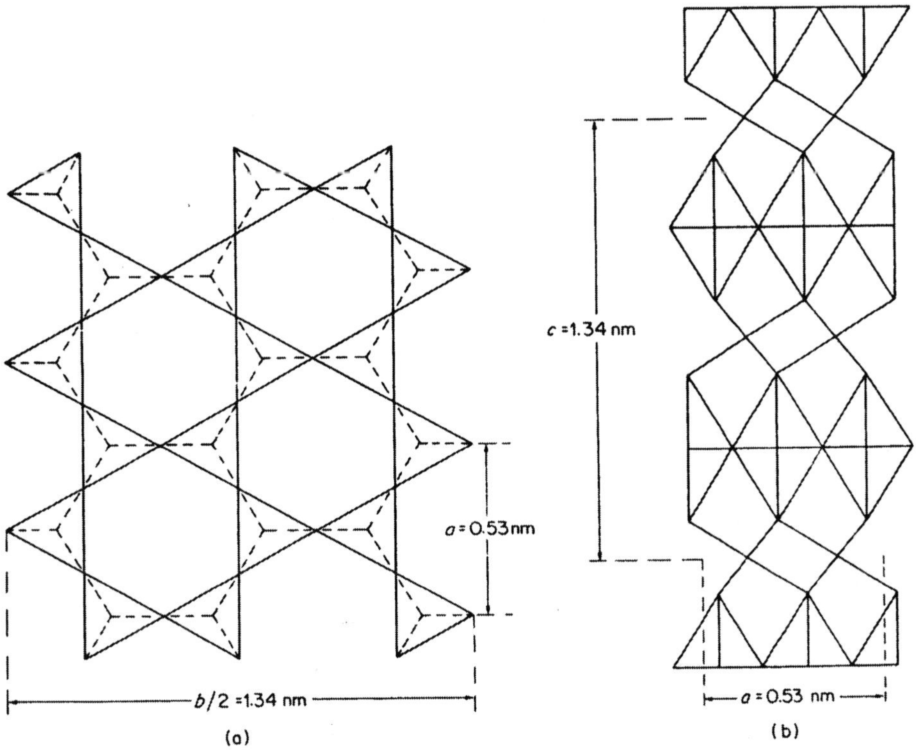

Figure 2.30 (a) The upper surface of a single 2:1 layer ribbon of the sepiolite structure; (b) b-axis projection of sepiolite (orthorhombic)

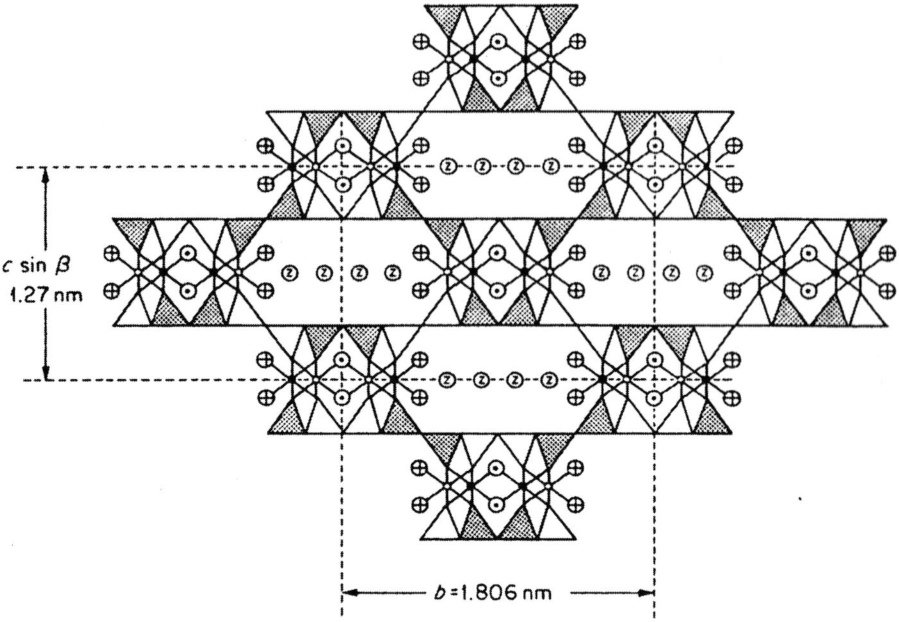

Figure 2.31 a-axis projection of palygorskite (after Bradley, 1940) (symbols as in Figure 2.29)

porous structure may be observed directly at high magnification (Plate 2.3c; the boxed areas coincide with the *bc* plane of the unit cell shown in Figure 2.29).

When sepiolite is viewed in the *ac*-plane (Figure 2.30b) the octahedral stagger of one 2:1 strip is reversed in those above and below it in the *c*-direction so that the overall symmetry is orthorhombic with cell constants $a = 0.31$, $b = 2.68$ and $c = 1.34$ nm. Powder diffraction data for sepiolites is given by Caillère and Hénin (1961). The minerals are normally recognized from their strong 011 reflection at 1.21–1.23 nm and by their fibrous form.

The structure of palygorskite is similar to that of sepiolite in that the mineral consists of 2:1 strips alternating with channels running the length of the fibres (Figure 2.31). According to Bradley (1940) the strips are one oxygen hexagon and two half hexagons wide in the *b*-direction (Figure 2.32a) and have vacancies for eight silicons and five octahedral ions, approximately four of which are normally filled (Drits and Aleksandrova, 1966). In the *ac*-plane, if the direction of octahedral stagger is the same for all 2:1 strips, the minerals have monoclinic symmetry (Figure 2.32b). However, Christ *et al.* (1969) have given powder data for some specimens with orthorhombic symmetry in which the octahedral stagger alternates between adjacent strips along *c* as in sepiolites. Although orthorhombic palygorskite has a very similar structure to sepiolite, it has *a* and *c* cell dimensions that are somewhat different: ortho-palygorskite, $a = 0.521$, $b = 1.789$, $c = 1.277$ nm; clino-palygorskite, $a = 0.524$, $b = 1.783$, $c = 1.278$ nm, $\beta = 95.8°$.

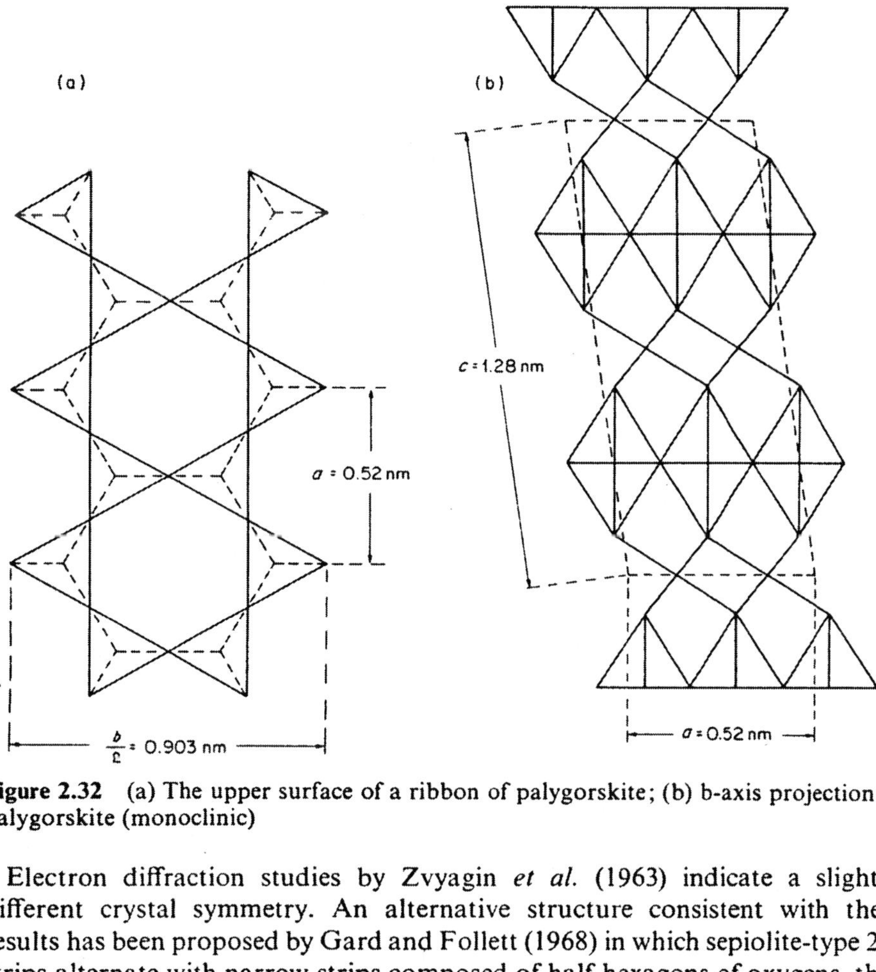

Figure 2.32 (a) The upper surface of a ribbon of palygorskite; (b) b-axis projection of palygorskite (monoclinic)

Electron diffraction studies by Zvyagin *et al.* (1963) indicate a slightly different crystal symmetry. An alternative structure consistent with these results has been proposed by Gard and Follett (1968) in which sepiolite-type 2:1 strips alternate with narrow strips composed of half hexagons of oxygens, thus giving rise to two different sizes of cavity. Neither of the proposed structures has been confirmed by detailed structural studies. Palygorskites are recognized by a strong 011 diffraction reflection at 1.02 to 1.05 nm and by their fibrous morphology.

Composition. Sepiolite has the ideal composition: $Si_{12}Mg_8O_{30}(OH)_4(OH_2)_4 \cdot 8H_2O$. This is equivalent to a sheet silicate with all the octahedral vacancies filled, as in trioctahedral minerals such as talc and saponite. The limited width of the 2:1 strips reduces the ratio of octahedral: tetrahedral cations from 9:12 to 8:12. In natural specimens some tetrahedral Si may be replaced by Al and Fe^{3+} and octahedral Mg by Fe^{3+}, Fe^{2+}, Al, Mn^{2+} and Ni. The iron-rich variety is called xylotile. A sodium-rich variety containing up to 3.7 ions of Na that can be replaced by Mg ions is termed loughlinite. Weaver and Pollard (1973) show that

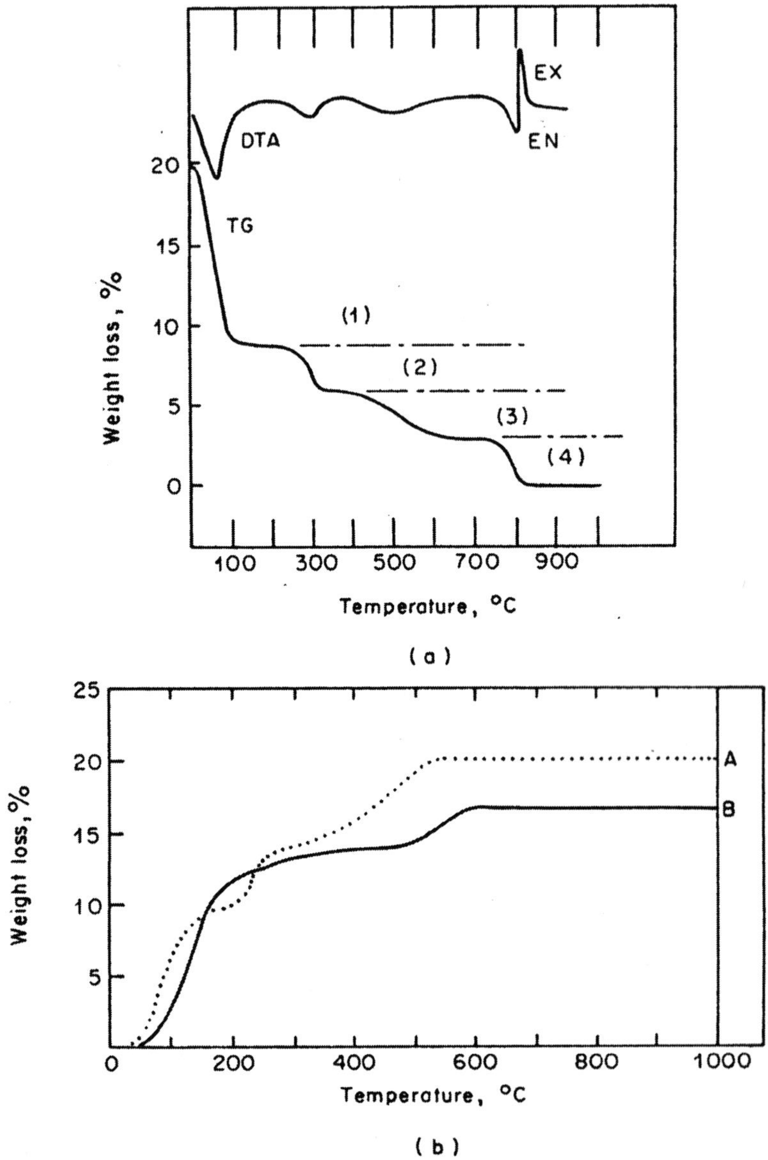

Figure 2.33 (a) Differential thermal analysis (DTA) and thermogravimetric (TG) curves for sepiolite. Regions of weight loss are: 1, zeolitic water; 2, half the Mg coordinated water (change from structure in Figure 6a to Figure 6b); 3, loss of other half of Mg coordinated water; 4, loss of hydroxyl and recrystallization as enstatite (after Nagata, Shimoda and Suda, 1974). Reproduced by permission of Pergamon Press Ltd. (b) Thermogravimetric curves for palygorskites from (A) Meyssonial en Mercoeur, Upper Loire, France; (B) Attapulgus, Georgia, U.S.A. (After Caillère and Hénin, 1957.) Reproduced by permission of the Mineralogical Society, London

for nine analysed sepiolites the total of octahedral cations varies from 6.96 to 8.14 and that six of the samples analysed contain less than 8.00 octahedral cations. The net negative charge on the structure is balanced by exchangeable cations, exchange capacities ranging for different samples from 20–45 meq 100 g^{-1}.

Palygorskite has an ideal composition: $Si_8(Mg_2Al_2)O_{20}(OH)_2(OH_2)_4 \cdot 4H_2O$ (Martin Vivaldi and Robertson, 1971). Al may substitute for Si in tetrahedral positions and Fe^{3+}, Fe^{2+} and Al for Mg in octahedral positions. For the fifteen samples quoted by Weaver and Pollard (1973) the total of octahedral cations varies from 3.40 to 4.27, a greater variation than is shown by the sepiolites. Exchangeable cations balance the permanent negative charge arising from isomorphous substitution. As palygorskite samples commonly contain smectite and other impurities, many of the published exchange capacity values are too large; pure palygorskites have capacities in the range 5–20 meq 100 g^{-1}, appreciably smaller than sepiolites.

Water content. The unit formulae and the diagrams of the structures of sepiolite and palygorskite show water present in three different forms: (i) free or zeolitic water occupying the central spaces of the channels between the 2:1 strips, (ii) bound water formed by the addition of protons to unsatisfied bonds of OH^- groups exposed at the edges of strips, and (iii) structural water from OH^- groups within the strips and those formed by the addition of protons to unsatisfied O^- bonds at the edges of strips. Nagata, Shimoda and Sudo (1974) studied the dehydration of sepiolite and showed that water is lost in discontinuous steps as temperature is increased (Figure 2.33). Step 1, from 20–200 °C covers the loss of zeolitic water, step 2, from 200–380 °C, half of the bound water, step 3, 380–680 °C, the other half of bound water and step 4, from 680–900 °C, hydroxyl water (Table 2.10). A change of structure occurs at step 2 when half the bound water is lost; the 2:1 strips rotate as the inter-strip channels collapse and there is a small change in the *bc*-dimensions of the strips themselves (Figure 2.34, Plate 2.3(d)). This crystalline form, which is characterized by a 011 reflection at 1.04 to 1.00 nm instead of 1.21 nm, is termed sepiolite anhydride. It persists through both stages 2 and 3 of dehydration, but when half the

Table 2.10 Observed water losses (per cent) compared with those for ideal sepiolite (after Nagata, Shimoda and Sudo, 1974). Reproduced by permission of Pergamon Press Ltd

	Step 1	Step 2	Step 3	Step 4
	$S^{a}(OH)_4 4H_2O \cdot 8H_2O \rightleftarrows S(OH)_4 \cdot 4H_2O + 8H_2O$	$S(OH)_4 \cdot 4H_2O \rightleftarrows S(OH)_4 \cdot 2H_2O + 2H_2O$	$S(OH)_4 \cdot 2H_2O \rightarrow S(OH)_4 + 2H_2O$	$S(OH)_4 \rightarrow SO_2 + 2H_2O$
Weight loss (%)				
Ideal	11.1	2.78	2.78	2.78
Obs.	11.1	2.90	2.90	3.39

^{a} S represents $Mg_8Si_{12}O_{30}$.

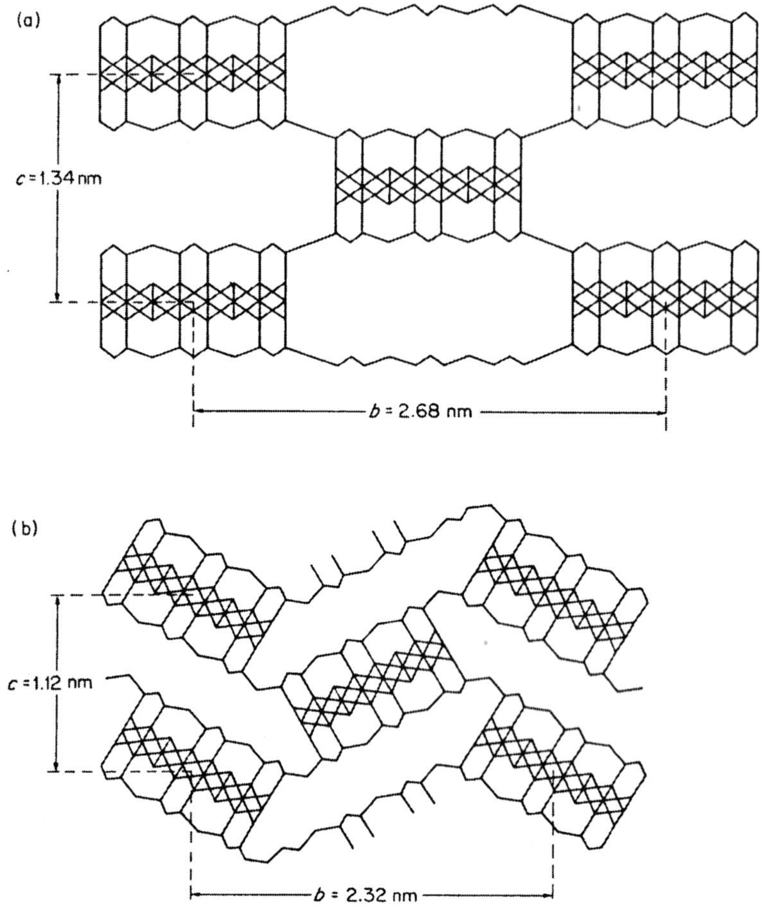

Figure 2.34 Structure of (a) Sepiolite; and (b) Sepiolite anhydride. ⊕ water loss in stage 3 of dehydration (see Figure 2.33a)

bound molecules are still present in the new structure the original sepiolite structure may be restored by the application of water at laboratory temperature and pressure, whereas after removal of all the bound water the sepiolite structure can only be restored by water vapour at elevated temperatures and pressures, if at all. Changes in the infrared spectra on dehydration of sepiolite have been studied by Serna, Ahlrichs and Serratosa (1975) and others. They interpret some features of the sepiolite anhydride spectrum as bonding between atoms in adjacent ribbons, brought closer by the rotation.

Palygorskite also loses water in discontinuous steps on heating (Figure 2.33b), but only stages corresponding to stages 1, 2, and 4 of sepiolite dehydration can be recognized. Table 2.11 taken from Caillère and Hénin (1961) shows that bound water in natural samples, lost between 250 and 400 °C, is less than that

Table 2.11 Observed water losses (per cent) compared with those for ideal palygorskite (from Caillère and Hénin, 1961). Reproduced by permission of the Mineralogical Society, London

	1	2	3
Zeolitic water, below 200 °C	10.8	9.8	8.60
Bound water, 250–400 °C	3.6	4.0	8.60
Hydroxyl water, above 400 °C	4.2	6.5	2.15
Total water	18.6	20.3	19.35

1. Attapulgus, Georgia, U.S.A.
2. Meyssonial en Mercoeur, Upper Loire, France.
3. Ideal palygorskite, $(Si_8)(Mg_5)O_{20}(OH)_2(OH_2)_4 \cdot 4H_2O$.

predicted by the ideal composition and that hydroxyl water, lost above 400 °C, is more than predicted, but the total water is approximately what is predicted. Thus it appears that dehydration of palygorskite may follow a similar course to that of sepiolite, but that the loss of stage 3 of bound water may overlap that of hydroxyl water of stage 4 and be indistinguishable from it by thermogravimetry.

Conditions of formation. Palygorskite is formed hydrothermally in veins in igneous rocks and from solution where the pH is high and there are abundant Mg and Si ions and some Al available, in lagoons, playa lakes and evaporitic basins. It is found extensively in calcareous soils of arid lands and has been shown by Singer and Norrish (1974) to form in soils.

Sepiolite is commonly associated with palygorskite, smectite, dolomite and magnesite. It is formed by the hydrothermal alteration of serpentine, the dissolution of phlogopite in the presence of calcite and in sedimentary environments similar to those in which palygorskite is formed. Sepiolite appears to be formed in preference to palygorskite where the concentration of Si and Mg ions is higher and that of Al lower.

2.3.8 Interstratified Minerals

The phyllosilicate clay minerals are characterized by planes of atoms joined in tetrahedral and octahedral sheets to form layers (Bailey *et al.*, 1971a). Layers usually have cations associated with them, either separate and hydrated, or as part of hydroxide sheets; *2:1 layers plus cations* may be referred to as unit structures or *units* (Bailey *et al.*, 1971b), a concept that is referred to frequently in this section.

The clay mineral groups, kaolinite–serpentine, pyrophyllite–talc, smectite, vermiculite, mica, brittle mica and chlorite, consist of minerals that are formed of one type of unit repeated throughout all the constituent particles. In addition to these groups, however, there is a further category of clay minerals termed interstratified minerals. In these, different types of unit alternate in either a regular or irregular manner. Interstratified minerals may contain units of two, three, or more different types, smectite with kaolinite, vermiculite with chlorite, mica with smectite and chlorite, for instance, so that a very wide range of compositions can occur.

Regular interstratification

The most common form contains equal proportions of two units. For mica (M) and smectite (S) the sequence would be as follows:

MSMSMSMSMS MSMSMS MSMSMSMS

where these three groups of letters represent a simplified sample containing equal proportions of particles with ten, six and eight units per particle. In a sample from a clay deposit consisting of thousands of particles the number of units per particle may vary from four or five to twenty or more, the mean value increasing in general as the mean particle diameter increases.

Other forms of regular interstratification have units in the ratio of 2:1, 3:1 etc. A mica smectite would then contain sequences like MMSMMS or MMMSMMMS. Such sequences have been reported from electron optical studies but not from bulk samples by X-ray diffraction. It therefore seems that such interstratifications occur as short sequences within samples that have a different overall interstratification.

Plate 2.4 illustrates interstratification in clay minerals in the 'craie à silex' from the Paris basin. The minerals are mixtures of 0.7 nm kaolinite units and 1.1 nm smectite units (impregnation with resin followed by heating in the electron beam gives smectite a 1.1 nm basal spacing). The electron diffraction pattern in Plate 2.4b is of a random interstratification of 0.7 and 1.1 nm units found in areas of flakes illustrated end-on in Plate 2.4a. Areas with strong fringes spaced at 1.8 nm in (c) gave the diffraction pattern (d) that is thought to represent regular 0.7, 1.1, 0.7, 1.1 nm or KSKS interstratification. Other areas, not illustrated, gave KKS and KKKS sequences (Eberhardt and Triki, 1972).

Irregular interstratifications

Many different types of irregular interstratification are possible, but four that have been used to describe two-component arrangements are termed random, ordered, segregated and zoned. Using the same convention as for regular interstratification, but with samples having on average 3S and 5M units per particle, the different sequences may be shown as follows:

Random	MSSMSMMMSM	SMMMSM	MMSMSMMS
Maximum order	MSMSMSMSMM	MSMSMM	MSMSMSMM

Segregated			
(partial)	MMMMMMSSSS	MMMMSS	MMMMMSSS
(complete)	SSSSSSSSSS	MMMMMM	MMMMMMMM
Zoned	SSMMMMMMSS	SMMMMS	SMMMMMSS

In random interstratifications the sequence of M and S units is haphazard, as obtained after shuffling playing cards. Maximum order resembles the regular interstratification described above, but has unmatched M units because the proportions of M and S units are unequal. 'Segregated' is the special case of a mixture of two different types of particle. 'Zoned' has a core of M units in the centre of each particle surrounded by an outer rim of S units.

Nomenclature and Description

The nomenclature sub-committee of the Association Internationale Pour l'Étude des Argiles (AIPEA) has not yet recommended a system for interstratified minerals. In its absence the system proposed by the Clay Minerals Society of America (Bailey *et al.*, 1971a) modifying the proposals of the British Clay Minerals Group (Brown, 1955) will be followed. In this system component units are named: kaolinite, mica (not illite), talc, vermiculite, smectite or montmorillonite and saponite, chlorite. These are listed in order of abundance with the proportions of units, if known, preceded by the words regular or irregular and terminated with the word interstratification (not mineral), e.g. an irregular 30:70 mica–chlorite interstratification. Names may be used for regular interstratifications, but are not preferred.

Essential features of interstratified minerals may be described by specifying three properties: (1) the nature of the component units, (2) their relative proportions, and (3) the rules of succession of the units.

The *component units* of interstratified minerals resemble those of minerals of the same name: mica, talc, vermiculite, smectite, chlorite, kaolinite. They are normally identified from their basal spacings by X-ray diffractometry and, if the specimens are sufficiently pure, from their chemical compositions. The simplest minerals to recognize are regular interstratifications, because their spacings are sums of those of the component units, usually in the ratio 1:1. Table 2.12 lists basal spacings for some of these. For each the left column records spacings given in the listed publications and the right a sub-division into component spacings. The results are not quite consistent with the spacings expected for pure minerals, but it is not clear whether the differences are due to experimental errors or are real, i.e. due to genuine differences between units of the same name in minerals and interstratifications. Kaolinite units are assumed to have a spacing of 0.715 nm up to a temperature of 500 °C; thereafter the structure is destroyed. Halloysite has a spacing of 1.0 nm when hydrated, 0.72 nm when dehydrated and is also destroyed above 500 °C.

There is a further group of units with properties intermediate between those of chlorite, vermiculite and smectite. The name swelling chlorite (Stephen and

Table 2.21 Basal spacings, nm, of regular or nearly regular interstratifications and their components:

	1		2		3		4	
	Spacing	Components	Spacing	Components	Spacing	Components	Spacing	Components
In air	2.47	1.00+1.47	2.38	1.44+0.94	2.91	1.40+1.51	3.00	1.40+1.60
Glycerol	2.46	1.00+1.46	2.77	1.83+0.94	3.20	1.40+1.80	3.20	1.40+1.80
500/600 °C	1.00	1.00+1.00	0.94	0.94+0.94	2.37	1.40+0.97	2.85	1.40+1.45

	5		6		7	
	Spacing	Components	Spacing	Components	Spacing	Components
In air	2.50	0.96+1.54	3.00	1.45+1.55	2.89	1.42+1.47
Glycerol	2.74	0.96+1.78			3.22	1.42+1.80
Ethylene glycol	2.65	0.96+1.69	3.19	1.45+1.74	3.11	1.42+1.69
500/600 °C	1.93	0.96+0.97	2.35	1.40+ .95	2.34	1.39+ .95

1. Regular biotite–vermiculite interstratification (hydrobiotite), Palabora, S. Africa; Veniale and van der Marel (1969).
2. Regular talc–saponite interstratification (aliettite), Monte Chiari, Italy; Veniale and van der Marel (1969).
3. Regular chlorite–montmorillonite interstratification, Texas, U.S.; Earley, Brindley, McVeagh and Vanden Heuvel (1956).
4. Regular chlorite–swelling chlorite interstratification (corrensite), Perigny, France; Martin Vivaldi and MacEwan (1957).
5. Regular sodium mica–beidellite interstratification (rectorite), Baluchistan, Pakistan; Brown and Weir (1963a).
6. Regular dioctahedral chlorite–beidellite interstratification (tosudite), Takatama, Japan; Shimoda (1969).
7. Nearly regular ditrioctahedral chlorite–smectite interstratification, Huy, Belgium; Brown, Bourguignon and Thorez (1974).

MacEwan, 1950) is given to units that expand to 1.8 nm when solvated with glycol, but collapse to chlorite spacings of approximately 1.4 nm when heated. A second group of units that are unnamed swell like smectites, but show partial or hindered collapse. Some appear to collapse completely if heated strongly, but the degree of collapse at intermediate temperatures is always less than that of smectites. Others collapse to a mean spacing that is intermediate between those of chlorite and collapsed smectite even after heating to 600 or 700 °C. A third group is described in detail in the section on clay vermiculites (page 88). A fourth group of units show no expansion following attempted solvation, but hindered collapse with heat treatment. Because they have properties intermediate between those of vermiculite, chlorite and mica these last two groups have been described generally as intergrade or intergradient. They are of great complexity because different treatments produce different degrees of intergradient behaviour and because an interstratification containing them, instead

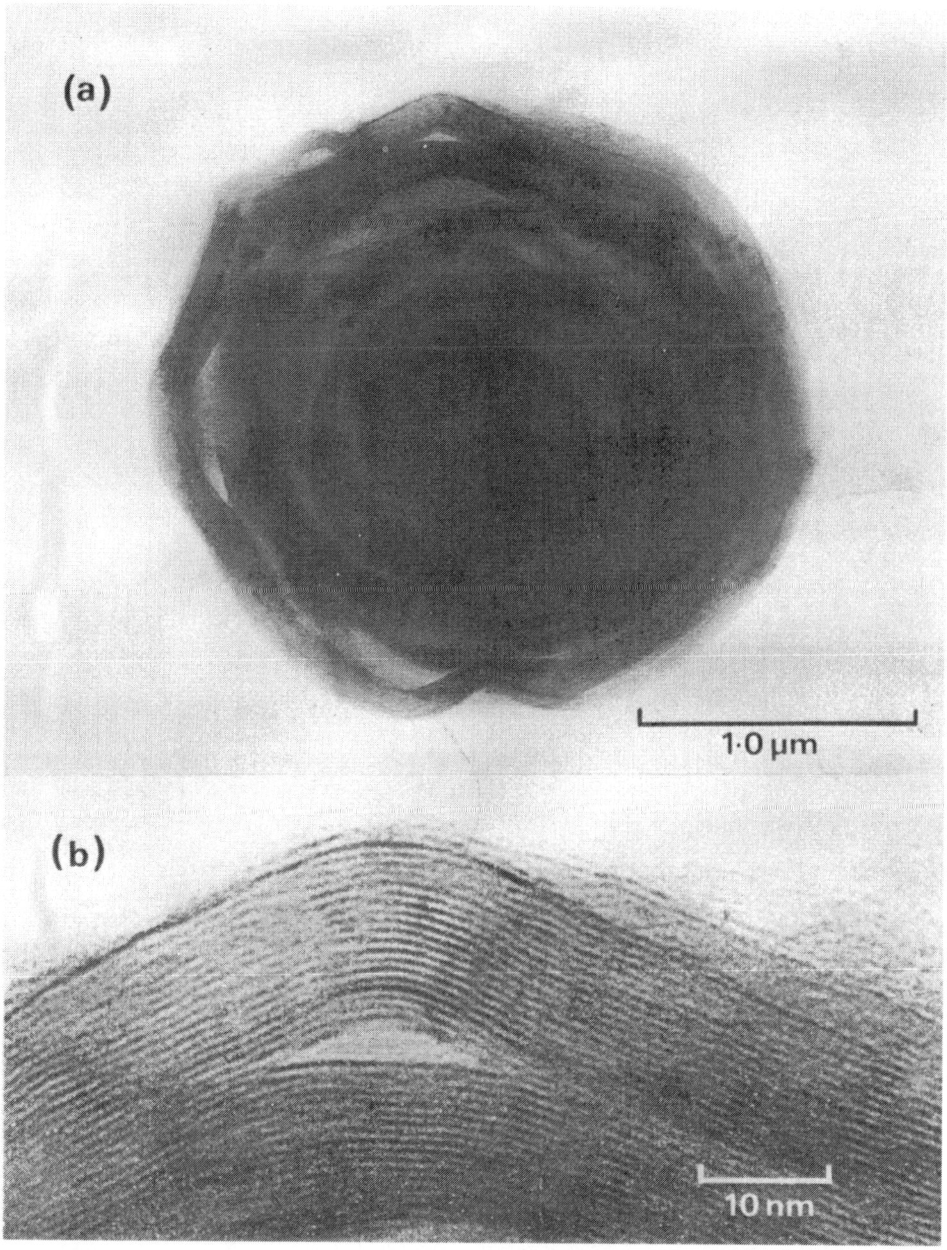

Plate 2.1 Transmission electron micrograph of a polygonal halloysite particle from weathered pumice beds at Yawata, Japan. (a) whole particle; (b) direct imaging at high magnification of the same particle showing lines corresponding to the basal unit repeat distance of 0.72 nm (Sudo and Yotsumoto, 1977). The more common tubular form of halloysite is seen in Plate 2.11(c)

(a)
(b)

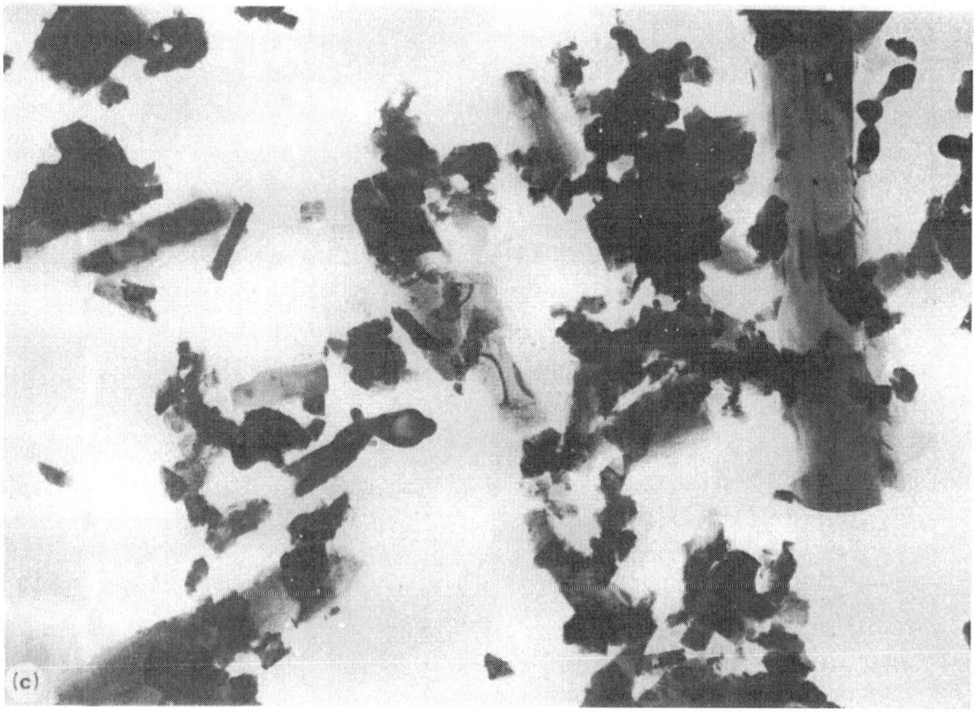

Plate 2.2 Transmission electron micrographs of chrysotile and antigorite illustrating the effect on particle morphology of different structures: (a) tubes of chrysotile (asbestos) from Mijas, Malaga, Spain (I.G.S. No. MR 25928) × 3250; (b) part of the sample at higher magnification (× 16,250); (c) platy particles of antigorite from Antigoria, Piedmont, Italy (JGS No. Ludlam 6550)

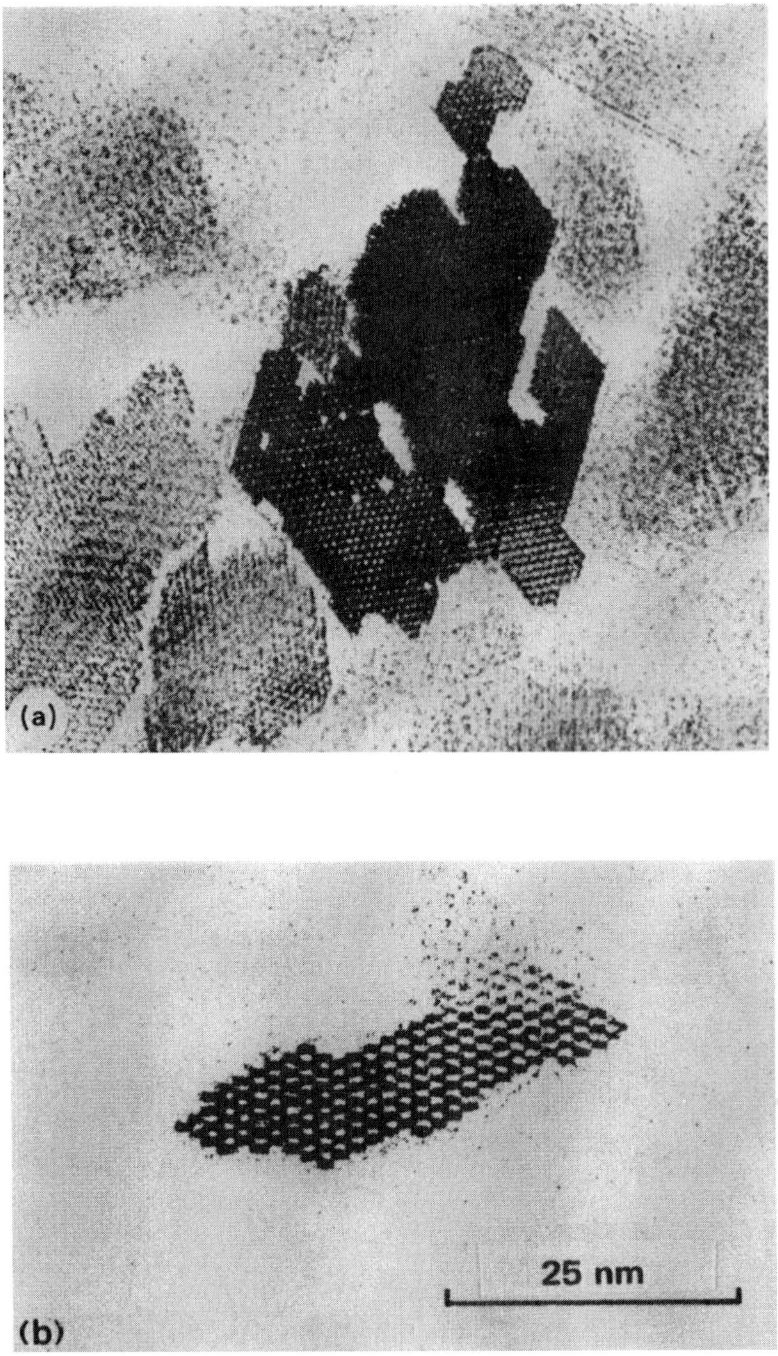
(a)
25 nm
(b)

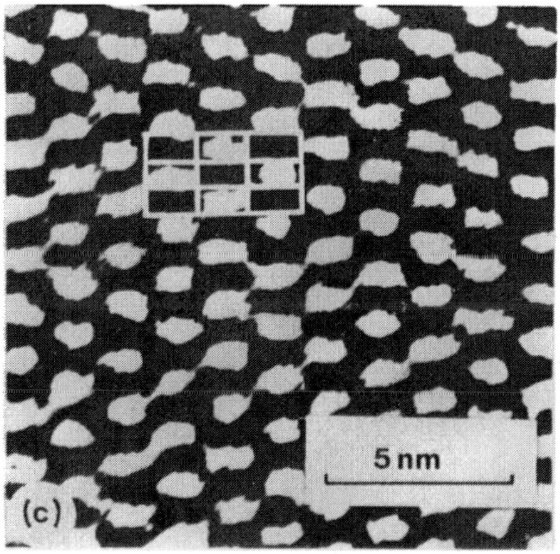

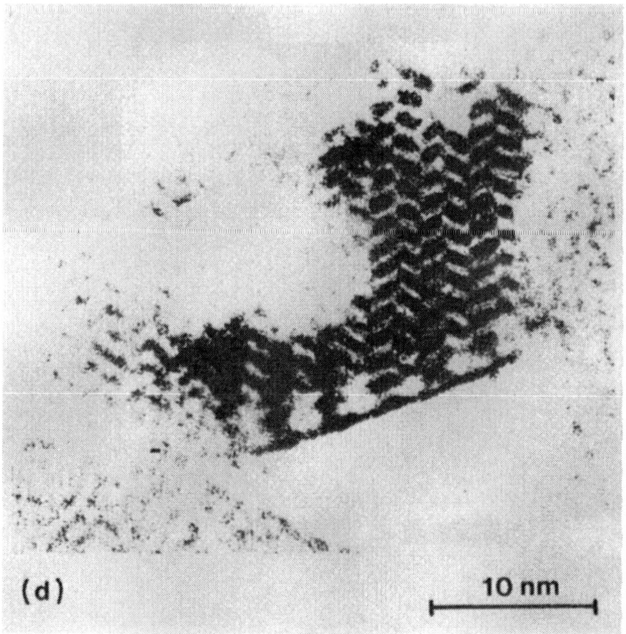

Plate 2.3 Cross-section of a fibre of Ampandrandava sepiolite (a, b and c) and its anhydride (d) at high magnification, showing development of 011 crystal faces (a and b, cf. Figure 2.29) and the occurrence of channels in the structures (transmission electron micrographs from M. Rautureau). Sections (a), (b), and (c) reproduced by permission of Pergamon Press Ltd.; section (d) reproduced by permission of the Mineralogical Society, London

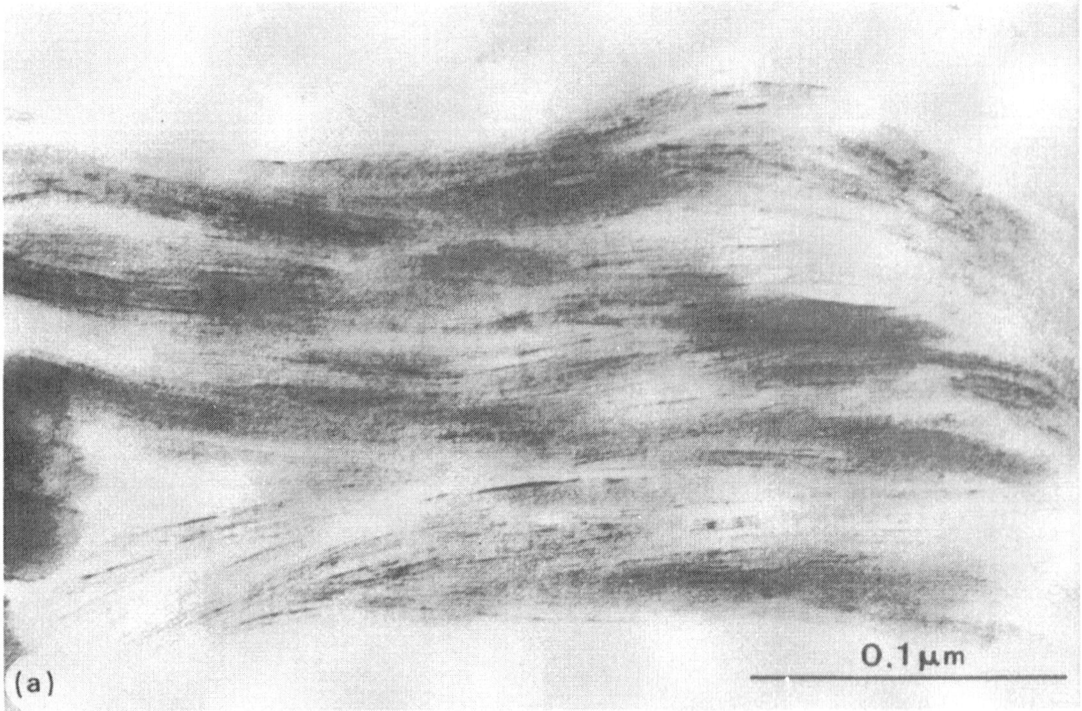
0.1 µm
(a)

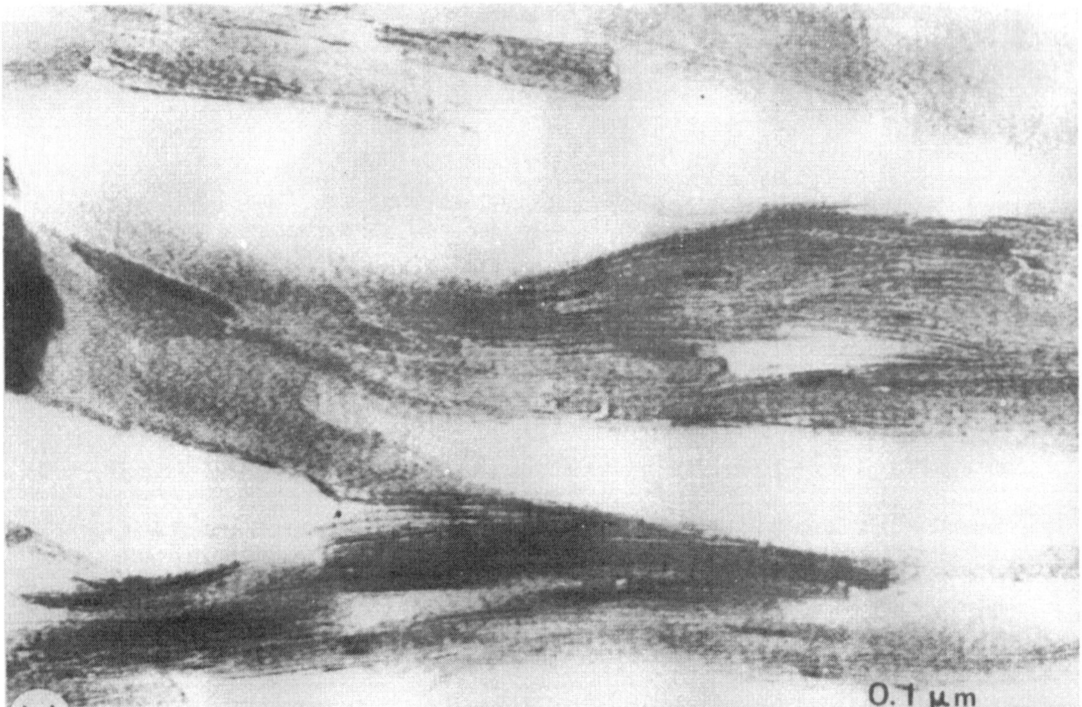
0.1 µm
(c)

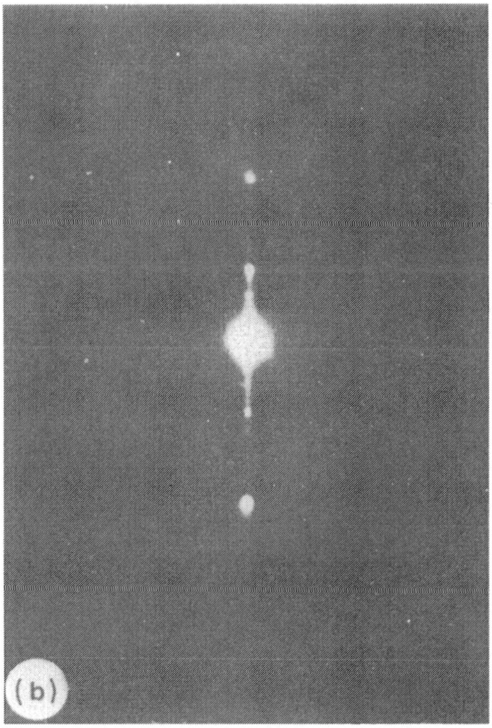

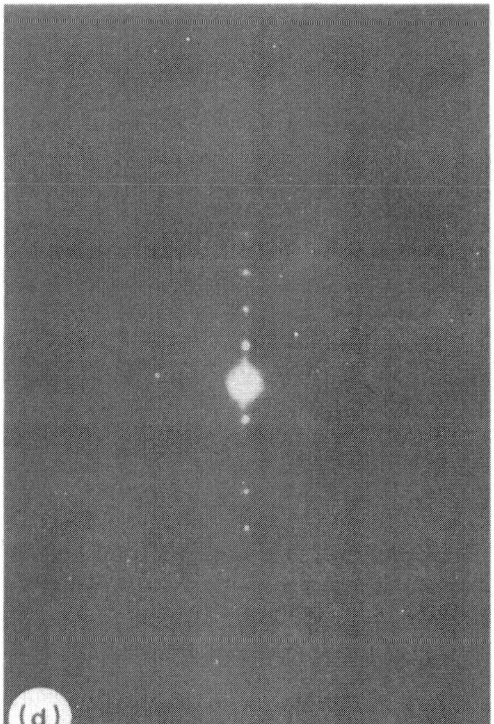

Plate 2.4 Transmission electron micrographs and selected area electron diffraction patterns of kaolinite–smectite interstratifications from the 'craie à silex' of the Paris Basin (Eberhardt and Triki, 1972): (a) area of mainly random interstratifications viewed perpendicular to the basal planes of the components; (b) selected area diffraction pattern of part of the same field; (c) area showing 1.8 nm fringes related to regular 50:50 kaolinite–smectite interstratification 0.7 + 1.1 nm—the 1.1 nm spacing for smectite results from partial collapse under electron bombardment; and (d) the selected area diffraction pattern of part of the same field. Reproduced by permission of the Société Française de Microscopie Électronique

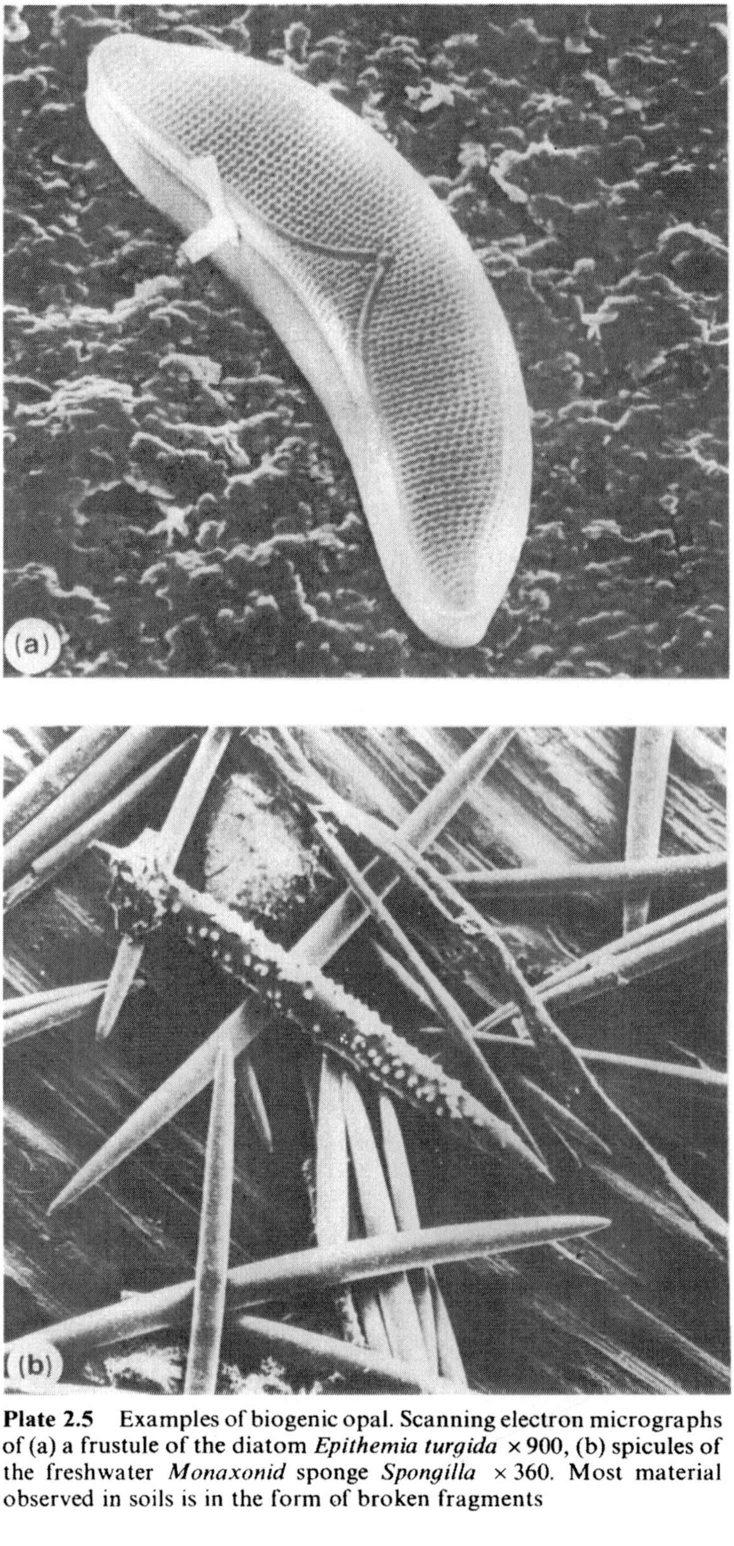

Plate 2.5 Examples of biogenic opal. Scanning electron micrographs of (a) a frustule of the diatom *Epithemia turgida* × 900, (b) spicules of the freshwater *Monaxonid* sponge *Spongilla* × 360. Most material observed in soils is in the form of broken fragments

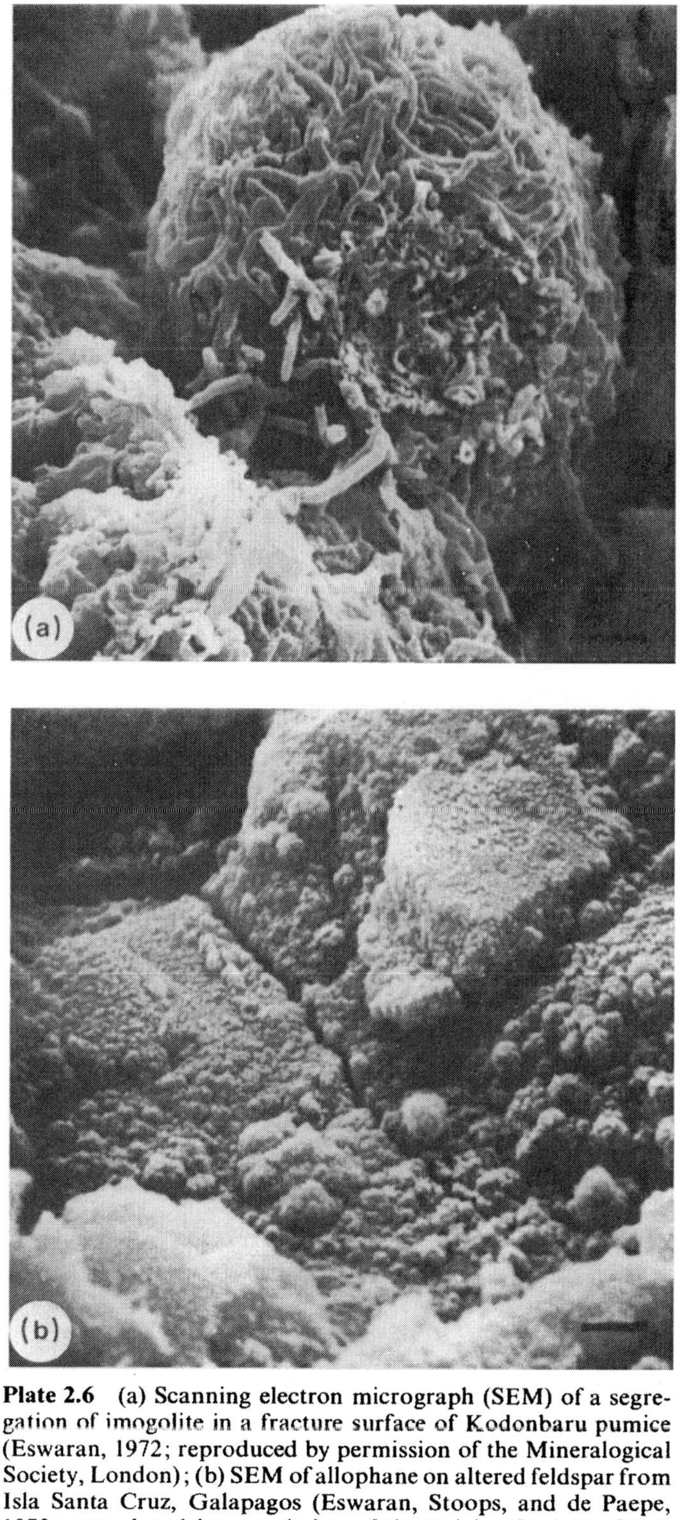

Plate 2.6 (a) Scanning electron micrograph (SEM) of a segregation of imogolite in a fracture surface of Kodonbaru pumice (Eswaran, 1972; reproduced by permission of the Mineralogical Society, London); (b) SEM of allophane on altered feldspar from Isla Santa Cruz, Galapagos (Eswaran, Stoops, and de Paepe, 1973; reproduced by permission of the Belgian Society of Soil Science)

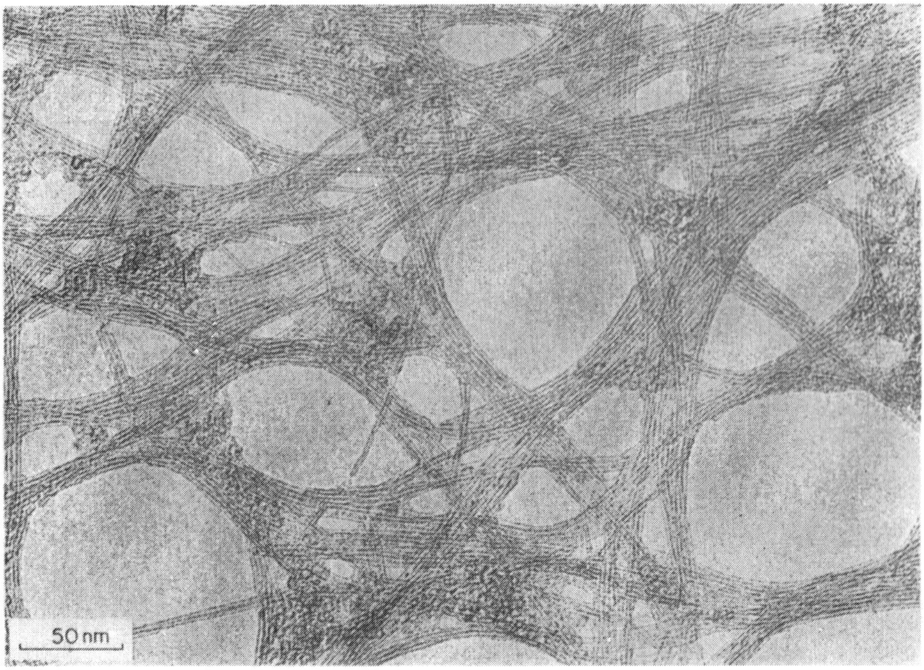

Plate 2.7(a) High resolution (× 348,000) transmission electron micrograph of imogolite showing the tubular form of individual threads (Wada *et al.*, 1970; reproduced by permission of the Mineralogical Society, London)

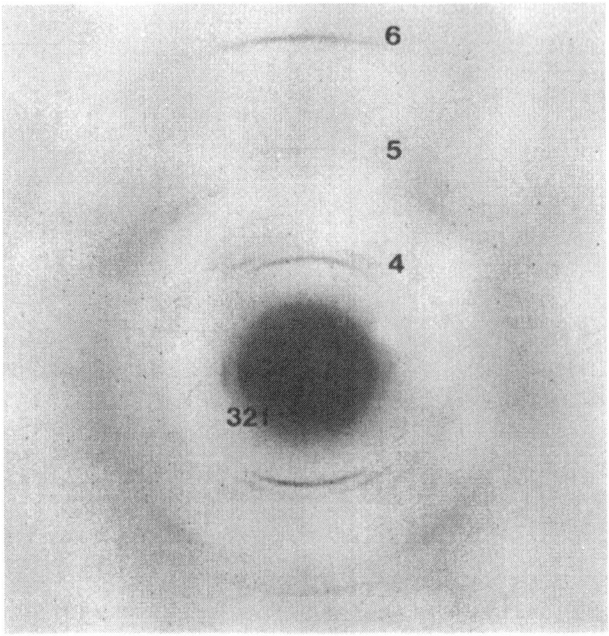

Plate 2.7(c) Selected area electron diffraction pattern of sub-parallel imogolite tubes. Reflections 1, 2 and 3 are related to the structure of tubes normal to the tube axis. Reflections 4, 5 and 6 along the tube axis are 2nd, 3rd and 4th orders respectively of the 0.84 nm spacing (Russell

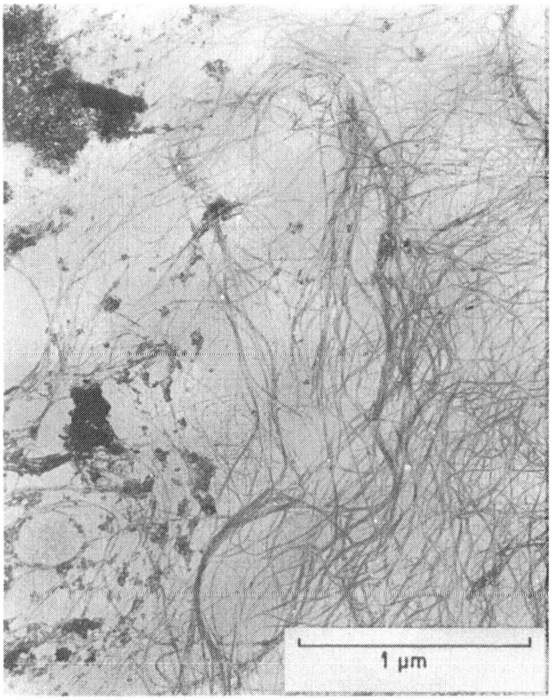

Plate 2.7(b) Transmission electron micrograph (× 30,400) of a less well dispersed sample of imogolite showing areas with many sub-parallel tubes (Russell, McHardy and Fraser, 1969). Reproduced by permission of the Macaulay Institute for Soil Science

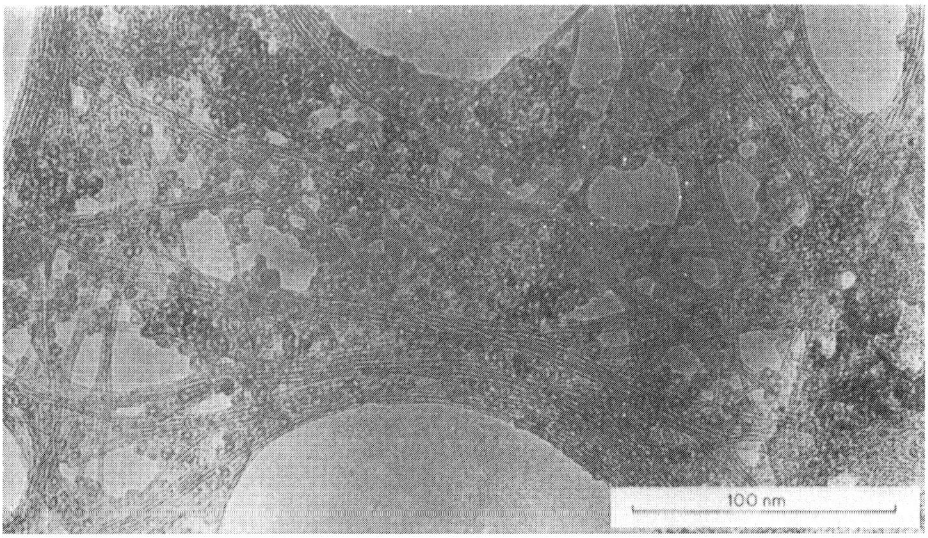

Plate 2.7(d) High resolution transmission electron micrograph of a mixture of imogolite threads and polygonal allophane particles (Henmi and Wada, 1976; reproduced by permission of the Mineralogical Society of America)

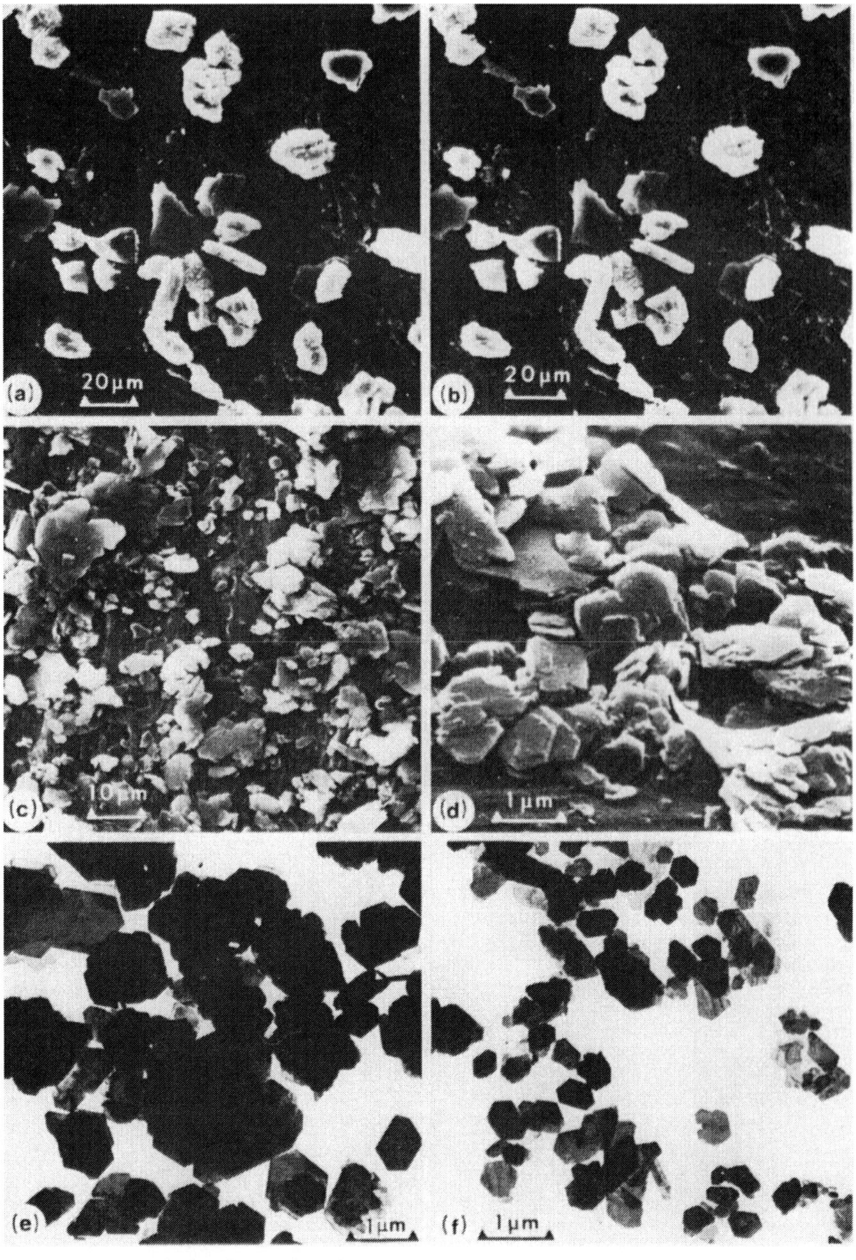

Plate 2.8 Kaolinite from the Blackpool Pit, St Austell, Cornwall; (a)–(d) SEM, (e) and (f) TEM. (a) and (b) Stereo pair of photographs taken by SEM, >10 µm fraction, (c) 10–2 µm fraction, (d) 2–0.5 µm fraction, (e) 2–0.5 µm fraction, sample shadowed with Pt metal at 45°, and (f) <0.5 µm fraction, unshadowed

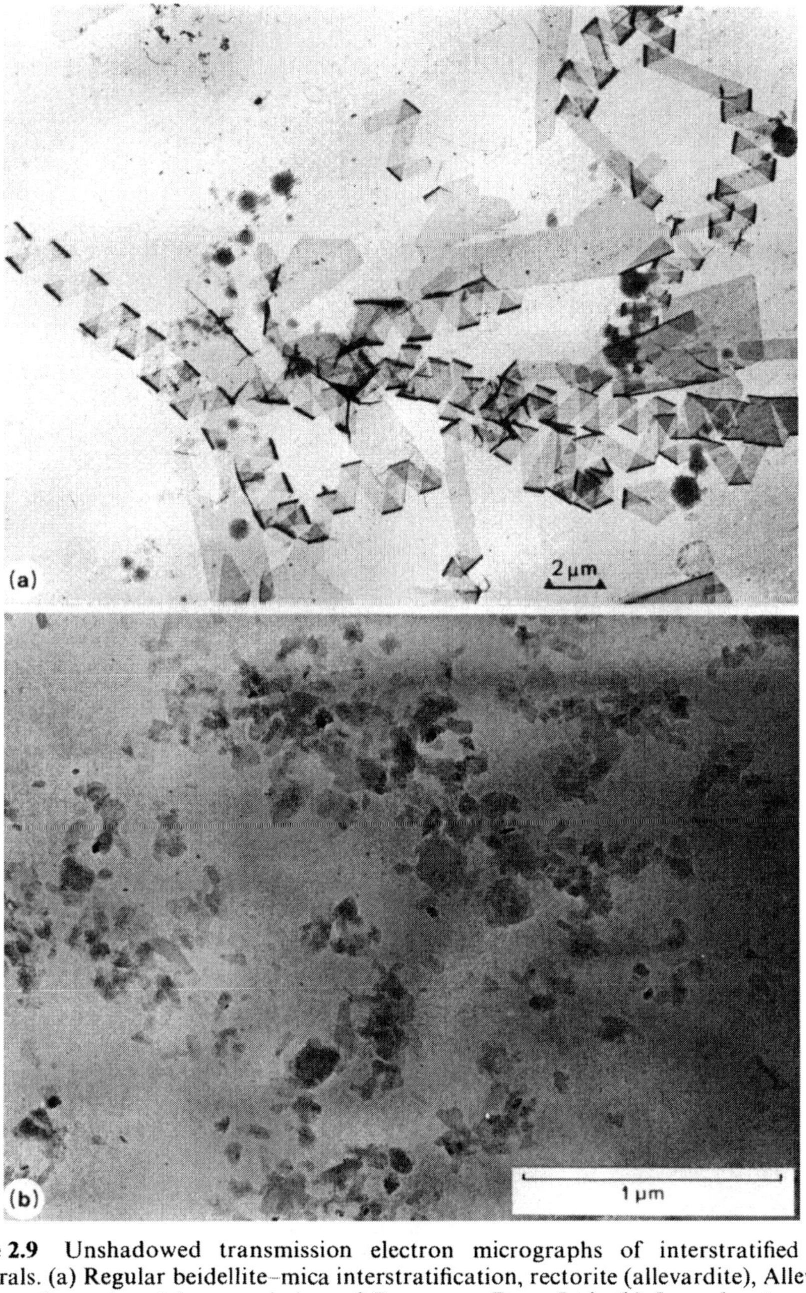

Plate 2.9 Unshadowed transmission electron micrographs of interstratified clay minerals. (a) Regular beidellite–mica interstratification, rectorite (allevardite), Allevard, France. Reproduced by permission of Pergamon Press Ltd. (b) Irregular (random) mica–smectite interstratification, <0.04 µm fraction of subsoil horizon of Denchworth series soil in weathered Oxford Clay, England

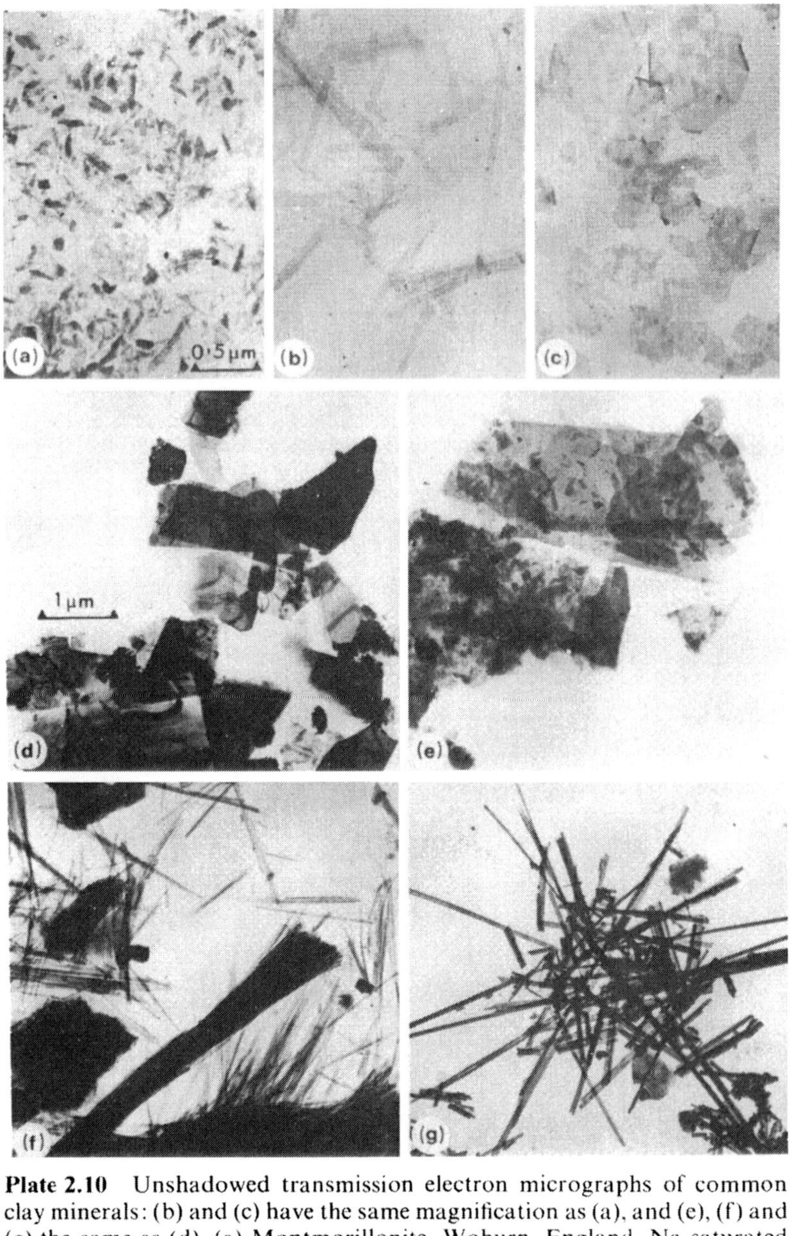

Plate 2.10 Unshadowed transmission electron micrographs of common clay minerals: (b) and (c) have the same magnification as (a), and (e), (f) and (g) the same as (d). (a) Montmorillonite, Woburn, England, Na-saturated <0.04 μm fraction; (b) Hectorite, Hector, California, U.S.A., Na-saturated, <2 μm fraction; (c) Montmorillonite, Clay Spar, Wyoming, U.S.A., Na-saturated, <0.04 μm fraction; (d) Chlorite, Snowdonia, Wales, <2 μm fraction; (e) Vermiculite, Transvaal, South Africa (World vermiculite), Li-saturated, 0.2–0.04 μm fraction; (f) Sepiolite, Vallecas, Spain, unfractionated; and (g) Palygorskite from Enderby, Leicestershire, England

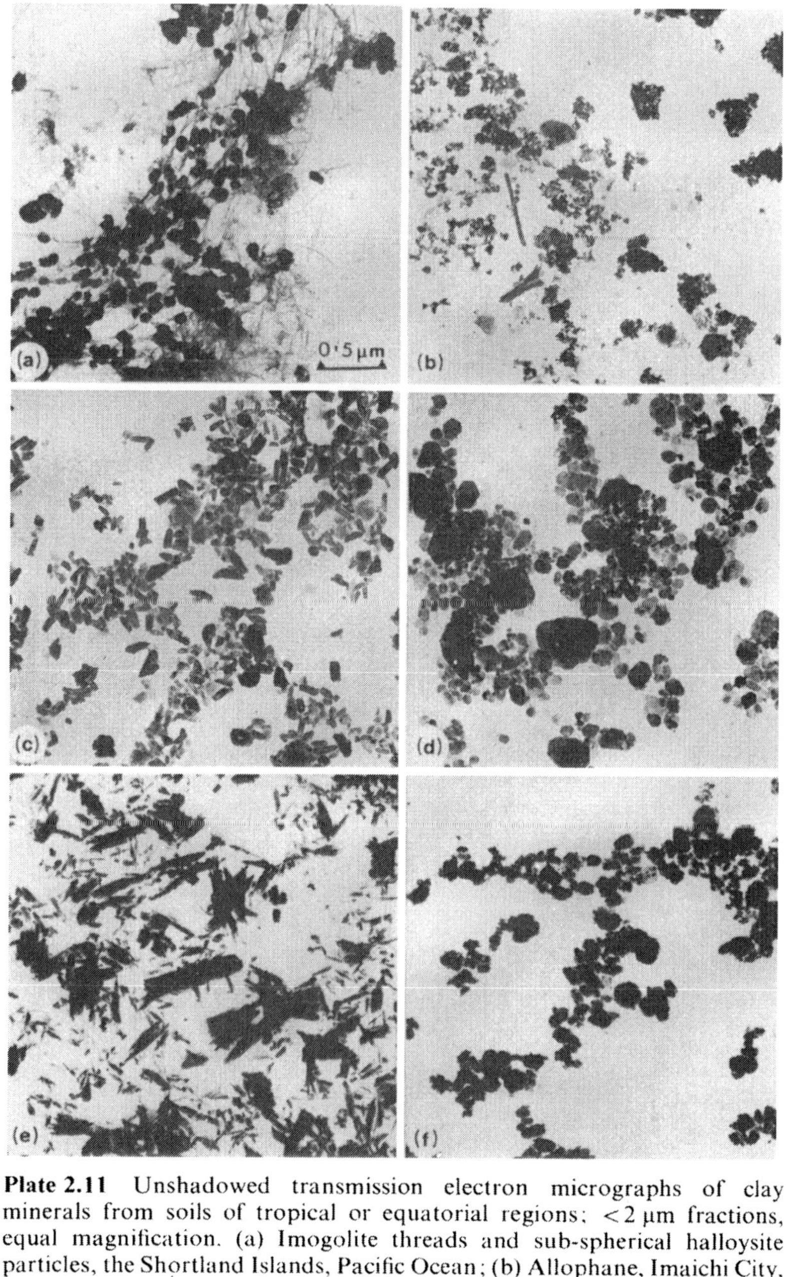

Plate 2.11 Unshadowed transmission electron micrographs of clay minerals from soils of tropical or equatorial regions; <2 µm fractions, equal magnification. (a) Imogolite threads and sub-spherical halloysite particles, the Shortland Islands, Pacific Ocean; (b) Allophane, Imaichi City, Japan; (c) Tubes and rolled flakes of halloysite, Mexico; (d) Kaolinite and iron oxide concretions, Ikenne, Nigeria; (e) Goethite, Santa Isobel, Solomon Islands, Pacific Ocean; and (f) Gibbsite partly coated with iron oxide, Rennell Island, the Shortland Islands, Pacific Ocean

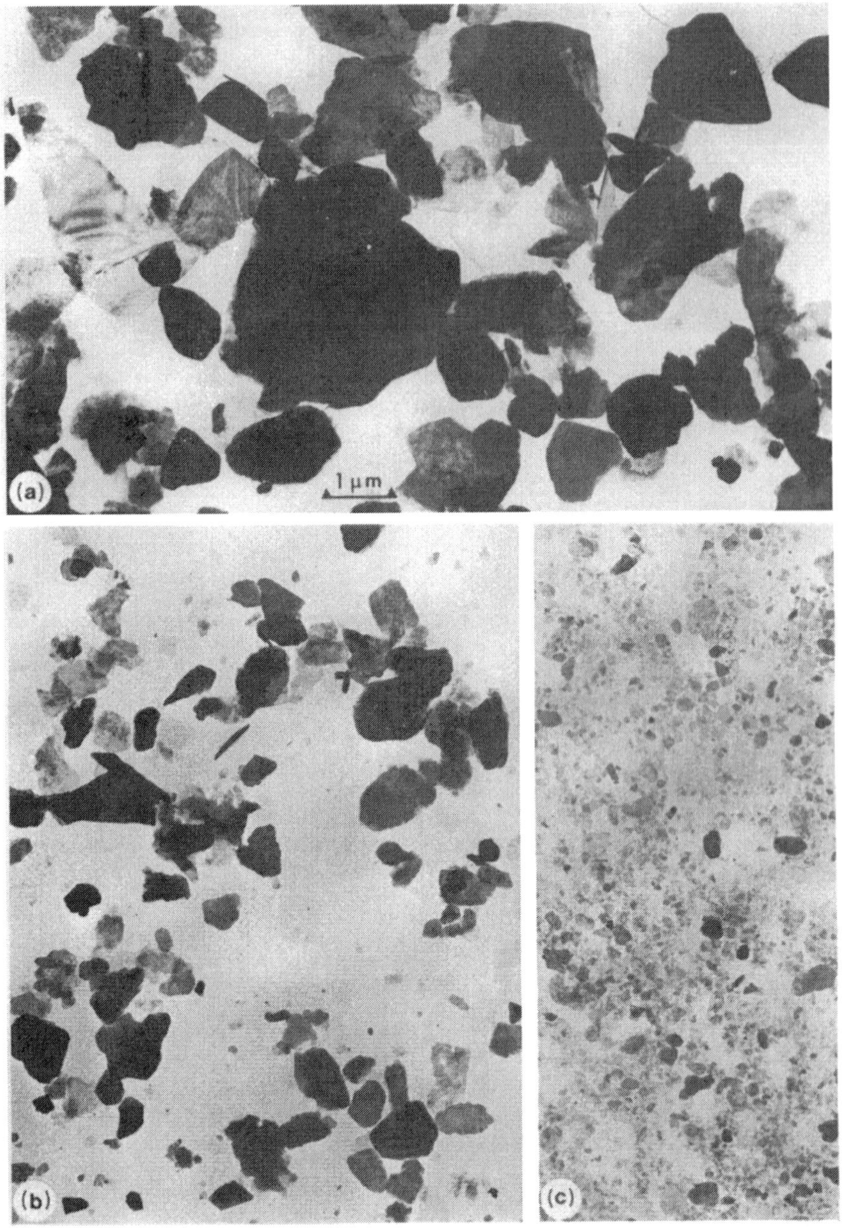

Plate 2.12 Unshadowed transmission electron micrographs of size fractions of clay from the B horizon of Swanmore series soil in weathered Reading Beds, Sussex, England, equal magnification. (a) 2–0.5 μm fraction, (b) 0.5–0.1 μm fraction, and (c) <0.1 μm fraction

of having two or three clearly defined units mixed in a simple or complicated manner, may have a whole series of units with gradational properties. They are of great importance, however, particularly in the weathering of clay minerals and much research effort has been devoted to them (Rich, 1968).

Table 2.13 gives unit formulae that have been calculated from the chemical compositions of a selection of interstratified minerals. Methods for determining proportions of units, as given for samples 6, 7, 8 and 9, form the subject of the next subsection. The compositions of minerals containing vermiculite or smectite units are most complete if exchangeable cations and exchange capacity are given. As natural specimens frequently contain exchangeable K, Na and Mg, this enables non-exchangeable K, or exceptionally Na, ions to be allocated to mica units and non-exchangeable Mg ions to be allocated to octahedral sites in 2:1 layers or hydroxide interlayers. This was not done by Boettcher (1966) for the hydrobiotite, sample 1, so 0.22 Mg ions have been placed in the interlayer position in order that the sum of octahedral cations should exactly equal 6.00. In all these interstratifications there are four tetrahedral and at least two octahedral sheets; in samples 3 and 4, containing chlorite units, there are three octahedral sheets. Substitution of Al for Si and Mg for Al can be switched to make units look more saponite- or montmorillonite- or beidellite-like, but as Weaver and Pollard (1973, p. 116) point out, in the absence of direct evidence that favours one composition rather than another, such subdivisions are quite arbitrary.

The *relative proportions of the component units* of an interstratified mineral may be deduced from the positions and intensities of the basal reflections obtained by X-ray diffraction and from the chemical composition. In favourable circumstances additional data may be obtained from other techniques, such as differential thermal analysis, thermogravimetry or infrared spectrometry. X-ray data should always be used, but as the determination of the proportions of units cannot be separated from the identification of the units themselves and the rules of succession, discussion of its application is deferred to the next subsection.

Chemical composition can sometimes be used if components are recognized but the rules of succession not understood. It was successfully done by Shimoyama, Johns and Sudo (1969) with a halloysite–montmorillonite interstratification from acid clays of Japan. These authors converted the per cent oxides of Si, Al and Mg from the analysis to molecular proportions and then assumed ideal compositions for kaolinite (Ka)$-4SiO_2 \cdot 2Al_2O_3 \cdot 4H_2O$ and montmorillonite (Mo)$-0.67M \cdot 8SiO_2 \cdot 3.33/2Al_2O_3 \cdot 0.67MgO \cdot nH_2O$, where M represents the exchangeable cations taken as monovalent. The following equations were then solved:

$$8\text{Mo} + 4\text{Ka} = \text{mol. prop. } SiO_2$$
$$3.33/2\text{Mo} + 2\text{Ka} = \text{mol. prop. } Al_2O_3$$
$$0.67\text{MgO} = \text{mol. prop. MgO}$$

Table 2.13 Unit formulae of regular and irregular interstratifications

		1	2	3	4	5	6	7	8	9
Anions $O_{20}(OH)_4$ or $O_{20}(OH)_{10}$	Charge (e.s.u.)	−44	−44	−50	−50	−44	−44	−44	−44	−44
Tetrahedral cations	Si^{4+}	5.64	7.58	6.86	6.52	6.56	6.61	7.20	7.38	7.10
$\Sigma = 8.00$	Al^{3+}	2.36	0.42	1.14	1.48	1.44	1.39	0.80	0.62	0.90
Maximum charge: +32.00	Charge	+29.64	+31.58	+30.86	+30.52	+30.56	+30.61	+31.20	+31.38	+31.10
Octahedral cations	Al^{3+}	—	—	6.03	5.44	4.02	3.75	2.82	3.16	1.94
	Fe^{3+}	1.22	—	0.03	0.12	0.02	0.21	0.68	0.10	1.44
	Fe^{2+}	0.10	0.34	—	0.38	—	—	—	—	0.02
If dioctahedral, $\Sigma = 4-6$	Mg^{2+}	4.54	5.70	0.02	0.51	0.03	0.04	0.52	0.80	1.04
If trioctahedral, $\Sigma = 6-9$	$Ti^{4+}_{(1)}, Li^{1+}_{(4)}$	0.14	—	—	0.33	—	—	—	—	—
	Σ	6.00	6.04	6.08	6.78	4.07	4.00	4.02	4.06	4.44
	Charge	+12.50	+12.08	+18.22	+18.79	+12.18	+11.96	+11.54	+11.38	+12.26
Interlayer cations	K^{1+}	0.64	—	0.29	0.13	—	0.90	0.58	1.08	0.30
	Na^{1+}	0.06	—	0.04	0.04	0.77	—	—	—	—
	Ca^{2+}	0.26	—	0.29	0.01	0.02	—	—	—	—
	Mg^{2+}	0.22	—	—	—	—	—	—	—	—
	X^{1+} (exchangeable)	—	0.26	—	0.48	0.45	0.48	0.66	0.18	0.40
	Charge	+1.72	+0.26	+0.91	+0.67	+1.26	+1.38	+1.24	+1.26	+0.70
Total cation charges		+43.86	+43.92	+49.99	+49.98	+44.00	+43.95	+43.98	+44.02	+44.06

1. Regular biotite–vermiculite interstratification (hydrobiotite), Montana, U.S.A.; Boettcher (1966).
2. Regular talc–saponite interstratification (aliettite), Taro, Italy; Alietti (1959).
3. Regular chlorite–beidellite interstratification (tosudite), Takatama, Japan; Shimoda (1969).
4. Nearly regular, ditrioctahedral chlorite–smectite interstratification, Huy, Belgium; Brown, Bourguignon and Thorez (1974).
5. Regular sodium mica–beidellite interstratification (rectorite), Baluchistan, Pakistan; Brown and Weir (1963b).
6. Irregular, partially-ordered, 64:36 potassium mica–beidellite interstratification, Surges Bay, Tasmania; Cole (1966).
7. Irregular, random, 55:45 mica–smectite interstratification, Denchworth series soil, Oxford Clay, England; Weir and Rayner (1974).
8. Irregular, partially-ordered, 76:24 mica–smectite interstratification, Kalkberg, Sweden; Hower and Mowatt (1966).
9. Irregular, random, 65:35 smectite–mica interstratification (glauconite), Eocene, Texas; Thompson and Hower (1975).

The proportions obtained were 25 % montmorillonite and 75 % kaolinite. An X-ray investigation by Cradwick and Wilson (1972) confirmed these proportions and showed that the interstratification is random. A similar but more sophisticated use of chemical data was made by Schultz, Shepard, Blackmon and Starkey (1971) on three kaolinite–montmorillonite interstratifications from Mexico. In this work they assumed the ideal composition for kaolinite and then adjusted the proportions of the components to obtain the most probable composition for the montmorillonite units. They obtained 45, 50 and 55 % kaolinite for the three samples. Cradwick and Wilson (1972) and Sakharov and Drits (1973) suggested on the basis of X-ray data that Schultz *et al.* had underestimated the proportions of kaolinite units. The samples contained interstratified mica–smectite that had not been identified by Schultz *et al.* in addition to the kaolinite–smectites.

In the introductory section on interstratified minerals sequences of mica and smectite units were given to represent random, ordered and segregated interstratifications. Methods used for calculating these sequences are based on probability mathematics, and rules of succession were developed for use with interstratified clay minerals by MacEwan. The following account is based on descriptions in MacEwan, Ruiz Amil and Brown (1961) and Sato (1965).

Consider a mica (M) and smectite (S) interstratification in which the succession of units is affected by nearest neighbours, referred to subsequently as nearest-neighbour interaction, but not by those further away. Let p_M be the probability of finding a mica unit and p_S the probability of finding a smectite unit. p_M and p_S also represent the proportions of mica and smectite in the mineral, since the probability of finding a mica layer is the same as its proportion of the whole sample. Thus

$$p_M + p_S = 1 \tag{1}$$

Further, let p_{MM} be the probability of finding a mica unit next to a given mica unit when passing through the particle in an agreed direction. p_{MS}, p_{SS} and p_{SM} are similarly defined. Then, because either mica or smectite must follow any particular unit,

$$p_{MM} + p_{MS} = 1 \tag{2}$$

$$p_{SS} + p_{SM} = 1 \tag{3}$$

and

$$p_M p_{MM} + p_S p_{SM} = p_M \tag{4}$$

The left side of equation (4) represents the probability of mica following mica added to the probability of mica following smectite, which is the same as the probability of a mica unit occurring at all. Equations (4) and (3) may be combined to eliminate p_{SM}, as follows

$$p_M p_{MM} + p_S - p_S p_{SS} = p_M \tag{5}$$

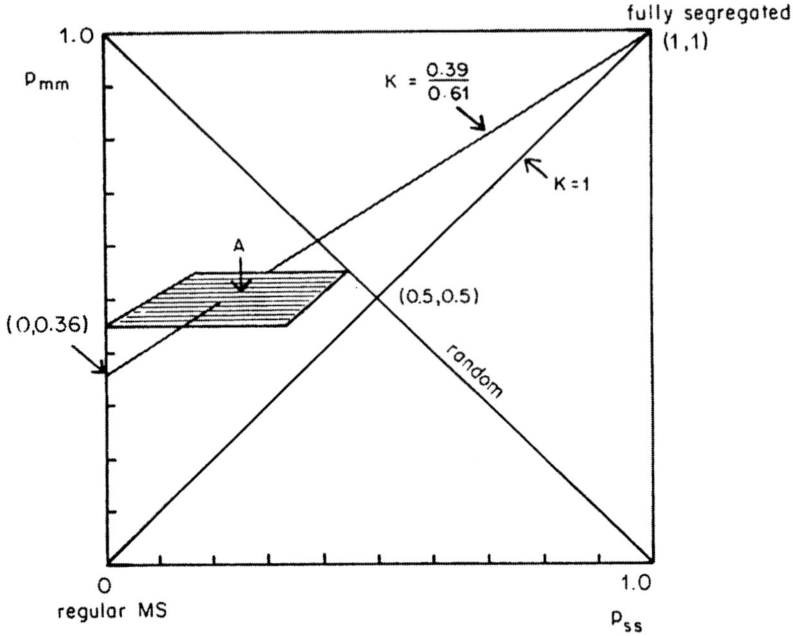

Figure 2.35 Sato plot for two-component interstratifications with nearest neighbour interactions only. For full explanation see the text (pages 120 and 121). Point A represents the composition of the mica–smectite HS-124 Figure 2.38b) as determined by the direct Fourier transform method (page 130 and Figure 2.38d). The shaded area shows the limits of the composition of HS-124 obtained using the peak position method (page 130 and Figure 2.38c)

and equation (5) rearranged to give

$$p_{MM} = p_{SS}K + (1 - K) \tag{6}$$

where

$$K = p_S/p_M$$

Thus a knowledge of p_M and p_{MM} completely specifies the sequence of units found in two-component interstratifications when only nearest neighbour interactions are considered. Given these two probabilities the other four may be obtained by the use of equations (1) to (4). The object of a structural analysis is therefore to discover the values of p_M and p_{MM} that apply to a particular interstratification. Techniques for doing this are given below.

Values of p_M and p_{MM} may be plotted on a graph with p_{MM} as ordinate and p_{SS} as abscissa (Figure 2.35). On this graph, equation (6) represents a family of straight lines of slope K that pass through the point (1, 1). At (1, 1) p_{MM} and p_{SS} have their maximum values and the probabilities of having MS or SM pairs are zero, so that the interstratifications are fully segregated. Along the axes $p_{MM} = 0, p_{SS} = 0$, the probabilities of finding MS or SM for any value of p_M are maximum and those for MM and SS pairs minimum, and so the interstratifications have maximum order. The line joining (0, 0) and (1, 1) has the slope $K = 1$, and as $K = p_S/p_M$, $p_S = p_M = 0.5$ for this line and the point (0, 0) represents the composition of the regular 50:50 mica–smectite interstratification. The mid-point of the same line, the point (0.5, 0.5) has values of $p_{MM} = p_M = 0.5$. This means that the probability of finding an MM pair is the same as the probability of finding an M unit at all, and there is no interaction between neighbours, i.e. the sequence is random. The relationship $p_{MM} = p_M$ holds for all points on the line joining (1, 0) and (0, 1); this is the line $p_{MM} = 1 - p_{SS}$, which represents all possible random interstratifications.

Figure 2.35 may be termed a 'Sato plot' after Sato (1965). It is a useful way of displaying results for interstratified minerals when p_M and p_{MM} or their equivalents are known. Points falling within the triangle bounded by (0, 0) and the line of random interstratification are partially ordered and those in the triangle above the line partially segregated. It is generally easier to visualize the relationships between minerals from their positions on the plot than from a comparison of their p_M and p_{MM} values.

Although two-component interstratifications with nearest-neighbour interaction are the sequences that have been most commonly looked for in clay mineral work, they are not the only ones to have been considered.

Investigation by X-ray diffractometry

Figure 2.36 shows diffractometer traces of Mg-saturated rectorite from Baluchistan, Pakistan (Brown and Weir, 1963a and b), (a) solvated with ethylene glycol and (b) heated to 250 °C on a heating stage. The trace in Figure 2.36a shows sharp, evenly spaced basal reflections corresponding to fifteen submultiples of the basal spacing. Table 2.14 lists the spacings and shows that $n \times d_n$—the spacing multiplied by the order of the reflection—varies from 2.629 to 2.6625 nm, a range of 0.0335 nm, the mean spacing being 2.6517 nm. Such a sequence of reflections that all give effectively the same $n \times d_n$ values is said to be rational and indicates that the specimen producing it is ordered, i.e. the reflections are produced by the repeat of a particular spacing, 2.6517 nm in this instance. Figure 2.36b shows the collapsed form of the same mineral. For the eight reflections measured, $n \times d_n$ varies from 1.949 to 1.9233 nm, a range of 0.026 nm, so that the sequence is again rational. The mean spacing is 1.932 nm. Assuming that the two components have approximately the same spacing when collapsed, Table 2.13 indicates that the permanently collapsed component is a sodium and not a potassium mica—this gives components of 0.96 nm for mica and 0.97 nm for the other component when collapsed, 1.69 nm when solvated with ethylene glycol, 1.78 nm solvated with glycerol and 1.54 nm solvated with water at room humidity (column 1, Table 2.12). Thus the expanding component

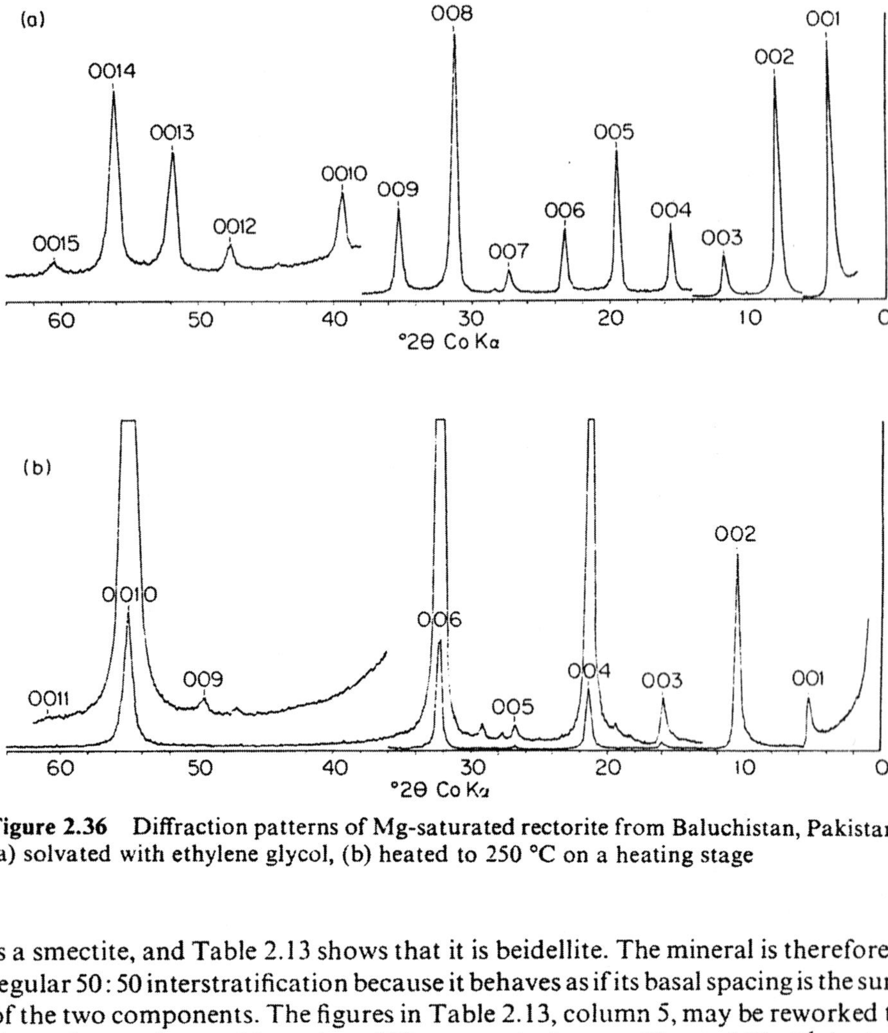

Figure 2.36 Diffraction patterns of Mg-saturated rectorite from Baluchistan, Pakistan: (a) solvated with ethylene glycol, (b) heated to 250 °C on a heating stage

is a smectite, and Table 2.13 shows that it is beidellite. The mineral is therefore a regular 50:50 interstratification because it behaves as if its basal spacing is the sum of the two components. The figures in Table 2.13, column 5, may be reworked to show that exchangeable cations X^+ are equivalent to 58 meq 100 g^{-1} ignited weight for Na^+ or Ca^{2+} ions for the whole mineral, or 116 meq 100 g^{-1} for the beidellite component, i.e. a value that is well within the range found for smectites.

Figure 2.37 shows traces for similar treatments of a mineral from a soil of the Denchworth series in weathered Oxford Clay from England. In Figure 2.37a some reflections are sharp, but others are broad and poorly resolved. The list of spacings in Table 2.14 shows that $n \times d_n$ varies from 1.596 to 1.864 nm, a range of 0.268 nm. This large range coupled with the broadening of certain of the reflections indicates that the basal reflections are produced by an irregular interstratification and that the mean spacing does not give a reliable guide to the spacings of the components. In Figure 2.37b, however, where the specimen is

Table 2.14 Spacings, nm corresponding to basal reflections of Mg-saturated rectorite from Baluchistan and Denchworth series soil clay, <0.04 μm fraction (illustrated in Figures 2.36 to 2.37)

Mg-saturated rectorite

(a) ethylene glycol solvated

n	001	002	003	004	005	006	007	008	009	0010	0011	0012	0013	0014	0015
d_n	2.629	1.3151	.8821	.6633	.5303	.4422	.3798	.3316	.2953	.2656	—	.2215	.2043	.1899	.1775
$n \times d_n$	2.629	2.6302	2.6463	2.6532	2.6515	2.6532	2.6586	2.6528	2.6577	2.6560	—	2.6580	2.6559	2.6586	2.6625

(b) heated to 250 °C

n	001	002	003	004	005	006	007	008	009	0010
d_n	1.949	.9651	.6440	.4822	.3862	.3212	—	—	.2137	.1935
$n \times d_n$	1.9490	1.9302	1.9320	1.9288	1.9310	1.9272	—	—	1.9233	1.9350

Mg-*saturated Denchworth series soil clay*, <0.4 μm *fraction*

(a) ethylene glycol solvated

'n'	1[a]	2	3	4	5	6	7	8	9
d_n	1.68	.932	.532	—	.3351	.2794	—	.2012	—
'n' $\times d_n$	1.68	1.864	1.596	—	1.6755	1.6764	—	1.6096	—

(b) heated to 250 °C

n	001	002	003	004	005
d_n	.996	.4946	.3295	.2496	.1987
$n \times d_n$	.996	.9892	.9885	.9984	.9935

[a] As the sequence is irrational the reflections are not orders of a series; they are therefore numbered 1–9, to be understood as 1st reflection, etc.

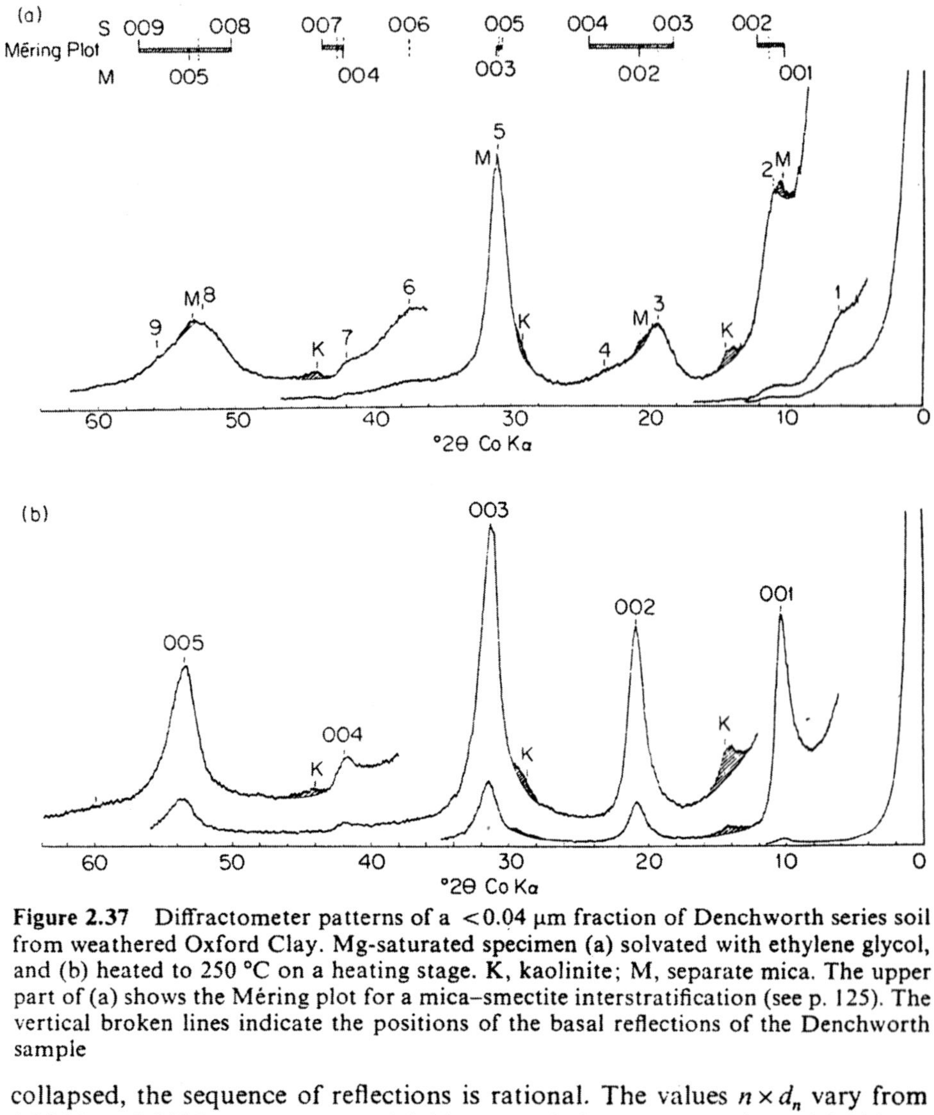

Figure 2.37 Diffractometer patterns of a <0.04 μm fraction of Denchworth series soil from weathered Oxford Clay. Mg-saturated specimen (a) solvated with ethylene glycol, and (b) heated to 250 °C on a heating stage. K, kaolinite; M, separate mica. The upper part of (a) shows the Méring plot for a mica–smectite interstratification (see p. 125). The vertical broken lines indicate the positions of the basal reflections of the Denchworth sample

collapsed, the sequence of reflections is rational. The values $n \times d_n$ vary from 0.9984 to 0.9885 nm, a range of 0.01 nm, and the mean spacing of 0.993 nm indicates that the components when collapsed resemble those of mica, vermiculite or smectite. In fact, according to Weir and Rayner (1974), the mineral is a random 55:45 mica–smectite interstratification. With such an interstratification there is no simple way to deduce the components, their proportions, or the rules of succession. The d_{06} spacing shows that the mineral is dioctahedral, and this is confirmed by the unit formula (Table 2.13), which has been calculated assuming that the components are mica, smectite or vermiculite, but not chlorite or kaolinite, i.e. the total anion charge was set to -44. The non-exchangeable *K* content confirms the presence of mica units and the exchange capacity that of vermiculite or smectite units, but more than this cannot be deduced without further analysis of the basal reflections.

Two approaches have been made to the problem of solving structures of interstratified minerals by the analysis of basal diffraction reflections. In the first (I), a model is assumed for the interstratification and used to calculate a diffraction pattern for comparison with that of a clay mineral. In the second (II), a model is assumed for the structures of the constituent units and used to transform the observed diffraction intensities into a function. This function gives the principal sequences of units occurring in the mineral and their relative abundance. Thus in (I) different values of p_M and p_{MM}, or the corresponding probabilities of other units, are tried until the best fit is obtained, whereas in (II) values of p_M and p_{MM} are obtained from the analysis.

The calculation of diffraction patterns from model structures (Model I). A mica unit has eight planes of atoms parallel to its basal plane and an expanded smectite may have four more, so that a diffracting unit composed of five mica and five smectite units has 100 atom planes separately diffracting the incident beam; the resulting intensity is the complex sum of all the individual diffracted beams. An oriented aggregate clay sample contains thousands of such diffracting units, in which the number and sequence of planes of atoms vary with the sizes of the particles and the type of interstratification. Hendricks and Teller (1942), Méring (1949) and Kakinoki and Komura (1952) who have sought to calculate diffraction from such a complicated structure have stated the problem in general terms, but have then devised ways of simplifying the calculations in order to reduce the amount of computation involved. For example, they all introduced a form factor, or structure factor (F), to represent diffraction from the groups of planes of atoms forming a unit or part of a unit.

Following work by Hendricks and Teller on infinitely large diffracting particles, Méring worked out the scattering to be expected from two-component random interstratifications of limited particle size, and showed how peak positions could be used to give the proportions of constituent units. On such a 'Méring plot' the positions of the basal reflections of pure phases of the constituent units are marked, as illustrated in Figure 2.37a for 1.69 nm smectite and 1.0 nm mica. Shaded blocks joining these positions show regions where the diffraction maxima of the interstratified minerals may be expected. Maxima migrate along the blocks as the proportions of their components vary, and if the migration is assumed to be linear, the proportions, p_S and p_M in this instance, can be estimated. The average results from Figure 2.37 is $p_M = 0.70$, $p_S = 0.30$, which may be compared with the values of $p_M = 0.55$, $p_S = 0.45$ obtained by the more elaborate method involving matching traces (Weir and Rayner, 1974).

MacEwan (1958) developed Méring's work by making the further simplifying assumption that a single structure factor may be used to represent either component. The diffracted intensity from an interstratified mineral may then be represented as the product of the squared modulus of the structure factor and a mixing function Φ, and an angular factor Θ.

$$I \rightarrow |F|^2 \, \Phi\Theta$$

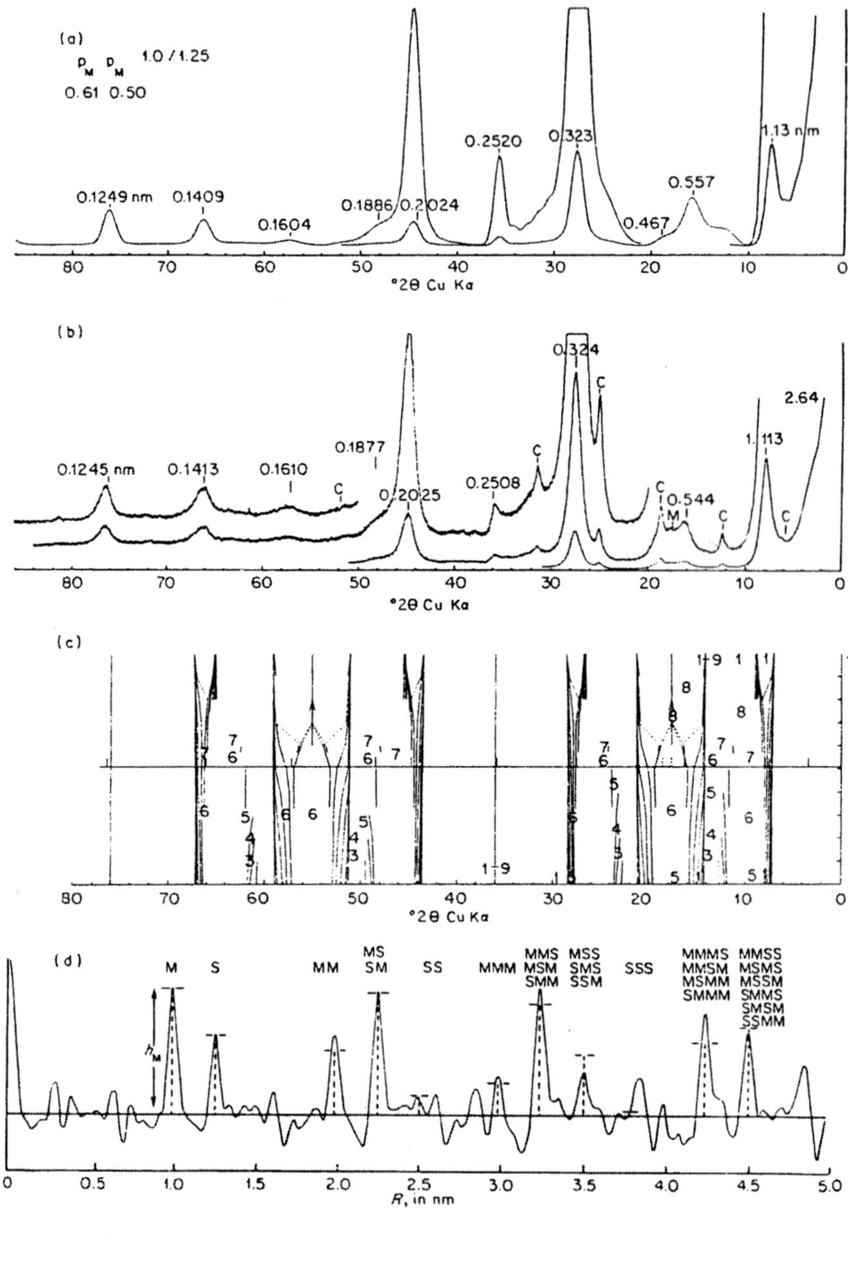

Φ is a function of layer spacing, layer succession and particle size. F, Φ and Θ vary with the angle of scattering, θ. Ruiz-Amil, Ramirez and MacEwan (1967) published a book of data for two-component interstratification that contains plots of this mixing function for a wide range of unit spacings, proportions of units and degrees of order or segregation (all values of p_M and p_{MM}). A set of peak migration curves, redrawn from the mixing function data from this book is shown in Figure 2.38c, in which the components are smectite, containing one layer of water molecules, and mica, with unit thicknesses of 1.25 and 1.0 nm respectively. The curves show peak positions of basal reflections in °2θ (Cu Kα) of mixtures of 1.25 and 1.0 nm units in all proportions varying from fully segregated to maximum order. Individual lines represent selected proportions of components, thus curve 1 represents 10% mica (p_M = 0.10) and 90% smectite (p_S = 0.90), curve 2, 20% mica (p_M = 0.20) and 80% smectite (p_S = 0.80), etc. This method of plotting places all the peak positions of fully segregated interstratifications (p_{MM} = 1.00) at the top of the graph and those of maximum order (p_{MM} = 0.00) at the bottom. Peaks of random interstratification occur on curves numbered x at the value p_{MM} = 0.x, e.g. p_{MM} = 0.50 on curves numbered 5 etc. Many of the curves are so close together that only the outside ones have been numbered in the figure. An example of the use of the curves to obtain values of p_M and p_{MM} for an interstratification is given on page 129. A similar set of peak migration curves was produced by Cesari, Morelli and Favretto (1965) from the equations of Allegra (1961), who had applied simplifying assumptions to the fundamental equations for diffraction intensities derived by Kakinoki and Komura (1952). These methods of obtaining values of p_M and p_{MM} from peak positions are an improvement on the Méring plot method because they are extended to cover interstratifications other than random, and also because they do not assume that the migration of peaks is linear with change in p_M.

MacEwan also calculated whole diffraction traces, using single F values and simple ratios of unit spacings, to reduce the amount of calculation involved (see MacEwan, Ruiz-Amil and Brown, 1961). As computers became available more detailed and realistic structures could be calculated. Reynolds (1967)

Figure 2.38 Diagrams to illustrate the structural analysis of an interstratified clay mineral: (a) calculated diffractometer trace of HS-124 for p_M=0.61, p_{MM}=0.51, using computer program details as given in Weir and Rayner (1974); (b) diffractometer trace of oriented aggregate of HS-124, from tuffaceous mudstone, Gwent, Wales, using a Na-saturated, air-dry <0.2 μm, oriented aggregate specimen, Cu Kα radiation, C, chlorite, M, separate mica (c) plot of peak positions for 1.25/1.0 nm interstratifications (from Ruiz-Amil *et al.*, 1967). Values of p_M (proportions of mica units) are written against individual curves so that curve 1 represents p_M=0.1, curve 2p_M=0.2, etc. The horizontal line at p_{MM}=0.51 represents the best fit position for the sample HS-124 illustrated in (b), as obtained by the direct Fourier transform method illustrated in (d). The short vertical lines show the positions of the basal reflections of HS-124, (d) function of W_R against R representing the transform of the basal reflection data obtained from the diffractometer trace illustrated in (b) above (see page 129) introduced separate F values for 2:1 layers and interlayer material and Wright (1975) derived complex functions to simulate naturally occurring particle size distributions. Cradwick (1975), using expressions derived from Kakinoki and Komura's equation, employed different F values to represent 2:1 layers of differing composition and, from the same equation, Cradwick and Wilson (1978) derived an expression for diffraction from random and segregated structures of three-component mixtures. Reynolds and Hower (1970) used Reynolds' computer programmes to match experimental and calculated traces by altering p_M and p_{MM} values.

The sequence of units derived for all these methods were given in terms of nearest neighbour interaction described above. This does not allow complex regular sequences like MMSMMS, or the concentration of one layer type in the interior of particles, to be treated. The first of these, non-nearest-neighbour ordering, was studied by Sato (1969), who calculated diffraction patterns for 70:30 mica–smectite interstratifications using an equation derived from that of Kakinoki and Komura. The second type of sequence, that of zoned interstratification, was investigated by MacEwan (1968), who used a distribution function to set the number of mica units that separated smectite units in particles. Other methods of obtaining unit sequences have been suggested by Ross (1968) and Tettenhorst and Grim (1975). Ross calculated the diffraction from a whole particle of 30 units after setting the sequence of units in the particle by the use of random numbers. Tettenhorst and Grim investigated diffraction by particles containing 5, 10 and 20 units in which M and S units could be placed in particular positions. They showed *inter alia* that diffraction from particles with one S unit at the edge is quite different from that with the S unit at the centre.

The direct Fourier transform method (Model II). The development and application of this method is largely the work of MacEwan (1956). MacEwan, Ruiz-Amil and Brown (1961) showed that measured positions and intensities of basal reflections of interstratified minerals may be transformed to give a distribution function of interlayer distances from which values of p_M and p_{MM} may be deduced. Examples of the application of the method are given by MacEwan *et al.* and the method has been widely used.

Cole (1966) extended the application of the method to investigate non-nearest-neighbour ordering for the sample from Surges Bay by determining the most common layer sequence in the sample, MS, and then repeating the analysis using a composite MS unit as one component.

Example of identification of an interstratification by X-ray diffractometry. Figure 2.38b shows the diffractometer trace of an air-dry, Na-saturated, <0.2 μm clay fraction from a tuffaceous mudstone from Gwent, Wales (sample HS-124). In the original trace nineteen orders of the basal reflection could be measured, of which twelve are shown in the figure. Analysis of the pattern will be discussed in terms of the use of the mixing function curves of Ruiz-Amil *et al.*, the direct Fourier transform method of MacEwan and the matching of computed diffractometer patterns with the observed pattern.

1. Matching of peak positions. Inspection of the traçe suggests that the mineral has components of smectite (basal spacing in air of 1.25 nm with interlayer Na ions) and mica (1.0 nm)—note the well-resolved first reflection intermediate in position between the 1.25 and 1.0 nm peaks and the similarly well-resolved 0.250 nm reflection where the smectite (005) and mica (004) reflections coincide. The horizontal line on Figure 2.38c represents the best fit values of $p_{MM} = 0.5$, $p_M = 0.6$, but it would be difficult to be more precise than, say, $p_M = 0.45-0.55$ and $p_M = 0.55-0.65$. The short vertical lines on the horizontal line represent experimental peak positions transferred from Figure 2.38b.

2. The direct Fourier transform method. The trace of Figure 2.38b was used by Brown and Machajdik (unpublished) to investigate this method. They used the expression from MacEwan *et al.* (1961)

$$W_R = (a/\pi\Sigma_x(x) \cos (\mu_x R)$$

where

$$i(x) = \frac{I_x}{L_{p_x}|F_x|^2} \text{ and } \mu_x = \frac{2}{d_x} = \frac{4 \sin \theta}{\lambda}$$

W_R is the probability of finding a layer at a perpendicular distance R from another layer, a is the thickness of the layer, I_x and d_x are the measured intensities and apparent spacings of the basal reflections, L_{p_x} is the Lorentz polarization factor and F a structure factor applicable to both component units. $|F_x|^2$ was calculated using the coordinates for a 2:1 layer after Reynolds (1967). The summations were done by a computer programmed to print the results in graphical form (Figure 2.38d). In the figure the first two major peaks give the spacings of the components, 1.00 and 1.25 nm. Having established these, the remaining peaks can be identified as aggregate spacings, for example, the peak at 2.25 nm arises from pairs of units, 1.0 and 1.25 nm, that are nearest neighbours. The heights of the peaks are proportional to the probabilities of the sequences occurring, e.g. $h_M = Cp_M$, $h_{MM} = Cp_M p_{MM}$, $h_{MS,SM} = C(p_M p_{MS} + p_S p_{SM})$ etc. From a set of simultaneous equations built up in this way the best fit values of p_M, p_{MM}, etc. may be obtained. The result obtained by Brown and Machajdik was $p_M = 0.61$, $p_{MM} = 0.51, \ldots$; the horizontal bars above the dotted lines on Figure 2.38d indicate the measure of agreement between the calculated and experimentally derived values of the peak heights.

3. Matching of diffractometer traces. The trace in Figure 2.38a was calculated using the computer programme developed by Rayner from one devised by Reynolds. The programme used Reynolds' silicate layer coordinates with *K* ions for the mica interlayer and Na ions with one layer of water molecules (after the structure for allevardite proposed by Brindley, 1956) for the smectite interlayer.

The trace was computed for $p_M = 0.61$, $p_{MM} = 0.51$, the values obtained by the direct Fourier-transform method. Comparison of Figures 2.38a and b shows that the overall agreement between the experimental and calculated curves is good, although an improvement in the fit of the 0.55 and 0.25 nm peaks is desirable.

Because $p_{MM} < p_M$ it is clear that there is a tendency towards alternation of M and S units in this sample. If $p_M = 0.61$ and $p_{MM} = 0.51$, point A Figure 2.35 represents the sample on a Sato plot. The shaded area around A represents the uncertainty in the positioning of A arising from the possible range in values of p_M and p_{MM} obtained by the simplest method, that of matching peak positions only. Point A (and the shaded area) lies approximately half-way between the line representing random interstratifications and the p_{MM} axis—which represents the maximum amount of alternation obtainable in samples with an excess of M units. Although it was not possible to make an estimate of the accuracy of the other two methods, the uncertainty of the result is thought to be considerably less than the shaded area around A.

In assessing the relative merits of the three methods, (1) is rapid and simple and the basic data easily accessible, (2) may give information about the type of ordering that cannot be obtained from the other two, but needs a number of basal reflections (more than 10) for the W_R peaks to be interpreted with confidence, (3) is a development of (1) and capable of greater precision and versatility.

Interstratified clay minerals in soils. Soils form a weathered surface layer on rocks wherever they are exposed to the atmosphere. It follows that any type of interstratified clay mineral, whatever its mode of origin or resistance to weathering, will be incorporated into soil at some place and for some time. Millot (1973) described the geological processes involved in the formation and weathering of sediments, and described an equatorial zone in which the soil clay minerals are predominantly kaolinitic with iron and aluminium hydrated oxides, flanked by tropical zones in which the soil clays are mainly montmorillonitic. Thereafter surface chemical weathering is not so dominant as to impose a uniform clay mineralogy on to soils, and the inheritance of clay minerals from parent rocks becomes increasingly important. In the temperate zone of the northern hemisphere the loess belt that extends around much of the world gives the near surface layers of the soils a more uniform clay mineralogy than they would have if derived entirely from underlying solid rock formations.

Most soil clay fractions contain mixtures of minerals, including, for some, more than one type of interstratified mineral. It is therefore rarely possible to study the structure of interstratified soil clay minerals in detail, so that in spite of the vast literature on soil clay mineralogy, few detailed structural studies have been made. As an alternative, clay-containing rocks, many of hydrothermal origin, have been searched for interstratified minerals free from impurities and the knowledge of structures gained from these applied to soil clays.

Trioctahedral clay minerals are found in soils derived from ferromagnesian igneous and metamorphic rocks and in rocks such as some of the Triassic rocks of Europe formed in sedimentary basins in which magnesium ions are concentrated. They may be short-lived in soils, however, as they tend to weather rapidly. Regular trioctahedral interstratifications include mica-vermiculite (hydrobiotite), mainly formed hydrothermally but also found in soils as a weathering product of biotite, talc–saponite, chlorite–saponite, chlorite–vermiculite and chlorite-swelling chlorite (Tables 2.12 and 2.13). In addition to regular interstratifications there exists a range of irregular ones, the description and nomenclature of which has led in the past to great confusion (Marel, 1964).

Stephen and MacEwan (in Martin Vivaldi and MacEwan, 1960) described a mechanism by which swelling chlorite may form from chlorite or vermiculite. They envisaged the interlayer material in swelling chlorite as being composed of islands of polymerized magnesium hydroxide together with hydrated and exchangeable magnesium ions. Changes in the pH of the external solution and supply of magnesium ions could either precipitate more brucite until complete chlorite units were formed or remove brucite completely to form vermiculite units.

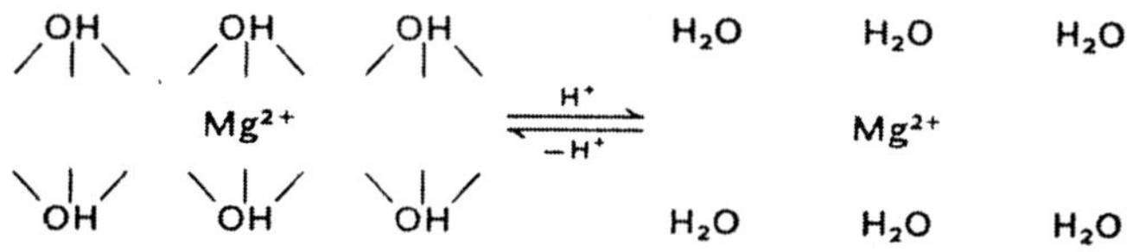

Although an oversimplification, this is thought to be the basis of the chloritization and subsequent stripping that occurs in soils both for trioctahedral minerals, as here, and for dioctahedral, mentioned below and on page 88.

Regular dioctahedral interstratifications include sodium mica–beidellite (rectorite), potassium mica–beidellite or -montmorillonite (Nemecz, Varju and Barna, 1965; Cole, 1966; Heystek, 1954) and chlorite–montmorillonite (tosudite). Unit formulae and basal spacings are in Tables 2.13 and 2.12, respectively. The minerals are of restricted occurrence, although partially-ordered dioctahedral mica-smectite interstratifications are common in older sediments.

Probably the most common interstratified minerals in sediments and the soils on them are dioctahedral mica–smectites. Weaver (1956) and Perry and Hower (1970) have described great thicknesses of them in the sedimentary column. Weir, Ormerod and El Mansey (1975) showed that clays from sediments in the Nile Delta that were formerly thought to be montmorillonites are in fact random mica–smectite interstratifications containing approximately 70% smectite units. A more detailed examination of clays of the tropical belt described by Millot would probably show that many of these are not strictly smectites, but interstratifications.

Interstratified minerals have also been described from soils in glacial till and lacustrine sediments of the Canadian Prairie Provinces (Kodama and Brydon, 1966). In a detailed structural study they showed that the fine fractions of the clays that are inherited from sediments of Upper Cretaceous age contain

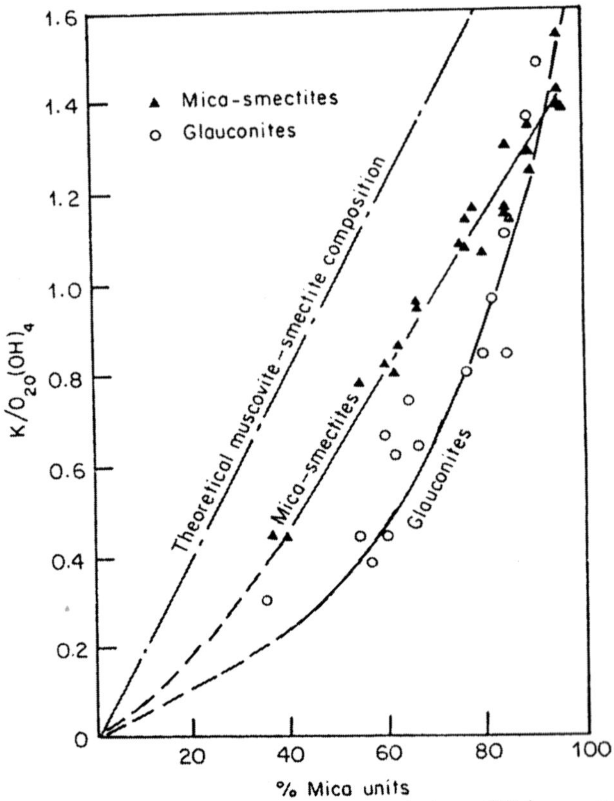

Figure 2.39 Graph showing the variation of K ions per formula unit with proportion of mica layers in (▲) mica–smectite interstratifications and glauconites (O) (after Thompson and Hower, 1975). Reproduced by permission of Pergamon Press Ltd

dioctahedral 55:45 random montmorillonite–mica interstratifications. A similar mineral was described from a soil developed in weathered Oxford Clay of Jurassic age from near Oxford, England by Weir and Rayner (1974). Some details are given in Table 2.13 and Figure 2.37. The mineral has fewer K ions in the mica units than are found in mica minerals. A detailed study of mica-smectites and glauconites from sediments ranging in age from the Cambrian to the Recent (Hower and Mowatt, 1966; Thompson and Hower, 1975) shows that the results found for the soil mineral mentioned above hold for both mica-smectites and glauconites, particularly those from Mesozoic and more recent sediments. Figure 2.39, reproduced from Thompson and Hower's paper, shows clearly that the interstratified minerals with the most expanding units have the least number of K ions per mica unit. Their work also shows that as the K content of the minerals increases the type of interstratification changes from random to partial nearest-neighbour ordering and then to non-nearest-neighbour ordering.

The minerals discussed in the previous paragraph contain smectite units that show unhindered collapse on heating. They occur in soils, but frequently in the deeper horizons, where they have been inherited from the parent rock. Clays found in upper horizons, and these include minerals found throughout many loess deposits, contain interstratified minerals with units that are dioctahedral intergradient chlorites. These minerals have been referred to as dioctahedral vermiculite (Brown, 1953; Rich and Obenshain, 1955), chloritized montmorillonite (Sawhney, 1958; Frink, 1965), aluminium interlayered vermiculite (Douglas, 1965), intergrade-chlorite–vermiculite–montmorillonite (Jackson, 1960) and chlorite-like intergrade (Weed and Nelson, 1962). The units are formed when aluminium, iron, magnesium and, perhaps, other ions are precipitated in the interlayer space or absorbed into the interlayer space as large hydroxy ions and polymerize there. After studying soils of the North Carolina Coastal Plain, U.S.A., Malcolm, Nettleton and McCracken (1969) suggested that in these soils polymers containing mainly aluminium ions were precipitated when the soil solution was in the pH range of 5–6. If the pH becomes more acid than 4.5 the polymers become unstable and are eventually replaced by aluminium and hydronium ions. The parent material of these soils contains montmorillonite, which in the pH range of 5–6 becomes transformed to an interstratification containing intergradient chlorite units and in upper acid horizons with a pH of less than 4.5 is again transformed, to an Al–smectite. Very similar sequences of mineral changes, although from different parent clay minerals, have been described in the upper horizons of podzols in Norway by Kapoor (1973) and Canada by Kodama and Brydon (1968).

An interstratification that has been increasingly reported in recent years contains units of kaolinite and smectite. It occurs in both hydrothermal deposits and soils. In the acid clays of Japan where it is formed from Al-montmorillonite, the kaolinitic component is sometimes halloysite and the interstratification gives such diffuse reflections that at first it was thought to be amorphous (Sudo and Hayashi, 1956). However, other occurrences are easier to analyse and progressively more complete accounts of these minerals have been given (Shimoyama, Johns and Sudo, 1969; Schultz, Shepard, Blackmon and Starkey, 1971; Wiewióra, 1973). Cradwick and Wilson (1972) and Sakharov and Drits (1973) calculated patterns to match diffractometer traces of the mineral. The model devised by Sakharov and Drits was the most detailed in that they assumed that where a smectite unit was adjacent to another smectite, two layers of glycerol would be absorbed, and where the basal oxygens of a kaolinite unit were adjacent to a smectite one layer of glycerol would be absorbed, but that the contact between the basal OH layer of kaolinite and the basal oxygens of either another kaolinite unit or a smectite unit would be held closed by hydrogen bonding so that no glycerol could penetrate. This model was used to match satisfactorily diffraction patterns of two randomly interstratified kaolinite–smectites from the U.S.S.R., one formed hydrothermally and the other by weathering.

A striking example of the formation of interstratified kaolinite–smectite by weathering was given by Altschuler, Dwornik and Kramer (1964). The transformation takes place in beds of clayey sands of Eocene age in Florida, U.S.A. Where the beds are exposed, the percolation of slightly acid ground water produces a surface weathering zone in which clay of the parent material, montmorillonite, is transformed into kaolinite through the intermediary of a regular 50:50 kaolinite–montmorillonite interstratification.

Randomly interstratified kaolinite–smectite should be suspected if a Ca-saturated specimen gives a 0.73 to 0.75 nm reflection that increases to 0.77 to 0.79 nm after glycol solvation and *increases* further to 0.80 to 0.83 nm following heating to 300 °C (Cradwick and Wilson, 1972). Infrared absorption spectra confirm the presence of kaolinite units if they show three or four peaks in the 362–370 nm region (Farmer 1974, p. 335).

2.4 OXIDES AND HYDROUS OXIDES

2.4.1 Iron and Aluminium Oxides and Hydrous Oxides

The oxides (including for brevity oxyhydroxides and hydroxides in the term) of iron and aluminium are almost universal constituents of soil clays occurring in several mineralogical forms (Table 2.15) as discrete particles or associated with surfaces of other minerals. Even though they are usually present in lesser amounts than the aluminosilicate clay minerals they may be as important in their effect on soil properties. Iron oxides are found in most soils and are responsible for the red, orange, yellow and brown colours that are so widely used to distinguish soils and soil horizons. Pedogenetic processes are also frequently inferred from colours and the distribution of colours that iron oxides impart to soils. Iron can exist in ferrous and ferric forms and so affects and is affected by the oxidation–reduction state of the soil. It is a transition metal, as are many important trace elements, and transition metals may be expected to enter the crystal structures of iron minerals. Both iron and aluminium are trivalent cations that are hydrolyzed at pH's common in soil. The hydroxy polymers formed by hydrolysis have important effects on aggregation and flocculation, soil pH and surface charge on soil particles. They also react with added phosphate and therefore modify the availability of applied phosphate fertilizers to plants.

Iron oxides in soils normally are the product of weathering of iron-bearing silicates but the magnetic iron oxide, magnetite, is usually inherited from the parent rock. The aluminium oxides are the ultimate product of intense weathering of aluminosilicate minerals and are the major constituent of bauxite; in soils, aluminium oxides are abundant only in certain tropical soils. Boehmite, γ-AlOOH, and gibbsite, γ-$Al(OH)_3$, are the only aluminium oxides and hydroxides common in soils. Both are products of intense weathering and are therefore found in many strongly weathered soils of tropical regions. Gibbsite is also occasionally found in soils of humid regions in some situations in which

Table 2.15 Naturally occurring hydroxides, oxyhydroxides and oxides of iron, aluminium, manganese, titanium and silicon; those common in soils are italicized

Iron		Aluminium	
Goethite	α-FeOOH[a]	Diaspore	α-AlOOH
Lepidocrocite	γ-FeOOH	*Boehmite*	γ-AlOOH
Akaganeite	β-FeOOH	Corundum	α-Al_2O_3
Hematite	α-Fe_2O_3	*Gibbsite*	$Al(OH)_3$
Ilmenite	$FeTiO_3$	Nordstrandite	$Al(OH)_3$
Maghemite	γ-Fe_2O_3	Bayerite	$Al(OH)_3$
Magnetite	Fe_3O_4		
Ferrihydrite	$Fe_5HO_8 \cdot 4H_2O$		
	Manganese		Titanium
Pyrolusite	MnO_2	Rutile	TiO_2
Birnessite		Brookite	TiO_2
Lithiophorite		*Anatase*	TiO_2
Hollandite			
Todorokite			
	Silicon		
Quartz	SiO_2		
Cristobalite	SiO_2		
Tridymite	SiO_2		
Opaline silica	SiO_2		

[a] Greek letters are used to denote different polymorphs of the same chemical constitution. The γ-forms have structures based on cubic close packing of anions whereas the α-forms are based on hexagonally close-packed anions.

cations and silica are rapidly removed from the site of weathering by strong leaching, as in some young, acidic volcanic ash deposits in humid areas.

The metal cations are in octahedral coordination in all the iron and aluminium oxides and hydroxides listed in Table 2.15 except on magnetite and maghemite, in which some of the iron is tetrahedrally coordinated. Figure 2.40 summarises schematically the arrangement of octahedra in goethite, lepidocrocite, hematite, gibbsite and also for comparison the magnesium hydroxide brucite, $Mg(OH)_2$.

Goethite, α-FeOOH

Goethite is the commonest iron oxide in soils and is found in temperate, sub-tropical and tropical regions. It is isostructural with diaspore (AlOOH), groutite (MnOOH), and montroseite (VOOH), and so isomorphous replacement of some of the Fe by Al, Mn and V may be expected. Synthetic goethites have been prepared in which as much as 30 per cent of the Fe is replaced by Al. Replacement of Fe^{3+} by the smaller Al^{3+} ion decreases the size of the unit cell and alters the X-ray powder pattern. Natural goethites in some iron ores

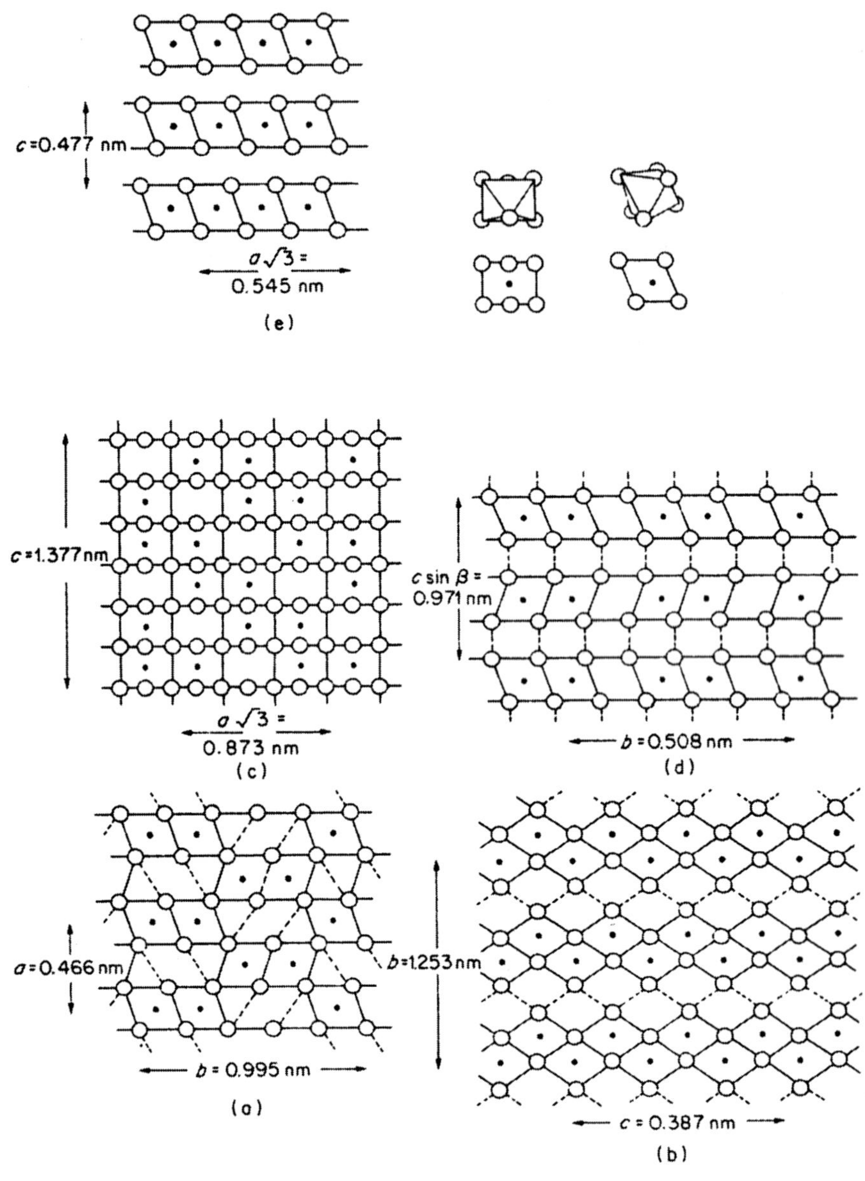

Figure 2.40 Schematic diagram illustrating packing of octahedra formed by six-coordinated oxygens or hydroxyls (O) surrounding Fe, Al or Mg ions (•) in (a) goethite, α-FeOOH, (b) lepidocrocite, γ-FeOOH, (c) hematite, α-Fe_2O_3, (d) gibbsite, γ-$Al(OH)_3$ and (e) brucite $Mg(OH)_2$. In these figures all the octahedra are shown as regular, and of the same size; in actual structures the octahedra are irregular and the size of the octahedra differs in different minerals. Hydrogen bonds are depicted by broken lines

have also been found that were shown by chemical analysis to have up to about 10 per cent replacement of Fe by Al and decreased cell dimensions. Goethite in soils is usually fine-grained and almost invariably occurs intimately mixed with major amounts of other minerals and cannot be separated for chemical analysis. Determination of the amounts of Fe and Al in extracts of soil materials containing much goethite suggests that in many soil goethites, 10 to 20 per cent of the Fe is substituted by Al. Differences in the X-ray powder patterns related to the Al content confirm that the Al is in the goethite structure.

The structure of goethite is shown in Figure 2.40a and in more detail in 2.41. The oxygens lie in close-packed planes perpendicular to the a-axis, the length of which is twice the distance between the oxygen planes. Iron atoms occur in distorted octahedra which, by sharing edges, form strips two octahedra wide which run parallel to the c-axis. These strips of octahedra occupied by Fe alternate with similar strips of unoccupied octahedra. The hydrogens, which are approximately tetrahedrally coordinated, have been located by neutron diffraction; their positions are shown in Figure 2.41 where it is seen that they link oxygens by hydrogen bonds across the 'vacant octahedra'.

The predominant surface of goethite crystals is corrugated with grooves and ridges running parallel to the *c*-axis as seen in the upper part of the diagram, Figure 2.41. Exposed on the ridges are three kinds of oxygens linked to one, two and three iron atoms in the crystal and in the grooves one kind of oxygen linked to three atoms. The surface structure provides important anion adsorption sites, discussed fully in Chapter 10.

Lepidocrocite, γ-FeOOH

Lepidocrocite is a common minor constituent of soil clays in humid temperate regions. It is found in non-calcareous seasonally waterlogged soils in which oxidizing and reducing conditions alternate and is frequently seen as strong orange coloured mottles. Lepidocrocite is rare in soils of tropical regions in which maghemite, γ-Fe_2O_3, appears to take its place. Both lepidocrocite and maghemite appear to form via an intermediate unstable mixed ferrous ferric hydroxide ('green rust') produced by oxidation of ferrous iron, provided the silica concentration is small (Schwertmann and Thalmann, 1976).

Laboratory experiments have shown that many factors influence the product formed by oxidation of the 'green rust' intermediate. More rapid oxidation, lower pH, smaller total Fe concentrations, lower temperature and absence of ferric iron have been shown to favour the formation of lepidocrocite, but it is not clear why lepidocrocite should be limited to humid temperate regions.

Although the corresponding Al mineral, boehmite, γ-AlOOH, is isostructural with lepidocrocite, substitution of Al for Fe has not been reported.

Figures 2.40b and 2.42 clearly show that lepidocrocite is a layer structure. The layers, which are perpendicular to the b-axis, consist of pairs of condensed ribbons of octahedra, that run parallel to the c-axis and are linked by sharing

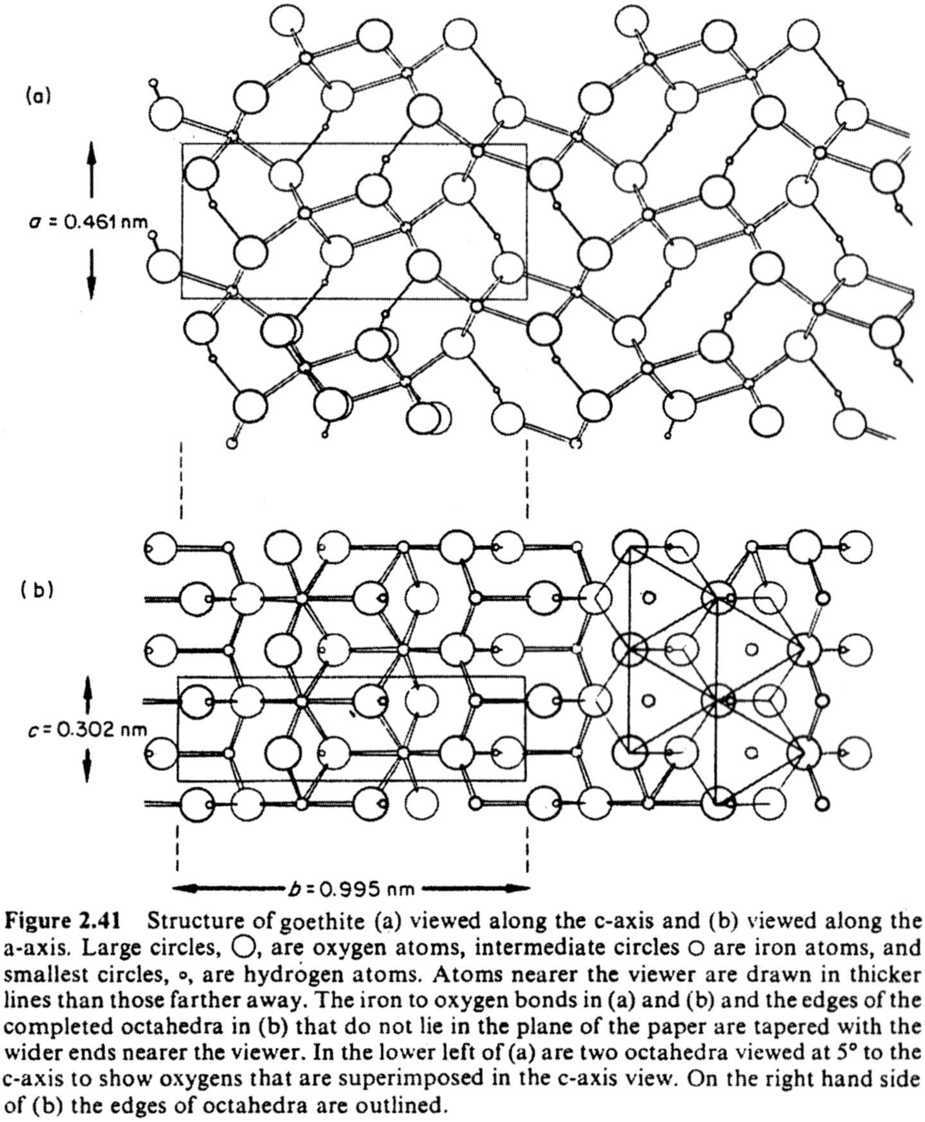

Figure 2.41 Structure of goethite (a) viewed along the c-axis and (b) viewed along the a-axis. Large circles, ○, are oxygen atoms, intermediate circles ○ are iron atoms, and smallest circles, ∘, are hydrogen atoms. Atoms nearer the viewer are drawn in thicker lines than those farther away. The iron to oxygen bonds in (a) and (b) and the edges of the completed octahedra in (b) that do not lie in the plane of the paper are tapered with the wider ends nearer the viewer. In the lower left of (a) are two octahedra viewed at 5° to the c-axis to show oxygens that are superimposed in the c-axis view. On the right hand side of (b) the edges of octahedra are outlined.

opposite corners. The chains are linked to other parallel chains by sharing the edges of octahedra, thus forming infinite two-dimensional sheets one octahedron thick. The layers are composed of two such sheets of octahedra that share oxygens so that each oxygen common to the two sheets is shared by four octahedra (Figure 2.42). The octahedra are distorted, but because the structure was determined in 1938 and has not been redetermined, accurate atomic coordinates are not available. Although the positions of the hydrogens have not been established it is almost certain that they form zig-zag strings of hydrogen

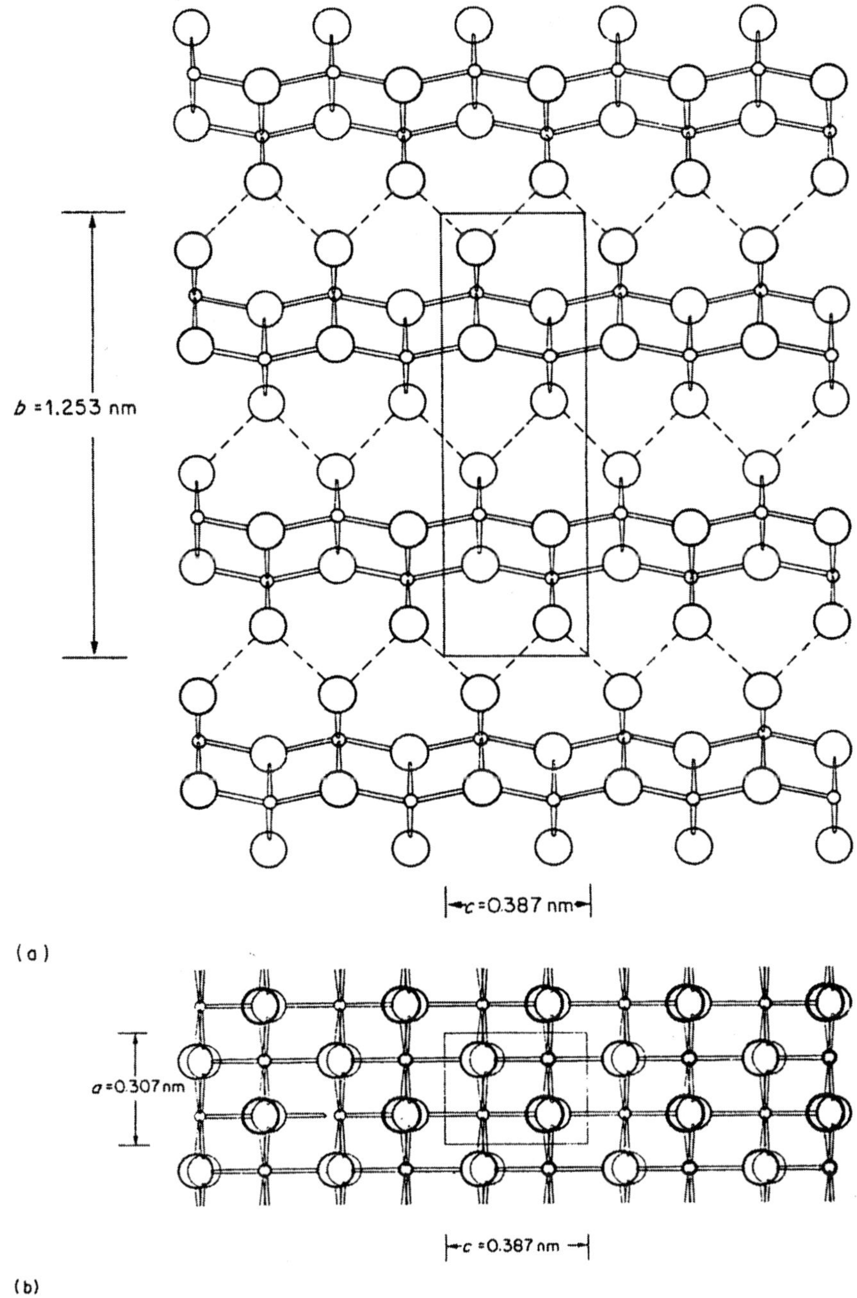

Figure 2.42 Structure of lepidocrocite (a) viewed along the a-axis showing the layers; the upper layer in the unit cell, outlined in (a), is shown in (b) viewed at 5° to the b-axis. Drawing conventions and symbols are as in Figure 2.41. Hydrogen bonds denoted by broken lines - - - -

bonds parallel to the c-axis between oxygens on opposed surfaces of the layers (Figure 2.42b, lower).

On heating at about 300 °C, lepidocrocite is transformed topotactically to γ-Fe_2O_3, maghemite.

Hematite, α-Fe_2O_3

Hematite is also a common mineral in soils. It is often inherited from parent materials but is also formed pedogenically in soils in warm regions. It occurs in Mediterranean soils, soils of warm deserts and in many strongly weathered tropical soils. There is evidence that hematite is very slowly replaced by goethite in cool humid regions.

Most analyses of hematite show that it is almost pure Fe_2O_3, but small amounts of TiO_2, Al_2O_3 and SiO_2 are often found. Much of the SiO_2 and some of the alumina probably arises from contamination with other minerals; titanium however can replace iron isomorphously in the hematite structure and the mineral ilmenite, $FeTiO_3$, is isostructural with hematite. The amount of titanium that is taken up depends on the temperature at which the hematite is formed; about 5 % can enter at 800 °C and less at lower temperatures. Fe^{2+} can also replace Fe^{3+} in small amounts but less than 1 % substitution occurs at 1000 °C. Similarly, hematite formed at 100 °C can contain up to 10 % by weight of Al_2O_3. It is clear therefore that although hematite can contain appreciable amounts of other elements, isomorphous substitution will occur only to a small extent in hematites formed in soils.

The structure of hematite is illustrated in Figure 2.40c and 2.43. It consists of planes of hexagonally close packed oxygens perpendicular to the c-axis. Iron atoms are sandwiched between every pair of oxygen planes and occupy two thirds of the octahedral sites. Neighbouring sheets have a common plane of oxygens and are linked by octahedra sharing faces. Iron atoms repel each other across shared faces and lie in two planes between every pair of oxygen sheets. The resulting octahedra are distorted because the Fe atoms do not lie in the middle of their coordinated octahedra of oxygens.

Ilmenite, $FeTiO_3$

Ilmenite is not a common mineral in soils; when it is found it is usually inherited from igneous or metamorphic parent rocks. Although the positions of the oxygens and the cations in ilmenite are almost identical to those in hematite, the iron in ilmenite is almost entirely divalent and the titanium quadrivalent. Although considerable substitution of Mg^{2+} and Mn^{2+} for Fe^{2+} can occur, normally only small amounts of these elements are found.

The atomic coordinates of the oxygens and metal ions in ilmenite are almost identical to those of hematite. The iron and titanium atoms are distributed in an ordered way on the cation sites. Between one pair of planes of oxygens all the

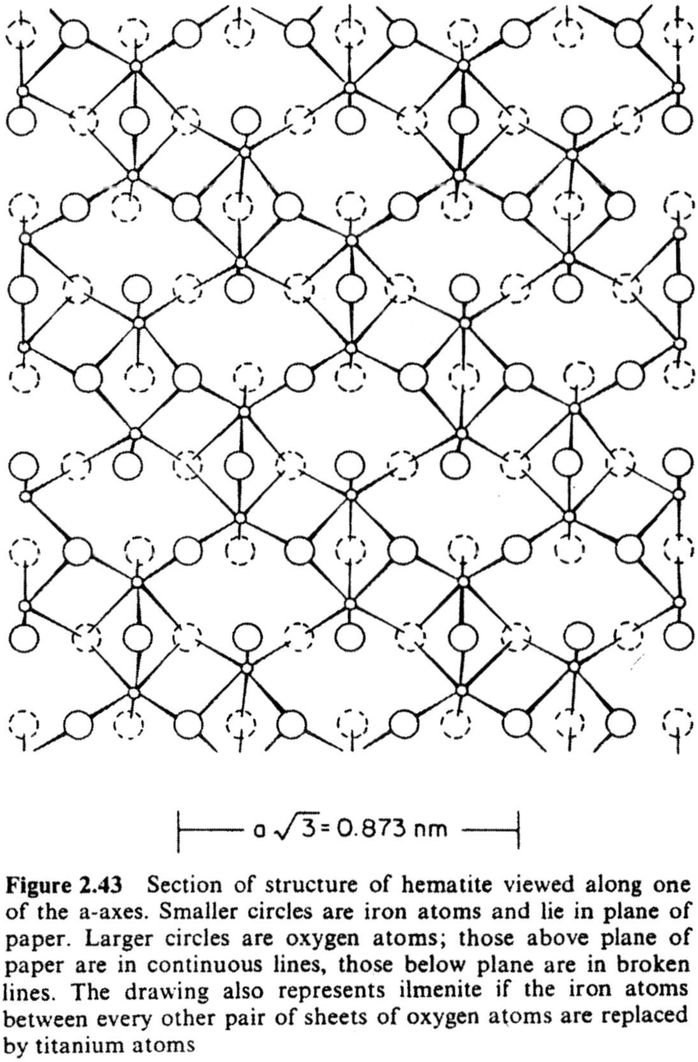

Figure 2.43 Section of structure of hematite viewed along one of the a-axes. Smaller circles are iron atoms and lie in plane of paper. Larger circles are oxygen atoms; those above plane of paper are in continuous lines, those below plane are in broken lines. The drawing also represents ilmenite if the iron atoms between every other pair of sheets of oxygen atoms are replaced by titanium atoms

cations are Fe^{2+} and between the next pair all the cations are Ti^{4+} and thus an alternating sequence is followed through the structure. Figure 2.43 therefore serves to illustrate the structures of hematite and ilmenite if it is remembered that in the latter Fe^{2+} and Ti^{4+} occur between alternate sheets of oxygens.

Ferrihydrite, $Fe_2O_3 \cdot nH_2O$

When ferric solutions are rapidly hydrolysed in the laboratory a brown colloidal precipitate often referred to as 'hydrous ferric oxide', or 'brown amorphous ferric hydroxide' is formed. These materials are poorly crystalline and of ill-defined composition. Some of them give X-ray powder diffraction patterns consisting of a few broad reflections indicating there is some degree of ordering in their atomic arrangement. Natural materials giving somewhat similar diffraction effects have been found and named ferrihydrite (Chukrov *et al.*, 1973). Most of them appear to be products of oxidation of ferrous iron catalytically accelerated by iron bacteria. Because they are so poorly crystalline it is not possible to determine their crystal structures. However, consideration of the few broad reflections shown in their diffraction patterns suggest that they consist of planes of oxygens hexagonally close-packed as in hematite but so frequently interrupted by stacking faults that no three-dimensional regularity exists. Within the structured regions sufficient Fe^{3+} ions to hold the oxygen sheets together occupy some but not all of the octahedral sites. The deficiency in positive charge caused by deficiency of Fe^{3+} is made up by additional protons associated with some of the oxygens. Absence of infrared absorption bands characteristic of OH groups leads to the conclusion that most of the protons are in the form of H_2O groups. The most commonly quoted composition is $5Fe_2O_3 \cdot 9H_2O$ but clearly the composition could show wide variations between $Fe_2O_3 \cdot 3H_2O$ and Fe_2O_3 depending on the Fe:O ratio.

Ferrihydrite is considered to be unstable and on ageing is said to transform to hematite in warmer areas or to goethite in humid temperate regions. The presence of finely divided silica or silica in solution appears to stabilize the ferrihydrite structure and some of the silica found by analysis may represent silica closely associated with the iron oxide structure.

It is probable that ferrihydrite-like material is more widespread in soils than has been reported. Its indefinite composition and poor crystallinity would make its recognition difficult when it occurs, as iron oxides usually do, as a minor component of soil clay. It has been identified as the major component of the 'iron ochre' deposits found in drains and ditches where it undoubtedly originates by the oxidation of ferrous iron by bacteria.

Maghemite, γ-Fe_2O_3 *and Magnetite*, Fe_3O_4

The remaining iron oxides that are found to any extent in soils are maghemite and magnetite, both of which are magnetic. They have very similar crystal structures. In both the anion framework consists of cubically close packed oxygens. Although some maghemites produced synthetically have a tetragonal unit cell arising from a particular ordered arrangement of the iron atoms on the available sites, soil maghemites and magnetite have a cubic unit cell containing 32 oxygens. The cell content of ideal magnetite, Fe_3O_4, is $Fe_8^{2+}Fe_{16}^{3+}O_{32}$; $9Fe^{3+}$ ions are in tetrahedral coordination and $8Fe^{3+} + 8Fe^{2+}$ occupy octahedral sites. In end-member maghemite, γ-Fe_2O_3, there are 64/3 Fe^{3+} ions distributed randomly over the 8 tetrahedral and 16 octahedral sites. Materials with amounts of Fe^{2+} intermediate between Fe_2O_3 and Fe_3O_4 are common. The composition of the Fe_2O_3–Fe_3O_4 series can be represented by $Fe_{2+n}O_{3+n}$ where $n = 2p/(1-p)$ when $p = Fe^{2+}/(Fe^{2+} + Fe^{3+})$; for the ideal end-member magnetite, $n=1$ and for the maghemite end-member, $n=0$. Natural minerals have n between 0 and 1. All the pedogenic minerals in this group that have been studied, 'soil maghemites', contain some Fe^{2+} with n between 0.1 and 0.5 while in natural magnetites from magmatic and metamorphic rocks n ranges from about 0.9 to 1.0 (Taylor and Schwertmann, 1974).

The magnetite–maghemite minerals have the well-known spinel (or rather inverse spinel) structure and can be represented by the formula XY_2O_4 in which X and Y are divalent and trivalent cations respectively. A wide variety of divalent cations, Mg, Fe^{2+}, Mn^{2+}, Zn and Ni, and trivalent cations Al, Fe^{3+} and Cr^{3+} have been found to occupy the X and Y sites in spinels. Ti^{4+} can also occupy some Y sites accompanied by an equivalent amount of a divalent cation, usually Fe^{2+} or Mg^{2+}, to maintain charge balance. Minerals of the magnetite–maghemite group can therefore be expected to take up any of these elements if they are available when the crystals are formed.

On heating in air magnetite is first oxidized to maghemite which in turn is converted to hematite, the iron oxide stable at high temperatures, as the temperature is increased.

Maghemite is formed pedogenically especially in highly weathered soils of tropical and sub-tropical regions. It frequently occurs in the form of concretions, usually accompanied by hematite and sometimes by goethite.

Gibbsite, boehmite and other hydrous oxides of aluminium

In gibbsite, the Al^{3+} ions occupy two thirds of the available octahedral interstices between two close-packed planes of hydroxyls (Figure 2.40d). These layers are stacked so that the OH groups on opposing surfaces of the layers do not key into hollows between three hydroxyls on the opposite surface as they do in brucite ($Mg(OH)_2$) (Figure 2.40e) but are almost directly opposite hydroxyls of the adjacent layer. The layers are held together by hydrogen bonds between the opposed OH groups. Hydrogen bonds also link some neighbouring hydroxyls in the plane of the layers resulting in some distortion of the octahedra,

The structure of boehmite is very similar to that of lepidocrocite (Figure 2.42); the extent to which Fe^{3+} may replace Al has not been established.

Nordstrandite and bayerite are two natural polymorphs of gibbsite. Both are occasionally found in soils, usually accompanied by much gibbsite. Structurally they consist of octahedra in which Al is coordinated by hydroxyls. Diaspore, α-AlOOH, isostructural with goethite, and corundum, isostructural with hematite, are not often found in soils.

2.4.2 Manganese Oxides

Manganese oxides occur widely in soils usually as a minor constituent but they can have a considerable effect on some properties of the soil (McKenzie, 1972). Although manganese oxides exist with a continuous range of compositions between MnO and MnO_2, only the more oxidized forms have been found in soils. Manganese oxides are chemically complex; in soil minerals, manganese is mainly quadrivalent but there is substitution of Mn^{2+} and Mn^{3+} for Mn^{4+}, some of which is balanced by replacements of O^{2-} by OH^-. Frequently a small proportion of large cations, Na, K, Ba, Ca, forms an essential part of the structures. Manganese oxides in soil occur, often associated with much iron, mainly as dark brown to black segregations on the surfaces of particles, in cracks or veins, or in the form of nodules. The commonest minerals in soils are birnessite and lithiophorite; hollandite, todorokite and pyrolusite have also been found. Birnessite occurs in both acid and alkaline soils whereas lithiophorite seems to be favoured by neutral to acid conditions. The composition of a soil birnessite has been given as $(Na_{0.7}Ca_{0.3})Mn_7O_{14}\cdot 28H_2O$, but synthetic materials containing potassium and barium or free of large cations, have been prepared. The composition of soil birnessites is therefore likely to vary considerably from the above formula. The structure of birnessite has not been determined but, as in all the soil manganese oxides, the cation is almost certainly octahedrally coordinated. Lithiophorite, $(AlLi)MnO_2(OH)_2$, has a structure in which sheets of MnO_6 octahedra and $(Al_{2/3}Li_{1/3})(OH)_6$ octahedra alternate; some of the Mn^{4+} is replaced by manganese of lower valency. Hollandite, $Ba_2Mn_8O_{16}$, and the related mineral cryptomelane, $K_2Mn_8O_{16}$, have a structure consisting of double chains of MnO_6 octahedra linked into a three-dimensional framework which forms channels in which the large cations are located. Only one occurrence has been reported in soils; it is a chemically complex manganese oxide that can contain Mg, Ca, Ba, Na and K. Pyrolusite, MnO_2, has also been found once in soils (Taylor, McKenzie and Norrish, 1964; Taylor, 1968).

Manganese oxides are noted for their strong absorption capacity for Mo, Fe, Co, Ni, Cu and Zn; other elements such as Ba in hollandite, Al and Li in lithiophorite, Na and Ca in birnessite and K in cryptomelane, respectively, are essential components. It seems that most of the Co in soils is associated with manganese oxide minerals, probably incorporated in octahedral sites in the crystals. The concentration of Ni, Mo, Cr, V, Cu and Zn is also much higher in manganese oxides than in the remainder of the soil. The affinity of manganese oxides for Co is so marked that most of the native Co in soils is unavailable to plants and the presence of moderate amounts of Mn (>1000 ppm) renders normal ameliorative additions of Co unavailable to plants (Adams, Honeysett and Norrish, 1969).

2.4.3 Titanium Oxides

In addition to ilmenite, $FeTiO_3$ (see page 140), titanium occurs in nature as the dioxide, TiO_2, in three polymorphic forms, rutile, brookite and anatase. In all, the titanium is octahedrally coordinated to six oxygens and each oxygen is surrounded by three titanium ions. Rutile, the most common mineral, is a high-temperature form and occurs widely in igneous rocks and as an accessory mineral in metamorphic rocks. It is also common as a detrital mineral in the sand fraction of sediments and soils. Though less common than rutile, the occurrence of brookite is similar and in soils it is found only as a detrital mineral. Anatase is the low temperature form of TiO_2 and usually occurs as an alteration product of other titanium-bearing minerals such as sphene and ilmenite. It is a common minor constituent of soil clays particularly those rich in kaolin minerals, and is probably formed by weathering in the course of soil formation.

Although all these minerals are essentially TiO_2 some less common elements can occur replacing Ti^{4+}. Tantalum, Ta^{5+}, and niobium, Nb^{5+}, have ionic radii very similar to Ti^{4+} and quite large amounts (10–20%) of Ta_2O_5 and Nb_2O_5 can occur in rutile. Tin, Sn^{4+}, and vanadium, V^{3+}, are of similar size and it appears that they can be incorporated in rutile crystals although in smaller amounts. Few analyses of brookite are available but usually 1 to 2 per cent of Fe_2O_3 is reported. Analyses of anatase show similar amounts of Fe_2O_3.

2.4.4 Silicon Oxides

Silica occurs in soils in non-crystalline and crystalline forms. Of the several crystalline forms that exist, only quartz and a partially-disordered low temperature form with structural affinities to the high temperature forms tridymite and cristobalite, occur to any extent in soils, although cristobalite is sometimes found in soils derived from volcanic material. The non-crystalline form, commonly referred to as opaline silica, is probably largely of biological origin (Plate 2.5). Opaline silica is formed in many grasses and in the leaves of some deciduous trees from silica taken up in solution from the soil. When the plant dies and decays the silica is left in the upper layers of the soil. Particles of opaline silica derived from the skeletons of sponges and diatoms are also found in some soils and are usually recognizable as such from their distinctive shapes. Although most of the non-crystalline silica, especially in surface soils, is probably of direct biological origin, there is some evidence that non-crystalline silica of inorganic origin, either alone, or associated with iron and aluminium also occurs in many soils (Mitchell, Farmer and McHardy, 1964).

Some but not all, silica of biological origin remains in the non-crystalline state in soils. A form of silica that gives X-ray powder diffraction patterns that resemble those of the high temperature crystalline forms, tridymite and cristobalite, is often found in association with biogenic opaline silica. The X-ray patterns have been interpreted as arising from structures in which elements of the structures of tridymite and cristobalite are stacked in a partially disordered manner. Some of this material, which is referred to as low temperature disordered tridymite–cristobalite or opal–CT, is clearly of biological origin, as it maintains the characteristic shape of the organism which deposited the silica, e.g. diatoms or sponge spicules (Plate 2.5). It seems to develop at low temperatures from the non-crystalline opaline form and there is some evidence for its eventual transformation to quartz.

Quartz, one of the commonest minerals in rocks, is practically 100% SiO_2, only very small amounts of other elements being found by chemical analysis. It is almost ubiquitous in soils and is often the most abundant mineral. Much of the quartz occurs as individual crystals of sand and silt size derived from the parent rocks. Quartz in microcrystalline form also occurs in the form of nodules of chert and flint.

In common with other crystalline forms of silica, quartz is a framework structure. Each silicon atom is linked to four oxygens to form SiO_4 tetrahedra in which the oxygens form the vertices. Each tetrahedron is linked by sharing the vertices to four surrounding tetrahedra via Si—O—Si links; each oxygen is shared by two Si. This results in a compact chemically inactive structure which is strongly bonded equally in three dimensions; the bonding forms a three-dimensional framework. Because every oxygen is shared between two silicons and each silicon is linked to four oxygens, the composition is $Si + 4(\frac{1}{2}O) = SiO_2$.

In addition to quartz– and opal–CT, silica may occur in two other crystalline forms in soils, tridymite and cristobalite. Classically they are high temperature forms but they appear to have the ability to form and persist at temperatures and pressures outside their normal stability equilibrium fields. Their crystal structures consist of different arrangements of SiO_4 tetrahedra linked by sharing all corners; the structures of both are more open than that of quartz. Both are typically minerals of volcanic rocks and they may occur in soils derived from volcanic rocks and in soils to which substantial amounts of volcanic debris have been added.

2.5 OTHER MINERALS

In addition to layer silicates and oxides many other minerals are often found in the clay fraction of soils. Occasionally they are the main component but usually they are present in minor proportions only. Nevertheless, their occurrence often provides valuable information about the history and development of soils and they can modify or control some soil properties. A brief account of some of the more common of these minerals is given in this section.

2.5.1 Feldspars and Zeolites

Feldspars and zeolites are two important mineral groups with framework structures that occur in soils. In them, the tetrahedra are, as in quartz, linked in three dimensions by sharing all their vertices with adjacent tetrahedra. In both groups some of the tetrahedra contain Si^{4+}, while others contain Al^{3+}. The aluminosilicate framework defined by the linked tetrahedra therefore has the composition $(Si_{1-x}Al_x)O_2^{x-}$. The negative charge on the framework is balanced by large alkali or alkaline earth cations that occupy cavities in the framework.

Feldspars (Smith, 1974)

Feldspars have the general formula A $Al_{2-y}Si_{2+y}O_8$, where A represents the large cations (usually Na^+, K^+ and Ca^{2+}) and y lies between 1 and 2. The principal end members are $KAlSi_3O_8$ (sanidine, orthoclase and microcline), $NaAlSi_3O_8$ (high and low albite) and $CaAl_2Si_2O_8$ (anorthite) and most natural feldspars can be classified chemically within this ternary system. Alkali feldspars have compositions between $KAlSi_3O_8$ and $NaAlSi_3O_8$ and plagioclases lie between $NaAlSi_3O_8$ and $CaAl_2Si_2O_8$.

The framework formed by the three-dimensionally linked tetrahedra is very similar in all feldspars. The essential structural feature consists of strings of linked tetrahedra shaped like a crankshaft. The strings are cross-linked in pairs to produce chains in which four-membered rings of tetrahedra are alternately horizontal and vertical. The chains of four-membered rings are cross-linked to other similar chain by sharing oxygens in a way that forms elliptical rings of eight tetrahedra. The large cations are completely enclosed in the cavities within the framework (Smith, 1974; Bragg and Claringbull, 1965).

Zeolites are also framework structures. They display a diversity of structural schemes. The tetrahedra are again linked by sharing all their vertices but there are a wide variety of ways in which they are arranged to produce the framework. They may be linked into four-, six-, eight- or twelve-membered rings which are joined together in a less compact way than in the feldspars. The open framework of the zeolites leaves channels of different sizes that run in different directions from the surfaces through the crystal. The channels are not of uniform cross-section but consist of wider cavities interrupted by periodic constrictions which can conveniently be called 'windows'. The wider cavities contain loosely-held water molecules and/or the charge-balancing cations which can readily move to adjacent cavities if the 'windows' are large enough. The way the different kinds of channels are interconnected and the sizes of the 'windows' determines the ease with which cations and absorbed molecules can move into, through, and out of the crystals.

The channels are frequently interconnecting and the sizes of the windows are larger than the common cations, and so both water and charge-balancing cations can diffuse readily through the crystals of most zeolites. The water in the channels is easily removed by mild heating and can be replaced by other molecules in great variety but often in a selective way. The selectivity depends on the relative sizes of the molecules and the 'windows' they have to pass through to penetrate the channels. If the molecules are bigger than the windows they cannot penetrate the crystals. This property has led to zeolites being referred to and used as 'molecular sieves' (Barrer, 1964; Breck, 1974).

In a similar way the open commonly interconnected channels in zeolites often allow the ready replacement of one species of charge-balancing cation by another, i.e. cation exchange. Cation selectivity is more marked in zeolites than in clays because the specific stereochemical configuration of adsorption sites defined by the particular aluminosilicate framework is sometimes extremely favourable for cations of a particular size and charge.

By contrast, the feldspars do not display cation exchange properties in aqueous solution. In them the charge-balancing cations are tightly held in completely enclosed cavities within the crystal and cannot be exchanged under normal conditions using aqueous solutions. Cations in the feldspar can, however, be exchanged by reaction with molten salts (Orville, 1967; Viswanathan, 1971). For example, the sodium in plagioclase can be replaced by potassium producing an equivalent K-feldspar by treating a small amount of the feldspar with a large excess of potassium chloride at 850 °C for several hours.

2.5.2 Calcium Carbonate

Calcium carbonate is an abundant substance in nature and is known to occur in at least five polymorphic varieties. The two common forms are calcite and aragonite. Calcite is the main component of most limestones where it occurs as a chemical precipitate and in the form of fossil shells. It is the stable form of $CaCO_3$ and, although about equal numbers of organisms make their shells of calcite and aragonite, the latter eventually reverts to calcite, so that calcite is much more commonly found in soils. Precipitated calcite usually arises from solutions of calcium bicarbonate which decompose losing CO_2 as the temperature rises at or near the earth's surface. The extensive hard indurated soil horizons called calcretes formed in arid regions and other deposits of calcite in soils probably arise in this way. The calcium carbonate precipitated in lakes and in seas is usually aragonite.

The essential structural feature of carbonates is the CO_3^{2-} anion which consists of an equilateral triangle of oxygens around the very much smaller carbon atom. The structures adopted by many carbonate minerals are governed by the radius of the divalent cation. Cations smaller than calcium form structures of the calcite type and those larger than calcium have aragonite-type structures. Thus magnesium, iron, zinc, manganese and cadmium carbonates have the calcite structure and strontium, lead and barium carbonates have the aragonite structure. Although many divalent cations may partially replace Ca in calcite, most calcites are fairly close to pure $CaCO_3$ but Mg, Fe and Mn can occur in substantial amounts.

Calcite is the only form of calcium carbonate widespread in soils, where it frequently occurs as a constituent of the clay fraction. The presence or absence of calcite has a major influence on the pH of soils and therefore controls many of the chemical processes involved in soil formation and in the provision of nutrients for plants.

2.5.3 Minerals of the Alunite and Plumbogummite Groups

There is a wide range of structurally similar minerals that have the general formula $MY(XO_4)_2(OH)_6$ or $MY_3(H_{1/2}XO_4)_2(OH)_6$ in which M is a large cation, often an alkali or alkaline earth metal, Y is a trivalent cation usually Al^{3+} or Fe^{3+}, and X may be S, P or As. These minerals, which can be regarded as basic sulphates, phosphates, or arsenates of aluminium or iron, can be divided into three groups according to the nature of the XO_4 anions and are listed in Table 2.16. Minerals of the alunite and plumbogummite groups have been found in soils but beudantite minerals have not.

Minerals of the plumbogummite group have been found in many soils (Norrish, 1968). In some of the Pacific coral atolls, the soil consists almost entirely of the mineral crandallite, probably formed by reactions between guano and volcanic debris. In the majority of soils in which they occur the plumbogummite minerals comprise only a small fraction of the whole soil but they can contain a considerable proportion, up to 60% of the total phosphorus in the soil. They have been identified in a wide range of soil types, ranging from the less weathered soils of the temperate zone (such as rendzinas) to deeply weathered soils of the tropics, and in soils developed from many different kinds of parent rocks. Their widespread occurrences in different soils and their absence in the parent materials indicates that they are formed pedogenically. The composition of these plumbogummites is intermediate between the crandallite (Ca), goyazite (Sr), gorceixite (Ba) and florencite (rare earth) end members; some trace elements, notably Pb, Cu and Cr, also appear to be associated with them. Plumbogummite minerals are significant because they are extremely stable in soils so that the phosphorus and trace elements they contain are not available to plants. Norrish (1968) has suggested that they may represent the ultimate sink for phosphorus added to the soil.

Although the minerals of these three groups have similar crystal structures, there is not a continuous range of substitution in the Y or X position. Y is either entirely Al^{3+} or Fe^{3+}, X is either S, P or As in the alunite and plumbogummite group or S with either P or As, in 1:1 proportions, in the beudantite minerals. In contrast the M position is frequently occupied by more than one element. In jarosites there is a continuous range between the K and Na end members, and varieties deficient in large cations are also found in which it is probable that extra protons balance the charge. In the plumbogummite minerals found in soils several cations are often present in the M site. Most analyses show that Ca, Sr, Ba, Ce, rare earths, and sometimes Pb are present in appreciable quantities.

The minerals of these groups are generally clay-sized in soils and resemble the aluminosilicate phyllosilicates in that a wide range of elements can be accommodated within a single structural scheme. All the minerals in these three groups have similar crystal structures. The dominant structural feature is a sheet of linked octahedral groups to both sides of which tetrahedral groups are attached by an arrangement whereby the apical oxygens of three adjacent octahedra become the basal oxygens of a single tetrahedron. The large cations are sandwiched in cavities between the composite octahedral–tetrahedral layers. As shown in Figure 2.44a, the octahedra form rings of six and each ring is linked to

Table 2.16 Minerals of the alunite, plumbogummite and beudantite groups

Alunite group	
Alunite	$KAl_3(SO_4)_2(OH)_6$
Natroalunite	$NaAl_3(SO_4)_2(OH)_6$
Jarosite	$KFe_3(SO_4)_2(OH)_6$
Natrojarosite	$NaFe_3(SO_4)_2(OH)_6$
Argentojarosite	$AgFe_3(SO_4)_2(OH)_6$
Ammoniojarosite	$(NH_4)Fe_3(SO_4)_2(OH)_6$
Plumbojarosite	$PbFe_6(SO_4)_4(OH)_{12}$
Plumbogummite group	
Plumbogummite	$PbAl_3(H_{1/2}PO_4)_2(OH)_6$
Crandallite	$CaAl_3(H_{1/2}PO_4)_2(OH)_6$
Goyazite	$SrAl_3(H_{1/2}PO_4)_2(OH)_6$
Gorceixite	$BaAl_3(H_{1/2}PO_4)_2(OH)_6$
Florencite	$CeAl_3(PO_4)_2(OH)_6$
Dussertite	$BaFe_3(H_{1/2}AsO_4)_2(OH)_6$
Beudantite group	
Beudantite	$PbFe_3(AsO_4)(SO_4)(OH)_6$
Corkite	$PbFe_3(PO_4)(SO_4)(OH)_6$
Hinsdalite	$(PbSr)Al_3(PO_4)(SO_4)(OH)_6$
Svanbergite	$SrAl_3(PO_4)(SO_4)(OH)_6$
Woodhouseite	$CaAl_3(PO_4)(SO_4)(OH)_6$

six identical rings producing a sheet consisting of triads of octahedra. Every octahedron shares four corners, one with each of four neighbouring octahedra; the shared corners are OH groups. The octahedra, all of which are occupied by the trivalent Y cations, are tilted (Figure 2.44b) so that the apical oxygens of triads are brought closer together on one side of the sheet and moved further apart on the other side of the sheet forming smaller and larger triangles of oxygens. On both sides of the sheet small and large triangles alternate. The smaller triangles of apical oxygens form the bases of XO_4 tetrahedra, each tetrahedron being linked to three different octahedra. Each triad of octahedra is linked to one tetrahedron and alternate triads have tetrahedra on either the upper or lower surfaces. On the other side of the sheet opposite each tetrahedron, the tilting produces triangles of more widely separated oxygens. Successive octahedral–tetrahedral layers are displaced laterally in such a way that the opposing large triangles of oxygens form an approximately octahedral cavity in which the large cations are located midway between the layers. In addition to these six oxygens, six hydroxyls, three from each layer, are at about the same distance as the oxygens. Thus the interlayer cation is surrounded by twelve anions in two groups of six (Figure 2.44b).

As well as the electrostatic attraction between the interlayer cations and the octahedral–tetrahedral layer, a system of hydrogen bonds between the hydroxyls on the surface of one layer and the apical oxygens of the tetrahedra protruding

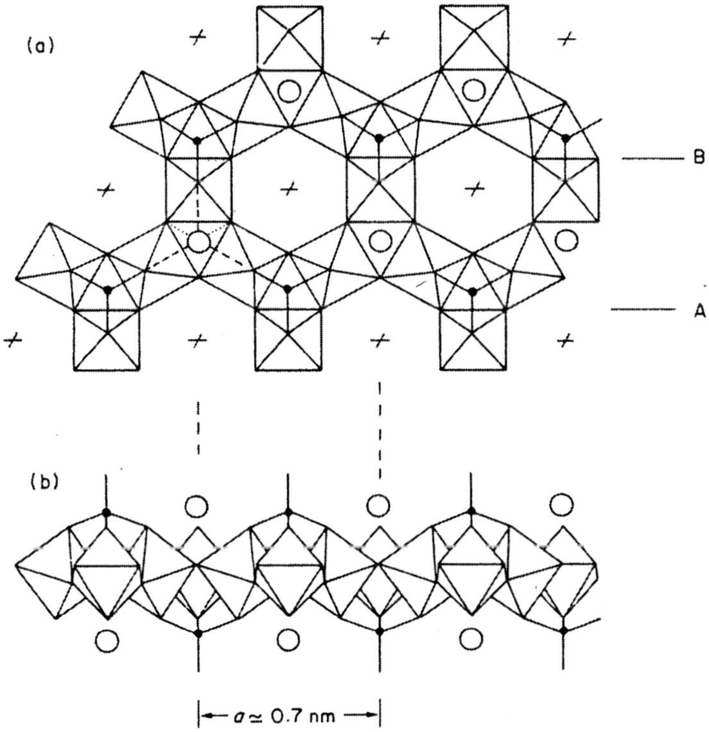

Figure 2.44 Structure of plumbogummite and jarosite minerals: (a) single layer viewed from above; tetrahedra and interlayer cations above the sheet of linked octahedra are shown; corners of base of hexagonal cell are marked, (b) section between A and B of (a) viewed from front; tetrahedra and interlayer cations on both sides of octahedral sheet are shown. This is a generalized structure; individual minerals differ in details of structure and dimensions of unit cell. The octahedral cations are not shown; octahedra are outlined. Tetrahedra are shown ⅄ in which • is tetrahedral ion. Anions are at the corners of the octahedra and at the end of the spokes of the tetrahedra. The equatorial anions of the octahedra are OH, apical anions of octahedra and apical anion of tetrahedra are oxygens. The large open circles are interlayer alkali or alkaline earth cations. There are three octahedral layers per unit cell with $c \approx 1.65$ nm. The a dimension is about 0.7 nm

from the adjacent layer holds the layers together, each apical oxygen being linked to three hydroxyl groups of the adjacent layer. In addition, in all the plumbogummite minerals except florencite, the divalent interlayer cation is insufficient to balance the negative charge on the layer. The extra proton required to balance the charge, represented by the $H_{1/2} \times 2$ in the $(H_{1/2}PO_4)$ group in the structural formula in Table 2.16, is thought to be shared between apical oxygens of tetrahedra of every other layer which point towards each other through the hexagonal holes in the intermediate layer.

Jarosite and plumbogummite minerals are formed pedogenically in certain soils in small but significant amounts. Jarosite minerals are found where sulphide-containing sediments are oxidized following uplift or drainage producing acid sulphate soils as, for example, when coastal mangrove swamps are drained. The sulphides are quite rapidly oxidized producing sulphuric acid; reaction of the sulphuric acid with other soil minerals releases Fe, K and Na in sufficient concentration to allow precipitation of the relatively insoluble jarosite minerals. Normally, although the jarosites found in soils contain both K and Na, they are deficient in alkalis. The jarosite minerals occur as pale yellow to pale buff mottles, or segregated as a pale powder in cracks, fissures, on surfaces and in old root channels. The presence of these pale yellow segregations is characteristic of acid sulphate conditions.

2.5.4 Apatites

Although they rarely occur naturally as soil minerals, members of the apatite group are the primary source of phosphate fertilizers. They are used in the manufacture of superphosphate and finely ground sedimentary apatite minerals can be applied directly as a slowly available source of phosphate.

Apatites are the commonest phosphorus-bearing minerals; they are represented by the isomorphous series with end-members: fluorapatite $Ca_{10}(PO_4)_6F_2$, chlorapatite $Ca_{10}(PO_4)_6Cl_2$, hydroxyapatite $Ca_{10}(PO_4)_6(OH)_2$, and carbonate–apatite $Ca_{10}((PO_4)_{6-x}(CO_3)_xF_x)F_2$. Fluorapatite is typically found in igneous and metamorphic rocks; chlor- and hydroxyapatite are both rare minerals. Sedimentary apatites, the commonest source of phosphate for fertilizer manufacturers and for direct application, are usually carbonate–apatites. McClellan and Lehr (1969) have shown that they can be conveniently represented by the structural formula, $(Ca_{(10-a-b)}Na_aMg_b)\cdot[(PO_4)_{(6-x)}\cdot(CO_3)_{(x)}\cdot F_{(y)}]F_2$. Neutrality is preserved when $a = x - y$. Ability of rock phosphates to provide phosphate to plants is related to the degree of replacement of PO_4^{3-} by CO_3^{2-}: solubility increases with increasing CO_3 content (Lehr and McClellan, 1972).

2.6 ALLOPHANE AND IMOGOLITE

The name allophane is derived from the Greek for 'to appear other'. It was first applied by Stromeyer and Hausmann (1816) to naturally occurring amorphous hydrous aluminosilicate materials that change appearance during drying from a glassy to an earthy lustre. Ross and Kerr (1934) described allophane as an amorphous material commonly associated with halloysite that has no structure and no definite chemical composition. They suggested that the name should be restricted to mutual solid solutions of silica, alumina, water and minor amounts of bases, but should include all such materials even though the proportions of these constituents might vary.

The name imogolite was first used by Yoshinaga and Aomine (1962) for a component of the clay fractions of soil (Imogo) derived from glassy volcanic ash. Imogolite differed from coexistent allophane in that it dispersed in acid rather than alkaline suspensions, gave several intense, but broad, X-ray diffraction reflections, a differential thermal analysis endothermic peak at 410–430 °C and possessed a thread-like habit. In these definitions imogolite is recognized as a crystalline or sub-crystalline material whereas allophane is defined as amorphous.

Allophanes, as defined, include natural materials of very diverse origins. They may be formed from basic igneous rocks by intense tropical weathering or from more acid rocks by podsolization, from physically comminuted minerals produced by glacial grinding and in cavities in limestones and other rocks from ground waters containing dissolved silica and alumina.

Much work on allophanes has been done on clay components of soils developed in weathered volcanic ash materials and other pyroclastic rocks. These soils, termed andepts and cryandepts in the U.S. classification (Soil Survey Staff, 1975) commonly have a low bulk density, $<0.85\ g\ cm^{-3}$ for moist soil, and much organic matter in the surface layer. They occur mostly in or near mountains containing active volcanoes, from the equator to high latitudes, and are known for phosphate deficiency in crops grown on them. Because of the economic importance of the soils and the resulting large amount of research into their mineralogy, much of the work reported here applies particularly to allophanes developed in weathered volcanic ash.

2.6.1 Occurrence and Morphology

Imogolite always occurs more or less intimately associated with allophane and it seems probable that some of the early reported occurrences of allophane contain imogolite. Wada and Matsubara (1968) have described the relationship between the two materials as they occur in weathered pumice (Figure 2.45). In this sample allophane occurs within weathered cores of pumice grains, whereas imogolite forms the major constituent of the gel material filling channels between the pumice and rock particles. Pockets of gibbsite form within the rock fragments. It thus appears that allophane and gibbsite and imogolite may coexist within one small part of a rock or soil, the allophane in a microenvironment found in close proximity to weathering pumice, the gibbsite in one associated with particular minerals in a rock fragment and imogolite in a third environment associated with drainage channels in the rock. Yoshinaga and Yamaguchi (1970) noted that in similar weathered pumice samples imogolite occurs where desilication has taken place, and halloysite lower in the profile where leaching is less pronounced, but that imogolite and halloysite never occur together.

Gel films of imogolite, separated by hand-picking, swell in water and become translucent. The mean refractive index of air-dry material is 1.47–1.49; films sometimes show birefringence in irregular areas and with very low colours.

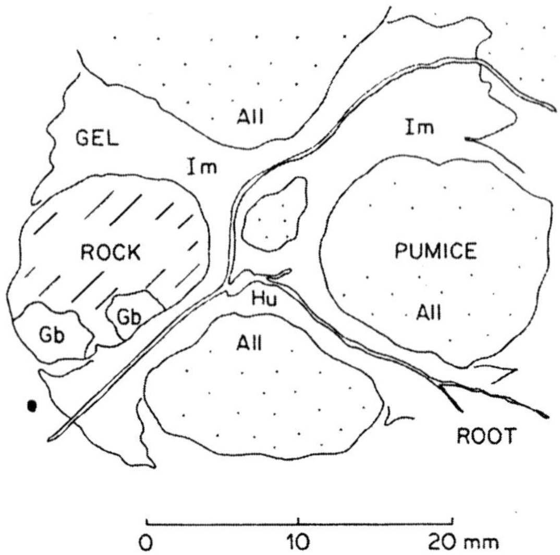

Figure 2.45 Sketch of a thin section of weathered pumice from the Kitakami pumice bed showing the spatial disposition of allophane, imogolite and gibbsite (after Wada and Matsubara, 1968). Abbreviations: All, allophane; Im, imogolite; Gb, gibbsite; Hu, humic materials. Reproduced by permission of the International Society of Soil Science

When viewed by scanning electron microscopy, imogolite in Kodonbaru pumice (Eswaran, 1972) occurs in subspherical masses of coiled tubes of diameter 2.5–3 μm and length at least 20–30 μm (Plate 2.6a). By contrast, allophane from Santa Cruz (Plate 2.6b) has the appearance of melted sugar with small protuberances, 0.1–0.5 μm across, covering the surface of irregular masses (Eswaran, Stoops, and de Paepe, 1973).

Fully dispersed imogolite may be seen by transmission electron microscopy to consist of very fine tubes with outside and inside diameters of 1.7–2.1 and 0.7–1.0 nm that are up to 1 μm long (Plate 2.7a). Incomplete dispersion may leave groups of nearly parallel fibres (Plate 2.7b) that give electron diffraction fibre patterns of a para-crystalline solid (Plate 2.7c). High resolution transmission electron microscopy of fully dispersed allophane (Plate 2.7d) shows it to be composed of sub-spherical particles of 2–5 nm diameter. Electron diffraction gives a broad ring or rings corresponding to an amorphous or very poorly crystalline solid.

2.6.2. Composition

The chemical compositions of eighteen allophane samples from weathered pumice are given by Yoshinaga (1966) and of four imogolite samples by Wada

Table 2.17 Selected chemical data for allophanes and imogolites (from Yoshinaga, 1966 and Wada and Yoshinaga, 1969). Reproduced by permission of the Society of the Science of Soil and Manure, Japan

Sample	Lab No	Oxides, weight % SiO_2	Al_2O_3	Other Oxides	Na_2O	H_2O^+	H_2O^-	Mol. proportions SiO_2/Al_2O_3	H_2O^+/Al_2O_3
Allophane									
1	1008	31.72	40.89	2.05	5.35	19.18	25.21	1.32	2.86
2	1010	34.32	40.94	2.14	5.08	17.65	22.16	1.42	2.44
3	1015	34.01	39.73	2.93	4.77	18.38	24.02	1.45	2.82
4	1033	37.16	38.41	3.17	5.69	15.64	23.03	1.64	2.31
5	1049	32.25	41.55	2.00	5.48	18.38	26.21	1.32	2.50
6	1057	37.90	36.18	2.58	6.09	16.54	20.12	1.78	2.59
7	1056	39.92	34.78	2.65	6.56.	15.62	19.08	1.95	2.55
Imogolite									
a	Ki-G(i)	31.10	48.22	—	0.97	19.74	—	1.09	2.31
b	KiG(ii)	30.90	48.77	—	0.57	21.96	—	1.11	2.65
c	Ka-G	29.12	45.19	—	1.82	23.86	—	1.09	2.98
d	Im-G	31.32	45.86	—	0.53	22.49	—	1.16	2.78
e	905-Ac	29.68	47.36	—	0.63	22.32	—	1.06	2.66

and Yoshinaga (1969). The allophanes were separated as fine clay (<0.1 or <0.2 μm) fractions from crushed whole pumice samples and the imogolites from hand-picked gel fragments. Both were treated with hydrogen peroxide to destroy organic matter and then with sodium dithionite–sodium citrate followed by sodium carbonate to remove free oxides. The allophanes were Na-saturated and both were washed with water and organic solvents before drying. These details are quoted because of the possibility that the pre-treatments needed to remove known impurities may have affected the separated product.

The imogolite compositions, recalculated as percent oxides, are given in Table 2.17 together with those of seven of the allophanes selected to show the range of variability of the original eighteen. Table 2.17 shows that the allophanes are less pure than the imogolites because they contain, in addition to silica, alumina, soda and water, 2–3% of oxides of iron, titanium, phosphorus, calcium, magnesium and potassium. They also contain significantly more soda, indicating that they retain more exchangeable cations (210–154 compared to 59–17 meq 100 g^{-1}) at the pH of the wash solutions used. Figure 2.46 shows a plot of the same data after converting them to molecular proportions by dividing by the appropriate molecular weights. For values of the $SiO_2:Al_2O_3$ ratio the imogolites are clustered about 1.1 whereas the allophanes spread evenly from 1.3 to almost 2. The $H_2O(+)/Al_2O_3$ ratio of both imogolites and allophanes varies from 2.3 to 3. Wada and Yoshinaga compared sample Ki-G with and without the Na_2CO_3 pretreatment and showed that there was an apparent change in $H_2O(+)$ content resulting from the method of preparation of the

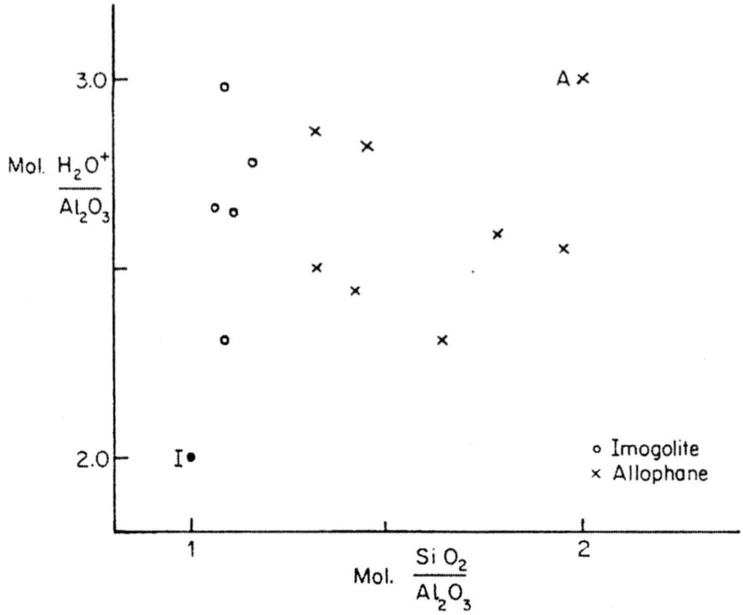

Figure 2.46 Plot of the chemical compositions of natural samples of imogolite and allophane (after Wada and Yoshinaga, 1969). I, composition of the unit formula of imogolite (Cradwick *et al.*, 1972); A, composition of end-member allophane (Wada, 1967). Reproduced by permission of the Mineralogical Society of America

samples. They concluded that the best empirical formula for imogolite is $1.1SiO_2 \cdot Al_2O_3 \cdot 2.3\text{–}2.8H_2O$. Wada (1967) had earlier proposed a formula of $2SiO_2 \cdot Al_2O_3 \cdot 3H_2O$ for a Si-rich end-member of an allophane series (point A on Figure 2.46).

Additional information on the status of water in allophane and imogolite may be obtained from infrared spectra and differential thermal analysis (DTA) traces (Figures 2.47 and 2.48). The infrared spectra show that adsorbed water (adsorbance at 1645 cm^{-1} from the angle deformation vibration of the water molecules) corresponding to the large DTA endothermic peak at 170 °C may be removed almost completely by evacuation at room temperature. Both allophane and imogolite contain structural OH, as shown by the OH-stretching absorbance at about 3500 cm^{-1}, but allophane starts to lose this when heated to only 150 °C whereas imogolite is stable to above 200 °C. The DTA trace of Kanumatsuchi allophane shows little evidence of loss of structural OH separate from that of adsorbed water, presumably because it is lost gradually over a range of temperatures, whereas the Imaichi imogolite shows a distinct endotherm at 415 °C corresponding to dehydroxylation.

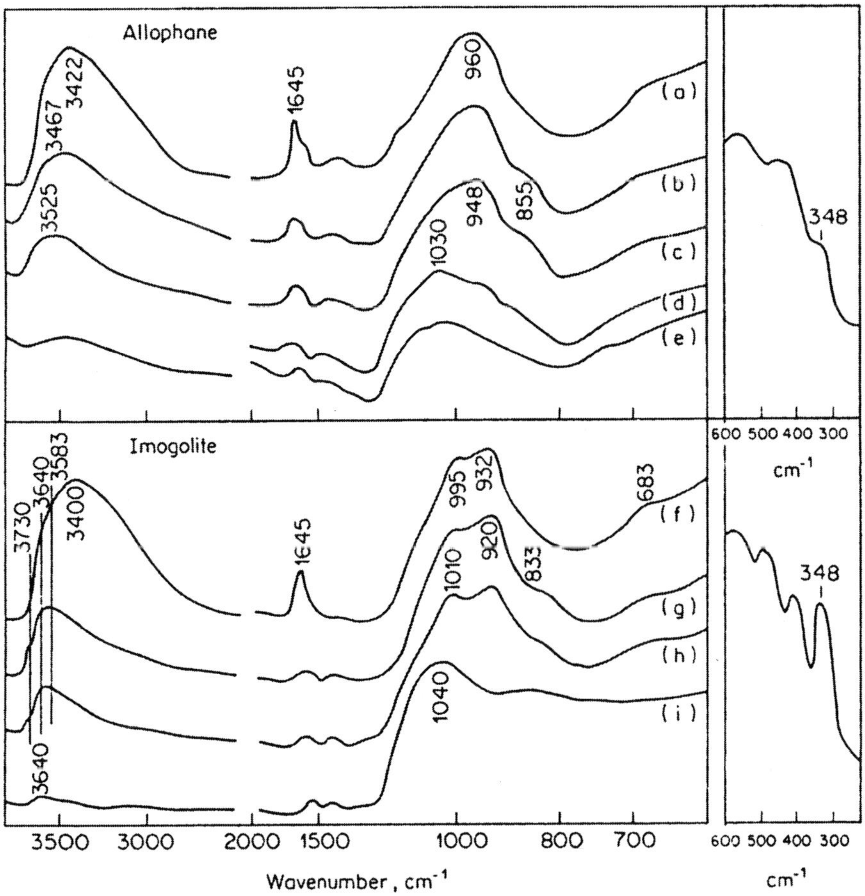

Figure 2.47 Infrared spectra of allophane and imogolite. Films normal to the incident radiation: (a) 20 °C; (b) 20 °C, 10^{-2} mmHg; (c) 150 °C, 10^{-2} mmHg; (d) 250 °C, 10^{-2} mmHg; (e) 350 °C, 10^{-2} mmHg; (f) 20 °C; (g) 20 °C, 10^{-2} mmHg; (h) 300 °C, 10^{-2} mmHg; (i) 475 °C, 10^{-2} mmHg. The weak OH vibration at 3640 cm^{-1} is due to a trace of micaceous impurity. The figure is of traces from Russell, McHardy and Fraser (1969) and Farmer Fraser and Tait (1977). Reproduced by permission of the Macaulay Institute for Soil Research

2.6.3 Structure

Allophanes give little evidence of structural order when investigated by either X-ray or electron diffraction techniques. Materials that give a few broad diffraction reflections have been described, however, and appear to be imogolite–allophane intermediates (Wada and Yoshinaga, 1969; Farmer, Fraser, Russell and Yoshinaga, 1977). These partially ordered materials—partially ordered by imogolite standards—have not yet been investigated. Perhaps when they are

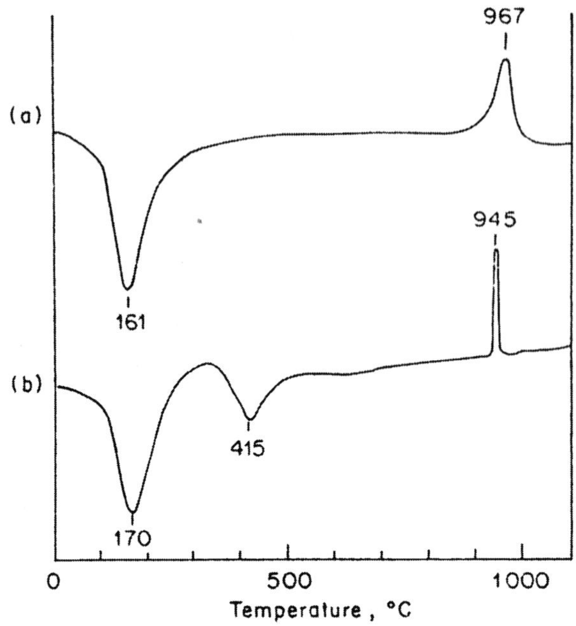

Figure 2.48 Differential thermal analysis traces of (a) Kanumatsuchi allophane (Udagawa *et al.*, 1969) and (b) Imaichi imogolite (Yoshinaga, 1968). (a) is reproduced by permission of KETER Publishing House Jerusalem Ltd, and (b) by permission of the Society of the Science of Soil and Manure, Japan

understood they will indicate how the fully disordered allophanes may be formed. For the present, however, structures put forward for allophane are speculative and largely based on the properties of synthetic aluminosilicate gels. For imogolite quite detailed diffraction evidence shows that it has strong sub-crystalline order. X-ray diffractometer traces (Figure 2.49) show four orders, at 0.14, 0.21 and 0.42 of an 0.84 nm spacing similar to the 0.86 nm cell spacing of gibbsite together, in the air-dry specimen, with a broad reflection for a 2.0 nm spacing reminiscent of certain interstratified clay minerals. On heating to 300 °C–i.e. the start of dehydroxylation—this reflection sharpens and becomes the first of four orders of a 1.7 nm spacing. Selected area diffraction of a group of nearly parallel tubes (Plate 2.7c) gives a fibre pattern in which the reflections of the 0.84 nm spacing run parallel to the tube axis and the broad 2.0 nm reflections normal to it. The latter thus appear to represent diffraction of separate tubes.

Imogolite. There have been several attempts to explain the structure of imogolite. Before it was realized that the threads were tubes Wada and Yoshinaga (1969) suggested that imogolite consists of ribbons of silica tetrahedra linked between

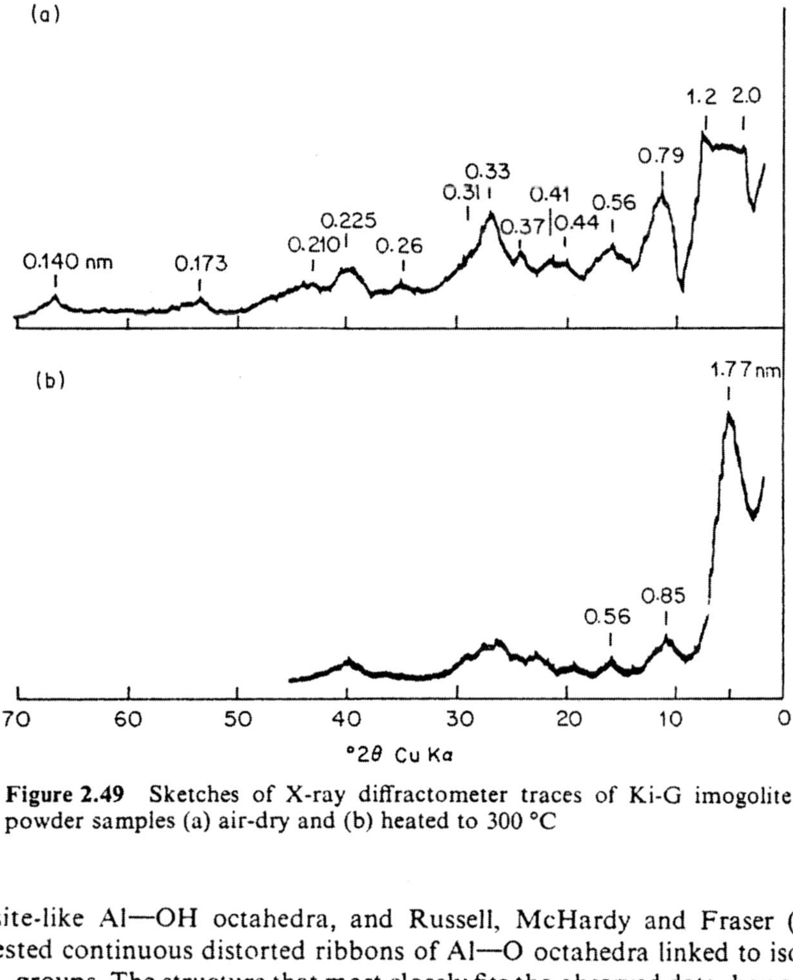

Figure 2.49 Sketches of X-ray diffractometer traces of Ki-G imogolite powder samples (a) air-dry and (b) heated to 300 °C

gibbsite-like Al—OH octahedra, and Russell, McHardy and Fraser (1969) suggested continuous distorted ribbons of Al—O octahedra linked to isolated Si_2O_7 groups. The structure that most closely fits the observed data, however, is one given by Cradwick *et al.* (1972). In arriving at their solution they first showed that the Si in imogolite is all or mainly in the form of orthosilicate—i.e. separated SiO_4—groups. They also knew that the threads are tubes and that the repeat distance along the fibre axis, 0.84 nm, is clore to the *a* cell dimension of gibbsite. They therefore considered paracrystalline cylindrical structures in which the OH groups of one side of a gibbsite octahedral sheet lose protons and bond to Si atoms, three such bonds spanning a vacant octahedral site, with the Si atoms out of the plane of the sheet. Because Si—O bonds are shorter than Al—O bonds the effect of introducing the Si atoms would be to bend the sheet around them in what in gibbsite would be the *b*-axis direction. The curvature could be made such that 10, 11, or 12 modified gibbsite unit cells complete a tube. The fourth Si bond would project radially towards the centre of the tube and be bonded to an OII group.

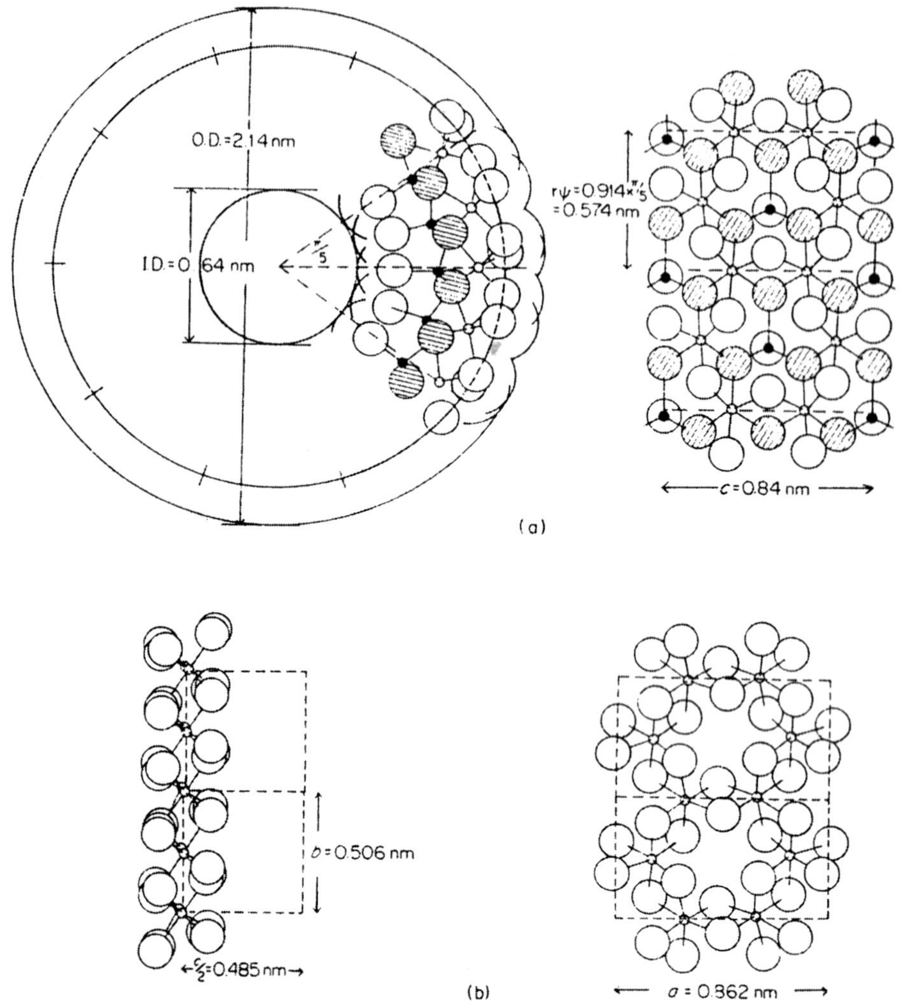

Figure 2.50 Structures of imogolite (Cradwick *et al.*, 1972, reproduced by permission of Nature and the Macaulay Institute for Soil Research) and gibbsite (Megaw, 1934). (a) Imogolite. Left, view down the tube axis showing atom positions in detail for two of the ten unit cells. The sizes of the circles illustrating atoms are arbitrary; the arcs of circles show the bond lengths of the inner and outer OH bonds, thus delimiting the inner and outer surfaces of the tube. Right, projection of the structure on a cylindrical surface through the centres of the outer OH groups, i.e. with a radius of 0.914 nm. Hatched circle, oxygen; large open circle, hydroxyl; small open circle, aluminium; small closed circle, silicon. (b) Gibbsite. One octahedral sheet from the gibbsite structure for comparison with the imogolite structure

This tubular structure of imogolite with ten unit cells forming the ring is shown in Figure 2.50 with part of the structure of gibbsite (Megaw, 1934) drawn to the same scale, for comparison. The imogolite structure has an empirical unit formula of $(HO)_3Al_2O_3SiOH$, with two formula units occurring in each unit cell. Written in this form it also represents the sequence of ions encountered in passing from the surface to the centre of the tube. The formula may also be written as $SiO_2 \cdot Al_2O_3 \cdot 2H_2O$ (Point I, Figure 2.46). In this form it can be seen to be very similar to Yoshinaga and Wada's empirical formula for natural imogolite samples.

In the gibbsite structure the repulsion of adjacent Al atoms and also hydrogen bonding between octahedral layers tends to distort the structure slightly so that Al and OH are not in exact planes. In imogolite the Al and Si show such displacement, but the O and OH ions do not. The view of imogolite on the right of Figure 2.50 is given by projecting atom centres along radii to a cylindrical surface through the centres of the outer OH ions. This surface may be visualized as having been cut along the z-direction and laid flat. The projection gives an accurate measure of the $r\psi$ side of the unit cell for the OH ions, but makes the dimensions appear the same for the other ions, whereas it is in fact 0.574 nm for the outer OH ions, 0.444 nm for the inner O atoms and progressively smaller as the centre of the tube is approached. It is the packing of the O and OH atoms in the octahedral sheet that determines the number of unit cells that can be accommodated in a tube of 2.1 nm diameter. With ten unit cells, as illustrated, the inner O atoms are as close-packed as in gibbsite, but the surface OH groups are much more open, particularly around cavities—0.33 or 0.37 nm apart compared with 0.29 nm. Although ten unit cells with a tube diameter of 2.1 nm gives the best fit with the diffraction data, Cradwick, *et al.* (1972) give reasons at the end of their paper why they would prefer ten cells in a tube of 2.4 nm diameter with a larger central hole to the tube. Such a structure would, however, require very long Al—OH bonds at the outer surface of the tube.

Synthesis of imogolite. Imogolite has been synthesized by Farmer, Fraser and Tait (1977). Optimum conditions are a dilute solution of SiO_2 monomer (0.08 %) and $AlCl_3$ (0.032 %). The pH is adjusted up to 5 with NaOH and then down to between 4.5 and 3.1 with acid, presumably to keep the Al ions in the six-coordinated form in contact with the monomeric silica. The solution is maintained just below boiling point for between one and five days, when the reaction is complete. The optimum product is said to be only slightly less well crystallized than the best natural imogolites. Variation of the solution conditions away from the optimum is said to produce materials with properties comparable to those found in many soils.

Allophane. As stated above, because allophane is amorphous or nearly so its structure cannot be verified directly from diffraction data. Cloos *et al.* (1969) used the properties of synthetic silica-alumina gels to suggest a model for the structure of allophane, involving tetrahedrally coordinated aluminium, as well as aluminium in octahedral coordination and silicon in tetrahedral coordination. Direct evidence that aluminium occurs in tetrahedral coordination in allophanes and amorphous aluminosilicate gels was obtained from X ray emission data by the method of White, McKinstry and Bates (1958). As when isomorphous replacement of Si by Al occurs in the tetrahedral sheets of micas, it may be expected to give rise to permanent negative charge in allophane. Thus the allophane dominated Sangara soil (Figure 4.4) has a negative charge of 12 meq 100 g^{-1} at pH 3 in 0.005 M CsCl. In addition many Si—OH and Al—OH groups will be accessible at the surface and give rise to pH and electrolyte concentration dependent charge (Chapter 4).

Under natural conditions imogolite and allophane can be formed inside soil crumbs within a millimeter of each other (Figure 2.45). Yet when they are synthesized the two minerals form in very different solution conditions. Allophane requires neutral or alkaline pHs to give maximum Al in fourfold coordination (Fripiat, 1965), whereas imogolite requires acid pHs to maintain Al in six-fold coordination and prevent the formation of an allophane core. Are similar solution conditions required in soils and, if so, can they occur? The formation of imogolite in open cavities and channels of the soil suggests that solution conditions there may be similar to those needed for synthesis. Drainage water, acidified with carbon dioxide and humus, can carry six-coordinated Al ions to react with dilute solutions of monomeric silica, H_4SiO_4. At grain contacts, however, solution conditions may be very different from those in free solution. Bases released from weathering feldspars and glass could both convert Al to fourfold coordination for inclusion in an allophane core and supply cations to satisfy the charges produced as the core forms. Thus it is not difficult to envisage processes by which both imogolite and allophane might be formed in the same soil, and also why natural samples can easily contain a mixture of both, together with any intermediate products that may occur.

2.6.4 Recognition and Estimation

Recognition and estimation of both allophane and imogolite is difficult, particularly if they are mixed with other clay minerals.

Imogolite may be recognized by its morphology (Plate 2.7b), diffraction patterns (Figure 2.49) and infrared spectrum (Figure 2.47). The latter, according to Farmer *et al.* (1977) has an absorbance at 348 cm^{-1} that in some instances can be used to estimate the amount present. The recognition of allophane is more difficult because, although the DTA trace is characteristic, the allophane may be missed when other clay minerals are present. Both allophane and imogolite contain Al(VI) ions that are highly reactive to fluoride solutions and this fact has been made the basis of qualitative and quantitative tests. Fieldes and Perrott (1966) give a simple rapid test for whole soils in which fluoride solution reacts with dry soil to release OH^- ions and the colour developed by an acid-base indicator gives a rough estimate of the number of OH groups released in the reaction. Methods of measuring this quantitatively using the rate of reaction have been given by Bracewell, Campbell and Mitchell (1970) and Perrott, Smith and Mitchell (1976). The reaction is not specific to allophane and imogolite, but a careful choice of pH and reaction time greatly reduces the reaction due to other phases.

Most quantitative estimations of allophane and imogolite in soils have been made by selective dissolution methods. That of Hashimoto and Jackson (1960), which employs heating the clay to disrupt the structure followed by boiling in 0.5N NaOH solution to dissolve and remove amorphous silica and alumina, has been the most widely used. It tends to overestimate allophane, however, because other clay minerals such as halloysite are attacked and partially dissolved. Kitagawa (1976) made a careful study of a number of these methods and concluded that the method of Segalen (1968), which employs alternate treatments with 8N HCl and warm 0.5N NaOH solutions, gives the most satisfactory method of estimating allophane in the clay fractions of soils.

Because of the dominant positive charge on imogolite at low pH (Figure 4.4) it disperses in acid solutions. This allows it to be separated from allophane and most other clay minerals (Yoshinaga and Aomine, 1962).

2.7 PARTICLE SIZE AND SHAPE

The size and shape of clay particles determines their ratio of surface area to volume and mass and thus directly to properties such as porosity, plasticity and adsorbency. Individual particles may be viewed by transmission or scanning electron microscopy (TEM or SEM) following dispersion, commonly in dilute aqueous soils, and subsequent drying; particles illustrated in Plates 2.8 to 2.12 were dispersed into soils diluted 1:1000–1:5000 w:w. Where particles have a recognizable habit direct viewing aids mineral recognition.

Plate 2.8 illustrates particles of kaolinite from the Blackpool Pit, St Austell, Cornwall, a nearly pure kaolin deposit in which the kaolinite was formed by hydrothermal alteration of feldspar. A sample was dispersed and fractionated by centrifugation and gravity settling to yield the following size fractions: >10 μm e.s.d. 21 %, 10–2 μm 55 %, 2–0.5 μm 16 % and <0.5 μm 8 %.

Plates 2.8a and 2.8b form a stereo pair of micrographs of the coarsest fraction, >10 μm, taken by SEM. The fraction consists of a mixture of platy and equidimensional particles, mostly kaolinite but with some mica. The largest particles are formed of stacks of thin plates joined by their plate surfaces.

Plate 2.8c illustrates particles of the 10–2 μm fraction. Most are platy, but some are equidimensional, including one of curved habit in the centre of the field that is similar in size and shape to many in the coarsest fraction. The maximum plate dimension of the particles varies from 15 to 2 μm.

Plates 2.8d and 2.8e illustrate particles of the 2–0.5 μm fraction. The micrograph in 2.8d was taken at an angle of 15° to the perpendicular in the SEM to show both the edges and plate surfaces of particles. Most have maximum plate dimensions of 2–0.3 μm, although a few much larger particles are present, but partially obscured, and edges have thicknesses of 0.2–0.1 μm. There is much overlap of platelets, probably because the suspension used was concentrated, but it is impossible to distinguish between platelets joined during crystal growth from those temporarily aggregated by drying. Plate 2.8e is a micrograph of the same size fraction taken by TEM after shadowing with platinum metal. A shadowing angle of 45° was used to make step and shadow the same length. The maximum plate dimension of the particles varies from 2.5 to 0.9 μm, average 1.5 μm, and their thicknesses from 0.5 to 0.025 μm, average 0.2 μm. The aspect ratio of plate breadth to thickness is thus 1.5:0.2 or 7.5:1. Variations in electron density across the particles show that even in this small particle size fraction most are composites of platelets of differing outline and thickness that are wholly or partially attached at their plate surfaces. The shadows thus represent the thickness of either the whole or part of a stack, depending on how the particle is lying. The particle on the right of the micrograph, thought to be single because of its even optical density, has dimensions of approximately $0.8 \times 0.6 \times 0.125$ μm. The platelet thus has a face area of approximately 1 μm^2 and an edge area of 0.04 μm^2, giving a ratio of face to edge of 25:1. The same particle also illustrates that a thickness of 0.125 μm (125 nm) is nearly the maximum that can be penetrated by electrons under the experimental conditions used.

Plate 2.8f and all subsequent plates are micrographs of unshadowed samples taken by TEM. Plate 2.8f shows particles of the finest fraction, <0.5 μm, of Blackpool kaolinite. The particles have maximum plate dimensions of 1.3–0.25 μm, average 0.7 μm, many are composite, and approximately one third too thick to be penetrated by the electron beam. Many platelets have straight edges subtending angles of 120°; they are thus hexagonal euhedral or subhedral in habit. This habit is very characteristic of kaolinite in both soils and sediments and may be used for tentative identification, although other soil minerals, such as gibbsite and jarosite, may occur as hexagonal platelets.

Plate 2.9 illustrates the two interstratified minerals that are described in Section 2.3.8. Plate 2.9a is of the regular beidellite–sodium–mica interstratification, rectorite. The one illustrated, from Allevard, France, is mineralogically similar to that from Baluchistan, Pakistan, but more photogenic. It consists of large plates, up to 10 μm by 3 μm, and long ribbons, up to 50 μm by 0.7 μm. The ribbons formed coils in suspension that have flattened on drying. The sample is not entirely salt-free and, as other preparations containing less salt contain uncoiled ribbons, it is thought that coiling may occur at certain salt concentrations. Although they show very good definition the plates and ribbons are very thin; Weir, Nixon and Woods (1962) concluded from measurements of shadowed micrographs that they are as thin as 1.8 nm, two 2:1 layers.

Plate 2.9b is of the irregular mica–smectite interstratification from the fine clay fraction, <0.04 μm, of Denchworth soil in weathered Oxford Clay. The particles are minute compared with those of the rectorite, the maximum platelet dimension ranging from 0.8–0.15 μm, average 0.4 μm. The thinnest soil particles show less contrast than the rectorite flakes and some so little that their outlines are difficult to distinguish. Close inspection shows that they are embedded in finely divided particulate material that covers most of the preparation. Material of this sort is commonly observed for soil clays—small amounts may be seen in Plate 2.12c—and for some montmorillonites, notably that from Camp Berteau, Morocco. It arises in part from the breakdown, or dispersion, of clay flakes in dilute slightly acid suspension, and its presence makes the observation and measurement of thickness of fine clay particles extremely difficult.

Plate 2.10 shows micrographs of clays from specially selected, monomineralic deposits. Plates 2.10a, 2.10b and 2.10c are of smectites that have been sized to the same range, <0.04 µm, as the Denchworth soil in Plate 2.9d. Plate 2.10b illustrates the smallest particles of hectorite, the trioctahedral lithium-containing smectite (see page 95) from Hector, California. They are lath-shaped, 2–0.2 µm long, 0.1–0.02 µm broad and very thin. A feature of hectorite, not shared with many smectites, is that when it is Na- or Li-saturated and dispersed the whole sample breaks into these tiny microaggregates, a fact which may account for its excellent gelling properties. Plate 2.10c is of montmorillonite flakes in Wyoming bentonite from Clay Spur, Wyoming. It consists of equidimensional, easily folded, very thin flakes, with a maximum platelet dimension of 1–0.1 µm, average 0.5 µm. The platelets have indented edges with certain commonly recurring angles that are similar to those on the rectorite flakes (Plate 2.9a); these probably represent the incipient development of crystal faces. This fraction of Wyoming montmorillonite has gelling properties similar to those of hectorite, but coarser fractions that consist of aggregates of platelets do not gel so well. Plate 2.10c illustrates the smallest particles of the montmorillonite from Woburn, England. The platelets are elongated with a long dimension of 0.25–0.05 µm, average 0.15 µm, and short dimension of 0.1–0.02 µm, average 0.05 µm; they are of variable thickness, with a proportion of quite thick particles. These particle dimensions are more similar to those of the two soil clays of Plates 2.9b and 2.12c than are those of either of the other two smectites. In general, the presence of very thin plate- or lath-shaped particles in a micrograph of a clay sample indicates the presence of an expanding mineral containing smectite layers. It is not possible to distinguish smectite from interstratified minerals by size and shape alone, however, and so the use of particle morphology as an aid to mineral identification is very limited for these two groups of minerals.

Plate 2.10c illustrates particles of chlorite from weathered pumice tuff, Snowdonia, Wales (Ball, 1966). The particles are platy with rounded outlines and are indistinguishable in habit from many micas or vermiculites.

Plate 2.10d is of particles of the 0.2–0.04 µm size fraction of World Vermiculite. The fraction was produced by lithium saturating a bulk sample of the vermiculite and using ultrasonic vibrations to disperse it. Thus some of the platelets are bounded by fracture surfaces produced by the ultrasonics rather than by natural flake edges. In finer fractions vermiculite occurs in very thin platelets that are difficult to distinguish from smectite; in coarser fractions, as in 2.10d it is in aggregates of platelets similar to those of mica or chlorite.

Plate 2.10f illustrates sepiolite from Vallecas, Spain. The smallest particles are laths of breadth 20 nm, and length 1–1.5 μm, although longer laths may occur in the large bundles.

Plate 2.10g illustrates palygorskite from Enderby Leicestershire, England. It consists of laths, 0.1–0.04 μm, broad and from 0.2–6 μm in length. Many of the laths are single, but some are joined edge to edge in small aggregates. Lath-shaped particles are a characteristic habit of both sepiolite and palygorskite and although the laths, particularly those of sepiolite, may not be as elongated as those illustrated here, the presence of laths in a micrograph should always lead to a search for confirmatory evidence for either of these two minerals.

Plate 2.11 illustrates clay minerals from tropical and equatorial soils, printed to the same magnification as those in Plates 2.10a, 2.10b and 2.10c. Plate 2.11a is of a soil from the Shortland Islands, a group near the Solomon Islands in the Pacific. It illustrates thread-like imogolite strands joining sub-spherical halloysite particles. The threads are very characteristic of imogolite and should be looked for in all acid soils, but particularly those developed in altered volcanic ash deposits.

Plate 2.11b illustrates allophane from Imaichi City, Japan. Most of the sample consists of tiny spheres about 5 nm in diameter that are joined in irregular aggregates, but includes a few imogolite threads, halloysite tubes and thin flat platelets that are probably layer silicate minerals.

Plate 2.11c is of clay containing much halloysite from Mexico. Most of the halloysite occurs as short rolled flakes or tubes, but some sub-spherical particles, similar to those in Plate 2.11a, are also present. The other constituent of the micrograph has flat flakes of irregular outline. This may be halloysite, as halloysite has been recognized in such a habit, but is more probably a 2:1 layer silicate mineral such as a smectite–chlorite interstratification, because such minerals commonly occur in soils containing halloysite. A tubular habit is common for halloysite and so the appearance of tubular particles in a micrograph should always lead to a search for confirmatory evidence. In addition to the tubular, sub-spherical and platy habits mentioned above, halloysite has been reported once as having elongated laths, similar to those of palygorskite. It is therefore clear that habit alone is unreliable for identifying halloysite, or indeed any other clay mineral.

Plate 2.11d illustrates the <2 μm clay fraction of a soil from Ikenne, S.W. Nigeria. It consists mainly of kaolinite and iron oxide. The kaolinite particles are platy, maximum plate dimension 0.5–0.05 μm, average 0.2 μm, hexagonal, euhedral to subhedral (cf. also Plate 4.3).

Plate 2.11e shows the <2 μm clay fraction from a soil on Santa Isobel in the Solomon Islands that consists of almost pure goethite. The smallest particles are irregular laths or needles 0.5–0.05 μm long and 5 nm broad aggregated into columnar-, fan- or spindle-shaped bundles. Particles of very similar morphology have been described for the mineral lepidocrocite by Schwertmann (1973).

Plate 2.11f is of the clay fraction of a soil from Rennell Island in the Shortland Islands. The clay is composed of gibbsite and iron oxide. The latter, which cannot be removed without destroying the gibbsite, obscures surface detail on many of the gibbsite particles. Those that can be seen bear a superficial resemblance to certain of the kaolinite particles of Plate 2.11d.

Plate 2.12 shows particles from the coarse, medium and fine clay fractions from the B horizon of a Swanmore series soil in weathered Reading Beds Clay from Sussex, England. The sample was pretreated by the method of Kittrick and Hope (1963) in which the iron oxide minerals, lepidocrocite and goethite, were dissolved before the separation of the clay fractions. The coarse clay fraction, Plate 2.12a, consists of quartz, feldspar, kaolinite, mica and a smectite-rich interstratified mineral. In the micrograph the equidimensional partly rounded grains are thought to be quartz or feldspar, the platy subhedral grains kaolinite and the platy particles of irregular outline mica or, if thin, smectite. The medium clay fraction, Plate 2.12b, contains less quartz and feldspar, but has otherwise a similar composition to the coarse fraction. The particles may thus be tentatively identified on the same basis. The fine clay fraction, Plate 2.12c, contains no detectable quartz or feldspar and small amounts of kaolinite and mica. Most of the fraction is composed of a complex interstratified expanding mineral containing a large proportion of smectite layers. The presence of kaolinite could be predicted from the subhedral platy particles in the micrograph, but the presence of mica could not, as the remainder of the sample consists of platy particles of irregular outline that cannot be sub-divided by size and shape alone.

To summarize, the shape of clay particles can be studied by electron microscopy of dispersed samples. Shape may be a valuable aid in identifying certain clay minerals, particularly kaolinite, halloysite, sepiolite, palygorskite and imogolite, but is of little value for mica, chlorite, vermiculite, smectite and interstratified minerals, which are so common in temperate soils. Particle size and shape may be investigated for most clay minerals by TEM of unshadowed specimens, measurement of the third dimension, the thickness, of platy particles being achieved by shadowing. However, measurements made by the latter method may lack precision if the particles being measured are composite, as in some kaolinites, or are very thin flakes surrounded by, and partly embedded in, very fine dispersed material, as happens with many smectites and the majority of fine clay fractions of soils.

REFERENCES

Adams, S. N., Honeysett, J. L. and Norrish, K. (1969), Factors controlling the increase of cobalt in plants following the addition of cobalt fertilizer, *Aust. J. Soil Research*, 7, 29–42.

Aleksandrova, V. A., Drits, V. A. and Sokolova, G. V. (1973), Crystal structure of ditrioctahedral chlorite, *Sov. Phys. Crystallogr.*, **18**, 50–53.

Alietti, A. (1959), Diffusione e significato dei minerali a strati misti delle serpentine mineralizzate a talche dell'Appenino Parmense, *Periodico Mineral.* (Roma), **28**, 65–110.

Allegra, G. (1961), Analysis of interstratified structures, *Nuovo Cim. Serie X*, **21**, 786; **22**, 661.

Allpress, J. G. and Sanders, J. V. (1973), The direct observation of the structure of real crystals by lattice imaging, *J. Appl. Crystallogr.*, **6**, 165–190.

Altschuler, Z. S., Dwornik, E. J. and Kramer, H. (1964), Genesis of kaolinite from montmorillonite by weathering: structural and morphological evidence of transformation, *Clays and Clay Min., 12th Nat. Conf.*, W. F. Bradley (ed.), Pergamon, Oxford, 1966.

Angell, B. K., Jones, J. P. E. and Hall, P. L. (1974), Electron spin resonance studies of doped synthetic kaolinite, *Clay Minerals*, **10**, 247–254.

Askenasy, P. E., Dixon, J. B. and McKee, T. R. (1973), Spheroidal halloysite in a Guatamalan soil, *Soil Sci. Soc. Amer. Proc.*, **37**, 799–803.

Azaroff, L. V. and Buerger, M. J. (1958), *The Powder Method in X-ray Crystallography*, McGraw-Hill, New York.

Bailey, S. W. (1963), Polymorphism of the kaolin minerals, *Amer. Mineral.*, **48**, 1196–1209.

Bailey, S. W. (1966), The status of clay mineral structures, *Clays and Clay Min.*, **14**, 1–23.

Bailey, S. W. (1969), Polytypism of trioctahedral 1:1 layer silicates, *Clays and Clay Min.*, **17**, 355–371.

Bailey, S. W. (1975a), Chlorites, in J. E. Gieseking (ed.), *Soil Components, Vol. 2, Inorganic Components*, pp. 191–263, Springer-Verlag, Berlin, Heidelberg, New York.

Bailey, S. W. (1975b), Cation ordering and pseudo symmetry in layer silicates, *Amer. Mineral.*, **60**, 175–187.

Bailey, S. W. and Brown, B. E. (1962), Chlorite polytypism: I. Regular and semi-random one-layer structures, *Amer. Mineral.*, **47**, 819–850.

Bailey, S. W. and Langston, R. B. (1969), Anauxite and kaolinite structures identical, *Clays and Clay Min.*, **17**, 241–243.

Bailey, S. W., Brindley, G. W., Johns, W. D., Martin, R. T. and Ross, M. (1971a), Summary of national and international recommendations on clay mineral nomenclature, *Clays and Clay Min.*, **19**, 129–132.

Bailey, S. W., Brindley, G. W., Johns, W. D., Martin, R. T. and Ross, M. (1971b), Clay Mineral Society report of nomenclature committee 1969–70, *Clays and Clay Min.*, **19**, 132–133.

Ball, D. F. (1966), Chlorite clay minerals in Ordovician pumice tuff and derived soils in Snowdonia, North Wales, *Clay Minerals*, **6**, 195–209.

Bancroft, G. M. (1973), *Mössbauer Spectroscopy: An Introduction for Inorganic Chemists and Geochemists*, McGraw-Hill, Maidenhead, England, 252 p.

Bancroft, G. M. and Brown, J. R. (1975), A Mössbauer study of coexisting hornblendes and biotites: Quantitative Fe^{3+}/Fe^{2+} ratios, *Amer. Mineral.*, **60**, 265–272.

Bancroft, G. M. and Maddock, A. G. (1967), Applications of Mössbauer effect to silicate mineralogy. I. Iron silicates of known crystal structure, *Geochim. Cosmochim. Acta*, **31**, 2219–2246.

Bancroft, G. M. and Stone, A. J. (1968), Applications of Mössbauer effect to silicate mineralogy. II. Iron silicates of unknown and complex crystal structures. *Geochim. Cosmochim. Acta*, **32**, 547–559.

Barrer, R. M. (1964), Molecular sieves, *Endeavour*, **23**, 122–130.

Barshad, I. and Kishk, F. M. (1968), Oxidation of ferrous iron in vermiculite and biotite alters fixation and replaceability of potassium, *Science*, **162**, 1401–1402.

Bassett, W. A. (1960), The role of hydroxyl orientation in mica alteration, *Bull. Geol. Soc. Amer.*, **71**, 449–456.

Bassett, W. A. (1963), The geology of vermiculite occurrences, *Clays and Clay Min.*, **10**, 61–69.

Bates, T. F. (1971), The kaolin minerals, in J. A. Gard (ed.), *Electron Optical Investigation of Clays*, p. 136, The Mineralogical Society, London.

Blount, A. M., Threadgold, I. M. and Bailey, S. W. (1969), Refinement of the crystal structure of nacrite, *Clays and Clay Min.*, **17**, 185–194.

Boettcher, A. L. (1966), Vermiculite, hydrobiotite, and biotite in the Rainy Creek igneous complex near Libby, Montana, *Clay Minerals*, **6**, 283–296.

Bracewell, J. M., Campbell, A. S. and Mitchell, B. D. (1970), An assessment of some thermal and chemical techniques used in the study of the poorly-ordered alumino-silicates in soil clays, *Clay Minerals*, **8**, 325–335.

Bradley, W. F. (1940), Structure of attapulgite, *Amer. Mineral.*, **25**, 405–410.

Bragg, W. L. (1933), *The Crystalline State, Vol. 1, A General Survey*, G. Bell and Sons Ltd., London.

Bragg, W. H. and Bragg, W. L. (1913), The reflection of X-rays by crystals, *Proc. Roy. Soc. A*, (London), **88**, 428–438.

Bragg, W. L. and Claringbull, G. F. (1965), in Sir Lawrence Bragg (ed.), *Crystal Structures of Minerals: The Crystalline State*, Vol. IV, G. Bell and Sons Ltd., London.

Brauner, K. and Preisinger, A. (1956), Structure of sepiolite, *Miner. Petrogr. Mitt.*, **6**, 120–140.

Breck, D. W. (1974), *Zeolite Molecular Sieves, Structure, Chemistry and Use*, John Wiley and Sons, New York.

Brindley, G. W. (1956), Allevardite, a swelling double-layer mica mineral, *Amer. Mineral.*, **41**, 91–103.

Brindley, G. W. (1961), Kaolin, serpentine and kindred minerals, in G. Brown (ed.), *The X-ray Identification and Crystal Structures of Clay Minerals*, pp. 57–131, Mineralogical Society, London.

Brindley, G. W. (1966), Ethylene glycol and glycerol complexes of smectites and vermiculites, *Clay Minerals*, **6**, 237–259.

Brindley, G. W. and Maksimovic, Z. (1974), The nature and nomenclature of hydrous nickel-containing silicates, *Clay Minerals*, **10**, 271–277.

Brindley, G. W. and Pedro, G. (1975), Meeting of the Nomenclature Committee of A.I.P.E.A., *Clays and Clay Min.*, **23**, 413–414.

Brochant, A. S. M. (1802), *Traité de Mineralogie* Vol. 1, Villier, Paris, p. 451.

Brown, B. E. and Bailey, S. W. (1963), Chlorite polytypism, II. Crystal structure of a one-layer Cr-chlorite, *Amer. Mineral.*, **48**, 42–61.

Brown, G. (1953), The dioctahedral analogue of vermiculite, *Clay Minerals Bull.*, **2**, 64–70.

Brown, G. (1955), Report of the Clay Minerals Group subcommittee on nomenclature of clay minerals, *Clay Minerals Bull.*, **2**, 294–300.

Brown, G. (1961), (ed.), *X-ray Identification and Crystal Structures of Clay Minerals*, Mineralogical Society, London.

Brown, G., Bourguignon, P. and Thorez, J. (1974), A lithium-bearing aluminian regular mixed layer montmorillonite-chlorite from Huy, Belgium, *Clay Minerals*, **10**, 135–144.

Brown, G. and Weir, A. H. (1963a), The identity of rectorite and allevardite, in I. Th. Rosenqvist and P. Graff-Peterson (eds.), *Proc. Int. Clay Conf.* (Stockholm), Vol. 1, pp. 27–35, Pergamon, Oxford.

Brown, G. and Weir, A. H. (1963b), An addition to the paper 'The identity of rectorite and allevardite', in I. Th. Rosenqvist and P. Graff-Peterson (eds.), *Proc. Int. Clay Conf.* (Stockholm), Vol. 2, pp. 87–90, Pergamon, Oxford.

Brydon, J. E., Clark, J. S. and Osborne, V. (1961), Dioctahedral chlorite, *Canad. Mineral.*, **6**, 595–609.

Caillère, S. and Hénin, S. (1957), The sepiolite and palygorskite minerals, in R. C. Mackenzie (ed.), *The Differential Thermal Investigation of Clays*, p. 242, Mineralogical Society, London.

Caillère, S. and Hénin, S. (1961), Sepiolite, in G. Brown (ed.), *The X-ray Identification and Crystal Structures of Clay Minerals*, pp. 325–342, Mineralogical Society, London.

Carr, R. M. and Chih, H. (1971), Complexes of halloysite with organic compounds, *Clay Minerals*, **9**, 153–166.

Cesari, M., Morelli, G. L. and Favretto, L. (1965), The determination of the type of stacking in mixed-layer clay minerals, *Acta Crystallogr.*, **18**, 189–196.

Christ, C. L., Hathaway, J. C., Hostetler, P. B. and Shepherd, A. O. (1969), Palygorskite: new X-ray data, *Amer. Mineral*, **54**, 198–205.

Chukrov, F. V., Zvyagin, B. B., Gorshkov, A. I., Ermilova, L. P. and Balashova, V. V. (1973), Ferrihydrite, *Akad. Nauk. SSSR*, 23–33; see *Amer. Mineral.*, 1975, **60**, 485.

Churchman, G. J. and Carr, R. M. (1975), The definition and nomenclature of halloysites, *Clays and Clay Min.*, **23**, 382–388.

Cloos, P., Leonard, A. J., Moreau, J. P., Herbillon, A. and Fripiat, J. J. (1969), Structural organisation in amorphous silico-aluminas, *Clays and Clay Min.*, **17**, 279–287.

Cole, W. F. (1966), A study of a long-spacing mica-like mineral, *Clay Minerals*, **6**, 261–281.

Cradwick, P. D. (1975), On the calculation of one-dimensional X-ray scattering from interstratified material, *Clay Minerals*, **10**, 347–356.

Cradwick, P. D. and Wilson, M. J. (1972), Calculated X-ray diffraction profiles for interstratified kaolinite-montmorillonite, *Clay Minerals*, **9**, 395–405.

Cradwick, P. D. and Wilson, M. J. (1978), Calculated X-ray diffraction curves for the interpretation of a three-component interstratified system. *Clay Minerals*, **13**, 53–65.

Cradwick, P. D. G., Farmer, V. C., Russell, J. D., Masson, C. R., Wada, K. and Yoshinaga, N. (1972), Imogolite, a hydrated aluminium silicate of tubular structure, *Nature Phys. Sci.*, **240**, 187–189.

Cullity, B. D. (1956), *Elements of X-ray Diffraction*, Addison-Wesley Publishing Company Inc., Reading, Mass.

De Lapparent, J. (1935), Sur un constituant essential des terres à foulon, *C.R. Acad. Sci. Paris*, **201**, 481–483.

De Waal, S. A. (1970), Nickel minerals from Barberton, South Africa: III. Willemseite, a nickel-rich talc, *Amer. Mineral.*, **55**, 31–42.

Deer, W. A., Howie, R. A. and Zussman, J. (1962), *Rock Forming Minerals, Vol. 3, Sheet Silicates*, Longman, London.

Douglas, L. A. (1965), Clay mineralogy of a Sassafras soil in New Jersey, *Soil Sci. Soc. Amer. Proc.*, **29**, 163–167.

Drits, V. A. and Aleksandrova, V. A. (1966), Crystallochemistry of palygorskite, *Zap. Vses. Miner. Obshch.*, **95**, 551–560.

Drits, V. A. and Karavan, Yu. V. (1969), Polytypes of the two packet chlorites, *Acta Crystallogr.*, **B25**, 632–639.

Earley, J. W., Brindley, G. W., McVeagh, W. J. and Vanden Heuvel, R. C. (1956), A regularly interstratified montmorillonite–chlorite, *Amer. Mineral.*, **41**, 258–267.

Eberhardt, J. P. and Triki, R. (1972), Description d'une technique permettant d'obtenir des coupes minces de minéraux argileux par ultramicrotomie: Application a l'étude de minéraux argileux interstratifiés., *J. de Microscopie*, **15**, 111–120.

Eberl, D. and Hower, J. (1975), Kaolinite synthesis: the role of the Si/Al and (alkali)/(H^+) ratios in hydrothermal conditions, *Clays and Clay Min.*, **23**, 301–309.

Eggleton, R. A. and Bailey, S. W. (1967), Structural aspects of dioctahedral chlorite, *Amer. Mineral.*, **52**, 673–689.

Eswaran, H. (1972), Morphology of allophane, imogolite and halloysite, *Clay Minerals*, **9**, 281–285.

Eswaran, H., Stoops, G. and De Paepe, P. (1973), A contribution to the study of soil formation on Isla Santa Cruz, Galapagos, *Pédologie*, **23**, 100–122.

Farmer, V. C. (1974), (ed.) *The Infrared Spectra of Minerals*, Mineralogical Society, London.

Farmer, V. C., Fraser, A. R. and Tait, J. M. (1977), Synthesis of imogolite: a tubular aluminium silicate polymer, *Chem. Commun.*, 462–463.

Farmer, V. C., Fraser, A. R., Russell, J. D. and Yoshinaga, N. (1977), Recognition of imogolite structures in allophanic clays by infrared spectroscopy, *Clay Minerals*, **12**, 55–57.

Farmer, V. C., Russell, J. D., McHardy, W. J., Newman, A. C. D., Ahlrichs, J. L. and Rimsaite, J. Y. H. (1971), Evidence for loss of protons and octahedral iron from oxidized biotites and vermiculites, *Mineral. Mag.*, **38**, 121–137.

Fieldes, M. and Perrott, K. W. (1966), The nature of allophane in soils, Part 3, Rapid field and laboratory test for allophane, *N.Z. J. Sci.*, **9**, 623–629.

Forbes, W. C. (1969), Unit-cell parameters and optical properties of talc on the join $Mg_3Si_4O_{10}(OH)_2$–$Fe_3Si_4O_{10}(OH)_2$, *Amer. Mineral.*, **54**, 1399–1408.

Foster, M. D. (1960), Interpretation of the composition of trioctahedral micas, *U.S. Geol. Surv. Prof. Pap.*, **354B**, 11–49.

Foster, M. D. (1962), Interpretation of the composition and a classification of the chlorites, *U.S. Geol. Surv. Prof. Paper*, **414A**, 1–33.

Foster, M. D. (1963), Interpretation of the composition of vermiculites and hydrobiotites, *Clays and Clay Min.*, **10**, 70–89.

Foster, M. D. (1964), Water content of micas and chlorites, *U.S. Geol. Surv. Prof. Paper*, **474F**, 1–15.

Frink, C. R. (1965), Characterisation of aluminium interlayers in soil clays, *Soil Sci. Soc. Amer. Proc.*, **29**, 379–382.

Fripiat, J. J. (1965), Surface chemistry and soil science, in E. G. Hallsworth and D. V. Crawford (eds.), *Experimental Pedology*, p. 5, Butterworths, London.

Frissel, M. J. and Bolt, G. H. (1962), Interaction between certain ionizable organic compounds (herbicides) and clay minerals, *Soil Sci.*, **94**, 284–291.

Gard, J. A. (1971), (ed.) *The Electron-optical Investigation of Clays*, Mineralogical Society, London.

Gard, A. and Follett, E. A. C. (1968), A structural scheme for palygorskite, *Clay Minerals*, **7**, 367–369.

Glocker, E. F. (1847), Synopsis, Halle, p. 190, in J. D. Dana (ed.), *A System of Mineralogy*, 5th edition, 1868, Tribner and Co., London.

Goodman, B. A., Russell, J. D., Fraser, A. R. and Woodhams, F. W. D. (1976), A Mössbauer and I.R. spectroscopic study of the structure of nontronite, *Clays and Clay Min.*, **24**, 53–59.

Greene-Kelly, R. (1955), Dehydration of the montmorillonite minerals, *Mineral. Mag.*, **30**, 604–615.

Greenland, D. J. (1965), Interaction between clays and organic compounds in soils, Pt. I. Mechanisms of interaction between clays and defined organic compounds, *Soils and Fertilizers*, **28**, 415–425.

Grim, R. E., Bray, R. H. and Bradley, W. F. (1937), Mica in argillaceous sediments, *Amer. Mineral.*, **22**, 813–829.

Grim, R. E. and Kulbicki, G. (1961), Montmorillonite: high temperature reactions and classification, *Amer. Mineral.*, **46**, 1329–1369.

Güven, N. (1974), Electron-optical investigations on montmorillonite, I. Cheto, Camp Berteaux and Wyoming montmorillonites, *Clays and Clay Min.*, **22**, 155–165.

Hashimoto, I. and Jackson, M. L. (1960), Rapid dissolution of allophane and kaolinite-halloysite after dehydration, *Clays and Clay Min. 7th Conf.*, pp. 102–113, Pergamon Press, London.

Hayes, M. H. B., Pick, M. E. and Toms, B. A. (1978), The influence of organocation structure on the adsorption of mono- and di-pyridinium cations by expanding lattice clay minerals. I. Adsorption by Na^+-montmorillonite. *J. Coll. Interf. Sci.*, **65**, 254–265.

Hazen, R. M. and Burnham, C. W. (1973), The crystal structures of one-layer phlogopite and annite, *Amer. Mineral.*, **58**, 889–900.

Hazen, R. M. and Wones, D. R. (1972), The effect of cation substitutions on the physical properties of trioctahedral micas, *Amer. Mineral.*, **57**, 103–129.

Helgeson, H. C. (1974), Chemical interaction of feldspars and aqueous solutions, in W. S. MacKenzie and J. Zussman (eds.), *The Feldspars*, University Press, Manchester.

Hendricks, S. B. and Teller, E. (1942), X-ray interference in partially ordered layer lattices, *J. Chem. Phys.*, **10**, 147–167.

Henmi, T. and Wada, K. (1976), Morphology and composition of allophane, *Amer. Mineral.*, **61**, 379–390.

Henry, N. F. M., Lipson, H. and Wooster, W. A. (1951), *The Interpretation of X-ray Diffraction Photographs*, MacMillan & Co. Ltd., London.

Hey, M. H. (1954), A new review of the chlorites, *Mineral. Mag.*, **30**, 277–292.

Heystek, H. (1954), An occurrence of a regular mixed-layer clay mineral, *Mineral. Mag.*, **30**, 400–408.

Hower, J. and Mowatt, T. C. (1966), The mineralogy of illites and mixed-layer illite–montmorillonites, *Amer. Mineral.*, **51**, 825–854.

Jackson, M. L. (1960), Structural role of hydronium in layer silicates during soil genesis, *Trans. 7th Int. Cong. Soil Sci.*, **2**, 445–455, Madison, Wisconsin.

James, R. W. (1954), in Sir Lawrence Bragg (ed.), *Principles of the Diffraction of X-rays: The Crystalline State, Vol. II*, G. Bell & Sons Ltd., London.

Jepson, W. B. and Rowse, J. B. (1975), The composition of kaolinite–an electron microscope microprobe study, *Clays and Clay Min.*, **23**, 310–317.

Joint Committee on Powder Diffraction Standards (1975), Search Manual, Inorganic Compounds—Alphabetical Listing SMA-25, published by Joint Committee on Powder Diffraction Standards, 1601 Park Lane, Swarthmore, Pennsylvania 19801, U.S.A.

Kakinoki, J. and Komura, Y. (1952), Intensity of X-ray by a one-dimensionally disordered crystal, *J. Phys. Soc. Jap.*, **7**, 30–35.

Kapoor, B. S. (1973), The formation of 2:1–2:2 intergrade clays in some Norwegian podzols, *Clay Minerals*, **10**, 79–86.

Kitagawa, Y. (1976), Determination of allophane and amorphous inorganic matter in clay fractions of soils, *Soil Sci. Plant Nutr.*, **22**, 137–147.

Kittrick, J. A. and Hope, E. W. (1963), A procedure for the particle-size separation of soils for X-ray diffraction analysis, *Soil Sci.*, **96**, 319–325.

Klug, H. P. and Alexander, L. E. (1974), *X-ray Diffraction Procedures for Polycrystalline and Amorphous Materials*, 2nd edition, John Wiley & Sons, New York.

Kodama, H. and Brydon, J. E. (1966), Interstratified montmorillonite–mica clays from subsoils of the Prairie Provinces, Western Canada, in W. F. Bradley and S. W. Bailey (eds.), 13th National Conference on *Clays and Clay Minerals*, pp. 151–173, Pergamon, London.

Kodama, H. and Brydon, J. E. (1968), A study of clay minerals in podzol soils in New Brunswick, Eastern Canada, *Clay Minerals*, **7**, 295–309.

Langston, R. B. and Pask, J. A. (1969), The nature of anauxite, *Clays and Clay Min.*, **16**, 425–436.

Lapham, D. M. (1958), Structure and chemical variation in chromium chlorite, *Amer. Mineral.*, **43**, 921–956.

Lehr, J. R. and McClellan, G. H. (1972), A Revised Laboratory Reactivity Scale for Evaluating Phosphate Rocks for Direct Application. Bulletin Y43, National Fertilizer Development Center, T.V.A., Muscle Shoals, Alabama.

Lipson, H. and Cochran, W. (1953), in Sir Lawrence Bragg (ed.), *The Determination of Crystal Structures; The Crystalline State* Vol. III, G. Bell & Sons Ltd., London.

Lyon, R. J. P. (1967), Infrared absorption spectroscopy, in J. Zussman (ed.), *Physical Methods in Determinative Mineralogy*, pp. 371–403, Academic Press, London and New York.

McCauley, J. W. and Newnham, R. E. (1971), Origin and prediction of ditrigonal distortions in micas, *Amer. Mineral.*, **56**, 1626–1638.

McCauley, J. W., Newnham, R. E. and Gibbs, G. V. (1973), Crystal structure analysis of synthetic fluorophlogopite, *Amer. Mineral.*, **58**, 249–254.

McClellan, G. H. and Lehr, J. R. (1969), Crystal chemical investigation of natural apatites, *Amer. Mineral.*, **54**, 1374–1391.

McConnell, J. D. C. (1967), Electron microscopy and electron diffraction, in J. Zussman (ed.), *Physical Methods in Determinative Mineralogy*, pp. 335–370, Academic Press, London and New York.

MacEwan, D. M. C. (1956), Fourier transform methods, I. A direct method of analysing interstratified mixtures, *Kolloidzschr.*, **149**, 96–108.

MacEwan, D. M. C. (1958), Fourier transform methods, II. Calculation of diffraction effects for different types of interstratification, *Kolloidzschr.*, **162**, 93–100.

MacEwan, D. M. C. (1961), Montmorillonite minerals, in G. Brown (ed.), *The X-ray Identification and Crystal Structures of Clay Minerals*, pp. 143–207, Mineralogical Society, London.

MacEwan, D. M. C. (1968), Type of interstratification in soil clay minerals., *Proc. 9th Int. Conf. Soil Sci.*, **I**, 1–10, Adelaide, Australia.

MacEwan, D. M. C., Ruiz Amil, A. and Brown, G. (1961), Interstratified clay minerals, in G. Brown (ed.), *The X-ray Identification and Crystal Structures of Clay Minerals*, pp. 393–445, Mineralogical Society, London.

McKenzie, R. M. (1972), The manganese oxides in soils—A review. *Zeits für Pflanzenebernährung und Bodenkunde*, **131**, 221–242.

Mackintosh, E. E., Lewis, D. G. and Greenland, D. J. (1971), Dodecylammonium-mica complexes, I. Factors affecting the exchange reaction, *Clays and Clay Min.*, **19**, 209–218.

Malcolm, R. L., Nettleton, W. D. and McCracken, R. J. (1969), Pedogenic formation of montmorillonite from a 2:1–2:2 intergrade clay mineral, *Clays and Clay Min.*, **16**, 405–414.

Marel, H. W. van der (1964), Identification of chlorite and chlorite-related minerals in sediments, *Beit. Z. Min. Pet.* **9**, 462–480.

Martin Vivaldi, J. L. and MacEwan, D. M. C. (1957), Triassic chlorites from the Jura and the Catalan coastal range, *Clay Min. Bull.*, **3**, 177–183.

Martin Vivaldi, J. L. and MacEwan, D. M. C. (1960), Corrensite and swelling chlorite, *Clay Min. Bull.*, **4**, 173–181.

Martin Vivaldi, J. L. and Robertson, R. H. S. (1971), Palygorskite and sepiolite, in J. A. Gard (ed.), *Electron-optical Investigation of Clays*, pp. 255–275, Mineralogical Society, London.

Mathieson, A. McL. and Walker, G. F. (1954), Crystal structure of Mg-vermiculite, *Amer. Mineral.*, **39**, 231–255.

Meads, R. E. and Malden, P. J. (1975), Electron spin resonance in natural kaolinites containing Fe^{3+} and other transition metal ions, *Clay Minerals*, **10**, 313–345.

Megaw, H. D. (1934), The crystal structure of hydrargillite $Al(OH)_3$, *Z. Kristallogr.* (*A*), **87**, 185–204.

Méring, J. (1949), X-ray diffraction in disordered layer structures, *Acta Crystallogr.*, **2**, 371–377.

Méring, J. and Oberlin, A. (1971), The smectites, in J. A. Gard (ed.), *The Electron-optical Investigation of Clays*, pp. 193–229, Mineralogical Society, London.

Millot, G. (1973), Data and tendencies in recent years in the field "Genesis and synthesis of clay and clay minerals", *Proc. Int. Clay Conf.*, Madrid, pp. 151–170.

Mitchell, B., Farmer, V. C. and McHardy, W. J. (1964), Amorphous inorganic materials in soils, *Advan. Agron.*, **16**, 327–383.

Mooney, R. W., Keenan, A. C. and Wood, L. A. (1952), Adsorption of water vapour by montmorillonite, II. Effect of exchangeable ions and lattice swelling as measured by X-ray diffraction, *J. Amer. Chem. Soc.*, **74**, 1371–1374.

Mortland, M. M. (1970), Clay–organic complexes and interactions, *Advan. Agron.*, **22**, 75–117.

Nagasawa, K., Brown, G. and Newman, A. C. D. (1974), Artificial alteration of biotite into a 14 Å layer silicate with hydroxy-aluminium interlayers, *Clays and Clay Min.*, **22**, 241–252.

Nagata, H., Shimoda, S. and Sudo, T. (1974), On dehydration of bound water of sepiolite, *Clays and Clay Min.*, **22**, 285–293.

Nagy, B. and Bradley, W. F. (1955), The structural scheme of sepiolite, *Amer. Mineral.*, **40**, 885–892.

Nemecz, E., Varju, G. and Barna, J. (1965), Allevardite from Királhegy, Tokaj Mountains, Hungary, *Proc. Int. Clay Conf.*, Stockholm, **II**, pp. 51–67, Pergamon, Oxford.

Newman, A. C. D. (1969), Cation exchange properties of micas, I. The relation between mica composition and potassium exchange in solutions of different pH, *J. Soil Sci.*, **20**, 357–373.

Newnham, R. E. (1961), A refinement of the dickite structure and some remarks on polymorphism in kaolin minerals, *Mineral. Mag.*, **32**, 683–704.

Noble, F. R. (1971), A study of disorder in kaolinite, *Clay Minerals*, **9**, 71–81.

Norrish, K. (1954), The swelling of montmorillonite, *Discuss. Faraday Soc.*, **18**, 120–134.

Norrish, K. (1968), Some phosphate minerals in soils, *Trans. 9th Int. Congress of Soil Science*, Adelaide, **2**, 713–723.

Norrish, K. (1972), Factors in the weathering of mica to vermiculite, *Proc. Int. Clay Conf.*, Madrid, 417–432.

Olejnik, S., Posner, A. M. and Quirk, J. P. (1970), The intercalation of polar organic compounds into kaolinite, *Clay Minerals*, **8**, 421–434.

Olphen, H. van (1965), Thermodynamics of interlayer adsorption of water in clays, I. Sodium vermiculite, *J. Colloid Sci.*, **20**, 822–837.

Olphen, H. van (1969). Thermodynamics of interlayer adsorption of water in clays, II. Magnesium vermiculite, *Proc. 3rd Int. Clay Conf.*, **1**, 649–657, Tokyo.

Orcel, J., Caillère, S. and Henin, S. (1950), Novel essai de classification des chlorites, *Mineral. Mag.*, **29**, 329–340.

Ovcharenko, F. D. (1964), *The Colloid Chemistry of Palygorskite*, Israel Program for Scientific Translations, Jerusalem.

Orville, P. M. (1967), Unit cell parameters of the microcline and sanidine-high albite solution series, *Amer. Mineral.*, **52**, 55–86.

Pauling, L. (1929), The principles determining the structure of complex ionic crystals, *J. Amer. Chem. Soc.*, **51**, 1010–1026.

Pauling, L. (1930), The structure of the chlorites, *Proc. Nat. Acad. Sci.* (Washington), **16**, 578–582.

Perrott, K. W., Smith, B. F. L. and Mitchell, B. D. (1976), Effect of pH on the reaction of sodium fluoride with hydrous oxides of silicon, aluminium, and iron, and with poorly ordered aluminosilicates, *J. Soil Sci.*, **27**, 348–356.

Perry, E. and Hower, J. (1970), Burial diagenesis in Gulf Coast pelitic sediments, *Clays and Clay Min.*, **18**, 165–177.

Pézerat, H. and Méring, J. (1954), Influence des substitutions isomorphes sur les parametres de structure des phyllites, *Clay Mineral Bull.*, **2**, 156–161.

Phillips, F. C. (1955), *An Introduction to Crystallography*, 2nd edition, Longmans, Green and Co., London, New York, Toronto.

Rausell Colom, J. A., Sweatman, T. R., Wells, C. B. and Norrish, K. (1964), Studies in the artificial weathering of mica, Proc. Univ. Nottingham 11th Easter Sch. Agric. Sci., pp. 40–72.

Rautureau, M. and Tchoubar, C. (1976), Structural analysis of sepiolite by selected area diffraction—relations with physico-chemical properties, *Clays and Clay Min.*, **24**, 43–49.

Rayner, J. H. (1974), The crystal structure of phlogopite by neutron diffraction, *Mineral Mag.*, **39**, 850–856.

Rayner, J. H. and Brown, G. (1966), Structure of pyrophyllite, *Clays and Clay Min.*, **13**, 73–84.

Rayner, J. H. and Brown, G. (1973), The crystal structure of talc, *Clays and Clay Min.*, **21**, 103–113.

Reynolds, R. C. (1967), Interstratified clay systems: calculation of the total one-dimensional diffraction function, *Amer. Mineral.*, **52**, 661–672.

Reynolds, R. C. and Hower, J. (1970), The nature of interlayering in mixed-layer illite-montmorillonites, *Clays and Clay Min.*, **18**, 25–36.

Rich, C. I. (1968), Hydroxy interlayers in expansible layer silicates, *Clays and Clay Min.*, **16**, 15–30.

Rich, C. I. and Obenshain, S. S. (1955), Chemical and clay mineral properties of a red-yellow podzolic soil derived from muscovite schist, *Soil Sci. Soc. Amer. Proc.*, **19**, 334–339.

Robert, M. and Pedro, G. (1968), Influence de l'oxydation thermique des biotites sur l'extraction du potassium (vermiculitisation), *C.R. Acad. Sci.*, Paris, **267**, 1805–1807.

Rosenberg, P. E. (1974), Pyrophyllite solid solutions in the system Al_2O_3–SiO_2–H_2O, *Amer. Mineral.*, **59**, 254–260.

Ross, C. S. and Hendricks, S. B. (1945), Minerals of the Montmorillonite Group, *U.S. Geol. Surv. Prof. Pap.*, **205B**, 23–79.

Ross, C. S. and Kerr, P. F. (1934), Halloysite and allophane, Prof. Pap. U.S. Geol. Surv., **185-G**, 135–148.

Ross, M. (1968), X-ray diffraction effects by non-ideal crystals of biotite, muscovite, montmorillonite, mixed-layer clays, graphite and periclase, *Z. Kristallogr.*, **126**, 1–3, 80–97.

Rothbauer, R. (1971), Untersuchung eines $2M_1$-Muskovits mit Neitronenstrahlen, *Neues Jahrb. Min. Mh.*, 143–154.

Ruiz-Amil, A., Ramirez, A. and MacEwan, D. M. C. (1967), X-ray diffraction curves for the analysis of interstratified structures, Volturna Press. Talleres Grafficos del C.S.I.C., Madrid.

Russell, J. D., McHardy, W. J. and Fraser, A. R. (1969), Imogolite: a unique aluminosilicate, *Clay Minerals*, **8**, 87–99.

Sakharov, B. A. and Drits, V. A. (1973), Mixed-layer kaolinite–montmorillonite: a comparison of observed and calculated diffraction patterns, *Clays and Clay Min.*, **21**, 15–17.

Sato, M. (1965), Structure of interstratified (mixed-layer) minerals, *Nature* (London), **208**, 70–71.

Sato, M. (1969), Interstratified structure with Reichweite $g=2$ and its X-ray diffraction pattern, *Proc. Int. Clay Conf.*, Tokyo, **1**, 207–214.

Sawhney, B. L. (1958), Weathering and aluminium interlayers in soil clay minerals, montmorillonite and vermiculite, *Nature*, (London), **182**, 1595–1596.

Schultz, L. G. (1969), Lithium and potassium absorption, dehydroxylation temperature and structural water content of aluminous smectites, *Clays and Clay Min.*, **17**, 115–149.

Schultz, L. G., Shepard, A. O., Blackmon, P. D. and Starkey, H. C. (1971), Mixed layer kaolinite-montmorillonite from the Yucatan Peninsula, Mexico, *Clays and Clay Min.*, **19**, 137–150.

Schwertmann, U. (1973), Electron micrographs of soil lepidocrocites, *Clay Minerals*, **10**, 59–60.

Schwertmann, U. and Thalmann, H. (1976), The influence of [FeII] [Si] and pH on the formation of lepidocrocite and ferrihydrite during oxidation of aqueous $FeCl_2$ solutions, *Clay Minerals*, **11**, 189–200.

Segalen, P. (1968), Note sur une methode de détermination des produits minéraux amorphes dans certain sol à hydroxydes tropicaux, *Cah. ORSTOM, sér. Pédol.*, **6**, 106–126.

Serna, C., Ahlrichs, J. L. and Serratosa, J. M. (1975), Folding in sepiolite crystals, *Clays and Clay Min.*, **23**, 452–457.

Shannon, R. D. and Prewitt, C. T. (1969), Effective ionic radii in oxides and fluorides, *Acta Crystallogr. B*, **25**, 925–946.

Sheppard, N. (1959), Infra-red spectra of adsorbed molecules, *Spectrochim. Acta*, **14**, 249–260.

Shimoda, S. (1969), New data for tosudite, *Clays and Clay Min.*, **17**, 179–184.

Shimoyama, A., Johns, W. D. and Sudo, T. (1969), Montmorillonite–kaolin clay in acid clay deposits from Japan, *Proc. Int. Clay Conf.*, Tokyo, **I**, 225–229.

Shirozu, H. (1958), X-ray patterns and cell dimensions of chlorites, *Miner. J., Sapporo*, **2**, 209–223.

Shirozu, H. and Bailey, S. W. (1965), Chlorite polytypism, III. Crystal structure of an orthohexagonal chlorite, *Amer. Mineral.*, **50**, 868–885.

Shirozu, H. and Bailey, S. W. (1966), Crystal structure of a two-layer Mg-vermiculite, *Amer. Mineral.*, **51**, 1124–1143.

Singer, A. and Norrish, K. (1974), Pedogenic palygorskite occurrences in Australia, *Amer. Mineral.*, **59**, 508–517.

Smith Aitken, W. W. (1965), An occurrence of phlogopite and its transformation to vermiculite by weathering, *Mineral. Mag.*, **35**, 151–164.

Smith, J. V. (1974), *Feldspar Minerals* Vol. 1 *Crystal Structures and Physical Properties* Vol. 2 *Chemical and Textural Properties*, Springer-Verlag, Berlin, Heidelberg, New York.

Smith, J. V. and Bailey, S. W. (1963), Second review of Al–O and Si–O tetrahedral distances, *Acta Crystallogr.*, **16**, 801–811.

Smith, J. V. and Yoder, H. S. (1956), Experimental and theoretical studies of mica polymorphs, *Mineral. Mag.*, **31**, 209–235.

Soil Survey Staff (1975), *Soil Taxonomy*, p. 230, Soil Conservation Service, U.S. Dept. of Agric., Washington.

Souza Santos, P. de, Souza Santos, H. de and Brindley, G. W. (1966), Mineralogical studies of kaolinite-halloysite clays. IV: A platy mineral with structural swelling and shrinking characteristics, *Amer. Mineral.*, **51**, 1640–1648.

Ssaftchenkov, T. V. (1862), Verh. Russich-Kaiser. *Gsell. Min. St. Petersburg*, 102.

Steinfink, H. (1958a), The crystal structure of chlorite, I. A monoclinic polymorph, *Acta Crystallogr.*, **11**, 191–195.

Steinfink, H. (1958b), The crystal structure of chlorite, II. A triclinic polymorph, *Acta Crystallogr.*, **11**, 195–198.

Steinfink, H. (1961), Accuracy in structure analysis of layer silicates: some further comments on the structure of prochlorite, *Acta Crystallogr.*, **14**, 198–199.

Stephen, I. and MacEwan, D. M. C. (1950), Swelling chlorite, *Geotechnique* (London), **2**, 82–83.

Stromeyer, F. and Hausmann, J. F. L. (1816), *Gottingische Geleherte Anzeigen*, **2**, 1251.

Sudo, T. and Hayashi, H. (1956), A randomly-interstratified kaolin-montmorillonite in acid clay deposits in Japan, *Nature* (London), **178**, 1115–1116.

Sudo, T. and Yotsumoto, H. (1977), The formation of halloysite tubes from spherulitic halloysite, *Clays and Clay Min.*, **25**, 155–159.

Suquet, H., de la Calle, C. and Pézerat, H. (1975), Swelling and structural organization of saponite, *Clays and Clay Min.*, **23**, 1–9.

Taylor, R. M. (1968), The association of manganese and cobalt in soils—further observations, *J. Soil Sci.*, **19**, 77–80.

Taylor, R. M., MacKenzie, R. M. and Norrish, K. (1964), The mineralogy and chemistry of manganese in some Australian soils, *Aust. J. Soil Res.*, **2**, 235–248.

Taylor, R. M. and Schwertmann, U. (1974), Maghemite in soils and its origin, I. Properties and observations on soil maghemites; II. Maghemite syntheses at ambient temperature and pH 7, *Clay Minerals*, **10**, 289–298, 299–310.

Tettenhorst, R. and Grim, R. E. (1975), Interstratified clays, I. Theoretical, *Amer Mineral.*, **60**, 49–59.

Theng, B. K. G. (1974), *The Chemistry of Clay–Organic Reactions*, Adam Hilger Ltd., London.

Thompson, G. R. and Hower, J. (1975), The mineralogy of glauconite, *Clays and Clay Min.*, **23**, 289–300.

Udagawa, S., Nakada, T. and Nakahira, M. (1969), Molecular structure of allophane as revealed by its thermal transformation. *Proc. Int. Clay Conf.*, **1**, 151–172, Jerusalem.

Veniale, F. and Marel, H. W. van der (1969), Identification of some 1 : 1 regular interstratified trioctahedral clay minerals, *Proc. Int. Clay Conf.*, 233–244, Tokyo.

Viswanathan, K. (1971), A new X-ray method to determine the anorthite content and structural state of plagioclases, *Contr. Mineral. and Petrol.*, **30**, 332–335.

Wada, K. (1967), A structural scheme for soil allophane, *Amer. Mineral.*, **52**, 690–708.

Wada, K. and Matsubara, I. (1968), Differential formation of allophane, imogolite and gibbsite in the Kitakami Pumice Bed, *Proc. 9th Int. Congr. Soil Sci.*, **3**, 123–131.

Wada, K. and Yoshinaga, N. (1969), The structure of imogolite, *Amer. Mineral.*, **54**, 50–71.

Wada, K., Yoshinaga, N., Yotsumoto, H., Ibe, K. and Aida, S. (1970), High resolution electron micrographs of imogolite, *Clay Minerals*, **8**, 487–489.

Walker, G. F. (1950), Trioctahedral minerals in the soil-clays of north-east Scotland, *Mineral. Mag.*, **29**, 72–84.

Walker, G. F. (1957), On the differentiation of vermiculites and smectites in clays, *Clay Minerals Bull.*, **3**, 154–163.

Walker, G. F. (1958), Reactions of expanding-lattice clay minerals with glycerol and ethylene glycol, *Clay Minerals Bull.*, **3**, 302–313.

Walker, G. F. (1961), Vermiculite Minerals, in G. Brown (ed.), *The X-ray Identification and Crystal Structures of Clay Minerals*, pp. 297–324, Mineralogical Society, London.

Walker, G. F. and Garrett, W. G. (1961), Complexes of vermiculite with amino acids, *Nature (London)*, **191**, 1389.

Wardle, R. and Brindley, G. W. (1972), The crystal structures of pyrophyllite, 1 Tc, and of its dehydroxylate, *Amer. Mineral.*, **57**, 732–750.

Warren, B. E. (1969), *X-ray Diffraction*, Addison-Wesley Publishing Company, Reading, Massachusetts, U.S.A.

Weaver, C. E. (1956), The distribution and identification of mixed-layer clays in sedimentary rocks, *Amer. Mineral.*, **41**, 202–221.

Weaver, C. E. and Pollard, L. D. (1973), Attapulgite and palygorskite, and sepiolite, in *The Chemistry of Clay Minerals*, pp. 118–130. Elsevier, Amsterdam.

Weed, S. B. and Nelson, L. A. (1962), Occurrence of chlorite-like intergrade clay minerals in Coastal Plain, Piedmont and Mountain soils of North Carolina, *Soil Sci. Soc. Amer. Proc.*, **26**, 393–398.

Weir, A. H. and Greene-Kelly, R. (1962), Beidellite, *Amer. Mineral.*, **47**, 137–146.

Weir, A. H. and Rayner, J. H. (1974), An interstratified illite–smectite from Denchworth series soil in weathered Oxford Clay, *Clay Minerals*, **10**, 173–187.

Weir, A. H., Nixon, H. L. and Woods, R. D. (1962), Measurement of thickness of dispersed clay flakes with the electron microscope, *Clays and Clay Min.*, **9**, 419–423.

Weir, A. H., Ormerod, E. C. and El Mansey, I. M. I. (1975), Clay mineralogy of sediments of the Western Nile Delta, *Clay Minerals*, **10**, 369–386.

Wenk, H. R. (1976), (ed.) *Electron Microscopy*, in *Mineralogy*, Springer-Verlag, Berlin, Heidelberg, New York.

White, E., McKinstry, H. and Bates, T. F. (1958), Crystal chemical studies by X-ray fluorescence, *Advances in X-ray Analysis*, **2**, 239–245.

Whittaker, E. J. W. and Muntkus, R. (1970), Ionic radii for use in geochemistry, *Geochim. Cosmochim. Acta*, **34**, 945–956.

Wiewióra, A. (1973), Mixed-layer kaolinite-smectite from Lower Silesia, Poland: Final report, *Proc. Int. Clay Conf.*, pp. 75–87, Madrid.

Wilkins, R. W. T. and Ito, J. (1967), Infrared spectra of some synthetic talcs, *Amer. Mineral.*, **52**, 1649–1661.

Wright, A. C. (1975), Closed form equations for X-ray diffraction by interstratified clay systems, 1. randomly occurring interlamellar species, *Clays and Clay Min.*, **23**, 278–288.

Yada, K. (1967), Study of chrysotile asbestos by a high resolution electron microscope, *Acta Crystallogr.*, **23**, 704–707.

Yoshinaga, N. (1966), Chemical composition and some thermal data of eighteen allophanes from Ando soils and weathered pumices, *Soil Sci. Plant Nutr.*, **12**, 47–54.

Yoshinaga, N. (1968), Identification of imogolite in the filmy gel materials in the Imaichi and Shichihonzakura pumice beds. *Soil Sci. Plant Nutr.*, **14**, 238–246.

Yoshinaga, N. and Aomine, S. (1962), Imogolite in some Ando soils, *Soil Sci. Plant Nutr.* **8**, 6–13.

Yoshinaga, N. and Yamaguchi, M. (1970), Occurrence of imogolite as gel film of the pumice and scoria beds of western and central Honshu and Hokkaido, *Soil Sci. Plant Nutr.*, **16**, 215–223.

Zvyagin, B. B., Mishchenko, K. S. and Shitov, V. A. (1963), Electron diffraction data on the structures of sepiolite and palygorskite, *Soviet Phys. Crystallogr.*, **8**, 148–153 (translated from *Kristallografiya*, **8**, 201–206).

CHAPTER 3

The chemistry of soil organic colloids

M. H. B. Hayes
Department of Chemistry, University of Birmingham
R. S. Swift
Soil Science Department, Edinburgh School of Agriculture

3.1 INTRODUCTION

The complete soil organic fraction is made up of live organisms, and their undecomposed, partly decomposed and completely transformed remains. Soil organic matter is the term used to refer more specifically to the non-living components which are a heterogeneous mixture composed largely of products resulting from microbial and chemical transformations of organic debris. These transformations, known collectively as the humification process, give rise to humus, a mixture of substances which have a degree of resistance to further microbial attack.

It is well known that soil organic matter has a beneficial effect on soil conditions and fertility. Amongst the more important functions which have been attributed to it are:

(i) the formation and maintenance of good soil structure;
(ii) the improvement of water entry into, and retention by the soil;
(iii) the retention of plant nutrients as the result of its cation-exchange properties;
(iv) the release of nitrogen, sulphur, phosphorous and possibly some trace elements as the result of degradation and mineralization;
(v) increasing soil temperature by enhancing absorption of solar radiation due to its dark colour.

It is probable that most of these beneficial effects can be attributed to the humus fraction of soil organic matter. The synthesis and degradation of humus is a dynamic process which attains an equilibrium in a particular soil environment.

This equilibrium can be disturbed by environmental changes, such as variations in soil water levels, alterations in cultivation practices, etc. Substantial depletion of the organic matter content of a fertile soil can lead to the deterioration of the soil structure and the loss of fertility.

Since soil organic matter is a heterogeneous mixture it is necessary for purposes of discussion to subdivide it into groups which have similar morphological or chemical characteristics. Here we are adopting a classification based largely on proposals by Kononova (1966, 1975). In this system soil organic matter is separated into two major groups:

(I) Unaltered materials, which include fresh debris and non-transformed components of older debris;

(II) Transformed products, or humus, bearing no morphological resemblance to the structures from which they were derived.

These transformed components are often referred to as humified products, but in fact they consist of both humic and non-humic substances and can be therefore subdivided as follows:

(IIa) Amorphous, polymeric, brown coloured humic substances which are differentiated on the basis of solubility properties (see Section 3.3.1) into humic acids, fulvic acids, and humins;

(IIb) Compounds belonging to recognizable classes, such as polysaccharides, polypeptides, altered lignins, etc. These can be synthesized by microorganisms or can arise from modifications of similar compounds in the original debris.

Humic substances are widely distributed in soils and waters, and are probably the most abundant of the naturally occurring organic polymers. Reserves of soil humic acids (the most abundant of the humic substances), for instance, are estimated to be of the order of $2–3 \times 10^{12}$ tonnes (Bazilevich, 1974). Yet their structures and properties are not sufficiently well understood. However, significant advances can soon be expected in these areas because of recent progress in degradation techniques and in instrumentation.

In order to carry out meaningful investigations of their compositions it is important to be able to extract humic materials in representative yields from soil; to separate these from co-extracted soil minerals and from non-humified substances; to isolate the relevant gross fractions; and to fractionate these further into components which are reasonably homogeneous with regard to molecular sizes and charge densities. If this can be successfully achieved results from chemical and physical measurements will have greater significance.

It is convenient to describe polymers in terms of primary, secondary, and tertiary structures. Primary structures can be considered to be the monomer units or 'polymer building blocks' which are released into the digest during degradation reactions. Secondary structures define the sequential arrangements of these units in the polymer, and tertiary structures describe the sizes and three-dimensional shapes of the polymers. When the primary structures are known it is possible to distinguish the chemical units which govern the reactivities of the structures. Detailed knowledge of the secondary and tertiary structures would allow predictions of polymer behaviour in different media, and would be especially valuable for a more complete understanding of interactions with soil inorganic colloids, sorbed organic molecules, and inorganic ions. Thus, advances in structural studies can provide essential information to help resolve many questions such as the stabilisation of soil structure, and the binding of pesticides and of heavy metals in the soil environment.

Some aspects of the role of humic substances in soil are discussed in the companion volume and by Kononova (1966, 1975), and Schnitzer and Khan (1972). This chapter deals in part with the composition and properties of soil polysaccharides (Section 3.6), but is primarily concerned with the chemical and physical properties of humic substances. We begin with a discussion of the isolation and purification procedures which are a necessary preliminary to the determination of composition, structure, shape, size and charge properties. When dealing with these subjects some emphasis is placed on the scientific principles on which the techniques are based and on the interpretation of the results obtained.

3.2 EXTRACTION OF HUMIC SUBSTANCES

Ideal solvent systems would quantitatively isolate homogeneous components of humic substances from the soil. Such systems are not known, and are unlikely to be devised because of the inevitable overlap of solubility properties among the vast numbers of organic materials which are present in the soil environment. For that reason most effort has been concentrated on finding solvents that will extract high yields of humic materials without significantly altering the compositions of the natural polymers. Large scale extractions are often preceded by preliminary treatments with organic solvents, such as benzene and ether, in order to remove waxy materials. Such solvents inevitably remove many of the low molecular weight organic compounds which generally compose only a small fraction of the soil organic matter.

We shall be concerned here with procedures used to extract the brown humic polyelectrolytes. At the pH values of most fertile soils these polyelectrolytes are rendered insoluble through ion-exchange reactions with polyvalent metal cations. Thus soils are usually pretreated with mineral acid which removes metallic cations, helps to break links between humic materials and sesquioxides, and dissolves soil carbonates. The carboxyl and phenolic hydroxyl groups in the resulting H^+-ion saturated humic materials are not dissociated, and this allows the polymers to associate through hydrogen bonding and van der Waals forces mechanisms. Thus the extraction of H^+-humic substances can be regarded as a problem of polymer solubility. More commonly, however, these substances are dissolved by replacing the H^+ with solubilizing ions such as Na^+, K^+, or NH_4^+, by using neutral or basic solutions of the cations.

Whitehead and Tinsley (1964) have outlined the criteria for effective solvents for humic substances. These are:

(i) a high polarity and a high dielectric constant to assist the dispersion of the charged humic molecules;
(ii) a small molecular size to penetrate through the humus polymers;
(iii) the ability to disrupt the existing hydrogen bonds, and provide alternative groups to form humic–solvent hydrogen bonds;
(iv) the ability to immobilize metallic cations.

From the treatment which follows it will be seen that solvents or solvent systems which give good yields of humic products satisfy most or all of these criteria.

3.2.1 Extraction of Ionized Humic Substances

Polymers are dissolved in two steps. First solvation, or a swelling process takes place to form a gel which results from the absorption of solvent molecules by the surface causing the polymer to change its average dimensions. Secondly, if the polymer intermolecular attraction forces can be overcome through the formation of strong polymer–solvent interactions the gel will rapidly disintegrate to form a true solution.

As a result of strong intermolecular attraction forces and the low degree of dissociation of H^+-ion saturated humic substances, the higher molecular weight components can only swell in water. However, the Na^+, NH_4^+-polymer salts are dissociated in water and the resulting repulsive forces between the negatively charged groups on the polymer are generally far stronger than the intermolecular attractive forces. Such repelling effects cause flexible macromolecules to assume shapes, or conformations, which minimize the electrostatic free energy and lead to very large expansions of the polymer structures. Such expansions allow free access of solvent to the absorption sites and solution takes place.

As the pH of the extracting medium is raised increased ionization of the different acid functional groups (largely carboxylic acid and phenolic hydroxyls) gives greater repulsion and increased solubility. However, solution of high molecular weight polyelectrolytes (charged, or ionizable polymers) will only take place when sufficient ionized species are present to cause significant changes in polymer conformations. Also, highly cross-linked polymers, even if extensively charged, will not dissolve because they cannot assume expanded conformations.

3.2.2 Extractions with Base, Complexing Agents, and Neutral Salt Solutions

Achard (1786) was first to isolate humic acid when he extracted a peat soil with aqueous sodium hydroxide. This is still the most widely used solvent system for humic substances. However, a certain amount of oxidation of humate, as measured by increases in the oxygen contents of the extracts (Bremner, 1950; Swift and Posner, 1972), takes place in the alkaline media. Nowadays, when sodium hydroxide solutions are used, the soil-solvent medium is flushed with nitrogen, extraction is carried out in an atmosphere of nitrogen, and sometimes reducing substances such as stannous chloride (Choudri and Stevenson, 1957) are added. Such laboratory practices can be expected to decrease the formation of artefacts, and are encouraged by the chemical, spectroscopic, and gel filtration data of Schnitzer and Skinner (1968). These indicate that the composition and properties of fulvic acids (see Section 3.3.1) are not significantly changed by solution contact with the base in an atmosphere of nitrogen. However, fulvic acids are the most highly oxidized of the polymeric humic substances.

Bremner and Lees (1949), in their search for solvents which would produce the least amounts of humic artefacts, showed that sodium and potassium salts of inorganic and of organic acids could be used to extract significant yields of humic substances. Of these the 0.1M solution of sodium pyrophosphate ($Na_4P_2O_7$), neutralized by addition of phosphoric acid, was best. This was followed, in decreasing order of efficiency for the sodium salts of the inorganic series, by the fluoride, hexametaphosphate $(NaPO_3)_6$, orthophosphate (Na_3PO_4), borate ($Na_2B_4O_7$), chloride, bromide and iodide. The order of decreasing extraction efficiencies of the organic salts was oxalate $[(CO_2Na)_2]$, citrate $[(NaO_2CCH_2C(OH)(CO_2Na)CH_2CO_2Na]$, tartrate $[NaO_2CCH(OH)CH(OH)CO_2Na]$, malate $[NaO_2CCH(OH)CH_2CO_2Na]$, salicilate (*o*-hydroxy-$C_6H_4CO_2Na$), benzoate ($C_6H_5CO_2Na$), succinate ($NaO_2CCH_2CH_2CO_2Na$), Na-4-hydroxybenzenecarboxylate, and ethanoate (CH_3CO_2Na).

The best extractants in the foregoing list form complexes with polyvalent metals bound to humic materials, and allow sodium ions to associate with the polyanions. Also the solubility is enhanced by the breakdown of the pseudo-cross-linking effects of polyvalent metals bridging charged functional groups on polymer chains. After removal of these metal ions the extents to which humic substances dissolve depends especially on the numbers and the extents of dissociation of their acid functional groups, and on the molecular weights of the polymers. However, solution will be influenced by the amount of low molecular weight electrolyte present. Additional electrolyte increases the ionic strength outside the polymer coil relative to that inside it, and decreases the thickness of the electrical double layer (see Chapter 5) causing the structure to contract. At high salt concentrations polymer concentration can be such that solution will not take place (see Section 3.5.5).

Alexandrova (1960) has shown that sodium pyrophosphate complexed calcium to such an extent that pretreatment of serosems (soils with high calcium contents) with acid prior to extraction was not necessary. Indeed, pretreatment could be expected to have a depressing effect on the solubility of humic substances in aqueous solutions of complexing agents because of the low degrees of

Table 3.1 Yields (%) and electron spin resonance (e.s.r.) data for humic (HA) and fulvic (FA) acids extracted by different solvents from a H^+-saturated organic soil (from Hayes *et al.*, 1975a). Reproduced by permission of Elsevier Scientific Publishing Co.

Solvent	% Yield			e.s.r. data (Spins $g^{-1} \times 10^{-16}$)		
	HA	FA	Total	HA	FA	H+FA
2.5M EDA	49.0	14.0	63.0	15.0	27.5	
EDA (anhydrous)	2.0	3.0	5.0	6.4	12.8	
0.5N NaOH	58.0	2.0	60.0	4.6	0.4	
0.1M $Na_4P_2O_7$	13.7	0.8	14.5	4.5	1.9	
1M Na^+-EDTA	12.5	3.8	16.3	0.3	0.3	
Pyridine	34.0	2.0	36.0	n.d.	n.d.	2.1
DMF	16.0	2.0	18.0	n.d.	n.d.	1.4
Sulpholane	10.0	12.0	22.0	n.d.	n.d.	1.0
DMSO	17.0	6.0	23.0	n.d.	n.d.	4.2

EDA = ethylenediamine; EDTA = ethylenediaminetetraacetic acid; DMF = dimethyl formamide; DMSO = dimethylsulphoxide; n.d. = not determined.

ionization of hydrogen-ion saturated (H^+-) humic substances. This is illustrated for extractions with pyrophosphate ($Na_4P_2O_7$, neutralized to pH 7.0 with phosphoric acid) and the sodium salts of ethylenediaminetetraacetic acid (EDTA) in Table 3.1. The poor solvent effects of these compounds for a H^+-organic soil can be explained by the fact that most of the polyelectrolytes were not significantly expanded in the unbuffered, and hence acidic media. The extractability in sodium hydroxide is generally significantly improved by acid pretreatment and removal of the excess acid, but Kononova and Bel'chikova (1961) have presented evidence which is widely accepted, to show that prior decalcification of soil with acid can be avoided by using an aqueous solution of 0.1M pyrophosphate and 0.1M NaOH.

3.2.3 Extraction with Organic Bases

Dryden (1952) has drawn attention to the usefulness of primary aliphatic amines, and particularly ethylenediamine (1,2-diaminoethane) or EDA, as solvents for brown- and low rank vitrain-type coals. He reasoned that the unshared electron pairs on the nitrogen atoms can partially accept a hydrogen atom attached to another molecule to form a hydrogen bond, and this would be the basis for a good solvent effect, as outlined among the criteria of Whitehead and Tinsley (1964).

Hayes *et al.* (1975a) have investigated the solubilities in anhydrous EDA, aqueous 2.5M EDA, sodium pyrophosphate, EDTA, pyridine and the dipolar aprotic solvents, dimethylformamide (DMF), sulpholane, and dimethylsulphoxide (DMSO). Aqueous EDA, with a medium pH of 12.6 (Table 3.2) extracted the greatest total quantity of humic materials, and this can be attributed

Table 3.2 Sequential extraction of H^+-organic soil samples using water and 1/1 solvent–water mixtures (after Hayes *et al.*, 1975a). Reproduced by permission of Elsevier Scientific Publishing Co.

Solvent[a]	Yield[b] (% of organic matter)			pH values	
	HA	FA	Cumulative	Extractant	Extract
Water	0.0	2.8	2.8		
DMF	15.0	2.2	20.0	6.8	2.9
Sulpholane	4.1	1.0	25.1	3.7	1.2
DMSO	0.7	0.2	26.0	5.9	1.3
Pyridine	14.8	0.6	41.0	11.6	4.2
EDA	23.2	6.3	70.9	13.0	12.6

[a] Solvent symbols are the same as those used in Table 3.1.
[b] HA = humic acids; FA = fulvic acids.

to the basicity of the medium, and to the ability of the organic compound to link to the polymer. However, most of the fulvic acid (FA) materials precipitated during dialysis. This precipitate was soluble in sodium hydroxide and reprecipitated on acidification to pH 1.0, and it would normally be classed as humic acid (see Section 3.3.1).

Table 3.3 provides evidence that EDA significantly increases the nitrogen contents of humic substances, and this could not be lowered further by repeated washings with, or dialysis against dilute acid solutions. Dryden (1952), Rybicka (1959) and others have observed the same behaviour for coal extracts, and the nitrogen is thought to be incorporated by formation of covalent linkages between EDA and quinone-type structures in the humic polymers, as illustrated in reaction scheme (3.1).

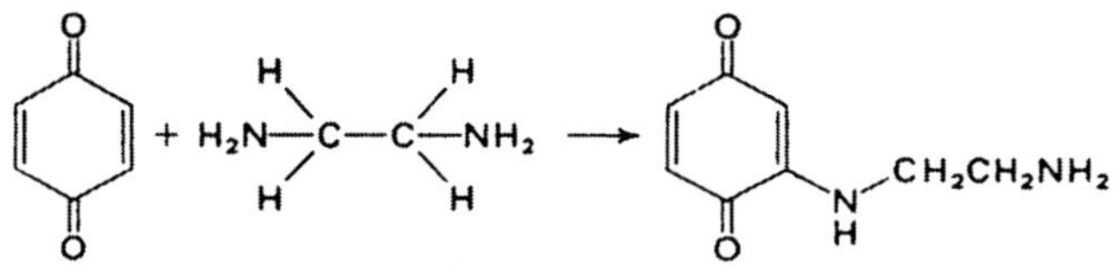

(3.1)

Anhydrous EDA is a poor solvent for dried organic matter (Table 3.1). This emphasizes the importance of hydrating the anionic species in the polyelectrolyte. Also this organic molecule would have difficulty in penetrating into the dry polymer matrix held together by van der Waals and hydrogen bonding forces.

Pyridine, with a pK_b value of 8.75 is a very much weaker base than EDA (pK_{b1} = 3.3, pK_{b2} = 6.44) and accordingly it is a poorer solvent than EDA for humic substances (note the pH of H^+-humic substances in pyridine, Table 3.2). Again, like EDA, it incorporated nitrogen into the polymers (Table 3.3), but it is very likely that this resulted from adsorption and not from the formation of covalent linkages. Adsorption of heterocyclic aromatic nitrogen compounds by humic substances is well known in the pesticide field (Hayes, 1970).

Table 3.3 Elemental composition (%) of humic acids (HA) and fulvic acids (FA) extracted with different solvents from an H^+-organic soil (after Hayes *et al.*, 1975a). Reproduced by permission of Elsevier Scientific Publishing Co.

Solvent[a]	C		O[b]		H		N		S	
	HA	FA	HA	FA	HA	FA	HA	FA	HA	FA
2.5M EDA	56.8	51.2	29.2	30.3	5.9	5.7	6.4	11.1	n.d.	n.d.
Pyridine	55.9	47.1	32.9	39.9	5.1	5.3	4.4	6.0	n.d.	n.d.
DMSO	55.0	55.0	35.5	37.1	4.2	4.4	3.3	2.2	2.0	1.3
Sulpholane	54.4	53.2	35.2	37.4	4.8	4.4	3.2	3.3	2.4	1.7
DMF	54.3	52.3	36.8	38.8	4.6	4.1	2.6	3.2	1.7	1.6
0.5N NaOH	53.1	45.0	36.3	43.0	6.0	6.0	2.9	4.3	n.d.	n.d.
Na^+ EDTA	52.1	48.4	—	—	4.1	4.2	n.d.	n.d.	n.d.	n.d.
0.1M $Na_4P_2O_7$	50.9	37.3	41.1	50.9	3.3	5.1	3.0	5.0	n.d.	n.d.

[a] Solvent symbols are given in Table 3.1.
[b] Where not determined, *S* values are assumed to be 1% for estimation of % O by difference; n.d. = not determined.

3.2.4 Extraction with Dipolar Aprotic Solvents

Because the acid groups of H^+-soil organic materials are largely undissociated, humic substances will have solubility properties in dipolar aprotic solvents (see pH values of extracts, Table 3.2) similar to those of a number of neutral polymers. For solution to take place a decrease must take place in the Gibbs free energy (G) of the system, and the free energy of mixing, expressed by the relationship

$$\Delta G_m = \Delta H_m - T\Delta S_m$$

will be negative (where T is the temperature, K; $\Delta H_m = -\Delta Q$ refers to the heat of mixing; ΔS_m is the change in entropy and is related to the increase in freedom of movement or disorder in the solution). The heat of mixing may be represented by the semiempirical Scatchard–Hildebrand equation (as outlined by Hildebrand and Scott, 1962)

$$\Delta H_m = V(\delta_1 - \delta_2)^2\phi_1\phi_2$$

where V is the total molar volume of solution, and δ_1 and δ_2, and ϕ_1 and ϕ_2 are solubility parameters and volume fractions, respectively, of solute and of solvent.

A negative value for ΔG_m is assured if the enthalpy value ΔH_m is small or negligible, and this will happen when the intermolecular interactions between solvent–solvent and polymer–polymer molecules are similar. This similarity in intermolecular interactions reflects similarities in δ values, and is an indication of what is known as the *cohesive energy density* (CED) value of the system. The CED measures the energy required to create a 'hole' in the solvent large enough to accommodate the solute and to promote solvation.

Enthalpies of mixing will approach zero when the solubility parameters are similar. Seymour (1971) has pointed out that, provided solvents are not dis-

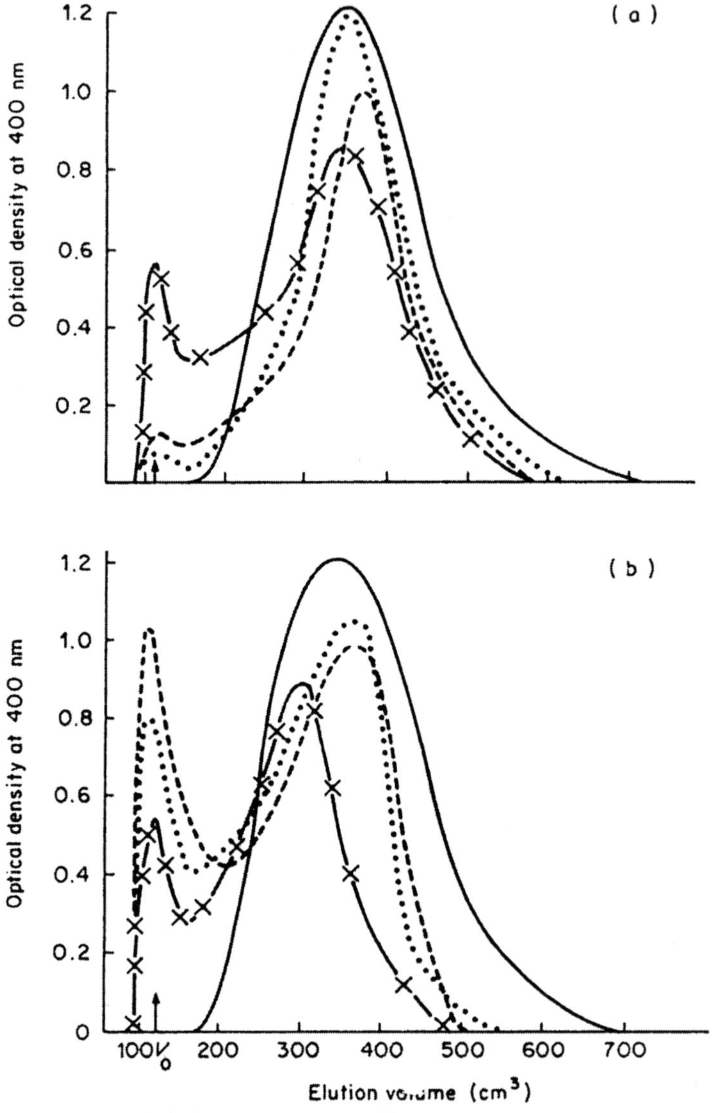

Figure 3.1 Gel chromatography of humic substances isolated with dimethyl formamide (———), pyridine (- - - - -), dimethylsulphoxide (· · · · · ·), and ethylenediamine (-×——×——×-) in single (a) and sequential (b) solvent extraction. (After Hayes *et al.*, 1975a.) Reproduced by permission of Elsevier Scientific Publishing Co.

similar in molecular structures, the solubility parameters of a solvent mixture will generally be equal to the sum of the products of the mole fractions and solubility parameters of each component. For that reason polymers which are insoluble in individual solvents may be soluble in mixtures of these.

The data in Table 3.1 shows that the total amounts of organic matter extracted by the three dipolar aprotic solvents were similar although the quantities of fulvic acids extracted were different. In addition the elemental compositions of the different humic and fulvic acids were very similar (Table 3.3) and these data indicate that the fulvic acids were less highly oxidized than those obtained by sodium hydroxide and pyrophosphate solutions. The solubility parameters (δ values) for DMF and DMSO (Tobolsky and Yannas, 1971) are nearly the same (12.4 and 12.93, respectively), which might partially explain the comparable solubilizing effects of the two solvents.

Figure 3.1a shows that, compared with EDA, only a trace of DMSO and none of the DMF extracted humic acids were of sufficiently high molecular weight to elute at V_0 (the void volume) during gel chromatography on Biogel P300 (see Section 3.5.2). Therefore it would appear that the dipolar aprotic solvents preferentially extracted the less highly oxidized and lower molecular weight fractions of humic substances. Such solvents may therefore be of value for selectively extracting relatively homogeneous humic materials.

3.2.5 Sequential Extraction Procedures

Table 3.2 provides data for yields of humic materials sequentially extracted with water, dipolar aprotic solvents, pyridine, and EDA. The technique involved exhaustively extracting with a single solvent system before proceeding to the next in the series. Similarities in the solution properties of the dipolar aprotic solvents is emphasized by the low yields extracted by DMSO in particular, which was preceded in the solvent series by DMF and sulpholane. Figure 3.1b shows that, as expected, the molecular weight distribution of extracted humic acid in the first organic solvent used was the same as for the single extraction (Figure 3.1a.) However, the amounts of high molecular weight materials from the subsequent extractions with DMSO and pyridine were increased. It is clear too that most of the lower molecular weight materials extractable in EDA were removed by the earlier solvents used. These data suggest that sequential extraction presents a further refinement of procedures for obtaining less heterogeneous fractions of humic materials.

3.2.6 Artefact Formation

Reference has been made to oxidations of humic substances during extractions with sodium hydroxide (particularly under aerobic conditions) and to the additions of nitrogen from EDA and pyridine solvent systems. Electron spin resonance (e.s.r) data (Table 3.1) suggest that free radicals were formed in the polymers (see Section 3.5.1) during contact with EDA. There is, however, no evidence for artefact formation during extractions with dipolar aprotic solvents and EDTA, and none would be expected during solution in pyrophosphate.

Oxidation of phenols to quinones can proceed by a free radical intermediate mechanism, but comparisons between e.s.r. data for the pyrophosphate and sodium hydroxide extract (Table 3.1) suggest that stable free radicals were not formed during contact with the base. Therefore it would appear that sequential extractions with non-polar solvents to remove waxes, followed by dipolar aprotic solvents, pyrophosphate in neutral media, and 0.5N sodium hydroxide would provide a good solvent series for the isolation of humic substances.

3.3 FRACTIONATION AND PURIFICATION OF HUMIC EXTRACTS

Brief reference has already been made to the heterogeneity of composition of soil organic matter extracts. Thus it is common practice to follow extraction by fractionation procedures in order to obtain materials which are less heterogeneous than the original extract and which can be characterized to some degree.

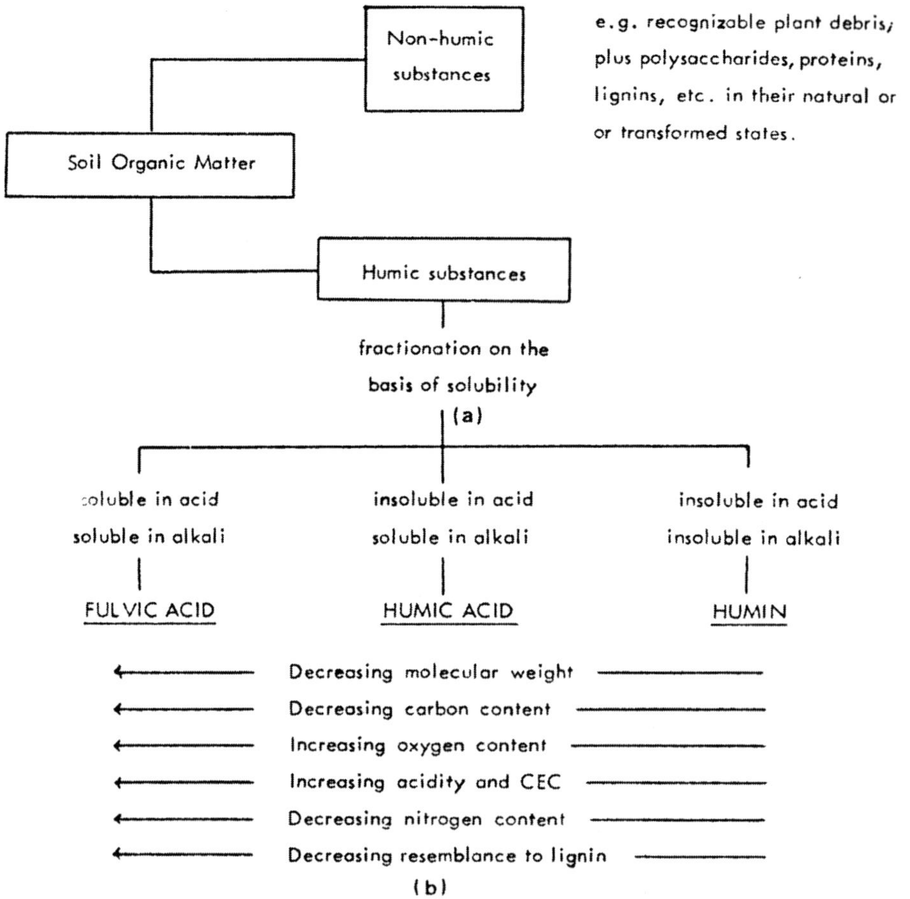

Figure 3.2 Fractionation of soil organic matter and humic substances (a), showing (b) how some properties vary in the fulvic acid–humin range

Such fractionations sometimes fulfil a dual purpose because the same techniques often allow the removal of some non-humic substances and hence permit a degree of purification to be achieved. In general the techniques used for purification and fractionation attempt to exploit differences in solubilities, adsorption behaviour, and molecular weight or charge characteristics either between the humic and non-humic substances, or between the different components of the humic substances.

3.3.1 Fractionation Based on Solubility Differences

Classically, the fractionation of the humic substances is based on differences in solubilities in acid and in alkaline media. The techniques involve extraction with an alkaline reagent, separation of the soluble extract from the soil residue, and then acidification of the extract. This procedure gives rise to three gross fractions (Figure 3.2a) called *fulvic acids*, soluble in acid and in alkali; *humic acids*, soluble in alkali but precipitated by acid; and *humins*, insoluble in acid and in alkali. Some workers further subdivide humic acids by addition of salt to give *gray* and *brown humic acids*, or by extraction of alcohol soluble *hymatomelanic acids*. The same nomenclature is used for the three main fractions even when the extraction is carried out using neutral salts, complexing agents, or organic solvents in place of the traditional alkali.

The classical fractionation procedures are discussed in detail by Kononova (1966) and by Stevenson and Butler (1965). Because of the nature of the extraction and fractionation processes it is inevitable that certain non humic impurities will accumulate in particular fractions. Thus, for example, some polysaccharides and numerous low molecular weight compounds are found in the fulvic acid fraction, highly lignified materials are found in the humic acid fraction, and inorganic contaminants, including colloidal materials and some salts are common to all fractions.

3.3.2 Purification of Humic Fractions

Humic extracts from mineral soils are generally contaminated with silicates and salts, and further quantities of the latter are generated during acidification of inorganic reagent extracts. Silicates can be dissolved by treatment with HF or with HCl–HF mixtures. Mortensen and Schwendinger (1963) effectively removed clay from their soil extracts in pyrophosphate by treating the fractions for 10 minutes at 4 °C with a solution 0.3 molar with respect to HCl and HF. The solution was then neutralized and dialysed. Dubach and Mehta (1963) have reviewed how preparations with very low ash contents can be obtained by repeated solution-centrifugation-precipitation techniques, or by treatment with complexing agents and ion-exchange resins, etc. Most of the non-polymeric organic materials, as well as salts, are removed during dialysis.

The fulvic acid fraction is composed of a variety of heterogeneous components, including polysaccharides (Section 3.6) which can be separated by adsorption on charcoal (Forsyth, 1947), or by use of gel and resin chromatography techniques (Barker *et al.*, 1967; Swincer *et al.*, 1968). In general humic acids are less contaminated by non-humic organic polymers and lend themselves more readily to purification procedures.

In many recent researches saccharide and protein or peptides are removed from humic acid by prolonged hydrolysis in 6M HCl. Decarboxylation takes place during hydrolysis (Deuel and Dubach, 1958a, 1958b), although Riffaldi and Schnitzer (1973) have indicated that charge, free radical and other properties of humic acids are not significantly affected by this treatment.

The three gross fractions of humic substances should not be regarded as distinct or discrete components, but should be considered as a spectrum of substances, or related molecules which display continuous variations with respect to certain properties. Amongst the properties which vary are molecular weight, cation–exchange capacity, elemental composition, etc. (Figure 3.2b). Each of these fractions can be refractionated to bring about further reductions in heterogeneity. The techniques generally employed are:

(i) electrophoresis, electrofocusing, and ion exchange—all of which exploit differences in charge and charge density;
(ii) gel chromatography which fractionates on the basis of molecular weight and size differences;
(iii) adsorption chromatography on a variety of media;
(iv) the further use of solubility differences, employing fractional precipitation with solvents, salting out and changes in medium pH.

Several of these procedures and their specific application to humic substances will be discussed in more detail in Section 3.5. Suffice it to say here that each of the techniques always produces a fractionation, but the different fractions can themselves be refractionated. Thus it will be difficult to isolate a single, homogeneous humic entity because such species are probably present in very low concentrations among related molecules in the polydisperse system. However, diligent application of fractionation techniques to humic substances enables the isolation of preparations with greatly reduced heterogeneity. This allows such preparations to be examined by a variety of physico-chemical techniques, but more important it permits more meaningful results to be obtained from structure (primary) and shape (tertiary structure) studies.

3.4 CHEMICAL COMPOSITION OF HUMIC SUBSTANCES

We define as chemical composition the elemental contents, and the numbers and types of functional groups (carboxyl, hydroxyl, etc.) and of organic mole- molecules or 'building blocks' which make up the humic polymer structures. Elemental analysis data can be readily obtained by standard procedures, and techniques have been developed which allow determinations to be made of many of the functional groups present. Difficulties are being experienced in finding suitable degradative reagents and conditions which allow the isolation of structurally meaningful organic molecules, although considerable progress has been made in this area of late.

Synthesis by chemical methods, and by microbial and enzymatic procedures offers alternatives to structural determinations by degradation reactions. Interest in non-enzymatic chemical synthesis has declined in recent times, although considerable research effort is now being directed to microbial and enzymatic procedures. These enable the disappearance of monomer units into the polymer structures to be followed, and the reappearance of the same units in digests after chemical (or enzymatic) degradation of the polymer is regarded as good evidence for their involvement in the 'humic-type' structures.

3.4.1 Elemental Analysis

Elemental or ultimate analysis involves the determination of the elements present in a compound. Dividing the percentage of each element in the compound by its atomic weight gives the ratios for the different elements present, and the compound's empirical formula can then be calculated by dividing all of the ratio values by the smallest of these. A knowledge of the molecular weight of the compound is required in order to determine its molecular formula, which will be some multiple of the empirical formula.

Elemental analysis and the data which it gives are used to check the purity of compounds and, together with chemical and spectroscopic data, to assign structures to organic compounds. In cases of complex structures, such as those of humic materials, ultimate analysis can only be interpreted in a very general way in structural terms, even when molecular weight values are available. They do, however, provide accurate values for the proportions of the elements in the different samples, and of course they allow comparisons to be made between samples.

Routine procedures are available for determinations of the carbon, hydrogen, oxygen, nitrogen, sulphur and ash contents; data for the elements are most useful when calculated on a dry, ash-free basis. Carbon and hydrogen are determined by combusting a weighed sample of material (usually in the presence of a CuO catalyst) in a stream of oxygen. The water and carbon dioxide given off are collected in anhydrous magnesium perchlorate [$Mg(ClO_4)_2$] and soda lime, respectively, and determined gravimetrically. Nitrogen is estimated from micro-Kjeldahl-, which uses selenium–mercury catalysts, or Dumas-digestion procedures, and the oxygen-flask method gives good determinations of the organic sulphur contents (MacDonald, 1965). Oxygen is determined by difference,

$$\%O = 100-(\%C+\%H+\%N+\%S),$$

or by direct methods. The best of these involve conversion of the organically bound oxygen to CO and its determination by instrumental procedures (Schöniger, 1969), especially by gas–liquid chromatography (Boos, 1964). The accuracy of such methods is impaired when various metals and phosphate and fluoride ions are present; consequently they should only be applied to H^+-saturated humates which have low ash contents. Ash can be determined gravimetrically by combusting the substrate in a stream of oxygen in a furnace at 400–500 °C.

The data in Table 3.3 show how the elemental compositions of humic and fulvic acids are influenced by the solvent used to extract the organic matter source. Reference has already been made to the different solvating effects of the extractants used and to the close similarity between the compositions of the humic and fulvic acid materials extracted by the dipolar aprotic solvents DMSO, sulpholane, and DMF. On the other hand, the elemental composition data suggest significant differences between the humic and fulvic acid materials extracted under basic conditions or in the presence of the chelating agents.

Thus close similarity should not be expected between humic materials extracted with different solvents from the same or from a variety of soils. Where the same extractant is used, however, elemental composition data allow interesting comparisons to be made between humic substances from different soils and sources. Table 3.4 provides such data for substances extracted in alkali from soils and coals. The fungal humic acids listed were obtained as precipitates at pH 2.0 from filtered culture media which had been incubated for seven weeks with the *Stachybotrys* fungal species, *S. chartarum* and *S. atra*.

The sod podzolic humates were from northern USSR, the chernozems were from the Ukraine, the inceptisols from Dominica in the West Indies, and Table

Table 3.4 Elemental composition (%) of humic acids (HA) and some fulvic acids (FA) extracted by alkali from different soils and materials

Sample	%C		%O		%H		%N	
	HA	FA	HA	FA	HA	FA	HA	FA
Sod podzolic[a]	57.6	42.6	35.3	44.6	5.2	5.0	4.8	4.1
Chernozem[a]	62.1	44.8	31.4	49.4	2.9	3.4	3.6	2.3
Inceptisol[b] (surface)	54.5	42.8	34.1	47.7	5.3	3.8	5.5	2.0
Inceptisol[b] (sub surface)	51.4	46.9	36.0	44.6	6.7	4.5	4.9	2.3
Brown coal[c]	63.1		(32.7)		2.8		1.4	
Brown coal[d]	65.4				4.0			
S. chartarum[e]	57.3		27.7		7.0		6.8	
S. atra[e]	53.1		34.8		5.7		5.4	

[a] From Kononova (1961).
[b] From Griffith and Schnitzer (1976).
[c] From Lawson and Purdie (1966).
[d] From Yokakawa *et al.* (1962).
[e] From Schnitzer and Neyroud (1975).

3.4 shows a carbon content maximum for the humates from the intermediate latitude. There was a close similarity in the composition of the chernozem humic acid and that of the brown coal from central England described by Lawson and Purdie (1966). Clearly there are only about half as many hydrogen as carbon atoms in these structures, and this could indicate extensive substitution or branching. The oxygen contents listed for all of the soil fulvic acids indicate that these compounds were highly oxidized.

Elemental analysis for the fungal 'humic' acids are presented for purposes of comparison. Inspection of the data for the *S. atra* product shows that its elemental composition is of the order of that for the inceptisol surface and subsurface soil humic acids. Those for *S. chartarum* are less comparable with the usual values for the soil humic substances.

Conclusions should not be reached about similarities between humic substances on the basis of ultimate analysis data alone. It would for instance, be highly erroneous to infer from the data in Table 3.4 that the humic acid material from *S. atra* is similar to that from either of the inceptisol samples, or that those from the chernozem and brown coal are the same. Such statements could only be made with any degree of conviction if the same functional groups were present in more or less the same abundances, if the numbers and types of degradation products from specific degradation reactions were the same, if the molecular weight distributions were closely similar, and if the shapes of the component molecules were comparable. It is very unlikely that all of these requirements for comparability would be met by any two highly polydisperse humic products from different soil or source types. For valid comparisons to be made it would be necessary to analyse products with decreased polydispersities by additional procedures which will be described in Section 3.5.2.

3.4.2 Determinations of Functional Groups

The numbers and the types of oxygen-containing functional groups greatly influence the reactivities of humic substances in the soil environment. Undoubtedly the carboxyl and phenolic structures are the most important of these because they are the major groups responsible for the contribution by organic matter to the cation-exchange capacity (CEC) of soils, and they can have chelating effects (Schnitzer and Skinner, 1965a) which might influence plant nutrition. Most of the carboxyl groups dissociate to the carboxylate anion at the pH values likely to be encountered in fertile soils. Phenol and hydroxybenzene derivatives containing activating substituents on the aromatic nucleus are not substantially dissociated when the pH is less than 9.0, but deactivating substituents can sufficiently lower the pK_a values to allow them to contribute to the soil CEC. Hydroxyquinone structures are also highly acidic although it is not certain to what extent, if any, they are present in humic substances.

It is not surprising, therefore, that substantial interest and research effort has been focused on investigations of the types and numbers of oxygen-containing functional groups in humic substances. It will be seen that accurate measurements and interpretations of data are sometimes difficult because of the complex chemical and steric environments of these groups.

Oxygen-containing functional groups

Carboxyl, phenolic hydroxyl, alcoholic hydroxyl, carbonyl and possibly quinone, and methoxyl groups are considered to be the principal oxygen-containing functional groups in humic substances. It is highly probable that additional ether groups, and in the light of recent researches some ester groups are also present.

Mention has been made of the complex chemical environments of the functional groups, and of the contributions of the carboxyl and of phenolic hydroxyl groups to the acidities of humic materials. Enolic structures can also be acidic and this adds a further complication to the analytical problem because none of these groups can be determined accurately by pK dependent techniques such as titration. This can be explained on the basis of the ranges of pK_a values for each group which result from the influences of neighbouring groups and from steric effects.

Determination of total acidity. Two methods have become generally accepted for this determination. The first involves the use of diborane (B_2H_6), which liberates gaseous hydrogen from even sterically hindered acidic groups in H^+-ion saturated preparations, and was used by Martin *et al.* (1963) to determine the total acidity in fulvic acid from a podzol B_h horizon. Their results were in good agreement with the second, or the barium hydroxide method.

The $Ba(OH)_2$ method was developed by Brooks and Sternhell (1957) to determine the acidity in preparations of brown coal, and it was first applied to soil humic materials by Wright and Schnitzer (1959a, 1960). When reacted with H^+-saturated humic substances the following reaction takes place:

$$Ba(OH)_2 + 2H^+\text{-humate} \longrightarrow Ba^{2+}(\text{humate})_2 + H_2O$$

A modification of the procedure by Brooks and Sternhell was described by Schnitzer and Gupta (1965) and is also presented in Schnitzer and Khan (1972). In practice an excess of $Ba(OH)_2$ is reacted with the humic materials and the unreacted base is back-titrated with standard acid. It is assumed that all of the exchangeable hydrogen ions are replaced by Ba^{2+}. This assumption need not always be valid, however. For instance Schnitzer and Gupta (1965) found that some substituted phenols did not quantitatively replace acidic hydrogen with barium ions, yet the $Ba(OH)_2$ procedure gave values for the total acidity of humic substances which were in close agreement with those for the diborane method. Similar low replacement effects might be expected for the humic substances because of steric effects which could prevent access to all of the exchange sites. On the other hand, barium taken up by chelation or complex formation would lead to falsely high values for total acidity.

It is, of course, possible to determine total acidity directly by titration of H^+-humic materials with strong base. The spread of pK_a values for the acidic groups of the polymers complicates this procedure, however, because it is difficult to locate the end point on the titration curve. Arbitrary end-points are taken at pH values of 7.0 or 7.5, and the acidities calculated from such titration data are reasonable indications of the sum of the contributions which the various acidic groups can make to the CEC of the soil. It is necessary to carry out titrations to higher pH values in order to estimate the total acitity of the sample being investigated.

Diazomethane reacts with functional groups containing acidic hydrogens to give ester products from carboxylic acids and methyl ethers from phenolic hydroxyls, as shown in reaction scheme (3.2)

$$\begin{aligned} RCO_2H + CH_2N_2 &\longrightarrow RCO_2CH_3 + N_2 \\ PhOH + CH_2N_2 &\longrightarrow PhOCH_3 + N_2 \end{aligned} \tag{3.2}$$

where Ph represents a phenyl group. The total acidity can then be estimated from the increase in the $-OCH_3$ content. Schnitzer (1974a) found that the acidity of seven humic acid samples measured by this procedure ranged from *ca.* 65–97% of that determined by $Ba(OH)_2$.

Determination of carboxyl groups. Carboxyl groups are the major contributors to the acidic properties of humic substances in what might be termed soils of normal pH values. The presence of the carboxyl group has been demonstrated by several workers who observed shifts in the maxima of the carbonyl bands from 1700–1750 cm^{-1} for the undissociated carboxyl to 1600–1650 cm^{-1} for dissociated carboxyl groups (e.g. Farmer and Morrison, 1960; Martin *et al.*, 1963; Theng and Posner, 1967), and by the disappearance of the carboxyl band after treatment with diborane (Martin *et al.*, 1963).

The amounts of carboxyl present in humic substances are generally determined by the calcium acetate method, described by Schnitzer and Gupta (1965), in which the carboxyl of H^+-humate reacts with excess calcium acetate, as shown in reaction scheme (3.3).

$$2RCO_2H + (CH_3CO_2)_2Ca \longrightarrow (RCO_2)_2Ca + 2CH_3CO_2H \tag{3.3}$$

The released ethanoic acid is measured by titration. To be effective the technique requires that Ca^{2+} ions should have access to and be able to exchange with all of the hydrogens on the carboxyl groups, and that exchange should not take place with hydrogen ions from the acidic phenolic groups. Because calcium acetate reacts with the more readily dissociated phenolic groups, as shown in experiments with model substances by Schnitzer and Gupta (1965), the method might be expected to give overestimates of carboxyl numbers. However, the bridging by calcium between carboxylate groups on different molecules (or randomly between such groups on the same polymer molecule) gives rise to aggregated and more compact structures and increases the difficulties of diffusion to and from the exchange site (see Sections 3.5.4 and 3.5.5). Such structural modifications would inhibit reaction with carboxyl groups in the interior of the aggregates, but these effects might be compensated by exchange with appropriate available phenolic or enolic structures to give, fortuitously, carboxyl values comparable with those obtained by other methods.

An alternative procedure uses $CuSO_4$–quinoline reagent (Hubacher, 1949) to decarboxylate the carboxyl groups. In theory this should be a more accurate method. Schnitzer and Gupta (1965) found that results from this method agreed well with those from the calcium acetate procedure.

Determination of phenolic hydroxyls. Despite inherent inaccuracies, the amounts of phenolic hydroxyl in humic substances is generally assessed as the difference between the total acidity and the carboxyl group contents. Wright and Schnitzer (1960) have described a modification of the procedure of Ubaldini and Siniramed (1933) which involves refluxing the humic material with alcoholic KOH. After removal of the excess base the residue is equilibrated with an 85% ethanol solution saturated with CO_2. Potassium released by the ethanol treatment, determined as K_2CO_3 by titration, is considered to be displaced from phenolic groups only. However, the technique cannot be relied upon for use with humic substances which have phenolic structures with a spread of pK_a values, and it is thus no more attractive than the difference method.

Determination of total hydroxyls. Total hydroxyl group content can be estimated following acetylation by means of acetic anhydride in pyridine (Brooks *et al.*, 1957), as shown in reaction scheme (3.4). Then the ester product formed can be separated, hydrolysed in base, and the ethanoic acid steam distilled from the acidified mixture and estimated by titration. Alternatively, a less favoured procedure involves determination of the ethanoic acid released during aqueous titration of the excess acetic anhydride (Wright and Schnitzer, 1960).

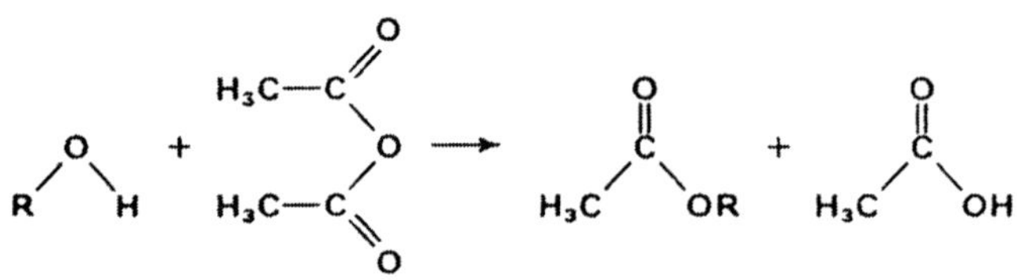

(3.4)

Determination of alcoholic hydroxyls. As yet there is no satisfactory procedure for the determination of alcoholic hydroxyl groups and these are usually unreliably estimated as the difference between the total hydroxyl and the phenolic hydroxyl contents.

Determination of carbonyl and quinone groups. Values for the carbonyl contents of humic substances can be obtained by determining the increases in nitrogen contents of humate derivatives formed by reactions with hydroxylamine (Flaig, Scheffer and Klamroth, 1955; Schnitzer and Skinner, 1965), phenylhydrazine (Schnitzer and Skinner, 1965), or 2,4-dinitrophenylhydrazine (Flaig *et al.*, 1955; Schnitzer and Skinner, 1965). The generalized reactions for hydroxylamine and 2,4-dinitrophenylhydrazine are outlined in reaction scheme (3.5) where R represents aliphatic or aromatic humic substituents and R′ would ideally be similar type structures, or hydrogen.

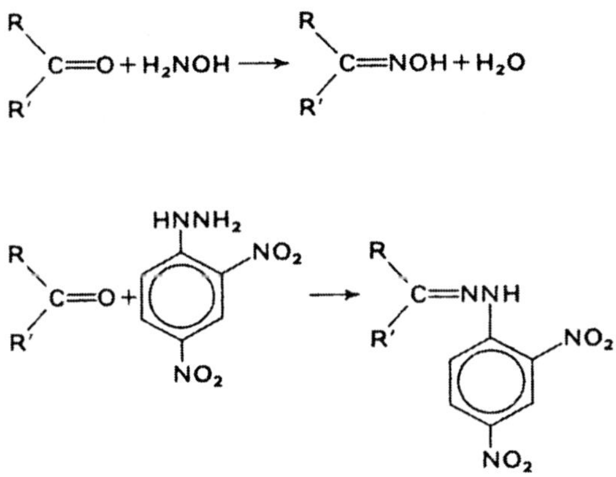

(3.5)

Other procedures determine the amounts of unreacted reagents (Brooks *et al.*, 1958; Schnitzer and Skinner, 1966). In this context Schnitzer and Skinner (1966) developed a polarographic procedure for the determination of carbonyl groups in humic substances. The 2,4-dinitrophenylhydrazone derivative was prepared, as outlined in scheme (3.5), by refluxing the substance in an excess of 2,4-dinitrophenylhydrazine reagent, and the carbonyl content was estimated from the amount of unreacted reagent, which could be measured from the polarographic reduction of the nitro groups.

Flaig and his colleagues (see Flaig, Beutelspacher and Rietz, 1975, pp. 75–95) have placed considerable emphasis on considerations of polyphenols and of polyquinones in humic structures. Quinones can form charge-transfer complexes, and this might provide a mechanism for the binding of some pesticides by humic substances (Hamaker and Thompson, 1972). They could also be involved in forming metal humate complexes and in hydrogen bonding. Foster and Foreman (1974) have reviewed the chemistry and the detection of various complexes of quinone structures.

It is unfortunate, in view of their probable presence and importance, that satisfactory procedures are not available for the unambiguous identification and the quantitative determinations of quinone groups in humic structures. A study of a review by Finley (1974) will show that reactions based on the formation of Schiff base structures with aldehyde and keto groups will also take place with quinone structures. Thus either the mono- or the dioxime can be formed from 1,4-benzoquinone as seen in scheme (3.6). However, the amount of oxime obtained will be governed by the extent of substitution of the quinone structure, and tetrasubstituted 1,4-benzoquinones will not form oximes for steric reasons. In addition substitution reactions, as shown in scheme (3.1), to give non-hydrolysable products can take place.

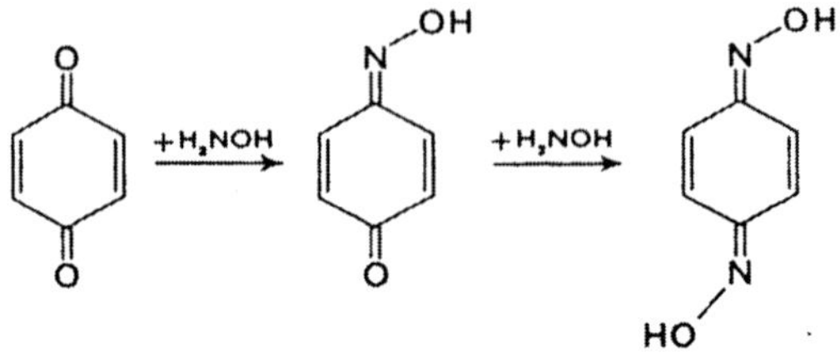

(3.6)

Obviously, from considerations of the foregoing reactions, it is not possible to differentiate between aldehyde-, keto- and quinone-carbonyl functional groups by estimating the increases in humic nitrogen contents from the formation of oximes or of derivatives of substituted hydrazines.

Berger and Rieker (1974) have stressed the usefulness of spectroscopic methods for the identification and quantitative determinations of quinones, and they have outlined in considerable detail the influences of adjacent carbon substituents on the quinone carbonyl vibration frequency in the 1630–1700 cm^{-1} region of the infrared spectrum. So far, however, spectroscopic techniques have failed to resolve the presence or otherwise of quinone structures in soil humic substances, although some success has been achieved with coal and coal humic acids (Brown and Wyss, 1955; Moschopedis, 1962; Fujii, 1963; Bailey *et al.*, 1965).

The best spectroscopic evidence for the presence of quinone structures in soil humic acids has been provided by electron spin resonance (e.s.r.) techniques (Section 3.5.1). Atherton *et al.* (1967) obtained spectra for a variety of acid-boiled humic acids and all had band breadths of the order of 1.75–1.9 gauss. The signals were eliminated when the samples were reduced with sodium dithionite but recovered when re-exposed to air. This could indicate the presence of semiquinone ion radicals, as was suggested by Steelink *et al.* (1963) and reviewed by Steelink and Tollin (1967). They suggest that polymers containing *ortho*- and *para*-quinhydrones would, under basic conditions give rise to the radical ions as indicated for *ortho*-quinhydrones in reaction scheme (3.7).

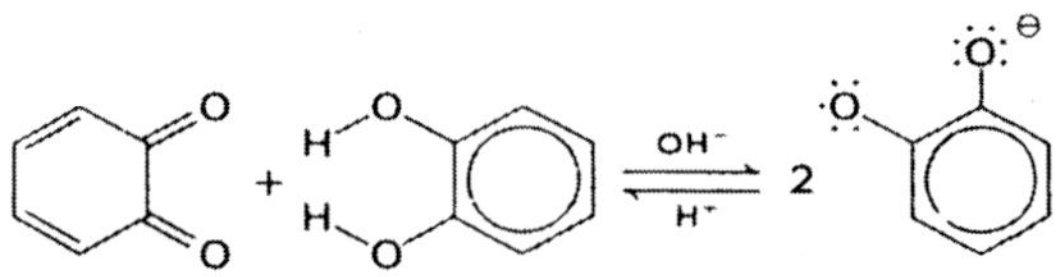

(3.7)

The reviews of Berger and Rieker (1974) and Finley (1974) draw attention to the usefulness of polarography for qualitative and quantitative studies on quinones. Flaig *et al.* (1969) have found linear correlations between the intensity of the infrared absorption of quinone carbonyls and the polarographic pulse heights at their half-wave reduction potentials. Unfortunately, as was pointed out earlier, only weak infrared absorptions indicative of soil humic acid quinones have been reported, and evidence for quinones from the polarography of humic acids is even weaker still. Should a variety of quinone structures be present, as seems likely, it is highly probable that these would have a range of half-wave reduction potentials and single, or well-separated polarographic reduction peaks would not be observed. Clearly considerable further experimentation, involving a range of buffers and a variety of treatments and of reaction conditions will be required before polarography can be properly evaluated for estimations of quinone structures in soil humic substances.

According to Vasilyevskaya *et al.* (1971) quinone structures can be quantitatively estimated from data obtained when humic substances are dissolved under N_2 in 0.1M NaOH, and solutions (2.5M) of NaOH and $SnCl_2$ are added until the final NaOH concentration is *ca.* 1.3M. If the reaction is allowed to take place for one hour at room temperature reduction of the quinone structures takes place and the unused $SnCl_2$ can be back-titrated with standard $K_2Cr_2O_7$ solution. Titanometric reduction (van Krevelen, 1961, pp. 160–176) has been used to determine quinones in coals and coal humic acids. This involves reduction, in a nitrogen atmosphere, with titanous chloride in acetic acid–hydrochloric acid (5:1), and oxidimetric titration of excess titanous chloride with ferric chloride.

Determination of methoxyl groups. Modifications of the Zeisel hydriodic acid procedure are generally used to determine methoxyl groups in lignins and in humic substances. The reaction involves boiling with HI. Iodomethane is liberated as indicated for an aromatic methyl ether structure in reaction scheme (3.8). Iodine, released from the iodic acid generated by oxidation of iodomethane, can be determined titrimetrically. Methoxyl content is highest in freshly humified plant tissue and it decreases as the humic substances become older.

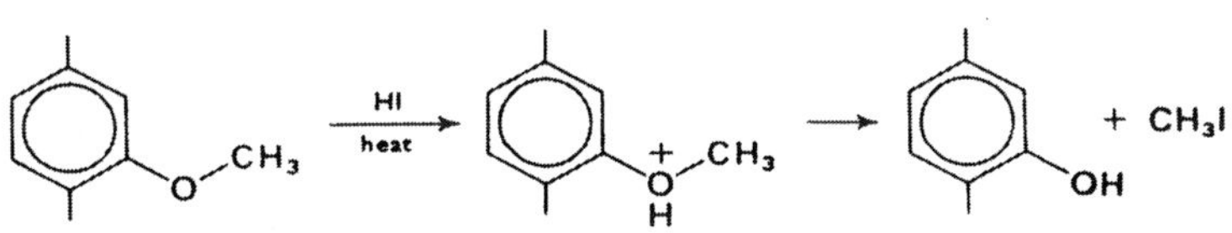

(3.8)

Distribution of oxygen among the functional groups in humic substances. Only in the case of fulvic acids does the sum of the oxygen in the carboxyl, phenolic and other hydroxyl, carbonyl, and methoxyl groups approach 100% of the oxygen contents of the humic fractions. Similar sums for brown coal- and soil humic acids, and humins range from 60–90 % of the total. Most of the oxygen of fulvic acids (the most highly oxidized component of humic materials) is in the carboxyl groups. This is generally true for humic acids also, but phenolic hydroxyls form an increased proportion of the total oxygen in these structures. Differences in the amounts of oxygen which can be assigned to detectable functional groups provide good evidence for the fact that humic structures formed under different environments can vary greatly. The unassigned oxygen might be bound, for instance, in heterocyclic structures or as phenolic ether linkages which would not be detected by the analytical procedures described. An ability to ascribe all of the oxygen to functional groups or organic structures would have an important bearing on predictions of humic polymer structures.

Nitrogen-containing functional groups

There are numerous reviews which deal with various aspects of the distribution of nitrogen in soil environments. Those by Bremner (1967) and Parsons and Tinsley (1975) are especially relevant to the contexts of this chapter.

Between 20 and 50 % of soil nitrogen is held as amino acids and 5–10 % as amino sugars (Bremner, 1967). Although amino sugars can be components of soil polysaccharides (Section 3.6) the amino acids cannot be accounted for as components of protein structures, but it is highly probable that peptide groups are present.

Hydrolytic procedures release amino acids and amino sugars, and these can be separated by use of buffers and ion-exchange resin techniques. There is an abundance of quantitative data, obtained by standard amino acid analyser techniques, for the amino acid composition of humic materials. One plausible theory proposes that much of the non-hydrolysable nitrogen is covalently linked to aromatic structures, introduced by electrophilic substitution of quinone structures as indicated by reaction scheme (3.1). Peptides could be linked in this way through the terminal amino group, or through the free-NH_2 group of diamino amino acids.

There is some evidence also for pyrrole-type structures and for some heterocyclic nitrogen in fused aromatic structures, e.g. acridine, which has been detected by zinc-dust distillation and fusion degradation processes (Section 3.4.3).

3.4.3 Degradation of Humic Substances

In order to determine the primary structure or the 'building blocks' of any naturally occurring polymer it must be degraded to the monomer units of which it is composed. This is relatively simple where the monomers are linked by labile bonds, such as the peptide bond in proteins, or glycosidic linkages in polysaccharides, since these are susceptible to hydrolysis in acid or alkaline

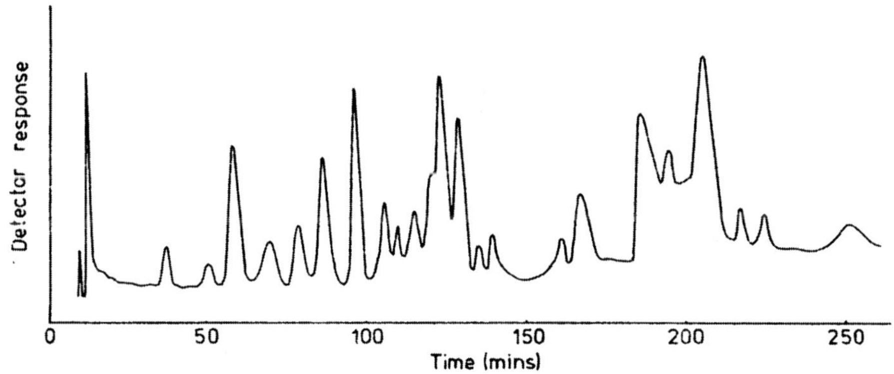

Figure 3.3 Gas–liquid chromatographic separation of low boiling products from the digest of a humic acid–sodium sulphide degradation reaction at 250 °C. (After Craggs, 1972)

conditions. However, labile bonds form only a small part of the structures in humic materials whose components are in the main linked by much stronger bonds, such as the C—C sigma bond, ether linkages, etc., cleaved only by highly energetic processes. These processes must be closely controlled to avoid degradation of the polymer to very simple but structurally meaningless products.

The most successful techniques used up to the present time for the degradation of humic substances yield large numbers of products. This is illustrated in Figure 3.3, which shows 24 compounds in the gas–liquid chromatography (g.l.c.) trace obtained by Craggs (1972) for methylated products in one fraction of a humic acid-sodium sulphide solution digest. Identification of such degradation products has been greatly accelerated by use of modern instrumental techniques such as preparative g.l.c., infrared, nuclear magnetic resonance (n.m.r.) and mass spectrometry. The direct coupling of gas chromatography and mass spectrometry (g.c.–m.s.), together with computer analysis of the output data, has provided a particularly powerful tool for identification of the many components in degradation products.

Frequently the products isolated from degradation reactions are regarded as the actual 'building blocks' of the humic polymers. Such interpretations can be false because, in many instances, the naturally occurring units are altered prior to release into the digests, and further alterations to the molecules can take place in the digests. For these reasons it is important to understand the reaction mechanisms which are likely to operate for the system being studied. This can be achieved by carrying out model studies with compounds of known structures which cleave, under the appropriate reaction conditions, to products of the types identified in the polymer degradation digest. It is important of course to use the same reagents and conditions for the polymer degradation and for model studies reactions.

Little attention will be given here to mild hydrolytic procedures which attack only the more labile components of humic substances, or to drastic conditions, such as those oxidation reactions which degrade the polymers into structurally meaningless products such as carbon dioxide, water, and short-chain aliphatic carboxylic acids. Emphasis will be placed on chemical reagents and reaction conditions that give rise to products which could reasonably be assigned to structures in the polymer molecule. As will become clear, this is often difficult because many of the compounds identified could be derived from different precursors.

Degradation by Oxidative Processes

Oxidative processes involve the loss of electrons from the substrate leading to the rupture of carbon–hydrogen and carbon–carbon bonds. Such rupture is energetically demanding; the cracking of the carbon–hydrogen bond, for instance, requires 300–400 kJmole^{-1}, and most of this energy must be supplied by a simultaneous bond making between oxidant and substrate.

Bond cleavage can be homolytic, giving rise to radical species, or heterolytic leading to the formation of ionic structures. Homolytic splitting of the carbon–hydrogen bond results from initiation and attack by molecular oxygen, although peroxides and ionizing radiation are better initiators. Once initiated the reaction can, like other free radical reactions, be readily perpetuated. Where formation of the radical involves hydrogen atom abstraction the reaction is facilitated by resonance stabilization, as in the cases of allylic [—C=C—C•] and benzylic [Ar—C•(H)(H), where Ar is the phenyl, C_6H_5 group] species. Oxidations of this type proceed more readily for hydrogen atoms on tertiary than on secondary and primary carbon atoms.

Proton and hydride (H^-) abstraction by chemical means are other well known mechanisms in oxidation reactions. Hydride extraction is well exemplified in base catalysed disproportionation reactions, such as the Cannizzaro reaction, for aldehydes lacking in α-hydrogen atoms. Benzenecarbaldehyde, for instance, when treated with base initially forms an addition product which then transfers a hydride ion to the aldehyde to give the reduction product (the alcoholate anion), and the residual oxidation product forms the acid. The reaction is completed by transfer of a proton from the acid to the alcoholate to form the alcohol and the acid anion.

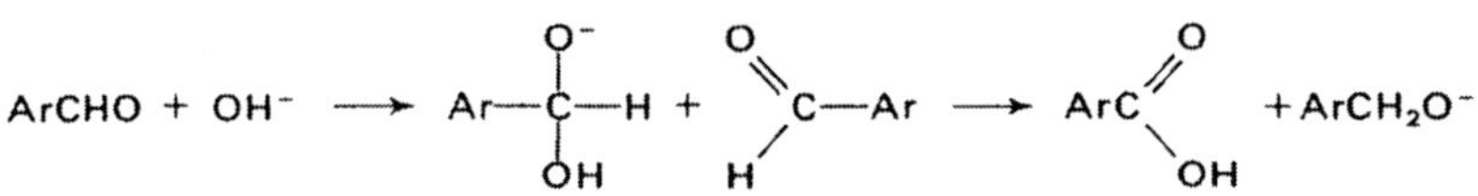

(3.9)

Other oxidation mechanisms will be dealt with as they arise in different oxidative degradation reactions.

Oxidation by permanganate. Solutions of potassium permanganate ($KMnO_4$) are extensively used for the oxidation of organic chemicals. There is extensive literature dealing with various applications of permanganate oxidation techniques and the reader is referred to reviews by Stewart (1964, 1965) for general background information. Here we shall be concerned primarily with reactions of manganese (VII) and (VI) but it is well to remember that manganese can have oxidation numbers from +1 to +7.

The permanganate species, Mn(VII), is commonly used as the starting oxidant for the degradation of humic substances. It is quite stable in acid solution, and and in 30–50 % sulphuric acid it forms the highly reactive permanganic acid ($HMnO_4$), although this decomposes to a less reactive soluble manganese (IV) species in concentrated sulphuric acid. In general the reactivity of the different species varies inversely with charge, and their oxidative effects follow the order: $HMnO_4 > MnO_4^- > MnO_4^{2-} > MnO_4^{4-}$

Because of the severe oxidation effects of $HMnO_4$ it is more usual to use a less drastic permanganate salt, e.g. $KMnO_4$ for the oxidation of humic substances. This gives a mildly alkaline solution which has the added advantage of being able to dissolve the humic substances to give a homogeneous reaction mixture. Under these conditions slow decomposition of permanganate occurs, and this is catalysed by MnO_2:

$$2MnO_4^- + H_2O \xrightarrow{MnO_2} 2MnO_2 + 2OH^- + \tfrac{3}{2}O_2 \qquad (3.10)$$

Since manganese dioxide is itself the final reaction product of oxidations by Mn(VII) it is necessary to use excess permanganate. In strongly alkaline solutions permanganate decomposes slowly to give manganate (VI) i.e. the MnO_4^{2-} species, which, although less reactive than permanganate, is still an effective oxidizing reagent.

The rates and the extents of oxidation of the different substrates are governed by their compositions, stereochemistry, the functional groups present, and especially by their solubilities in the media used. As a general rule anions are more readily oxidized than neutral molecules, which in turn are more readily oxidized than cations.

We shall now consider reactions of permanganate with some functional groups and linkages which are likely to be important in humic structures, beginning with alcoholic hydroxyls. Primary and secondary alcohols react more readily in basic than in neutral media because of the formation of the soluble anionic or alkoxide species. This will rapidly transfer a hydrogen atom or a hydride ion to the oxidant. Primary alcohols are oxidized to carboxylic

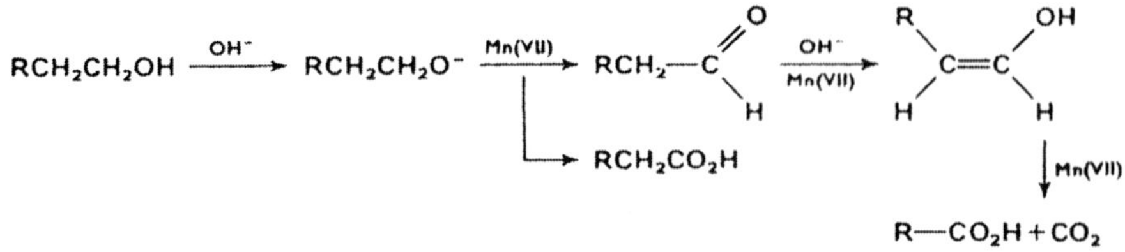

(3.11)

acids as shown in reaction scheme (3.11). Enolizable aldehydes and ketones (from secondary alcohols) can lose CO_2 to form the carboxylic acids with one carbon atom less than the original alcohol, as indicated by scheme (3.11). Tertiary alcohols resist oxidation, and are degraded only when very drastic conditions are employed.

Phenols (benzenols) and aniline (benzenamine), in which the aromatic nucleus is activated by the electron-donating substituents, are degraded to carbon dioxide when treated with permanganate (Stewart, 1964, p. 75). However, methylation confers a good degree of resistance to breakdown, and the importance of this will be evident in the later considerations of alkaline permanganate degradations of humic substances.

Saturated aliphatic hydrocarbons are relatively resistant to oxidation by permanganate. Some oxidation is observed when the molecule is rendered more soluble by the presence of a carboxyl group. When the chain is branched the molecule is oxidized to tertiary hydroxyacids (Kenyon and Symons, 1953). Alkenes are significantly more susceptible to oxidation than alkanes. Glycol formation from alkenes in controlled reactions with permanganate is well known, and although the reaction takes place with aqueous permanganate it proceeds best under alkaline conditions. Cyclic manganese esters are first formed and the hydroxyl groups add *cis* (see Stewart, 1965, p. 42).

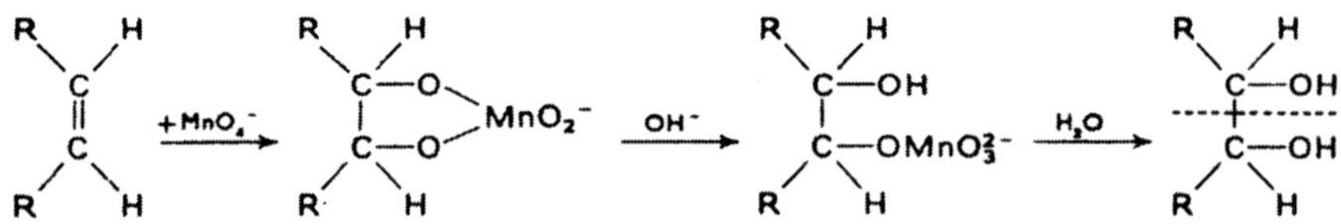

(3.12)

In hydrolysis of the Mn(V) ester to give the glycol in reaction scheme (3.12), Mn(VII) and Mn(IV) are formed by disproportionation. However, it should be remembered that further oxidation resulting in cleavage to carboxylic acids (as indicated by the broken line in a sequence (3.12)) always competes with glycol formation. Thus excess oxidant at higher temperatures will normally give rise

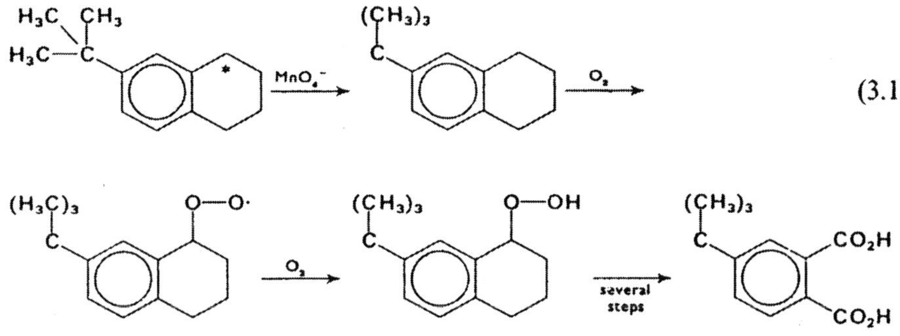

(3.13)

to cleavage. Alkynes are even more readily oxidized than alkenes and they form diketo, or more likely carboxylic acid cleavage products.

Permanganate oxidation of aromatic hydrocarbons is very relevant to degradation studies on humic substances. Reaction scheme (3.13), from Stewart (1964, p. 67) and Lee (1969, p. 37), illustrates products formed when an aromatic ring containing two different types of substituents is reacted with alkaline permanganate. This scheme shows that, in the absence of hydrogen on the α-carbon substituent, alkyl side chains resist degradation. Attack on the benzylic substituent (marked with an asterisk) is thought to cause expulsion of a hydrogen atom and the residual radical reacts with oxygen to give the peroxide, and degradation proceeds to the carboxylic acids. Some degradation of the aromatic ring occurs, particularly when excess permanganate is used, and, as previously indicated, complete degradation takes place when the ring has hydroxyl and amine substituents.

One or more rings of polycyclic aromatic structures are degraded in oxidations by permanganate. For instance 1,4-dimethylnaphthalene (**I**) gives 1,2,3,4-benzenetetracarboxylic acid (Bone *et al.*, 1935), naphthacene (**III**) gives 1,2-benzenedicarboxylic acid and 1,2,4,5-benzenetetracarboxylic acid (**V**), pyrene (**VI**) gives the 1,2,3-tricarboxylic acid (**VII**) and **II**. These reactions are summarized in reaction scheme (3.14) from Lee (1969, pp. 37–38).

Permanganate has been used for a long time for the oxidation of coal (Bone *et al.*, 1930; Bone and Himus, 1936; Ward, 1947; van Krevelen, 1961, pp. 225–231), lignin and wood (Lai and Sarkanen, 1971, p. 217; Chang and Allan, 1971, pp. 452–457). Piret *et al.* (1957) brought the permanganate degradation technique closer to the interests of soil chemists when they showed that peat degradation could yield 12 to 22% of aromatic polycarboxylic acids. More recently Schnitzer and his colleagues have extensively applied permanganate oxidation reactions to soil organic matter and humic substances. Their early work used potassium permanganate in dilute potassium hydroxide to oxidize the organic matter from the A and B_h horizons of an imperfectly drained podzol (Wright and Schnitzer, 1959b; Schnitzer and Wright, 1960a) and later to oxidize a humic acid preparation from the B_h horizon (Hansen and Schnitzer,

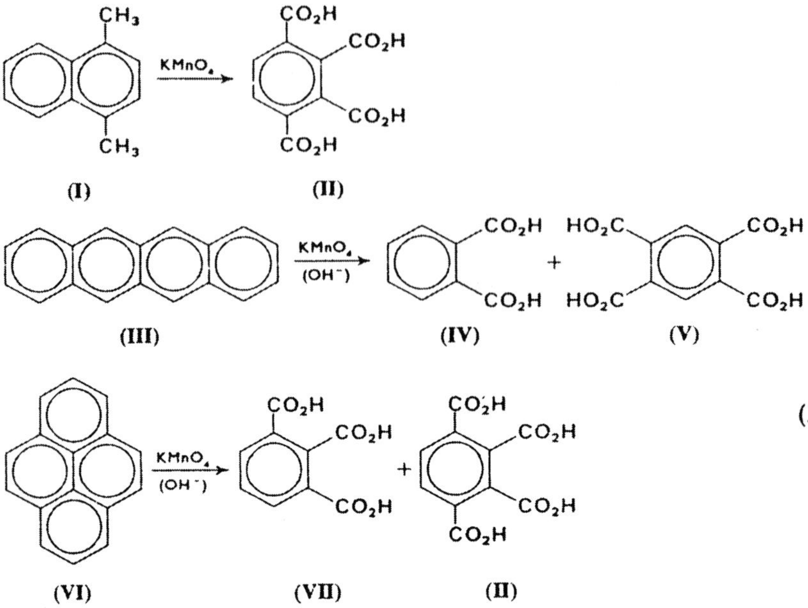

(3.14)

1966). Potassium hydroxide was omitted from subsequent work, and the humic substances were methylated with methyl iodide and silver oxide (Barton and Schnitzer, 1963), or dimethyl sulphate and anhydrous potassium carbonate in acetone (Briggs and Lawson, 1970), or with an excess of diazomethane, regarded by Schnitzer (1974a) as the best of the techniques.

The work of Matsuda and Schnitzer (1972) clearly illustrates the importance of premethylation. Substrate methylation had long been used to stabilize lignins prior to oxidation with permanganate (see Chang and Allan, 1971), because, as has been pointed out, the conversion to methoxybenzene derivatives could be expected to stabilize the phenolic constituents against degradation to meaningless products. Yields of oxidation products in the permanganate digests of humic acids from different soils were 160 to 250 % higher than those for the unmethylated materials after the polymers (1 g) had been refluxed for 8 hours in 250 cm^3 of *ca* 0.25M aqueous $KMnO_4$, pH 10 (Matsuda and Schnitzer, 1972). However, even methylated humic substances are degraded by prolonged reaction with excess permanganate (Schnitzer and Desjardins, 1970).

Figure 3.4 (assembled from selected data from Khan and Schnitzer, 1971 and 1972; Matsuda and Schnitzer, 1972; Schnitzer and Ortiz de Serra, 1973a Neyroud and Schnitzer, 1974a) includes compounds isolated from refluxed (8 hours) permanganate digests of humic and fulvic acids. These compounds were extracted in organic solvents, methylated, and fractionated and identified by g.l.c., g.c.–m.s. and microinfrared techniques. All of the carboxylic acids and phenols were identified as methyl esters and ethers, respectively. These

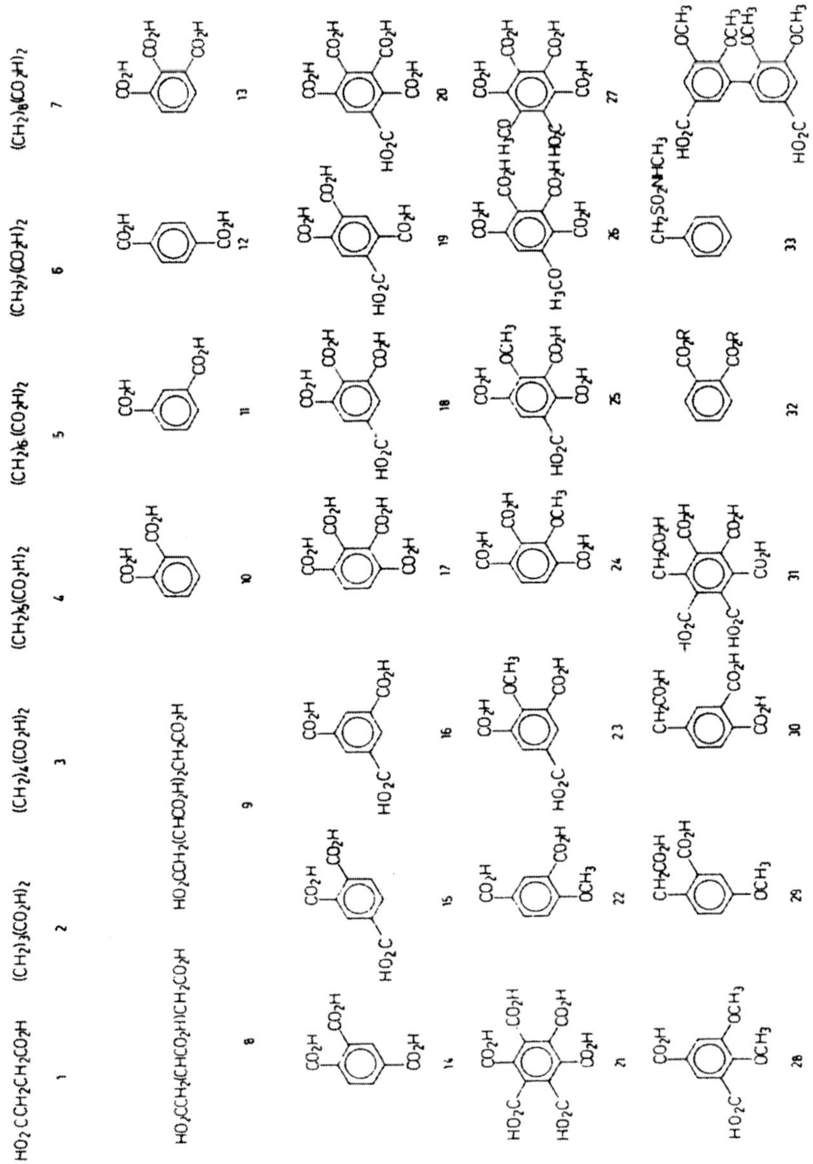

Figure 3.4 Products identified in the digests from permanganate oxidations of humic and fulvic acids

esters are represented by the free carboxylic acids. Some of the carboxyl groups originated from the humic substances, but many were formed during the degradation. By contrast few, if any, phenolic groups were produced during the degradation, and because it is impossible to say whether these were present as the ethers or hydroxyls in the polymer they are represented as the methyl ethers in Figure 3.4.

Aliphatic di-, tri-, and tetracarboxylic acids (compounds **1** to **9**) amounted to 20 % of the 0.194 g of digest products from the oxidation of humic acid (1 g) from the O horizon of the Armadale profile, and pentanedioic (glutaric, **2**), hexanedioic (adipic, **3**), heptanedioic (pimelic, **4**), and 1,2,3-propanetricarboxylic (**8**) acids were the most abundant of those present. Only octanedioic (suberic, **5**) and nonanedioic (azealic, **6**) acids were identified in the humic acid digests from the A_1 horizons of Japanese Volcanic and Diluvial soils (Matsuda and Schnitzer, 1972), and of the considerable amount of data inspected only the Armadale O horizon humic acid contained a substantial proportion of aliphatic di- and polycarboxylic acids in the permanganate digests. These were also present, but in relatively lesser abundance in the B_h humic acid of the same profile.

Unsaturated carbon to carbon bonds separated by three to nine single bonds could give rise to structures **1** to **7** in Figure 3.4. The appropriate mechanisms are outlined in reaction scheme (3.12), and the glycols formed would be expected to cleave to the dicarboxylic acids. The yields of products from alkene or alkyne functional groups would not be influenced by humic acid premethylation. Thus it is difficult to explain how the yields of di- and polycarboxylic aliphatic acids were less in the digest of unmethylated B_h humic acid than in the methylated sample.

A simple explanation for the occurrence in the digest of the aliphatic tri- and tetracarboxylic acids (compounds **8** and **9**) might involve carboxylic substituents in the chain, or the presence of hydroxymethyl side chains, or of branched chain olefinic structures, such as **VIII** and **IX**. These would degrade to the 1,2,3-propanetricarboxylic acid, and the mechanism for oxidation of alcohols is presented in scheme (3.11).

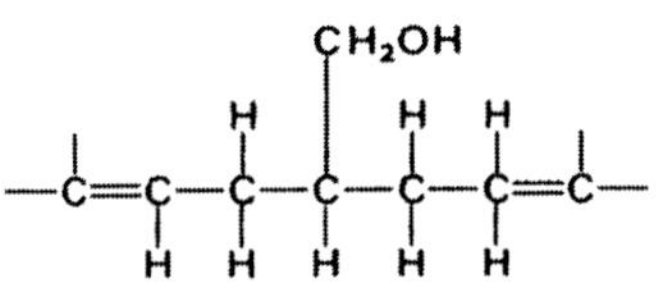

(VIII)

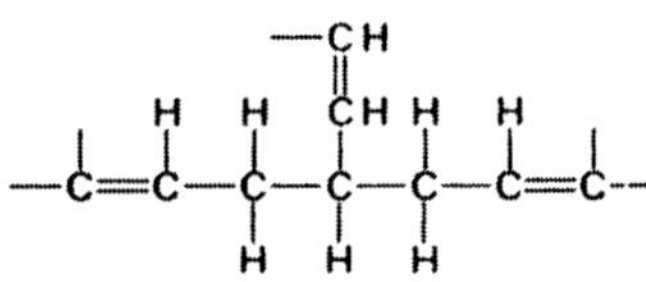

(IX)

All of the possible benzenecarboxylic acids, except benzoic acid, have been isolated and are shown in Figure 3.4 (structures **10**–**21**). These could have been formed by the oxidation of straight- and branched-chain substituents on a single benzene ring, or of aromatic and cyclic aliphatic structures of the type indicated in scheme (3.13), or they could arise from fused aromatic structures. Some appropriate reactions are outlined in schemes (3.13 and 3.14).

The methoxy-substituted aromatic structures indicate the presence of phenols or of methoxybenzene derivatives in the humic structures. Compound **28**, 1,3-dicarboxy-4,5-dimethoxybenzene, would suggest that the carboxyl groups were derived from aliphatic straight- or branched-chain substituents, or from a cyclic aliphatic structure linked 1,3- to the benzene nucleus. One or both of the methoxy substituents could have been present as phenolic hydroxyls. Compounds **29**, **30**, and **31**, give further evidence for aliphatic side chains, and the α-substituted ethanoic acid derivatives could indicate the presence of unsaturation in one of the side chains, as represented in hypothetical structure **X**, where the R's could be hydrogens, aliphatic side chains, or a naphthalene structure in the case of **30** (3,4-dicarboxyphenyl) ethanoic acid, or a fused aromatic structure in the case of **31** (2,3,4,5,6-pentacarboxyphenyl) ethanoic acid. The 2-carboxy substituent in **29** (2-carboxy-4-methoxyphenyl) ethanoic acid was most likely derived from an aliphatic side chain, or from a cyclic aliphatic structure linked 1,2- to the aromatic nucleus, and containing an appropriately placed unsaturated bond.

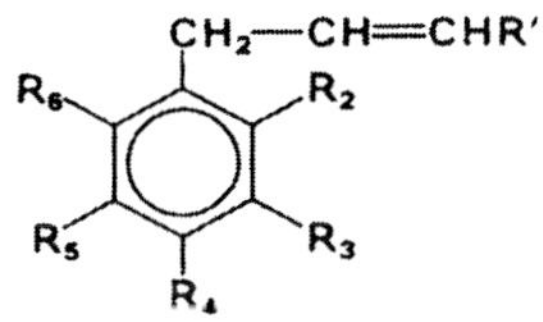

(X)

A number of dialkylphthalate structures of the general formula **32** were isolated, and it is tempting to infer that these were derived from plastic contaminants. However, Khan and Schnitzer (1971) were aware of this, and their exhaustive tests eliminated the possibilities of contamination during the isolation, degradation, and the preparation of products for analysis. It is also unlikely that such compounds were formed by esterification when the permanganate digest was acidified because, if this were so, similar esters could also be expected for the other benzenecarboxylic acids, and such esters were not reported. In one unoxidized fulvic acid fraction, dicyclohexyl phthalate accounted for 74 % of the products identified, and the total dialkylphthalate content was 91 % of the 382.4 g identified (Khan and Schnitzer, 1971). Such structures in permanganate digests could arise from incomplete oxidation of phthalate structures in humic substances.

The presence of bis(3-carboxy-5,6-dimethoxy)biphenyl, compound **34** (or dehydrodiveratric acid) is thought to be indicative of the presence of some lignin type structures in humic substances. However, the possibility exists that compounds of this type could be formed by coupling through a free radical mechanism (Lai and Sarkanen, 1971) during the oxidative degradation process.

The full potential of permanganate oxidation techniques for structural studies on humic substances has not yet been realized. Considerable progress can undoubtedly be made from extended studies with aqueous solutions of $KMnO_4$. It will, however, be necessary to investigate the products formed by the use of different amounts of permanganate over varying conditions of time and temperature. For instance, studies with less than the theoretical amounts of permanganate required for complete oxidation would lead to enhanced decomposition of permanganate in the medium, as the result of catalysis by accumulating MnO_2 (reaction scheme 3.10). This would allow milder conditions to prevail in the digest and avoid over-oxidation of the most readily released compounds. By increasing the amounts of permanganate in separate experiments it would be possible to follow the alterations to these initially released compounds and to observe the release of new products in the digest. In this way it might be possible to obtain a better indication of the structures in the original polymer.

Use of excess permanganate, as developed by Schnitzer's group is particularly useful for comparisons between humic substances from different soil environments. One example of this has shown that podzol B_h humic substances have a higher enrichment of phenolic constituents than the same substances in the O horizon of the same soil profile (Matsuda and Schnitzer, 1972). This would substantiate podzolization theories which involve the movement of phenolic humic substances down the soil profile. The same work showed significant differences between the degradation products of Japanese volcanic soils and those of the podzol.

Permanganate oxidation can also give an indication of the degree of humification of soil organic constituents. High yields of degradation products are thought to be indicative of lower extents of humification. Lower yields might result from increased blocking of the 4-hydroxy substituents on the benzene ring (the formation of phenylether cross links would be an example) to give rise to more complex degradation products not yet identified. Some evidence of the effects of humification on such yields is provided by Khan and Schnitzer (1972). They showed that amounts of degradation products were lowered by cultivation practices; these presumably had caused oxidation of some of the humic substances. It is interesting to note that many of the products from the permanganate oxidation of fulvic acids were present in the non-degraded materials (Khan and Schnitzer, 1971; Ogner and Schnitzer, 1971). This might indicate that such products were released by air and/or biological oxidation processes.

Oxidation by alkaline nitrobenzene. Following the pioneering work of Freudenberg *et al.* (1940), it has been established that when lignin is heated in the presence of 2M sodium hydroxide and nitrobenzene in a stainless steel pressure vessel for 2 to 2.5 hours at 170–180° C, vanillin (3-methoxy-4-hydroxy-benzenecarbaldehyde, **XI**), syringealdehyde (3,5-dimethoxy-4-hydroxybenzenecarbal-

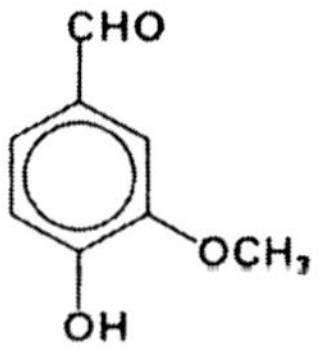

(XI)

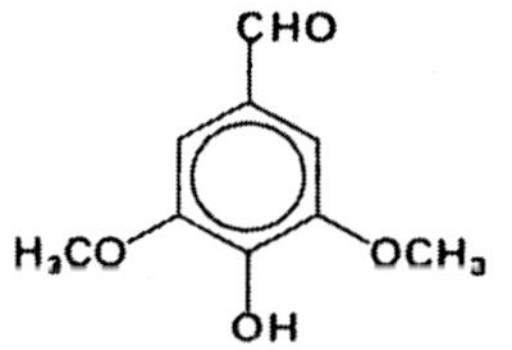

(XII)

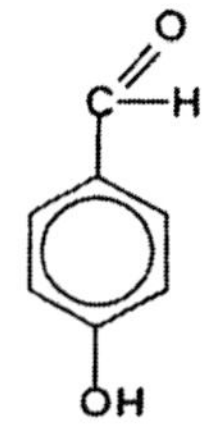

(XIII)

dehyde, **XII**), and 4-hydrobenzenecarbaldehyde (**XIII**) or *para*-hydroxybenzaldehyde are the most abundant degradation products. Coniferous lignin degrades chiefly to **XI** with some of **XII** (Hurst and Burges, 1967). Woody dicotyledonous plants yield **XI** and **XII**, but **XI**, **XII**, and **XIII** are isolated from the degradation of the lignin of monocotyledonous plants.

Leopold (1952) has reviewed appropriate mechanisms and the oxidation of the model substance 2-methoxy-4-(1-propenyl)benzenol, or isoeugonol (compound **XIV**) is cited as an example in reaction sequence (3.15).

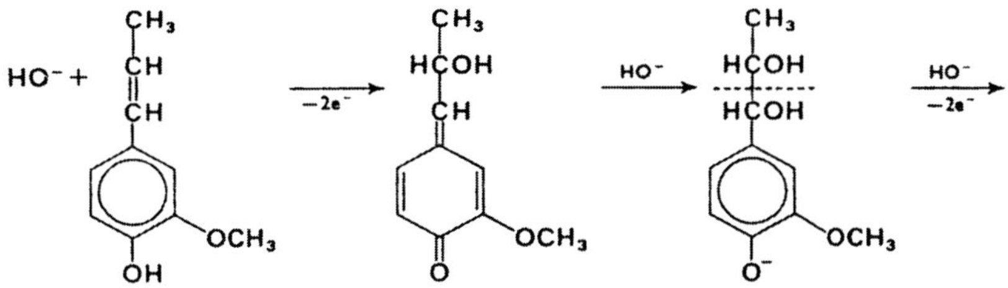

(XIV) (XV) (XVI)

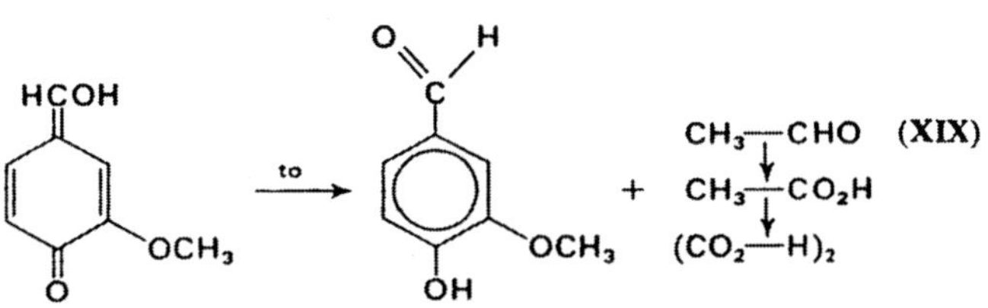

(XVII) (XVIII) (XIX) (XX)

(3.15)

In this sequence the oxidant, nitrobenzene acts as a two electron acceptor, and it undergoes stepwise reduction through nitrosobenzene and *N*-phenylhydroxylamine to aniline. Secondary condensation reactions can take place in alkaline media between the various products (Roberts and Casserio, 1965, pp. 869–870) to give rise to artefacts such as azoxybenzene, azobenzene, and 4-, and 2-hydroxyazobenzene, all of which have been isolated from lignin degradation reactions in the alkaline nitrobenzene medium. This focuses attention on the need for caution where interpreting the origins of compounds in degradation reactions.

Reaction sequence (3:15) illustrates the importance of the quinonemethide type of structure as an intermediate in the oxidation of phenols under alkaline conditions. Under such conditions the phenolate anion of **XIV** is readily formed and this transfers two electrons to the nitrobenzene to allow the formation of the quinonemethide **XV** which then adds a hydroxide ion to give the aromatic structure **XVI**, the phenolate anion of 2-methyoxy-4(propane-1,2-diol) benzenol. Further oxidation of **XVI** by the same mechanism would lead to cleavage along the broken line to yield **XVII** and ethanal (**XIX**) which further oxidizes to ethanedioic (**XX**) or oxalic acid. The quinonemethide **XVII**, would then rearrange to **XVIII**. The importance of the phenolic hydroxyl group and of the quinonemethide intermediate is emphasized by the fact that the methyl ether of **XIV**, which does not form the quinone methide structure, is not oxidized under the reaction conditions. The reaction will also not take place when the hydroxyl group is in the *meta* position.

Gottlieb and Hendricks (1945) were first to apply the alkaline nitrobenzene technique to the degradation of soil organic matter. They failed to isolate 3,5-dimethoxy-4-hydroxybenzenecarbaldehyde (**XII**) and could only detect traces of the odour of 3-methoxy-4-hydroxybenzenecarbaldehyde (**XI**) in the digest of a New York muck soil. However, they isolated significant yields of ethanedioic acid (**XX**). Later excellent work by Morrison (1958, 1963), which illustrated the value of establishing the best reaction conditions, showed very good agreement between the total amounts of **XI**, **XII**, and **XIII** isolated from the alkaline nitrobenzene oxidation at 170 °C of a phragmites humic acid and of phragmites straw. By column and paper chromatography he identified and isolated 5.9 and 6.42 % of the carbon contents of the acid and straw, respectively, as the aldehydes **XI**, **XII**, and **XIII** and a further 1.95 and 1.06 % as the acid oxidation products of these aldehydes. At temperatures of 150 °C significant quantities of 3-(3-methoxy-4-hydroxyphenyl) propenoic acid (**XXII**) and of 3-(4-hydroxyphenyl) propenoic acid (**XXI**) were detected, but the amounts present were significantly less at 170 °C. These compounds would be expected to oxidize to **XI** and **XIII** in the reaction and their presence indicates a non-lignin source.

Other compounds detected included 1-(4-hydroxyphenyl)-1-ethanone (**XXIV**), 1-(3-methoxy-4-hydroxyphenyl)-1-ethanone (**XXV**), 1-(3,5-dimethoxy-4-hydroxyphenyl)-1-ethanone (**XXVI**), methanoic acid, ethanoic acid, and 9.8% of the peat carbon was lost as carbon dioxide.

The work of Morrison and of Wildung *et al.* (1970) has shown that the yields of aldehydes decreased in the order, plant material > peat and eutro-amorphous peat (muck) > mineral soils. This suggests that the alkaline nitrobenzene degradation procedure could be used for the classification of soil organic matter. The more highly humified organic matter would be expected to provide low

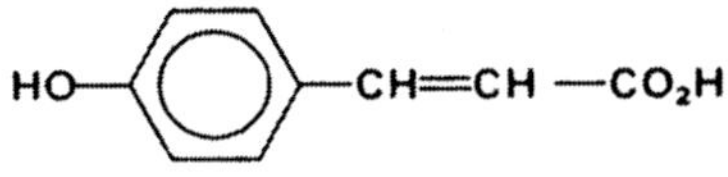

(XXI)

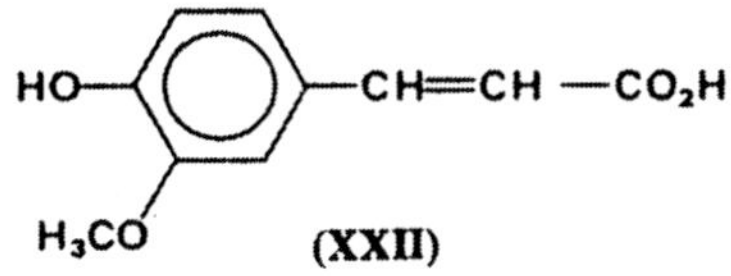

(XXII)

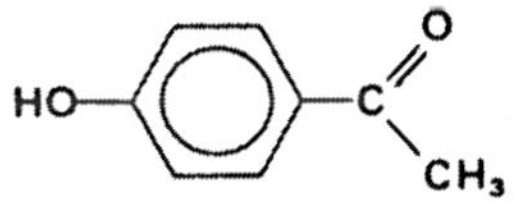

(XXIII)

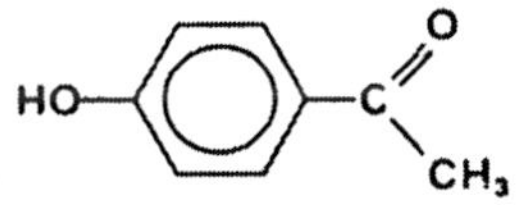

(XXIV)

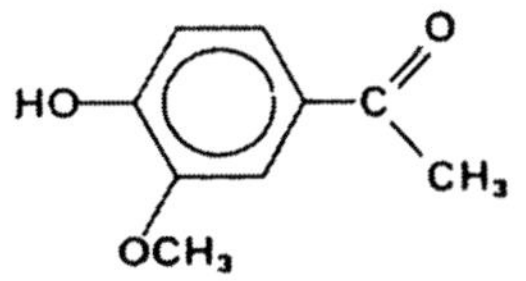

(XXV)

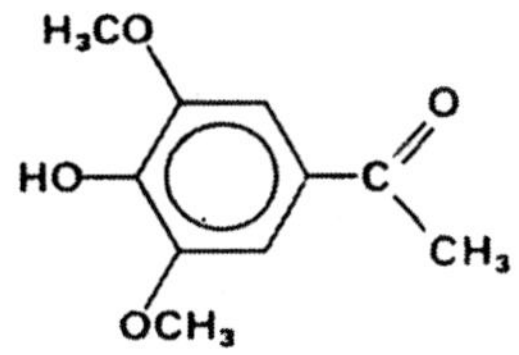

(XXVI)

yields of 4-hydroxybenzenecarbaldehyde derivatives, possibly because of the involvement of the 4-hydroxyl groups in ether or other linkages. Less humified material, with perhaps fewer cross-linkages or derivatives involving the 4-hydroxyl group, would give higher yields of benzenecarbaldehyde derivatives. It is tempting to infer that the less humified material might contain lignin residues which have only been partially modified.

In order to determine the full potential of the alkaline nitrobenzene procedure for the degradation of humic substances it will be necessary to degrade at different concentrations of alkali and nitrobenzene, at a range of temperatures, and in controlled atmospheres. The isolation for instance of **XXI**, and **XXII** at 150 °C is more significant than the carbaldehyde derivatives which would be formed at 170 °C. Model studies are required in order to make meaningful predictions of how such compounds could have been released, and of the types of functional groups from which they were released.

Oxidation by alkaline and alkaline cupric oxide solutions. Sodium hydroxide solutions at elevated temperatures are used for lignin degradation (Wallis, 1971) and it is therefore not surprising that similar procedures have been used extensively to degrade humic substances. Studies with models representing structural units in lignins have shown how ether linkages in the aliphatic side chain of phenylpropane structures can be broken. Compound **XXVII**, 1-(3,4-dimethoxyphenyl)-2-(2 methoxyphenoxy)propan-1,3-diol, when reacted for 2 hours at 170 °C with 2M sodium hydroxide cleaved (Gierer and Norén, 1962)

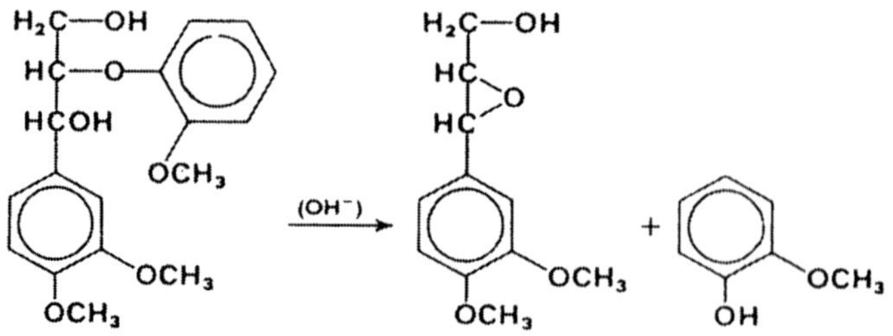

(XXVII) (XXVIII) (XXIX) (3.16)

as indicated in reaction scheme (3.16). This is known as β-ether cleavage, and it requires the presence of a hydroxyl group on the α- or γ-carbons of the aliphatic side chain. The scheme shows that the phenoxy substituent is cleaved in base to the 2-methoxybenzenol (**XXIX**) or guaicol, and that an epoxide structure (**XXVIII**) is formed which hydrolyses to the triol. When the *para*-substituent in **XXVII** is hydroxyl the epoxide structure can still be formed which can hydrolyse to the triol as before. However, the hydroxyl substituent also allows the formation of a quinone methide, and the elimination of water and methanal from the side chain.

Ethers, in which the oxygen linkage is attached to the carbon of the propane side-chain alpha to the benzene ring (α-ethers) are also cleaved, but the studies of Gierer and Norén (1962) indicate that such cleavages require the presence of a hydroxyl group *para* to the propyl side chain. This emphasizes the importance of the quinonemethide intermediate in the reaction.

Alkaline solutions of cupric, mercuric, silver, and cobalt oxides have been used to degrade lignin to aromatic aldehydes and aromatic carboxylic acids as the major products. The acid to aldehyde product ratio is governed by the oxidizing potential of the oxidant. Thus silver oxide, which has the highest oxidizing potential of the four, produces mainly acids, mercuric oxide gives a mixture of acids and aldehydes, and cupric oxide, a relatively weak oxidant, yields mainly aromatic aldehydes (Chang and Allan, 1971). Of these oxidants only alkaline cupric oxide has been used extensively so far for the degradation of humic substances.

Nigh (1973) has reviewed the type of complexes formed by copper(II). He also presented a comprehensive list of oxidation products and provided mechanisms for reactions carried out for the most part in non-aqueous systems. However, this list is only of limited use for studies with humic materials where degradations generally involve aqueous media. Extensive model studies of substrate Cu(II) interactions will be required in order to appreciate the mechanisms of alkaline cupric oxide degradations of lignins and humic substances.

There are some similarities between alkaline nitrobenzene and alkaline cupric oxide degradations; the most important is that both yield aldehydes in the digest. Copper oxide has a significant advantage because artefacts, such as those found in the nitrobenzene system, are not formed. They differ also in the fact that cupric oxide is a one-electron transfer oxidant. In their review on lignin oxidation Chang and Allan (1971, pp. 444–446) suggest that the reaction is initiated by extraction of an electron to give a phenoxy radical. A second electron is then rapidly transferred to another cupric oxide molecule preventing coupling and allowing the formation of the quinonemethide structure as indicated in reaction scheme (3.17).

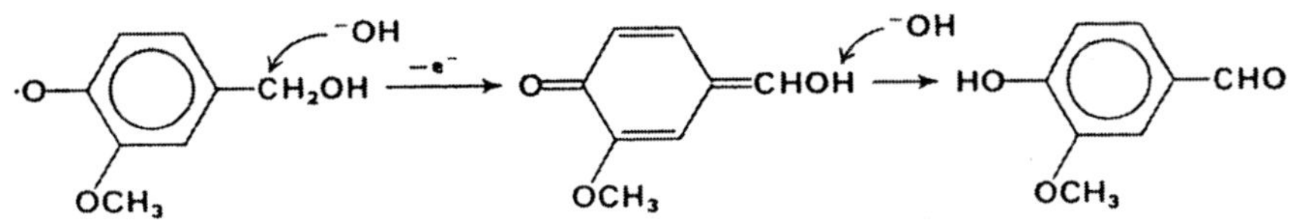

(XI) (3.17)

The high temperature and excess cupric oxide used in such reactions increases the rate of electron transfer but does not affect the coupling reaction, which requires no activation energy. The importance of using excess copper oxide is evident from the fact that the highest yield of 3-methoxy-4-hydroxybenzene-carbaldehyde (**XI**) was obtained when 13.5 moles of reagent was used per lignin building unit (referenced by Chang and Allan, 1971).

Figure 3.5 (which also refers back to compounds in Figure 3.4) illustrates the numbers and the types of products which have been separated after humic and fulvic acids were degraded in alkaline and in alkaline cupric oxide solutions. The list is compiled from data from Greene and Steelink (1962), Jakab *et al.* (1963), Schnitzer and Ortiz de Serra (1973a), Schnitzer (1974b), Neyroud and Schnitzer (1974a; 1974b, 1975), and Griffith and Schnitzer (1976). Degradation products reported by Greene and Steelink and by Jakab *et al.* were separated by paper chromatography, and those from Schnitzer's laboratories were identified by g.c.–m.s. and micro-infrared after they had been extracted with solvent from the digest and methylated with diazomethane. Again phenolic hydroxyls are represented as the methyl ether derivatives, whereas the identified methyl esters are presented as the appropriate carboxylic acids.

Comparisons, where possible, of yields of solvent-extractable materials have shown that hydrolysis in 2M NaOH produces at least as much soluble material as oxidation by the CuO–NaOH system (Neyroud and Schnitzer, 1974a). However, a higher proportion of the products from the oxidation reactions have been identified indicating that the oxidation produces simpler compounds, or more advanced degradation, than does the hydrolysis. Some aliphatic dicarboxylic (compounds **1**, **2**, **3**, **6**), tricarboxylic (**8**), and tetracarboxylic (**9**) acids were released by base from humic and fulvic acids, as was observed for permanganate oxidation. The release of aliphatic hydrocarbons (**35**), and of fatty acids (**36**; up to 10% of the weight of the humic substances) was, however,

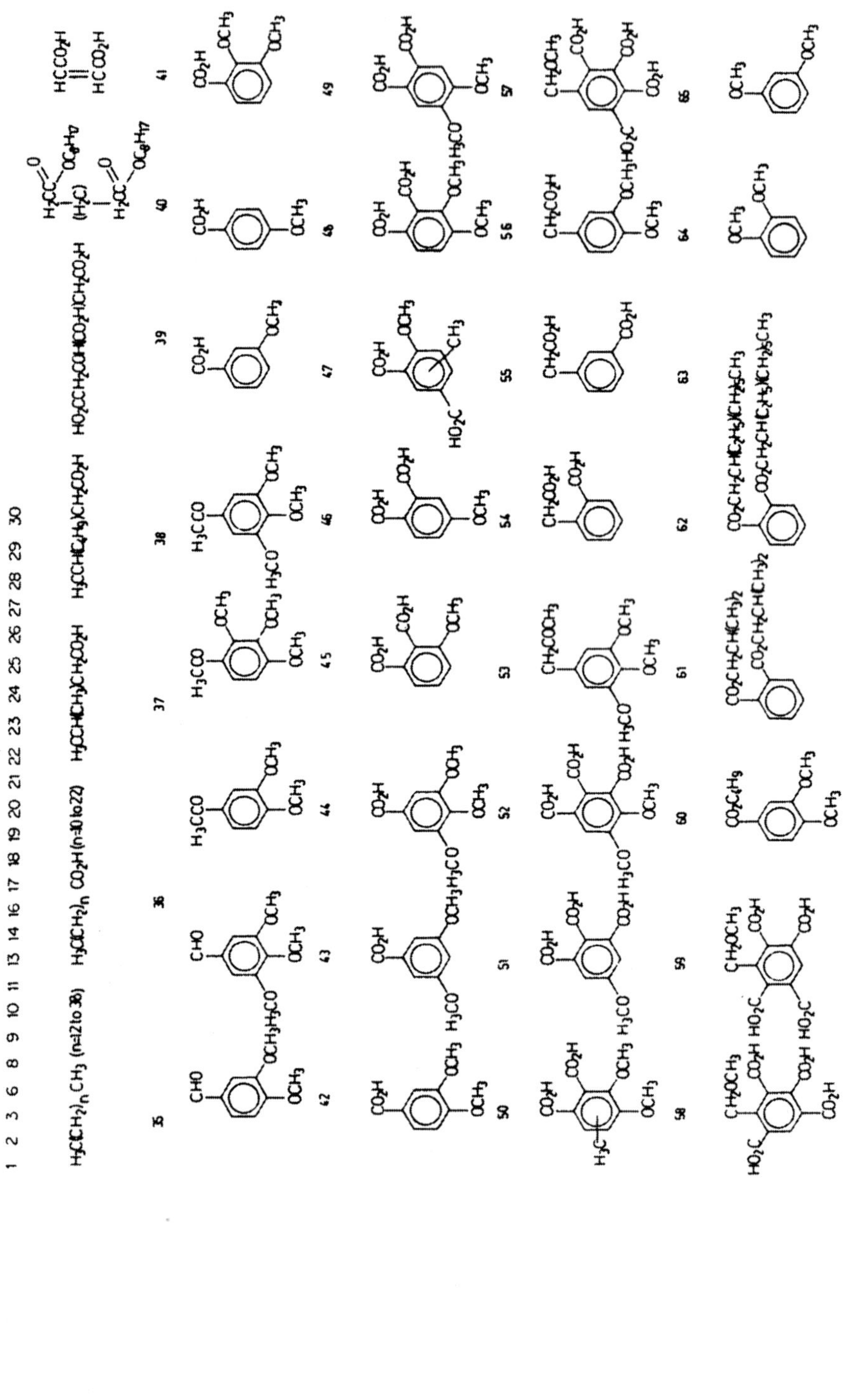

Figure 3.5 Products identified in the digests from alkaline cupric oxide degradations of humic and fulvic acids

surprising. Ultrasonication, followed by hydrolysis in water at 170 °C gave the maximum yield (1.7 % of the total weight) of alkanes from fulvic acids, but hydrolysis in 2M NaOH at 170 °C for 3 hours was required to obtain the high fatty acid yields reported (Schnitzer and Neyroud, 1975). It is difficult to explain, without additional experimental evidence, why the alkanes were not released in the basic medium. The authors observed that the majority of the alkanes were in the C_{18} to C_{36} range and that they had an odd to even carbon ratio of 1.0 which might indicate a microbial origin.

Considerations of the phenolic and fatty acid saponification products suggested to Schnitzer and Neyroud that these were largely held as phenolic esters as represented in structure **XXX**, where the R's represent hydrogen, aliphatic or aromatic substituents, carboxyl groups, or hydroxyl or ether derivatives.

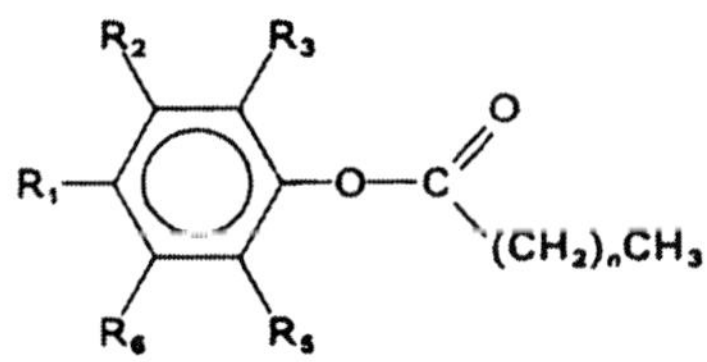

(XXX)

Reaction scheme (3.18) outlines mechanisms other than saponification which, according to Shemyakin and Shchukina (1956), would release acids and also hydrocarbons into the basic digests from alcoholic, ketonic, and alkene substituents in aliphatic structures. The reader can readily deduce how the positioning of such functional groups along the hydrocarbon components of humic substances would give rise to aliphatic dicarboxylic acids, monocarboxylic acids, and alkanes. Should compound **38** result from such release mechanisms it would indicate the presence of branched structures in the chain.

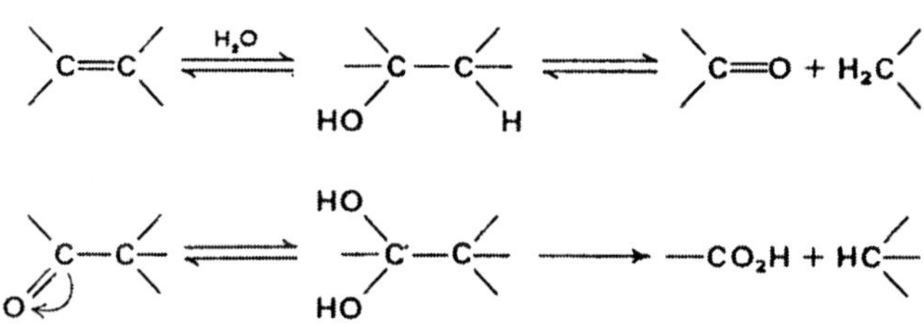

(3.18)

Benzenecarbaldehyde derivatives **42** and **43**, when present in CuO–NaOH digests, might suggest that they had origins in lignin type structures. The presence of ketones **44**, 1-(3,4-dimethoxyphenyl)-1-ethanone, **45** and **46** would substantiate this hypothesis because these could arise from the cleavage of

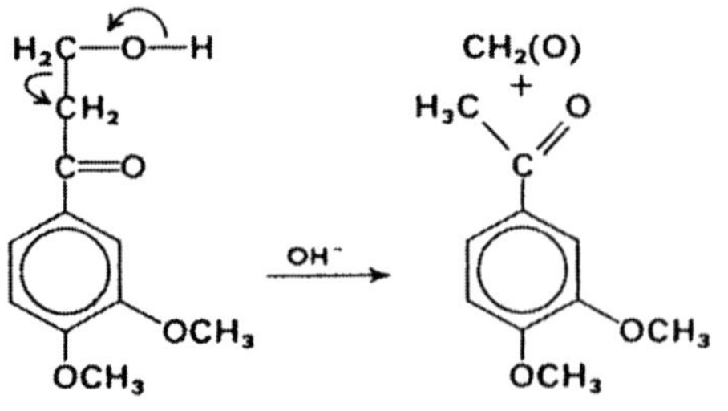

(3.19)

(XXXI) (XXXII)

β-ethers of the type shown in reaction scheme (3.16). Appropriate mechanisms for ketone formation are outlined (from Wallis, 1971) in reaction scheme (3.19).

Schnitzer and Ortiz de Serra (1973a) found that compounds **42**, **43**, **44** and **46** amounted to 26% of the weight of the structures identified in the CuO–NaOH digest of an unmethylated humic acid from the A_1 horizon of a Brunizem soil. The acids **50** and **52**, which might have been formed from more advanced oxidation of the same parent materials, amounted to an additional 14% of the identified mass. When the humic acids were methylated before degradation the values were 23% for aldehydes plus ketones and 20% for the related acids. Compound **45**, 1-(2,3,4-trimethoxyphenyl)-1-ethanone was also present in substantial yields in the degradation products of the unmethylated materials. These results were unusual because aldehyde and ketone derivatives were not detected in many of the digests of other humic substances although the appropriate acids were. Generally, when compounds **42**, and **43** (aldehydes), and **50** and **52** (acids) were present in the same digest the acids were significantly more abundant (Schnitzer, 1974*b*; Griffith and Schnitzer, 1976).

The information in Figures 3.4 and 3.5 indicates that the benzenecarboxylic acids released by permanganate oxidation are also released by NaOH hydrolysis and by CuO–NaOH oxidation. Also it will be obvious from comparisons between the compounds in the two figures that a greater variety of methoxybenzene- or of hydroxybenzenecarboxylic acids are preserved in the basic solutions than in the permanganate digests. The methoxy substituents detected were either present as such in the parent molecule, or more likely were derived from a combination of methoxy and of hydroxy substituents, phenolic esters, and less likely from *β*-ether structures. Jakab *et al.* (1963) have isolated 1,2- and 1,3-dihydroxybenzenes (shown as the methyl ethers in structures **71** and **72**) and we have referred to mechanisms, such as those in reaction scheme (3.16) which would release the methoxybenzenol derivatives from *β*-ether structures. Wallis (1971) has indicated that structure type **XXXIII** could also cleave in alkaline conditions to **XXXIV** and the aldehyde in reaction scheme (3.20).

It is almost certain that some carboxyl groups in the various benzenecarboxylic acid compounds were present as substituents on the aromatic structures of the parent polymers. However, it is probable that many or most of these were

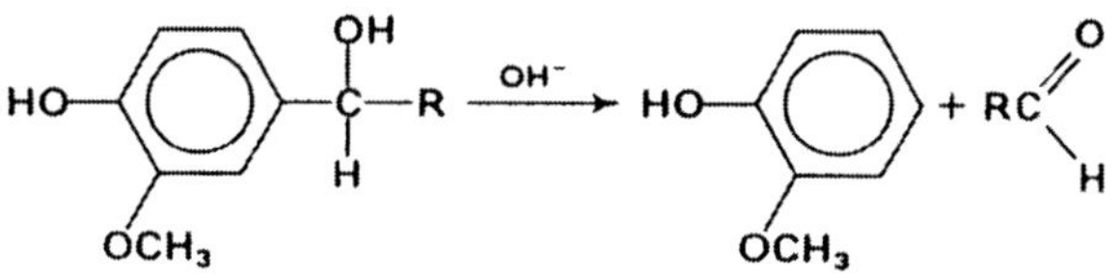

(3.20)

(XXXIII) (XXXIV)

generated under the alkaline oxidative conditions from side chains which contained appropriate alkene (or alkyne), hydroxyl, keto, and other substituents which would give reactions of the types outlined in scheme (3.18).

In the cases of degradations with permanganate reference was made to the possibility that fused aromatic structures gave rise to some of the polycarboxylic acids identified. Such fused structures would be resistant to degradations in 2M NaOH media; yet none were detected. Therefore it is likely that the polycarboxylic acids in Figure 3.4, which could have been formed from fused aromatic structures, were derived from another source.

The presence of phenylethanoic acid structures (**62** to **64**) provides direct evidence for aliphatic substituents in the aromatic nuclei, and mechanisms already discussed, particularly those in scheme (3.16), would allow such structures to be formed from appropriate functional groups located on carbon atoms β- to the aromatic nucleus. Reaction scheme (3.18) shows how the generation of a carbonyl group on the β-carbon could give rise to the methyl substituents in compounds **55** and **58**.

Benzylmethyl ethers **65**, **66** and **67** would not be expected to be formed as the result of methylation (with diazomethane) of the appropriate primary alcohol structures, but they could result from methylation of enols followed by cleavage reactions. It is, of course, possible that the ether structure existed in the parent macromolecular compounds. A number of aromatic esters were also present in the digests, but these should have been hydrolysed in the alkaline medium. Therefore it can be assumed that they formed on acidification of the mixtures. Also sulphonic acid and sulphonamide derivatives were isolated. Some of these might have been artefacts from reagents used to generate diazomethane, but others could indicate microbial origins (Neyroud and Schnitzer, 1975).

It can be concluded, particularly from the studies by Schnitzer and his colleagues, that alkaline hydrolysis and alkaline cupric oxide oxidation have provided useful techniques for the degradation of humic substances. Unmethylated humic and fulvic acids, when degraded in CuO–NaOH solution, yield ethylacetate soluble compounds which, after methylation, amount to 10–50 per cent of the weight of starting material. Although product identification has varied from 10 to 90 per cent of the weights of solvent soluble materials, it is clear that a broader range of compounds survive this reaction than they do permanganate oxidations. In particular CuO–NaOH preserves more of the aliphatic and phenolic structures than does $KMnO_4$, and it therefore allows better comparisons to be made between humic substances from different environments.

Oxidation by nitric acid. Nitric acid has been extensively applied in studies on lignins (Dence, 1971) and coal (Van Krevelen, 1961, pp. 220–224), but it has only been used to degrade soil humic substances (Schnitzer and Wright, 1960b; Scheffer and Kickuth, 1961a; 1961b; Hayashi and Nagai, 1961; Hansen and Schnitzer, 1967). This limited application may be partly attributed to earlier difficulties in quantitatively isolating the nitro derivatives formed, and to the problems of assigning origins in the polymers for these products. Digestions in nitric acid can give rise to a variety of aromatic substitution and replacement reactions in addition to oxidation processes.

Nitronium (NO_2^+) ions are usually generated for aromatic substitution reactions from concentrated nitric and sulphuric acids. They are present only in concentrations of about 4% as the result of self-dehydration reactions of concentrated nitric acid:

$$HONO_2 + H^+ \leftrightharpoons H_2ONO_2^+ \leftrightharpoons H_2O + NO_2^+$$

Thus concentrations of the nitronium species will be very low for the acid concentrations (e.g. the 2M and 7.5M nitric acid solutions applied by Hansen and Schnitzer, 1967) used to degrade humic substances. Therefore nitrosation (by NO^+) probably initiates the nitration reaction, as has been observed for lignins (Dence, 1971).

Nitroaromatic compounds can also arise from the electrophilic displacement by nitro groups of some ring substituents *ortho* and *para* to activating functional groups. Displacement will take place if the departing ionic species can be stabilized. For instance the leaving $^+CH_2OH$ group is stabilized by loss of a proton and rearrangement of the residue to methanal. The methyl ether derivative will not be stabilized in this way and will not be displaced. Neither will primary alkyl groups because their carbonium ions cannot be stabilized by simple proton transfer, but carboxyl and keto groups attached to the aromatic nucleus are readily displaced.

Nitric acid cleaves alkyl–aryl ethers by a solvolytic process and liberates the appropriate phenol and alcohol (Dence, 1971). The phenols are then nitrated, but are also highly susceptible to oxidative degradation by nitric acid to non-aromatic structures. Oxidation of polyhydroxy aromatic compounds goes through the corresponding *ortho* or *para* quinone intermediates and these can participate in addition reactions typical of conjugated dienone structures, or they can cleave to dicarboxylic acids and CO_2.

Yokakawa *et al.* (1962) when studying the degradation of coal humic acids investigated the degradation of a number of model aromatic compounds in boiling M HNO_3. Catechol and hydroquinone (2- and 4-hydroxybenzenol, respectively) lost 22% and 44% of their carbon contents as ethanedioic acid and CO_2, and some nitration of the aromatic structures was observed. Cinnamic, or 3-phenyl-propenoic acid, was resistant to degradation in the medium, but its phenol analogues were not. Methylation decreased the loss of CO_2 from the humic acid from 30 to 17%, although the authors were not satisfied that methylation had been complete.

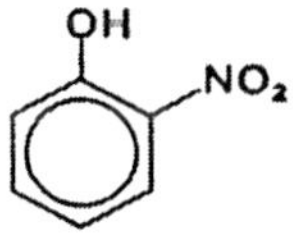

(XXXV)

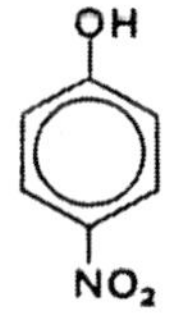

(XXXVI)

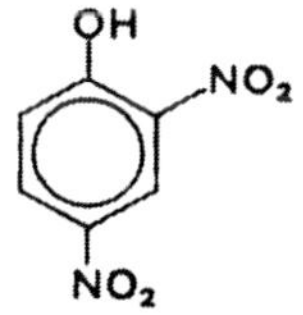

(XXXVII)

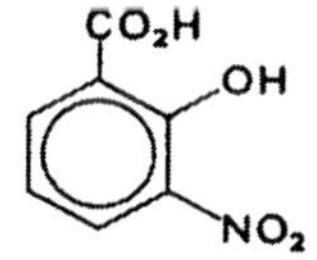

(XXXVIII)

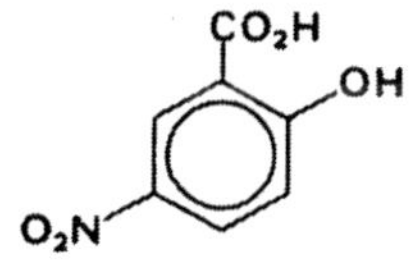

(IXL)

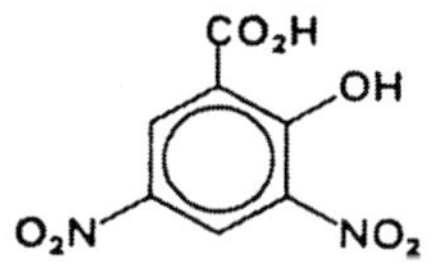

(XL)

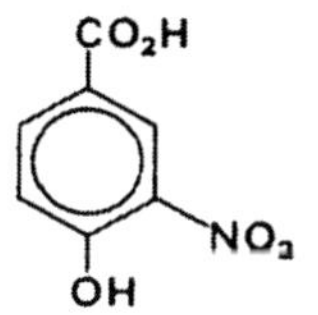

(XLI)

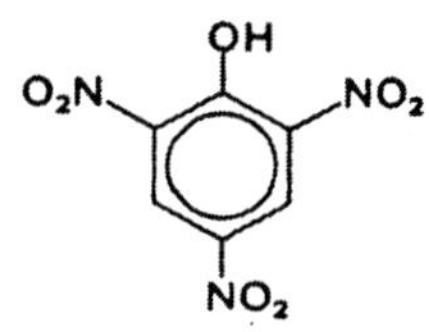

(XLII)

On the basis of R_f values from paper chromatography, and n.m.r. data, Hansen and Schnitzer (1967) identified the following nitrobenzene derivatives in the benzene soluble fraction of the 2M HNO_3 digest: small amounts of 2- and of 4-nitrobenzenols **(XXXV)** and **(XXXVI)**; possibly some 2,4-dinitrobenzenol **(XXXVII)**; 2-hydroxy-3-nitrobenzenecarboxylic acid **(XXXVIII)**; 2-hydroxy-5-nitrobenzenecarboxylic acid **(IXL)**; 2-hydroxy-3,5-dinitrobenzenecarboxylic acid **(XL)**; and 3-nitro-4-hydroxybenzenecarboxylic acid **(XLI)**. Only picric acid, or 2,4,6-trinitrobenzenol **(XLII)** was present in the benzene extract of the 7.5M nitric acid digest. Nitroaromatic structures accounted for 2.1% of the digests in 2M acid, and picric acid composed 5.30% of that in the 7.5M acid (5.93 and 9.38% of the weights of starting materials were identified as organic structures in extracts from the 2M and 7.5M acid digests, respectively). The nitrophenols shown can be explained by substitutions *ortho* and *para* to the hydroxyl groups and do not require the electrophilic displacement mechanism to be invoked.

Twelve benzenecarboxylic acids (compounds **10–21** in Figure 3.4) were detected, and amounted to 3.8% of the starting material in the 7.5M digest. These were probably formed by oxidation of carbon side chains. There was evidence for small amounts of pentanedioic (**2**), hexanedioic (**3**), and heptanedioic (**4**) acids in the 7.5M digests. The hydroxybenzenecarboxylic acids identified, or tentatively identified were: 2-, 3-, and 4-hydroxybenzenecarboxylic acids and 2,4-, 3,5-, and possibly 3,4-dihydroxybenzenecarboxylic acids, and these amounted to about 10% and 2.2% of the compounds identified in the 2M and

7.5M digests, respectively, which suggests that such compounds were degraded at the higher acid concentrations. Indeed the evolution of CO_2 from digests (Yokakawa *et al.*, 1962) shows that nitric acid treatment causes extensive decomposition of the degradation products, but the technique may have potential when the reaction is preceded by adequate methylation.

Degradation by reductive processes

Reactions which add hydrogen to a compound to form new covalent linkages are generally called reduction reactions. The most useful reducing agents in organic chemistry are hydrogen, generally used with a solid catalyst, metals such as zinc and sodium which transfer electrons to the substrate, and metal hydrides which transfer hydride (H^-) anions. In the transfer process the metals are oxidized to their cationic species. Hydride transfer reactions can also take place between certain molecules in basic media to yield oxidation and reduction products, as was described for the Cannizzaro reaction in reaction scheme (3.9). We will be mainly concerned here with reductions by active metals.

Reduction with sodium amalgam. Burges *et al.* (1964) introduced the sodium amalgam (Na/Hg) technique for the degradation of humic acids and isolated yields of 30–35% ether-soluble products from the digests. These products, composed for the most part of mixtures of phenols and phenolic carboxylic acids, were readily oxidized to a dark residue on standing. Since that time the technique has been used by a number of workers (Mendez and Stevenson, 1966; Stevenson and Mendez, 1967; Dormaar, 1969; Schnitzer and Ortiz de Serra, 1973a, 1973b; Martin *et al.*, 1974) who obtained lower yields of ether or solvent soluble materials. Amongst the many compounds identified, the most readily released were 3-methoxy-4-hydroxybenzenecarboxylic acid (vanillic acid **93**) and 3,5-dimethoxy-4-hydroxybenzenecarboxylic acid (syringic acid **94**), and these were common to all digests despite differences in the ratios of reagents and reactants used.

Piper and Posner (1972) studied the mechanisms of reaction, investigated the optimum conditions for Na/Hg degradations, and established the importance of the ratio of reactant to reagent and of the time of reaction. Their work showed that the same products were released from hydrolysed and unhydrolysed substrate, and illustrated the importance of using finely ground amalgam (prepared according to Vogel, 1956). It also showed the amounts of monomers released increased with time during 30 hours of reaction. The results suggested that under the experimental conditions employed, best results were obtained by using 25 mg of the more highly oxidized humic acids (such as those extracted with pyrophosphate) and 50 mg of less oxidized material (e.g. alkali extracted humic acid) in an alkaline medium (15 cm^3 of 0.5M NaOH) containing 30 g of 5% sodium amalgam. Whilst the reaction conditions used by Burges *et al.*

(1964) met these requirements, those used by other workers did not, and this might account for their lower yields.

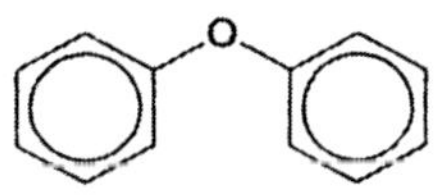

(XLIII)

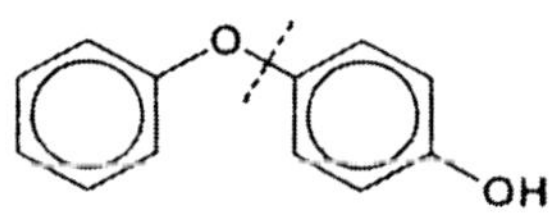

(XLIV)

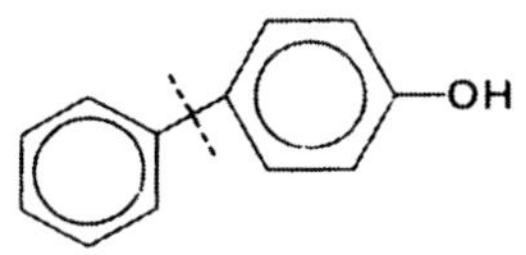

(XLV)

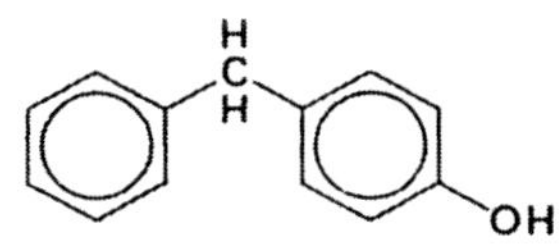

(XLVI)

Degradations on model compounds by Piper and Posner (1972) give indications about some relevant linkages which could be cleaved in Na/Hg digests. Although the mechanisms involved are not known, degradation is thought to result from attack by atomic hydrogen (H•). These atoms combine readily to form hydrogen gas (H_2), and this might account for the necessity to use excess amalgam. The hydrogen acts as an electrophile, as indicated by the products released during degradation of compounds **XLIII** to **XLVI**. Compound **XLIII** gave benzene and benzenol, **XLIV** gave equal amounts of benzenol because the electron density on the *para*-carbon is increased by the activating effect of the 4-hydroxy-group, and this would favour cleavage as indicated by the broken line. Again a similar activating effect caused **XLV** to cleave to equal amounts of benzene and benzenol, but **XLVI** did not cleave because excess electron density cannot reside on the bridging methylene carbon.

Figure 3.6 summarizes the products (Burges *et al.*, 1964; Piper and Posner, 1972; Dormaar, 1969; Schnitzer and Ortiz de Serra, 1973b; and Martin *et al.*, 1974) which have been identified in humic acid-Na/Hg digests. Model studies by Piper and Posner (1972) indicated that compounds **73**, **74**, **87**, **92**, **93**, **94**, and 4-hydroxybenzenol (hydroquinone) did not degrade when subjected to their reaction conditions, and this suggested that the identified phenolic structures in the humic acid digests could reasonably be assigned to structures in the polymer. Later, however, Martin *et al.* (1974) showed that varying extents of degradations occurred in their digests in the cases of compounds **75**, **76**, **88**, **89**, **91**, benzoquinone, 2-hydroxy-6-methylbenzenecarboxylic acid, and several hydroxynaphthenic acids. Decarboxylation was the principal reaction observed in the cases of the dihydroxybenzenecarboxylic acids, but the partial destruction of the trihydroxybenzene structures indicates that the yields reported for such compounds in humic acid digests were not representative of the amounts released as carboxylic acid derivatives. Variations in reaction conditions (e.g.

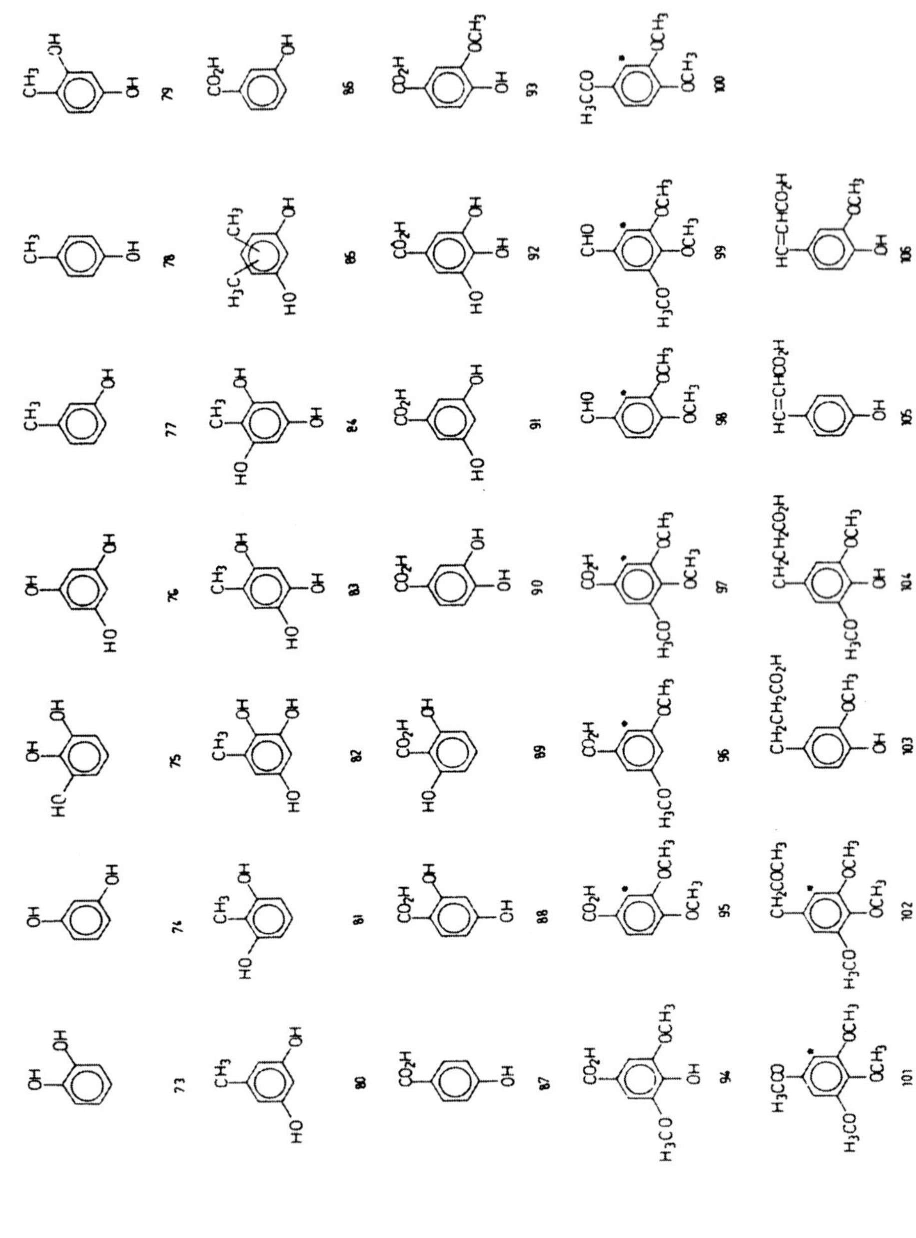

Figure 3.6 Products identified in the digests from sodium amalgam degradations of humic acids

the amounts of reductant used, and the maintenance of anaerobic conditions in the alkaline media) might account for some of the differences in the apparent stabilities of the compounds examined.

On the basis of the studies of degradations of model substances by Martin *et al.* (1974) we cannot be certain that all of the non-carboxylic phenolic structures in Figure 3.6 represent primary structures in the humic acid polymers. Also, the fact that none of the phenolic carboxylic acids listed have a methyl group attached to the aromatic ring suggests that such structural types did not exist in the original polymer, or (if they did) were decarboxylated when released into the digest. It would appear that the various cleaved structures were linked by aromatic ether linkages, or were present as biphenyl structures where activating (hydroxy or methyl) substituents were *ortho* or/and *para* to the linking bond. The model studies, and the compounds identified would suggest also that methylene and longer aliphatic bridging structures were not cleaved.

Detection of the phenolic compounds and their quantitative estimations generally used two-dimensional chromatography and co-chromatography with standards. Compounds marked with an asterisk in Figure 3.6 (taken from the data of Schnitzer and Oritz de Serra, 1973b), were methylated prior to g.c.–m.s. and micro-infrared analysis, and it is impossible to state which of the structures originally had free hydroxyls. However, it is highly likely that the distribution of —OH and —OCH_3 substituents were the same as for the other similar compounds in the Figure 3.6. It is interesting to note that these workers isolated the aldehyde and keto structures **98, 99, 100, 101** and **102**, and this is good evidence for the presence of the carbonyl group in humic substances.

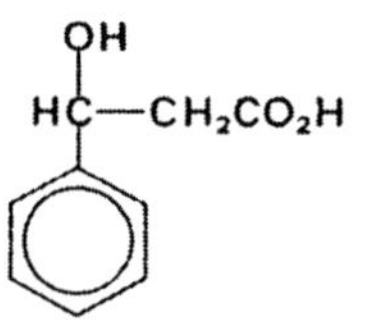

(XLVII)

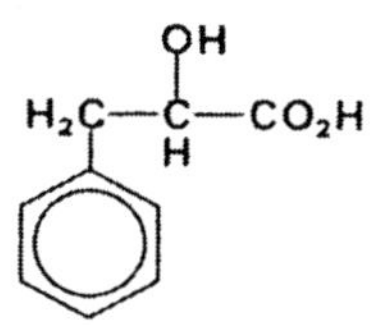

(XLVIII)

The side chain double bonds of the substituted propenoic acids **105** and **106** would be expected to rapidly hydrogenate, and Piper and Posner (1972) suggested that these might arise by dehydration of appropriate hydroxy derivatives **XLVII** and **XLVIII** during the drying of the ether extracts.

Despite the reducing conditions of the reaction many of the compounds in Figure 3.6 contain the carboxyl group. This provides strong evidence that this functional group, as well as hydroxyl, methoxyl, and carbonyl are present as substituents on aromatic nuclei of humic substances. This inference could not be made so strongly for carbonyl and carboxyl groups solely on the basis of oxidative degradation reactions.

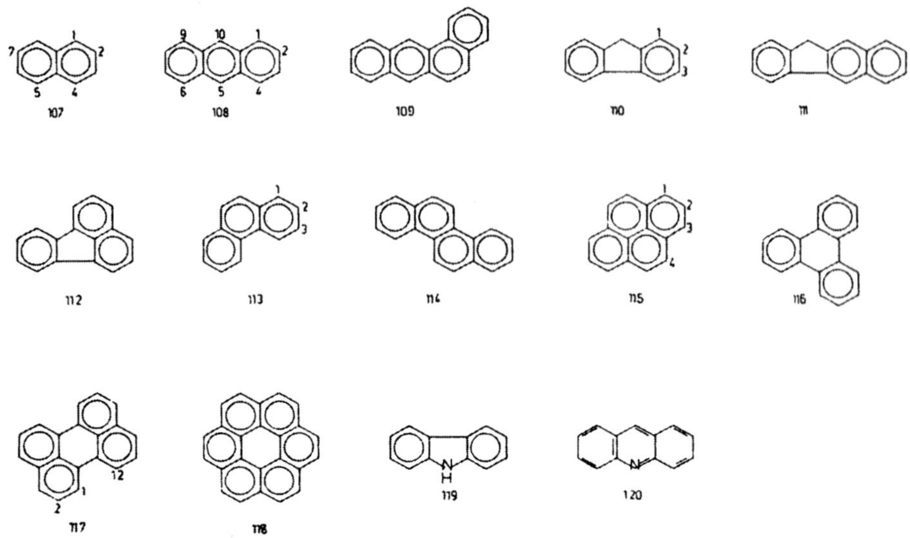

Figure 3.7 Products identified in the digests from degradations of humic materials in zinc-dust distillation and fusion reactions

Degradation by zinc-dust distillations and fusions. The zinc-dust distillation of humic substances has been described by Cheshire *et al.* (1967, 1968) and by Hansen and Schnitzer (1969a and 1969b). The latter authors (1969b) have also used zinc-dust fusion, and the products which have been obtained by both procedures are summarized in Figure 3.7. For the distillation process Cheshire *et al.* used 0.5 g quantities of acid (6M HCl) boiled humic acids mixed with 30 g purified zinc dust, and they placed an additional 10 g of the dust at the exit port of the pyrex tube used for the reaction. Distillation was carried out in a stream of H_2 gas at 500–550 °C. For fusion Hansen and Schnitzer combined 0.5 g of substrate with 3 g of Zn dust, 5 g $ZnCl_2$ and 1 g NaCl and heated at a temperature of 300 °C.

As Figure 3.7 shows the products isolated were composed of fused aromatic structures. The listed compounds are naphthalene (**107**), anthracene (**108**), 1,2-benzanthracene (**109**), fluorene (**110**), 2,3-benzofluorene (**111**), fluoranthene (**112**), phenanthrene (**113**), chrysene (**114**), pyrene (**115**), triphenylene (**116**), perylene **117**), coronene (**118**), carbazole (**119**), and acridine (**120**). In addition a number of derivatives of these were identified which include methylated naphthalenes (1-methylnaphthalene, 2-methylnaphthalene, and 1,2,7-trimethylnaphthalene), methylated anthracenes (1-methylanthracene and 9-methylanthracene), methylated phenanthrenes (2-methylphenanthrene and 3-methylphenanthrene), methylated pyrenes (1-methylpyrene and 4-methylpyrene), benzpyrenes (1,2-benzopyrene and 3,4-benzopyrene) and naphtho-(2′,3′:1,2)-pyrene, 1,12-benzoperylene, and homologues of carbazole and of acridine.

The yields of products from zinc-dust distillation reactions are invariably very low, often less than 1 %, and the recovery of 3 % of starting material as a pale yellow oil by Cheshire *et al.* (1968) was considered to be good. Hansen and Schnitzer obtained yields of the order of 0.6–0.7 % for humic acid, whilst those for fulvic acid were somewhat lower. Assuming a 10 % recovery of products and calculating on a functional group-free basis, they postulated that fused aromatic structures could account for 25 % and 12 % of the components in fulvic and humic acids, respectively.

Based on their data for zinc-dust distillation experiments, Cheshire *et al.* (1967) and Haworth (1971) proposed that humic substances are built around a polynuclear aromatic 'core'. Combining this with other data from hydrolysis, oxidation, reduction, and electron spin resonance (e.s.r.) they suggested that polysaccharides, simple phenols, proteins or peptides, and metals were attached to this 'core'.

The results from such drastic procedures must be interpreted with caution. In their later paper Cheshire *et al.* (1968) showed that 3,4- and 3,5-dihydroxybenzoic acids, furfural (from dehydration of pentose sugars), and quinone polymers gave polycyclic aromatic structures from zinc-dust distillation at 500–550 °C. At 400 °C only small yields of anthracene were obtained from the hydroxybenzenes, and no aromatic structures were detected in the distillates from furfural and polymers of *ortho*- and of *para*-benzoquinone. In addition, the same products were isolated in the same proportions from the distillations at the higher and lower temperatures and it was concluded that the products identified were largely released from the humic materials.

We conclude that judgement cannot yet be passed on the usefulness of zinc-dust distillation and fusion techniques for work with humic substances. An obvious experiment, which could help decide whether or not the fused aromatic structures detected were 'core' products or artefacts, would be to subject compounds released by other reductive techniques (such as the sodium amalgam method), and the undegraded residue to zinc-dust distillation at 400 °C, and to compare the yields of fused aromatic structures from both substrates.

Degradation with phosphorus and hydriodic acid. Hydriodic acid (HI) is a standard reagent for cleaving ether linkages, and red phosphorus provides reducing conditions to hydrogenate aromatic nuclei. A volatile and a nonvolatile oil, yielding 3% and 18% by weight, respectively, of starting material were formed when Cheshire *et al.* (1968) heated 1 g of acid-boiled humic acid in a sealed tube at 250 °C with 1 g of red phosphorus and 30 cm^3 of HI (density 1.74). When the oils were dehydrogenated with a palladium–carbon catalyst, or with sulphur, a number of fused aromatic structures were formed. This was thought to substantiate the zinc-dust distillation data, and indicated the presence of fused aromatic structures in the polymer. These promising techniques, initiated by Haworth and his group, are well worth further investigations.

Degradation by hydrogenation and by hydrogenolysis. Oils and in some cases cyclopentanols, phenols, and phenolcarboxylic acids have been reported from hydrogenation of humic acids under pressure (Ganz, 1944; Gottlieb and Hendricks, 1945; Kukharenko and Savel'ev, 1951 and 1952; Kukharenko and Kredenskaja, 1956; Murphy and Moore, 1960).

In carefully controlled work, which included investigations of possibilities for artefact formation, Felbeck (1965a; 1967) subjected an acid hydrolysed organic (muck) soil to hydrogenolysis at 350 °C and 41 Nm^{-2}. He isolated 50–60 % of the carbon in the starting material as hexane- and benzene-soluble materials. The products contained C11 to C35 n-hydrocarbons, and the C22 to C32 compounds were present in greatest abundance. Calculation of the carbon-preference index (CPI) for the data for the C12–C35 hydrocarbons gave an odd/even CPI value of 0.96 which contrasted with the values (3–8) for most biologically derived materials. His data would indicate that substantial fragments of the polymer cleaved at the weaker bonds, and he proposed that γ-pyrone structures in humic substances might account for the behaviour observed (Felbeck, 1965b).

Degradation with sodium and liquid ammonia. Maximov and Krasoskaya (1977) found that from 9 to 21 % of the substrates were degraded to water-soluble products when 6M HCl hydrolysed humic acids from a forest soil were reduced by metallic sodium in liquid ammonia at −33 °C. These amounts were increased to *ca.* 45 % when the reduction experiments were repeated five times on each substrate. However, only 10.5 % of the products were ether-soluble. By use of g.l.c. and g.c.–m.s. techniques, hydroxy- and dihydroxybenzene-carboxylic acids, hydroxybenzene polycarboxylic acids, and benzenecarboxylic acids were detected in amounts totalling 32.5 $mg\,g^{-1}$ of humic acid. They postulated that such compounds could have been released from ether linkages involving the foregoing structures.

Degradation with phenol

Herédy and Neuworth (1962) introduced this technique to coal science when they showed that 60 % of a bituminous coal (80.4 %C) was depolymerized at 100 °C by phenol and boron trifluoride catalyst. Later Ouchi *et al.* (1965) found that a Yubari coal (84.6 %C) was almost completely depolymerized when refluxed for 27 hours at 185 °C (the boiling point of phenol) in the presence of *para*-toluenesulphonic acid (PTS) catalyst.

In a reaction with the model substance 1-(4-hydroxy-3,5-dimethylphenyl)-2-(1-naphthyl)ethane (**IL**) Henédy and Neuworth (1962) proposed that clearages took place as outlined in reaction scheme (3.21). The products expected would arise from attack on bonds a and b. Their results showed that bond a was particularly susceptible to cleavage, as determined by the recovery of 2,6-dimethylbenzenol (**L**), but the small amount of naphthalene released suggested

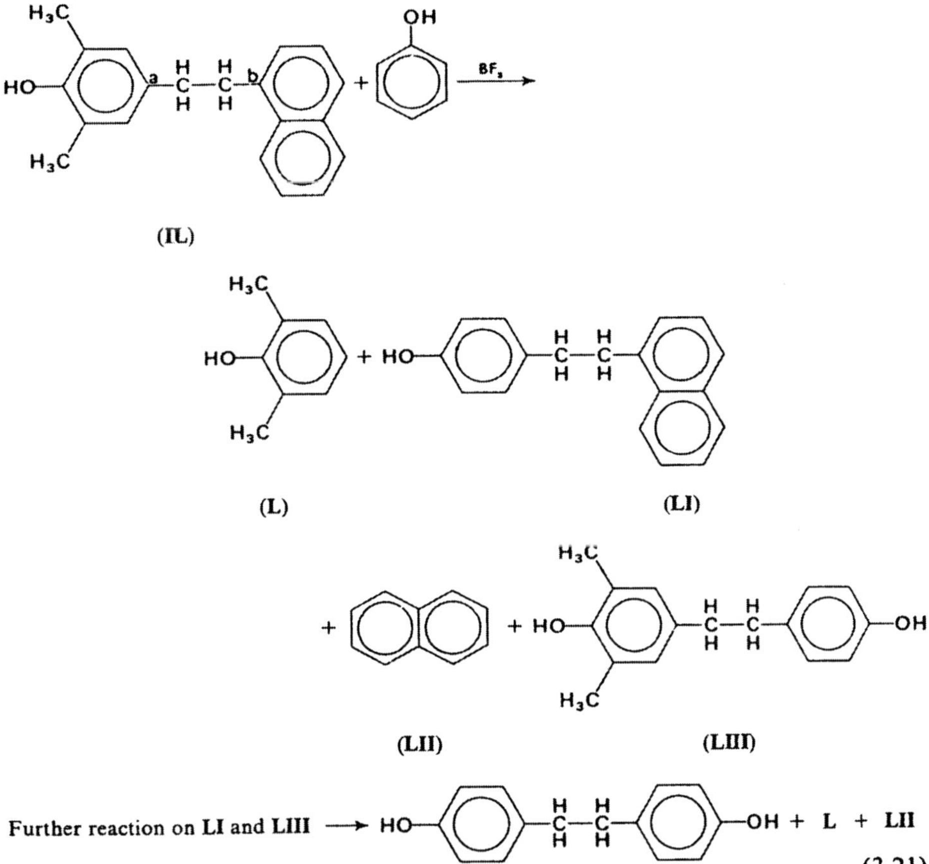

(IL) (L) (LI) (LII) (LIII)

Further reaction on **LI** and **LIII** → ... + **L** + **LII**

(3.21)

that bond b was relatively resistant to cleavage. However, when the ethylene linkage in **IL** was replaced by a methylene (—CH_2—) group 73% and 41% of the theoretical yields of **L** and **LII** were obtained, respectively, showing that in this instance significant cleavage of bond b had taken place.

Jackson, Swift, Posner and Knox (1972) introduced the phenol–PTS method, for the degradation of soil humic acids. After refluxing for 24 hours excess phenol and PTS were removed by steam distillation, and the petroleum ether-soluble fraction of the methylated residue was separated by column chromatography, preparative g.l.c. and thin layer chromatography. The separates were identified by chemical and spectroscopic methods. Figure 3.8 presents a list of the compounds identified. Compounds **122, 126, 127, 130,** and **131** were also identified (Ouchi and Brooks, 1967; Imuta and Ouchi, 1968, 1969) among the degradation products of coals.

From the few model studies which are available we can predict some polymer linkages which would cleave to give the structures in Figure 3.8. The xanthene

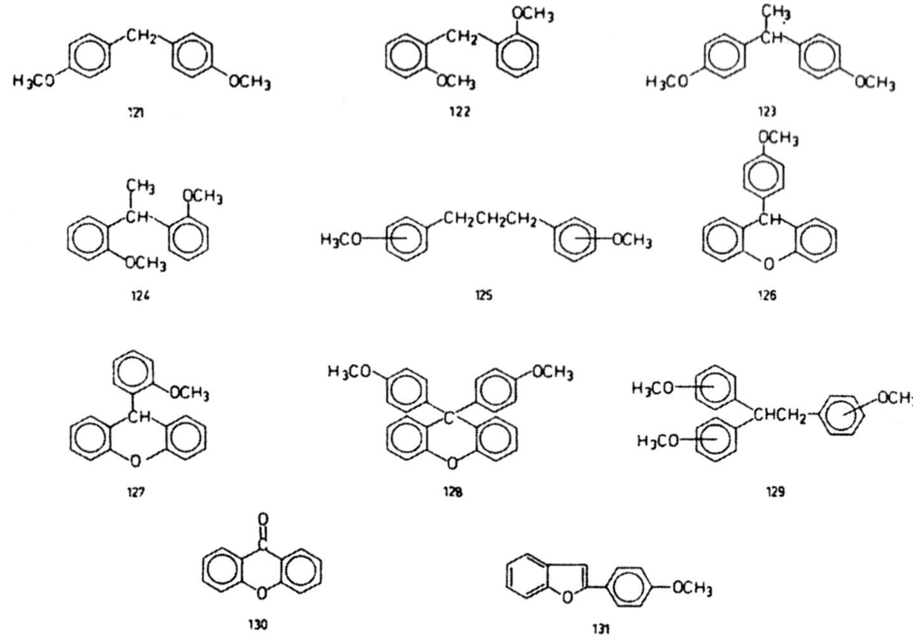

Figure 3.8 Products identified in the digests from phenol + *p*-toluenesulphonic acid–humic acid degradation reactions. (After Jackson *et al.*, 1972)

structures, 9-(4′-methoxyphenyl)xanthene (**126**), 9-(2′-methoxyphenyl)xanthene (**127**), 9,9-bis(4′-methoxyphenyl)xanthene (**128**) and xanthone (**130**) were formed by dehydration of appropriate 2-hydroxy benzene structures as indicated in reaction scheme (3.22), where R is a bridging (usually aliphatic) structure. Clearly the formation of xanthene and xanthone structures indicate that aromatic ether linkages are not broken. Also the 1-naphthylmethyl ether linkage remained intact in the model studies, and this suggests that aliphatic–aromatic ethers are not cleaved. Thus the bis(4-methoxyphenyl)methane (**121**), bis(2-methoxyphenyl)methane (**122**), 1,1-bis(4′-methoxyphenyl)ethane (**123**), 1,1-bis(2′-methoxyphenyl)ethane (**124**), and 1,3-bis(methoxyphenyl)propane (**125**) compounds could have formed from structures in which aromatic groups were linked by the appropriate aliphatic bridges.

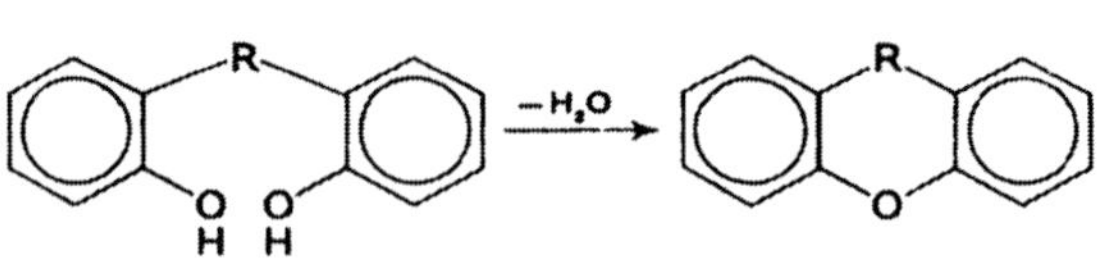

(3.22)

In a model study with biphenyl methanol, or benzhydrol (**LIV**), using the phenol–PTS system, Piper and Posner obtained, after methylation, the products

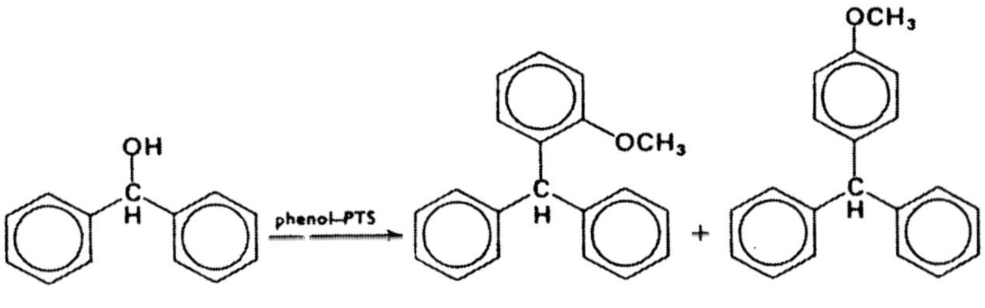

(LIV) 30 % 40 % (3.23)

in the yields indicated in reaction scheme (3.23). There it will be noted that benzene substituents were not cleaved and that the —OH was displaced by the phenol. This provides an alternative explanation for the origins of compounds **121–125**, which could also be formed from hydroxyl groups on terminal carbon atoms. Similarly 9-(4′-methoxyphenyl)xanthene (**126**), 9-(2′-methoxyphenyl)xanthene (**127**) and 1,1,2-tris(methoxyphenyl)ethane (**129**) could have been derived from secondary alcohols. The cyclic ether structures were formed from *ortho*-hydroxybenzenes in the polymer, or more likely from *ortho*-phenol groups substituted during the reaction. Alternatively these structures could be formed by displacement of triaromatic substituents on the tetrahedral carbon.

In another model study 4-hydroxy acetophenone (**LV**) cleaved to products in the ratios indicated by the methylated structures in scheme (3.24) (Piper and Posner, 1975). This reaction suggests that a methyl ketone origin is possible for structures **123** and **124**, and indicates, from compound **LVII**, that where enols can form, the hydroxy group is replaced by phenol.

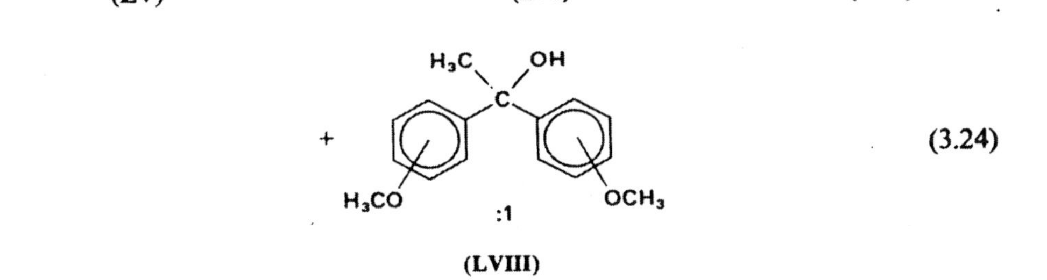

(LV) (LVI) (LVII) (LVIII) (3.24)

When 2,2′-dihydroxybenzophenone (**LIX**) was reacted with phenol in 1 : 10 ratio (Piper and Posner, 1975), four products were identified in the proportions

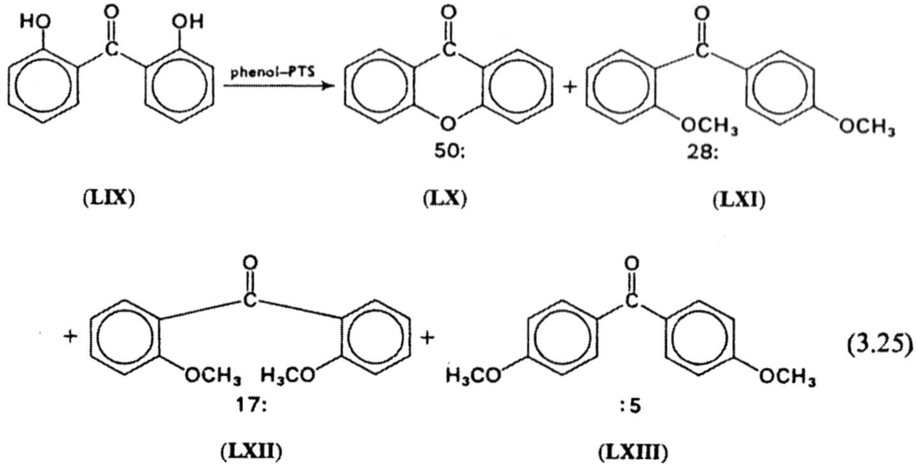

(3.25)

given in reaction scheme (3.25). This indicates that xanthone (**LX** and compound **130** in Figure 3.8) was derived from a diaromatic ketone structure in the humic acid molecule. Although some 9,9′bis(4-methoxyphenyl)xanthene (compound **128**) was formed when xanthone was reacted in a 1:110 ratio with phenol, this could not account for the amounts present in the humic acid digest. It would therefore appear more likely that **128** was formed from a triaryl tertiary alcohol structure, or less likely from a quaternary aryl group. In either case this would represent a potential branching point in the polymer structure. Other possible branching points could be indicated by structures **126**, **127**, and **129**, if it is assumed that hydroxyl was absent from the original (R) linkages.

No convincing explanation can be offered for the origins of **131**, but it could be formed from a lignin-type percursor.

The phenolic degradation technique differs from the others discussed because it helps to identify interaromatic linkages. As such it is complementary to other degradative techniques which are primarily used to identify monomer structures. Yield figures indicate that this reaction causes extensive degradation, giving approximately 50% ether-soluble components, and containing numerous, as yet unidentified products. There is obviously great scope for further research on this technique.

Degradation with sodium sulphide

Mixtures of sodium hydroxide and sodium sulphide have been used in the pulping industry since the end of the last century to delignify hardwoods and softwoods. These chemicals are the main ingredients of Kraft pulping liquor, and lesser amounts of sodium carbonate, sodium thiosulphate, and sodium polysulphide are also present. The potential value of sodium solutions for degradation of humic substances was examined by Swift (1968) and was further developed by Craggs (1972).

Hayes *et al.* (1972) isolated 60 % of the starting polymer mass as ether and ether–ethanol soluble products when H^+-humic acid was degraded in a 10 % sodium sulphide solution at 250 °C. Craggs *et al.* (1974) reported similar yields of ether-soluble components, and these were composed of materials in solution (45 %) and associated with the precipitate (15 %) formed when the digest was acidified. The average molecular weight values of the acid soluble and insoluble materials were 145 ± 15 and 245 ± 25, respectively, and these values indicated that the digestion procedure had caused extensive degradation of the polymer. More recently O'Callaghan (1977) has shown that yields of ether-soluble products are improved when cobalt molybdate (a dehydrodesulphurization catalyst) is added and the reaction is carried out in an atmosphere of hydrogen. It is thought that the reducing conditions inhibit oxidative coupling of phenols which give rise to artefact products.

Marton (1971) has summarized a number of model studies of reactions of sodium sulphide with compounds related to structures in lignins, and Craggs (1972), Burdon *et al.* (1974) and Craggs *et al.* (1974) have investigated some reaction mechanisms during the course of their humic acid degradation reactions.

$$S^{2-} + H_2O \overset{K_1}{\rightleftharpoons} HS^- + HO^-;\; HS^- + H_2O \overset{K_2}{\rightleftharpoons} H_2S + HO^- \qquad (3.26)$$

It is important to understand the solution properties of sodium sulphide before considering any of its mechanisms of degradation. The compound hydrolyses in water as shown in scheme (3.26) where the equilibrium constants in the two reactions, K_1 and K_2, are the first and second dissociation constants of hydrogen sulphide.

The values for pK_1 and pK_2 indicate that negligible amounts of S^{2-} are present in the alkaline medium ($pH = 12.5$) at high temperatures, and the HS^- and HO^- species are both present in similar concentrations. Nucleophilic strength increases in the order $HO^- < HS^- < CH_3S^-$, and this can be explained on the basis of the polarizability of the anions; the larger HS^- ion being more readily polarized than the smaller HO^- species. The ability to abstract a proton, or the basic strength, is in the reverse order.

Important features of sodium sulphide degradation reactions include:

(i) demethylation of methylphenyl ethers (a well known reaction in lignin chemistry);
(ii) the formation of quinone methide structures;
(iii) hydride transfer processes (Weedon, 1963);
(iv) the partial inhibition of condensation reactions (Marton, 1971).

Reaction scheme (3.27) incorporates the first three of these features. Attack by HS^-, in an S_N2 mechanism leads to the formation of methyl mercaptan (CH_3SH) and the anion species (**LXV**). This anion can lose OH as indicated to form the *ortho* quinonemethide structure (**LXVI**) which takes up hydride to give **LXVII**.

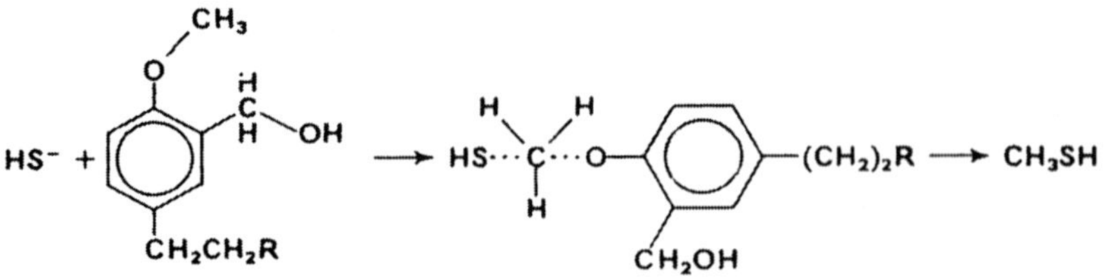

(LXIV)

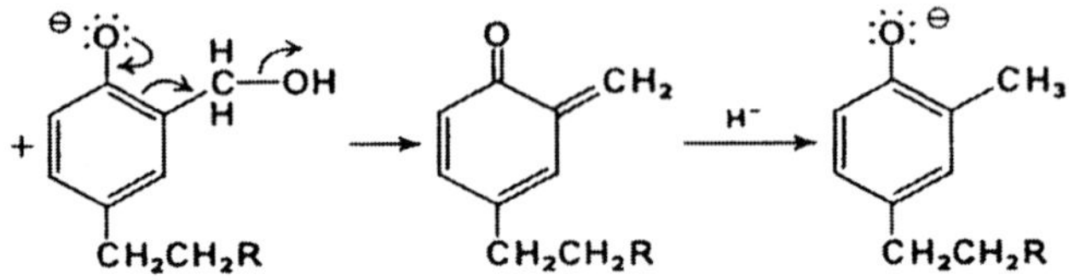

(LXV) (LXVI) (LXVII). (3.27)

A number of mechanisms can give rise to hydride formation; one of these is the Dumas–Stass reaction (3.28) where the base HX^- is HS^- or HO^-. The HS^- can also accept H^- to release H_2 and give the S^{2-} species. Methyl mercaptan is ionized to the mercaptide in the alkaline medium, and the highly nucleophilic mercaptide can attack another methyl phenylether group to form the phenate anion and dimethylsulphide (CH_3SCH_3).

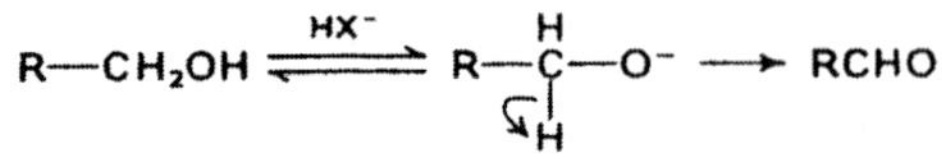

(3.28)

Several aliphatic compounds were identified among the more volatile components in the lower molecular weight fraction of the sodium sulphide digest of humic acid. These included aliphatic mono- and dicarboxylic acids, and some keto acids (Craggs *et al.*, 1974). Many of the products which had longer retention times during g.l.c. separation were aromatic and had spectral properties indicative of phenylpropane-type structures. It was considered unlikely that these were artefacts since no aromatic compounds were formed when cellulose was degraded by the same technique. By taking into account the glucose equivalent of the carbohydrate (7 %) associated with the humic acid material it was concluded that the short chain aliphatic acids identified were primarily derived from the humic polymer and not from associated carbohydrate.

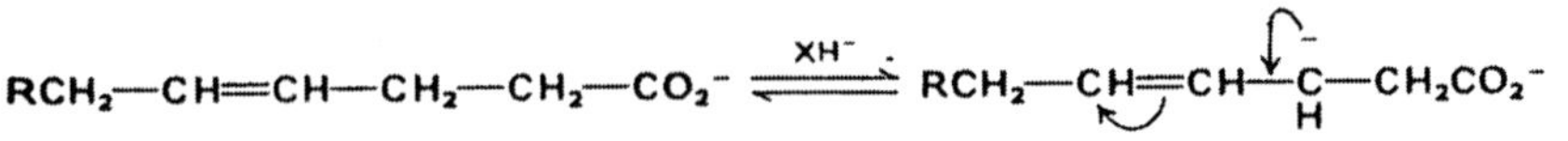

further migration to $RCH_2—CH_2CH_2CH=CHCO_2^-$

$\xrightarrow{\text{fission}} RCH_2CH_2CH_2CO_2^- + CH_3CO_2^-$ (3.29)

Ethanoic acid was present in greatest abundance in the digest. One plausible explanation would suggest that this and other acids might have formed by a Varrentrapp reaction as outlined in scheme (3.29).

This scheme illustrates the reversible migration of the double bond by a prototropic mechanism, and the mobility of the double bond increases as it approaches the carboxylate group. When the double bond reaches the α, β position fission takes place following attack by base to give the β-hydroxy or thiol acid.

Compounds **LXVIII** and **LXIX** are representative of a number of the methylated aromatic species isolated from the digest and identified or tentatively identified. On the basis of the demethylation illustrated in scheme 3.27 we can assume that the methoxyl substituents were present in the digest as phenolic hydroxyls prior to methylation. Thus quinone methide structures would be formed where appropriate substituents were located *ortho* and *para* to the hydroxyl groups. Reference again to reaction scheme (3.27) will show how the methyl substituent on the aromatic C_5 should have formed from $—CH_2OH$ (which could in turn have formed from the carbaldehyde structure). More significantly, perhaps, it could also have formed from elimination of an α-ether substituent, and the reduction of the resulting quinone methide.

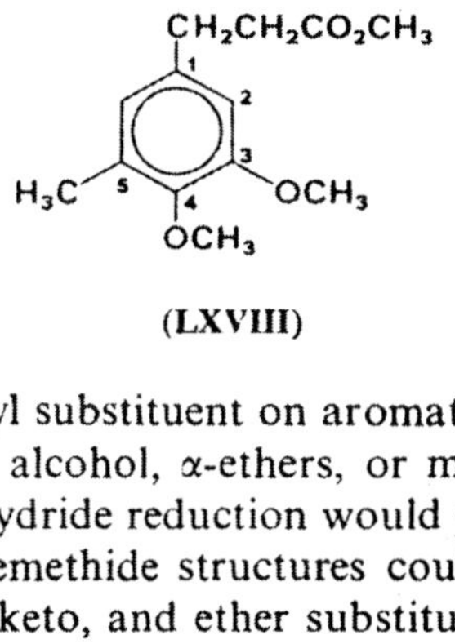

(LXVIII)

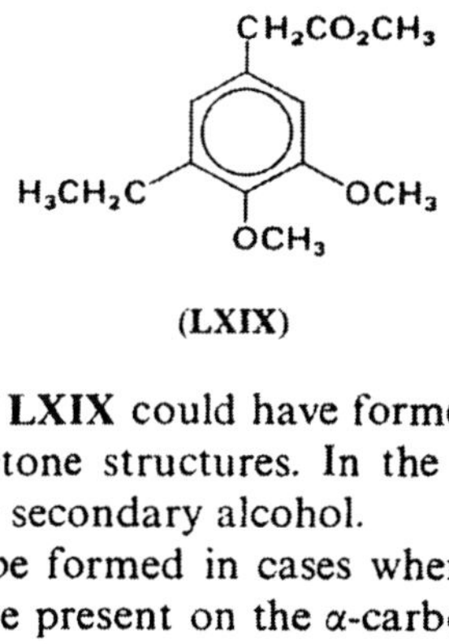

(LXIX)

The ethyl substituent on aromatic C_5 of **LXIX** could have formed from the secondary alcohol, α-ethers, or methyl ketone structures. In the case of the ketones, hydride reduction would yield the secondary alcohol.

Quinonemethide structures could also be formed in cases where alcoholic hydroxyl, keto, and ether substituents were present on the α-carbon attached to aromatic C_1 in **LXVIII** and **LXIX**. However, it is likely that the esters were formed after cleavage of aliphatic side-chains to yield the carboxylic acid structures. Again the Varrentrapp reaction, or reaction scheme (3.18) by Shemyakin and Shchukina (1956), could be invoked to explain the formation of these carboxyl groups, but further studies with model compounds will be required before definite conclusions can be reached with regard to their possible origins.

Clearly the sodium sulphide degradation technique can provide very useful information with regard to the primary structures of humic substances. At its present stage of development the technique strongly indicates that considerable amounts of aliphatic substituents are attached to the aromatic structures in these materials. Quantitative isolation and identification of digest products, coupled with extensive studies of reactions of model compounds are now required in order to be able to appreciate its full potential as a degradative procedure.

Additional degradation procedures

Peracetic acid oxidation, anodic oxidation, and pyrolysis procedures provide additional useful techniques for the degradation of humic substances. Although the first two procedures have extensive uses in organic chemistry they have only had limited applications to investigations of humic structures. There is, however, a growing interest in pyrolysis, and this developing technique appears to be especially useful for making comparisons between the compositions of humic materials of different origins.

Peracetic acid (CH_3CO_3H), when protonated dissociates to the reactive OH^+, and this species, as well as the protonated parent peracid, takes part in oxidation reactions. Well known reactions of peracids include the formation of oxiranes from alkenes, the conversion of ketones to esters and lactones, and the oxidation of aliphatic aldehydes to carboxylic acids.

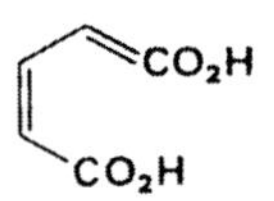

(LXX)

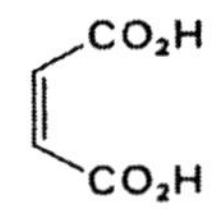

(LXXI)

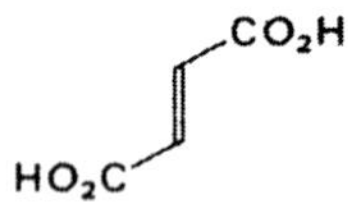

(LXXII)

Of special interest here is the fact that phenols and phenol ethers readily undergo electrophilic substitution when treated with peracetic acid. Substitution is *ortho* and *para*, and the dihydroxy benzenes form quinone structures which oxidize further to dicarboxylic aliphatic acids. For instance 2-hydroxybenzenol yields muconic acid (**LXX**) and 4-hydroxybenzenol gives the *cis* and *trans* butenedioic, or maleic (**LXXI**) and fumaric (**LXXII**) acids. Waters (1964), Behrman and Edwards (1967), and Lewis (1969) have reviewed the mechanisms of peracid oxidation reactions.

Chang and Allan (1971, pp. 458–459) have described the uses of peracetic acid for the degradation of lignins and wood, and they have presented some relevant mechanisms for reactions with model substances. Meneghel *et al.* (1972) oxidized a peat humic acid with peracetic acid for eight days at 40 °C and identified five amino acids and six sugars in the digest. Schnitzer and Skinner (1974a, 1974b) were more successful when they heated chernozem humic and fulvic acids for four hours in 10% peracetic acid at 80 °C. Among the compounds identified were numbers **1**, **2**, **3**, **8**, **9**, **10**, **11**, **13**, **14**, **16**, **17**, **18**, **19**, **20**, **21**, **23**, **24**, **25**, **26**, and **27** in Figure 3.4, and numbers **53** and **62** in Figure 3.5. Other compounds included propanedioic acid, 2-methoxy-1,5-benzenedicarboxylic acid, 3,4-dimethoxy-1,5-benzenedicarboxylic acid, 5-methoxy-1,2,3-benzenetricarboxylic acid, nC_{12}–nC_{20} fatty acids, methyl-3,4-furandicarboxylic acid, and branched chain butanoic acids.

Total yield of products and the types of structures identified were very similar to the results from degradations with permanganate. Again the results would suggest that aliphatic substituents on the aromatic structures were oxidized. It will be necessary to carry out model studies on possible precursors of the identified products before predictions can be made with regard to their parent structures in the humic polymers.

Ross *et al.* (1975) have extensively reviewed electrochemical oxidation reactions of organic compounds and illustrated numerous mechanisms of anodic oxidations. Strangely, despite the promising application by Belcher (1948) of anodic oxidation to coal fractions, the technique has not been adequately tested for the degradation of soil humic substances. Yields of up to 35 % ether-soluble products were obtained by Belcher when he oxidized vitrain coal materials at the copper anode. Some benzene polycarboxylic acids, similar to those produced by permanganate oxidation, were isolated and identified.

Modern separation and instrumental analysis procedures would allow identification of a high proportion of the products released during controlled anodic oxidation reactions. Thus, by step-wise increases of the electrode potential (and analysis of products released during each increase) it might be possible to observe differences in the strengths of bonding in the polymer of the products released at different anode potentials. The resistant residues could possibly be further degraded by chemical procedures.

Volatile degradation products are released when polymers are rapidly pyrolysed in an inert atmosphere. These products may be separated by g.l.c. and analysed by m.s. or g.c.–m.s. instrumentation (Meuzelaar *et al.*, 1974) and the traces obtained can be used to 'fingerprint' the polymers.

Pyrolysis technology has been greatly advanced by the development of Curie point pyrolysis procedures (Bühler and Simon, 1970). This allows samples in the microgram range to be heated at rates in excess of 1 °C $msec^{-1}$, and the possibilities of secondary reactions are minimized by the rapid removal of the degradation products from the reaction zone.

For more meaningful interpretations of results it is important to be able to predict the types of parent polymer structures which give rise to the degradation products. Thus considerable exploratory work is still required in order to evaluate fully data from the pyrolysis of humic substances. At this time interpretations are based on information related to non-humic systems, such as proteins, fats and waxes, carbohydrates, and lignins. According to Simmonds *et al.* (1969) the pyrolysis of proteins, peptides, and amino acids give nitriles, fats and waxes produce unbranched alkanes and alkenes, and carbohydrates give aliphatic aldehydes, ketones, and furan derivatives containing $—CH_2OH$, —CHO, and CH_3 substituents. Because the organic materials in ancient sediments and meteorites pyrolyse to predominantly hydrocarbon-type products compared to the more heteroatomic fragments which characterize present-day biologically derived materials, pyrolysis gas chromatography was chosen to examine Martian soils for evidence of materials from life systems during the Viking mission in 1975.

There are already a number of applications of pyrolysis techniques for studies of soil organic matter, and of soil and model humic substances (Nagar, 1963; Wershaw and Bohner, 1969; Kimber and Searle, 1970; Gomez Aranda *et al.* 1972; Murzakov, 1972; Bracewell and Robertson, 1973 and 1976; Martin, 1975; Nagar *et al.* 1975; Meuzelaar *et al.*, 1977; Martin *et al.*, 1977; Haider *et al.* 1977a). Bracewell and Robertson (1976) have given a good example of the uses of pyrolysis g.l.c. for 'fingerprinting' humus materials from different soils and in different horizons in the same soil. By use of a Curie-point pyrolyser linked to g.l.c. they were able to observe chromatogram differences between mull and mor humus of whole-soil samples. When the effluent from the g.l.c. column was analysed by mass spectrometry the identified products included furfural, pyrrole, 5-methyl-2-furfuraldehyde, benzenol, 2-methylbenzenol, 2-methoxybenzenol, and 4-methyl-2-methoxybenzenol. A lignin product gave significant yields of 2-methylbenzenol and 2-methoxybenzenol.

Mor humus produced considerable amounts of furfural derivatives, possibly indicating a carbohydrate source. By contrast pyrrole was the main component of mull humus and this could suggest an origin in proteins, or the amino acids proline and hydroxyprolines, porphyrin structures or soil microorganisms (Simmonds, 1970). The patterns for the relatively undecomposed surface organic matter suggested the presence of lignin, but the characteristic benzenol derivatives were not present to a significant extent in the humus in the mineral soil horizons.

By use of Curie-point pyrolysis in direct combination with a fast-scanning quadrupole mass spectrometer Meuzelaar *et al.* (1977) and Haider *et al.* (1977a) were able to show definite similarities and differences between humic acids from soils, peats, model phenolic polymers, fungal 'humic acids' and lignins. Again their results emphasize the value of pyrolysis techniques for 'fingerprinting' humic substances.

Further development of pyrolysis procedures could allow some interpretations to be made relative to the structures of humic substances. This will, of course, require extensive model studies with known polymers and chemicals pyrolysed under the same reaction conditions as used for the humic materials. Such an approach is encouraged by the work of Martin (1975) who identified 18 organic compounds among the low-boiling components in the chromatogram from pyrolysis g.l.c. of humic and fulvic acids. Furane derivatives were not detected when the samples were pre-hydrolysed, which indicates that carbohydrates were removed by the hydrolysis process. The 12 aliphatic alkanes and alkenes, with up to 10 carbon atoms, suggested aliphatic or/and alicyclic sources. Benzene and methylbenzene were present in all pyrolysates, but their amounts decreased after hydrolysis. On the other hand benzenol increased on hydrolysis, and this could have been derived from lignin and from humic structures.

General conclusions from degradation studies

Considerations of digest yields of ether- and ethylacetate-soluble products from degradations of humic substances by procedures employed until the present time emphasize the need for a continuing search for more efficient degradative reagents and conditions. It is unlikely that complete degradation to molecules which can be related to structures in the polymer will be achieved by use of a single reagent, and it is probable that a sequence of reactions will be required in order to obtain maximum yields of component products.

Hydrolysis in acid and base can be regarded as a preliminary treatment (Jakab *et al.*, 1962) which cleaves labile structures such as the glycosidic linkages of oligo- and polysaccharides (Section 3.2.6), ester groups from condensations between phenolic hydroxyls (as exemplified in structure **XXX**) or alcohols and carboxylic acids, and the peptide linkages bonding amino acid residues in proteins and polypeptides. Components held in these ways can be regarded only as peripheral to the structures of humic substances. Carbohydrates, for instance, are seldom present in excess of 10 per cent of the total mass of humic acids, and could be bound to them by physical adsorption forces, or more likely through phenolic glycoside linkages. Peptides or proteinaceous materials, which are present in lesser abundances than the carbohydrates, could also be physically adsorbed, but they are more likely to be bonded to quinone-type structures through terminal or free amino groups as indicated by scheme (3.1), or to carbonyl groups through Schiff-base type linkages as indicated by reaction schemes (3.5) and (3.6).

Cleavage of the 'core' or fundamental structures of humic substances is energetically demanding. This would not be true, as Maximov *et al.* (1977) have also pointed out, if the acid compounds shown in Figures 3.4 and 3.5 (or, for that matter, similar compounds among any of the digest products listed) were *per se*, the 'building blocks' of the 'core' structures. Although such acids could be held together when undissociated through hydrogen bonding they would be dispersed when solubilized under neutral or alkaline conditions and would not exhibit polymer properties. Because of the relatively high energy requirements for the degradation of humic polymers it is probable that carbon to carbon, carbon to oxygen (e.g. ether), and possibly some carbon to nitrogen bonds are important in linking together the monomer units or primary structures of humic materials.

Techniques which employ phenol plus *para*-toluenesulphonic acid and sodium sulphide at moderately elevated temperatures cleave humic acid polymers to ether-soluble components amounting to 50–60 per cent of the masses of the starting materials. However, the major proportions of the products released by these procedures have not yet been identified. Polymer degradations have not been as extensive from controlled oxidations with permanganate, alkaline cupric oxide, and peracetic acid, but the use of modern separation and spectroscopic techniques (as employed by Schnitzer and his colleagues) has helped to identify most of the ethylacetate and ether-soluble digest digest products which can amount to 20–30 per cent of the polymer masses. But even when allowances are made for the fact that product yields in organic reactions are generally depressed, it is obvious that the bulk of the polymer masses are not degraded to monomer components by the foregoing oxidative processes. Considerations of yields and of identities of degradation products do not provide sufficient information to allow predictions of meaningful structures for the polymers. Nevertheless, by taking into account the types of compounds identified from a variety of digest reactions in several laboratories, some predictions can be made about some of the humic polymer component molecules, and about the types of linkages which hold these together.

There is unquestionable evidence for the presence of substituted single aromatic structures in humic polymers. These are almost invariably di- or polysubstituted, because only traces of monosubstituted benzene derivatives, such as benzoic acid, have occasionally been detected in digests. Proof that fused aromatic compounds contribute to the structures is less convincing. The available evidence shows that such compounds, which are present only in small abundances in the digests, could have been formed as artefacts during the degradation (e.g. zinc-dust distillation and fusion) reactions. Some of the aromatic di- and polycarboxylic acids (Figure 3.4) from the permanganate degradations might also suggest fused aromatic parent structures, but it is noticeable that similar acids were formed or released in reactions (such as with alkaline cupric oxide, Figure 3.5) in which the fused structures would not be degraded to benzenecarboxylic acid derivatives. Thus we favour the interpretation that at least some of the substituent carboxyl groups were derived by oxidations of aliphatic side chains.

Functional group analysis data have shown that humic polymer structures contain carboxyl, phenolic and alcoholic hydroxyl, carbonyl and methoxyl groups. Some of the carboxyl groups are unquestionably present as substituents on the aromatic structures in the polymer, as can be inferred from compounds identified in digests of sodium amalgam reduction reactions (Figure 3.6). It is unlikely, however, that the aromatic nuclei in the polymer were polysubstituted with carboxyl groups, as seen in Figures 3.4 and 3.5, and it is more plausible to suggest that these were generated from aliphatic side-chains. There is evidence for the presence of such side-chain structures in the cases of some groups which resisted oxidation by permanganate and alkaline cupric oxide. However, the products from sodium sulphide degradation reactions give the strongest indications for the presence of more than one aliphatic side-chain substituent on aromatic nuclei.

There are striking differences between the types of products released by oxidative and reductive processes, and these differences must be related to the linkages which are cleaved in the parent polymer structures by the different reagents. Degradations with sodium amalgam released considerable amounts of di- and trihydroxybenzene derivatives, and these sometimes contained carboxyl (but never more than one group per aromatic structure), methoxyl, carbonyl, methyl and longer aliphatic chain substituents. It is possible that some (but perhaps not all) of the polyphenol derivatives identified had been decarboxylated during the degradation reactions. But clearly, with the exception of structure **85**, none of the products from the amalgam reductions in Figure 3.6 could be oxidized to aromatic structures containing more than one carboxyl group, and this contrasts with most of the structures isolated from oxidation reactions (listed in Figures 3.4 and 3.5). Therefore the available evidence indicates that different polymer structural units are released by the oxidation and reduction reactions, further indicating that different linkages are attacked although, of course, some overlaps could occur.

Diaryl ethers have been shown to be cleaved in sodium amalgam degradations, and all of the compounds listed in Figure 3.6 (where allowance is made for methylated products) could have been released from such ether linkages in the humic polymers. Additional model studies will be required in order to postulate other polymer origins for the various structures, but it can be stated with confidence that the carbonyl and carboxyl structures shown reflect active functional groups in the polymer, and do not represent artefacts formed in the digest mixtures. Other interesting amalgam digest products include phenylpropane units containing methoxyl and hydroxyl substituents on the aromatic nuclei, as such structures might originate from lignin-type materials.

Examination of the aromatic compounds obtained by the oxidative degradation processes shows that many of these are di- and polycarboxylic, and that several also contain methoxyl and/or hydroxyl substituents; it is impossible to estimate the free hydroxyl group contents in these instances because the digest products were methylated prior to analysis. Aromatic permanganate degradation products were especially rich in carboxyls, and it is plausible to assume that several of these were formed by the oxidation of aliphatic side chains. The same digests yielded considerable amounts of di- to tetracarboxylic acids which could have been formed from olefinic structures appropriately separated by $—CH_2$ groups, unsaturated branch chains, and/or aldehyde or primary alcohol substituents. Such structures could be bonded to the aromatic components through various ester and ether linkages, or they could of course be held as hydrocarbon substituents on the aromatic nuclei.

Oxidations with alkaline nitrobenzene and alkaline cupric oxide provide further evidence that some of the digest products might have been released from lignin-type units in the polymer structures. This is especially obvious in the cases of the compounds (see Figure 3.5) which contain carbonyl groups as well as methoxy substituents in the aromatic 3,4- or 3,4,5-positions. Like permanganate, alkaline cupric oxide digestion yielded di- to tetracarboxylic acids, but the latter reagent also produced significant numbers and yields of aliphatic hydrocarbons and monocarboxylic acids. Mechanisms, such as those in reaction schemes (3.18) and (3.29, the Varrentrapp reaction), show how both types of products could have been obtained in alkaline media from olefinic structures.

We cannot confidently predict whether or not the long-chain aliphatic hydrocarbons and acids were derived from polymer structures, or whether they should be regarded as humification products. They could merely represent unaltered products of microbial metabolism which were physically adsorbed to the humic structures and freed during the degradation processes. Alternatively (and very likely) the acids were released from ester structures, as has been discussed, by the hydrolytic action of the aqueous base media. It would appear unlikely that such long chain aliphatic structures are substituted into the aromatic structures, although we have no evidence that they were not.

Degradations with phenol plus *para*-toluenesulphonic acid suggest that aromatic groups in humic polymers could be linked by hydrocarbon bridges. The structures shown in Figure 3.8 represent only a small proportion of those released in the digests, but it can be inferred from the limited data available that straight-chain and branched hydrocarbon structures could be involved in such linking processes. Alcoholic hydroxyls could also have given rise to some of the structures identified.

On the basis of the information which is available at this time we conclude that most can be learned, in so far as humic structures are concerned, from degradations with alkaline cupric oxide, sodium amalgam, sodium sulphide, and phenol. It may well be worthwhile to investigate the effects of the different reagents applied in sequence. Such investigations would demand rigorous anaerobic conditions where alkaline media are used because of the possibilities of secondary polymerization reactions which are accelerated by the presence of oxygen.

3.4.4 Chemical and Biological Synthesis of Humic Substances

Most of the work directed towards understanding the structures of humic materials has concentrated on degrading the polymers and identifying the compounds released into the digests. It is obvious from the structures listed in Figures 3.4–3.8 that very many products, real or artefacts, can be detected by these procedures. An alternative approach would be to chemically, enzymatically, or biologically synthesize humic-type substances from known starting materials, and to compare their compositions and physical and chemical properties with those of natural humic polymers.

Chemical synthesis of 'humic' polymers

Maillard (1912, 1916, 1917) observed that brown polymers formed during the reaction of glucose with glycine (2-aminopropanoic acid) displayed some properties similar to those of soil humic acids. This reaction, classically known as the 'Maillard' or 'browning' reaction, is well recognized in food processing industries, and has been reviewed by Hodge (1953) and Ellis (1959).

Several reactions are known in which amino acids and amines react with reducing sugars and aldehydes to give melanoidins or 'humic-like' substances. Among the more interesting products are polymers formed by the reaction of propanedione (methylglyoxal) and glycine which Enders and his colleagues (Enders and Fries, 1936; Enders, 1943a, Enders *et al.*, 1948) have shown to possess properties comparable with soil humic acids. Enders (1943b) also postulated that methylglyoxal could be released from soil microorganisms in unfavourable growth conditions to form humic substances, and later showed (Enders and Sigurdsson, 1948) that the compound was present in ten of the sixteen soils tested.

Schuffelen and Bolt (1950) found that the C:N ratio of the products from the reaction of methylglyoxal and glycine varied with the relative concentrations of reactants used, and they isolated a polymer which compared favourably with a Dalgrund peat on the basis of this ratio, titration, and CEC data. Later Hayes (1960) showed that the polymer formed in an atmosphere of oxygen from equimolar proportions of the same reactants had C:N ratios, differential thermal analysis properties, paper electrophoresis mobilities, and infrared spectra similar to neutral pyrophosphate extracted humic acids from an organic soil. These comparabilities do not, of course, necessarily imply similarities in compositions, and more significant tests would involve investigations of the compounds released when the polymers are degraded by chemical or physical means.

Glucosamine (see Section 3.6), a component of chitin, has been shown by Drozdova (1959) and Young *et al.* (1977) to form humus-like products when heated in solution, but it would appear that browning was not as rapid and the polymer products were not as high in molecular weights as those from the glucose– and methylglyoxal–amino acid systems.

Flaig *et al.* (1975) have reviewed the extensive literature which deals with the formation of dark coloured 'humus-like' substances from phenolic precursors. Phenols are regarded as important components because: (a) plant residues, such as lignins, which contain them are relatively resistant to degradation; (b) phenols or quinones are synthesized by microorganisms; and (c) phenolic lignin degradation products can be isolated from the soil.

Eller (1925) and his coworkers were among the first to observe some similarities between soil humic acids and the polymeric products formed by alkaline oxidation of phenols. The polyoxybenzene theory, reviewed by Flaig (1950), assumes that hexose sugars might aromatize to hydroxybenzoquinone as outlined in reaction scheme (3.30), and this highly reactive quinone structure (**LXXIII**) could readily polymerize under alkaline conditions by a number of different mechanisms. One of these might involve formation of ether linkages between the quinone structures followed by rearomatization as indicated in scheme (3.31). A number of studies, including those by Flaig and his colleagues (e.g. Flaig, 1960, 1966; Flaig and Schulze, 1952; Flaig *et al.*, 1955; Flaig and Salfield, 1958, 1960a, 1960b) have investigated the alkaline polymerization of hydroxy-*p*-benzoquinone and of hydroquinone and compared the products

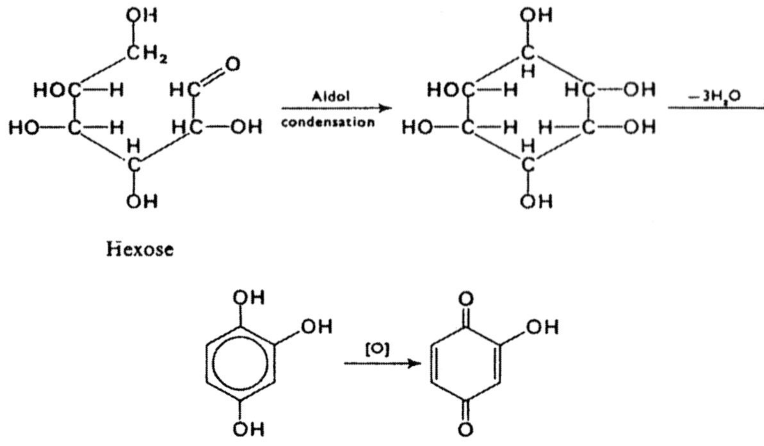

(3.30)

(LXXIII)

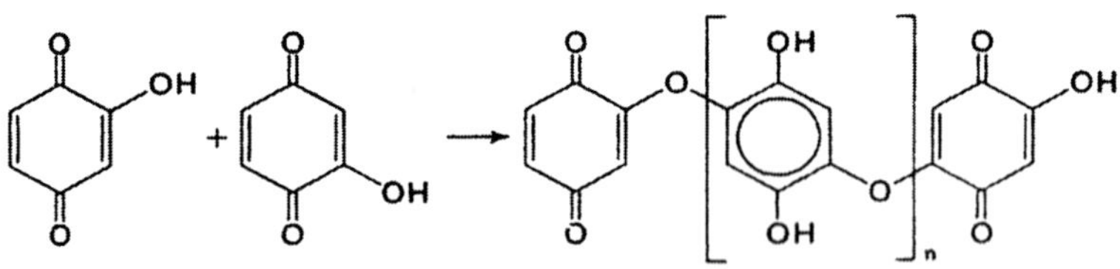

(3.31)

with natural humic acids. Best yields of polymer are obtained in the pH 8–10 range. Formation of carboxyl group and decarboxylation can result from oxidation of the quinone structures. Incorporation of nitrogen into the polymers can also take place by a variety of mechanisms, one of which is outlined in reaction scheme (3.1).

It is, of course, highly unlikely that naturally occurring humic polymers would be wholly composed of chemically synthesized polyphenol or polyquinone structures. However, without doubt chemical synthesis could contribute to humus formation where a variety of phenols are present in the appropriate chemical environment in the soil. In this way chemical processes might supplement extracellular enzymatically catalysed reactions.

Enzymatic synthesis of 'humic' polymers

Martin and Haider (1977) have drawn attention to species of *Basidiomycetes*, *Ascomycetes*, *Fungi imperfecti* and some bacteria which can degrade lignin. They suggested that phenols released from lignins, microbially synthesized phenols, partially degraded lignins, and other reactive compounds could undergo autoxidation, or be enzymatically polymerized to humic molecules either extracellularly or within microbial cells. Other possibilities for humus formation would link sugars and amino acids to reactive molecules by enzymatic or autoxidative processes.

Phenoloxidase enzymes are widespread and at least partially purified preparations can be isolated from the soil environment (Mayaudon and Sarkar, 1974). Mushroom phenolase, when added to mixtures of di- and tri-hydroxybenzenes, hydroxy- and methoxybenzenecarboxylic and -propenoic acids produce brown polymeric 'humic-type' structures (e.g. Martin and Haider, 1976, 1977; Verma and Martin, 1976; Haider *et al.* 1977b; Zunino and Martin, 1977). There is good reason to believe that such phenolic structures, released by degradation or biosynthesized, would be polymerized to humic materials by the phenoloxidase enzymes present, as well as by chemical reactions in the soil.

Microbial synthesis of 'humic' polymers

Numerous investigators have reported the formation of 'humic-like' substances by the activities of several fungal species, by bacteria, actinomycetes, and even by yeasts. Here we will concentrate on the products isolated by Martin and his associates from fungi grown on different media.

In 1967 Martin *et al.* cultured the fungus *Epicoccum nigrum* on a glucose–asparagine medium containing yeast extract and inorganic salts. After incubation for five weeks they isolated, at pH 2.0, mycelium-free 'humic acid' substances. Some of these were similar to a leonardite humic acid in elemental composition, total acidity, CEC, carboxyl and phenolic hydroxyl contents, and in molecular weight distributions, and the values reported were in the ranges quoted for soil humic acids. In soil incubation studies some of the synthetic polymers decomposed as little as 4% in eight weeks (compared with 1–7% in the same time for the leonardite preparations). The biosynthesized substances increased soil aggregation.

Subsequently Bondietti *et al.* (1971) showed that the addition of clays to similar media inoculated with the fungi *Hendersonula toruloidea* and *Aspergillus sydowi* accelerated growth, glucose utilization, and polymer synthesis. Montmorillonite was a more effective catalyst than kaolinite or illite. The catalytic effects of the clays were thought to result from (among other things) the concentration of enzymes and of substrates on the clay surfaces and the removal (also by adsorption) of metabolic wastes from the media.

Biosynthesis of fungal 'humic acid' precursors. It has been possible to follow the synthesis and degradation of phenols by various fungi by isolating at 4–7 day intervals the ether soluble products from fungal culture media. Figure 3.9, from Haider and Martin (1967), illustrates a plausible scheme for the genesis of 15 of the 20 or so phenols which were identified by two-dimensional thin layer chromatography from an *Epicoccum nigrum* culture. Compounds **132**, 2-methyl-3,5-dihydroxybenzenecarboxylic acid (cresorsellinic acid) and **133**,

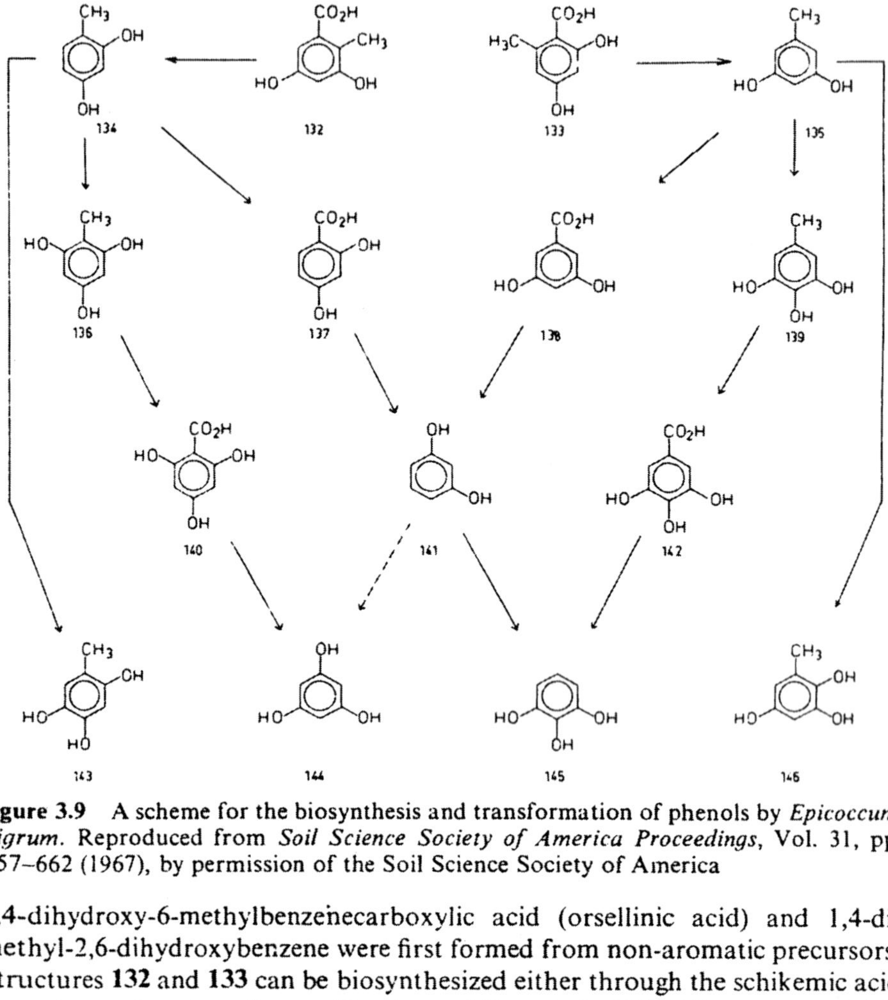

Figure 3.9 A scheme for the biosynthesis and transformation of phenols by *Epicoccum nigrum*. Reproduced from *Soil Science Society of America Proceedings*, Vol. 31, pp. 657–662 (1967), by permission of the Soil Science Society of America

2,4-dihydroxy-6-methylbenzenecarboxylic acid (orsellinic acid) and 1,4-dimethyl-2,6-dihydroxybenzene were first formed from non-aromatic precursors. Structures **132** and **133** can be biosynthesized either through the schikemic acid or acetate malonate pathways (Bentley and Campbell, 1968). These were quickly followed by **134** (2,4-dihydroxytoluene), **135** (3,5-dihydroxytoluene, or orcinol) and 1,4-dimethyl-2,6-dihydroxytoluene. Most of the remaining phenols in Figure 3.9 were formed during the later stages of incubation. By following the arrows it will be seen that some of the products identified could have been derived from precursors by oxidation of methyl substituents, by hydroxylation of the aromatic nuclei, and by decarboxylation of carboxylic acids. The unusual product 1,4-dimethyl-2,6-dihydroxybenzene might have been derived from non-aromatic precursors or from the introduction of a methyl group into the aromatic nucleus of orcinol (**135**).

Compound **133**, but not necessarily **132**, was found in the media in the early stages of incubation of other fungi. For instance the first products formed in a similar medium by *Stachybotrys atra* and *S. charatum* were **133**, *p*-hydroxycinnamic acid (4-hydroxyphenyl 3-prop-2-enoic acid), and 2-hydroxy-6-methylbenzenecarboxylic acid (Martin and Haider, 1969). In the case of *Hendersonula toruloidea* (Martin *et al.*, 1972) **133** and 2-hydroxy-6-methylbenzenecarboxylic acid were formed at pH 3.5–4.5 from non-aromatic precursors, followed by **136**, **138** (3,5-dihydroxybenzenecarboxylic acid), **142** (3,4,5-trihydroxybenzenecarboxylic acid or gallic acid), and later by **141** (3-hydroxybenenol or resorcinol). The substitution of additional hydroxyl groups into the various aromatic

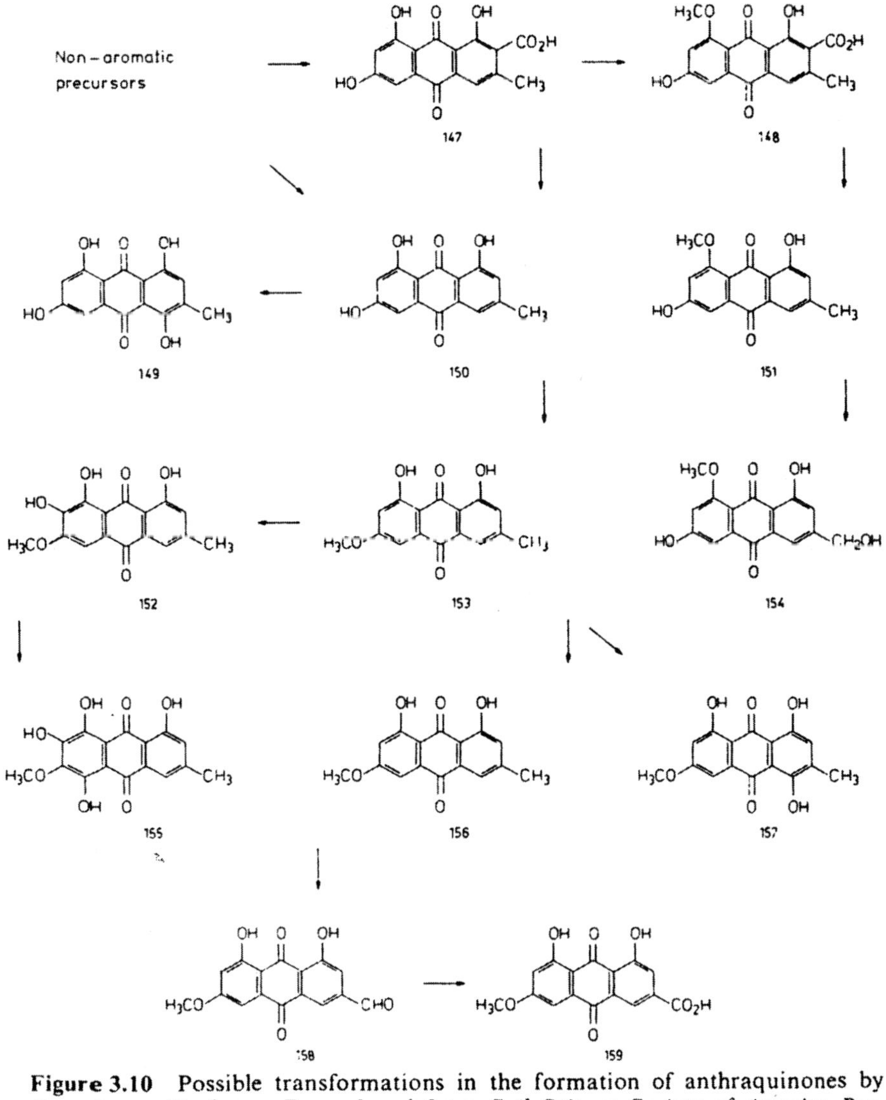

Figure 3.10 Possible transformations in the formation of anthraquinones by *Eurotium echinulatum*. Reproduced from *Soil Science Society of America Proceedings*, Vol. 39, pp. 649–653 (1975), by permission of the Soil Science Society of America

nuclei can result from strong mixed function oxidase enzyme systems (Haider and Martin, 1967).

Anthraquinones are synthesized by some fungi and Saiz-Jimenez *et al.* (1975) have identified the compounds shown in Figure 3.10, as well as several phenols in the ether extracts of cultures of *Eurotium echinulatum* grown on glucose–asparagine and glucose–nitrate media. Again the arrows lead from the possible precursors of the various anthraquinone derivatives identified. Compounds **147** (endocrocin) and **150** (emodin) can be formed from non-aromatic precursors by the acetate–malonate pathway (in Packter, 1973) and these can give rise to dermolutein (**148**), catenarin (**149**), questin (**151**), dermoglaucin (**152**), physcion (**153**), questionol (**154**), dermocybin (**155**), fallacionol (**156**), erythroglaucin (**157**), fallacinal (**158**) and parietinic acid (**159**). The first phenols formed, *p*-hydroxycinnamic acid and 4-hydroxybenzenecarboxylic acid were also synthesized from non-aromatic precursors by the shikimic acid pathway, and they were transformed by the appropriate oxidation, decarboxylation, and hydroxylation reactions into a range of phenolic compounds.

Resistance of precursors to microbial decompositions. By specific carbon-14 labelling of aromatic ring carbons, carboxyl groups, and aliphatic carbon side chains, Haider and Martin (1975), Martin and Haider (1976, 1977) and Haider *et al.* (1977b) showed that carboxyl groups in hydroxybenzenecarboxylic acids were readily decarboxylated in soil, that phenolic compounds with unsaturated aliphatic side-chains (models for constituents of lignin), such as 2(3-methoxy-4-hydroxyphenyl)propenoic acid (ferulic acid), 2(4-hydroxyphenyl)propenoic acid (cinnamic acid), 2(3,4-dihydroxyphenyl)propenic acid (caffeic acid), 3(3-methoxy-4-hydroxyphenyl)propen-1-ol (coniferyl alcohol), and 3(4-hydroxyphenyl)propen-1-ol (coumaryl alcohol) were more resistant to microbial decomposition in soil than glucose. The alcohols were significantly more resistant than the acids, and the aromatic nuclei had a relatively high degree of resistance and all could be substantially protected from biodegradation by incorporation into humic-type polymers. In the absence of polymerization, or of protection by polymeric materials, the various structures would degrade to ethanoic and butanedicarboxylic acids, and eventually to carbon dioxide and water. Thus it is probable that the primary structures are rapidly incorporated after synthesis into humic polymers in the soil.

Biosynthesis of fungal 'humic acid' polymers. Very little if any phenoloxidase activity and polymer formation takes place at the pH values of 3.5–4.5 characteristic of fungal culture media shortly after inoculation. Later, when media pH values rise to 5–6 pronounced phenolase activity takes place in the cells, and browning, indicative of polymerization takes place. Phenols and anthraquinones cannot be extracted from the medium during rapid polymer synthesis.

Raising the pH (to a limit of *ca.* 8.0) significantly increases the rate of polymerization, and also of autoxidation. Compounds such as 3,4,5-trihydroxytoluene (**139**), 3,4,5-trihydroxybenzenecarboxylic acid (**142**), 2,4,5-trihydroxytoluene (**143**), and 2,3,5-trihydroxytoluene (**146**) are capable of autoxidizing under alkaline conditions, and yield *ortho*- (and in the case of **146**, *para*-) quinone structures. Mention has already been made of chemical polymerization of quinone structures, and in particular of hydroxy quinones in alkaline media. Martin and his associates have drawn attention also to the possibilities of nucleophilic coupling of the quinones to give biphenyl and biphenyl ether structures.

General conclusions from humic synthesis studies

The information provided in this subsection has summarized how phenolic compounds in particular can be chemically or enzymatically polymerized to 'humic-type' structures. These phenols could arise from degradation of plant remains or from products from microbial metabolism. Peptides and various reactive structures could be incoporated in the polymeric materials.

Martin, Haider, and their associates have shown, in several of the studies referenced, how sodium amalgam reduction releases phenolic materials (and in one instance anthraquinone structures) from the polymers. These phenols were the same as those referred to as humic precursors, and many were identical to structures released in the sodium amalgam digests of soil humic substances.

Schnitzer and Ortiz de Serra (1973b) also identified similar sodium amalgam digest products from soil and fungal 'humic' materials, but their yields were low. Permanganate oxidation (Schnitzer *et al.*, 1973) yielded aliphatic mono- and dicarboxylic acids, benzenecarboxylic acids, phenolic acids, and aromatic compounds containing S and N (some of which may have been artefacts). Later, from alkaline cupric oxide digestion studies on 'humic acids' synthesized by *Stachybotrys chartarum*, *S. atra*, and *Epicoccum purpurescens* Schnitzer and Neyroud (1975) suggested that these compounds were, on average, composed of aliphatic materials (38 %), benzenecarboxylic acids (25 %), phenolic structures (21 %), and of materials which released dialkylphthalates (16 %). Such ratios were significantly different from those for soil humic substances. However, this information cannot be regarded as contrary to any thesis which proposes that fungal products contribute significantly to soil humic substances.

Soil humic materials must be considered in terms of highly heterogeneous mixtures formed as the result of numerous chemical and biological synthesis reactions. In addition these materials can be expected to consist of mixtures of old and newly synthesized polymers. For instance, humic substances have been estimated from $^{14}C/^{12}C$ ratio data to have residence times of 100 years in ferrallitic soils (Lobo *et al.*, 1974) and from 5540 years at 18 cm to 9300 years at 30 cm in volcanic soils (Otsuka, 1975). Thus significant changes can be expected to take place during these long residence times. These might involve further biological transformations, and chemical rearrangements and condensation reactions within the polymers. We therefore suggest that it would be highly exceptional to find a fungal humic structure whose composition would be comparable with any soil humic substance.

Thus there is strong evidence to indicate that fungi contribute significantly to the formation of soil humus. Also it would be hard to prove that extracellular enzymatically catalysed synthesis is not important. However, the evidence available at this time for the contribution of purely chemical synthesis is less convincing, and an extensive research programme will be required in order establish its importance in the soil environment.

3.5 PHYSICAL PROPERTIES

Whilst the physical properties of molecules are determined by their chemical composition these properties cannot necessarily be deduced from purely chemical data. This is particularly so in the case of polymeric molecules, and even more so for molecules with poorly defined structural characteristics such as humic acids. Consequently, direct measurements of properties such as molecular size and shape, charge and charge density, and of spectral characteristics are necessary. Knowledge of both the chemical structure and physical properties of humic substances, and the relationships between these is basic to our understanding of the role which they play in the soil.

3.5.1 Spectral Characteristics

Techniques are now available for the accurate recording throughout most of the electromagnetic spectrum of the spectral characteristics of compounds. These range from X-rays, through the ultraviolet, visible and infrared regions, to microwave frequencies. All of these techniques have been applied to humic substances with varying degrees of success and usefulness.

The accurate interpretation of spectral data obtained from polymeric materials is always problematical, and the complex nature of humic substances only serves to increase the difficulties. However, as Felbeck (1965b) pointed out, improvements in techniques of purification and fractionation will increase the usefulness of spectral data. But even so, there are limitations, imposed by the composition and structure of the humic materials, which should always be borne in mind when assessing and interpreting the results obtained from spectral measurements.

Ultraviolet and visible spectra

Humic substances, as has already been pointed out, contain phenolic and various aromatic and possibly aliphatic functional groups which act as chromophores in the visible and ultraviolet regions of the electromagnetic spectrum. However, with very few exceptions the visible and ultraviolet spectra recorded for humic substances tend to be rather featureless. The well defined peaks characteristic of simple organic compounds are absent and even the more diffuse peaks, such as those exhibited by some proteins, are not found. As a

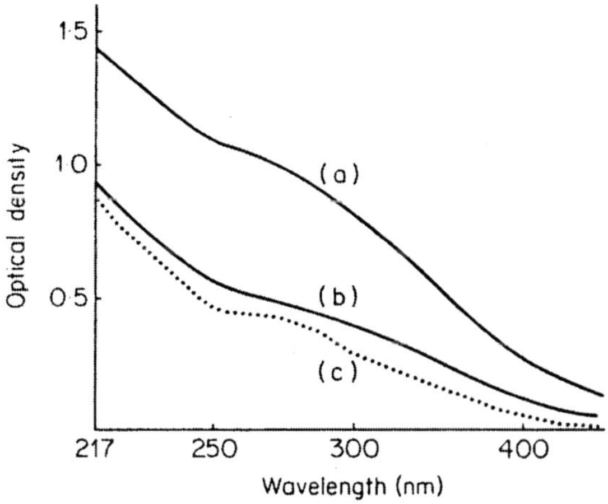

Figure 3.11 Ultraviolet spectra of low (a) and of high (b) molecular weight humic acids (extracted by 0.5N NaOH) from a lateritic podzol and of lignin (c, for comparison). Note the higher extinction coefficient for material (a). (Experimental conditions: sample concentration = 0.2 mg cm^{-3}, cell path length = 2 mm)

rule the spectra show a general increase in absorption with decreasing wavelength (Figure 3.11), although for some materials 'shoulders' or inflections can be detected which perhaps indicate the presence of particularly strong or abundant chromophores.

It is therefore reasonable to assume that humic substances contain numerous chromophores whose absorption bands overlap throughout this spectral region and absorb with growing intensity as the wavelength decreases. It is not possible to observe or measure one particular chromophore, or to derive definitive information with regard to chemical composition or structure. It is possible, however, to utilize the general features of the spectrum, namely the intensity of absorption and the gradient of the spectrum.

The intensity of absorption can be used for quantitative measurements by utilizing the Beer–Lambert law for monochromatic light, which is expressed in the equation

$$A = \log_{10}\frac{I_0}{I} = Kcl$$

where, A is the absorbance or optical density of the solution, I_0 and I are intensity of the incident and transmitted light, respectively, c is the concentration of the absorbing substance, l the path length of the solution, and K a constant which is called the absorptivity or extinction coefficient.

For a single well-defined material K is a constant at a given wavelength. When sample cells of the same path length (e.g. a 1 cm cell) are used the measurements of optical density will be directly proportional to the concentration of solute in solution. It is thus a simple matter to determine solution concentration from calibration curves of known concentrations.

This technique is commonly utilized for quantitative measurements of humic substances since it is simple, accurate, non-destructive, and requires very small amounts of material. Such measurements are usually made in the wavelength band of 400–500 nm, although other wavelengths have been used. However, extinction coefficients, on a weight for weight basis, vary sometimes by as much as a factor of two for humic materials extracted from different soils (Orlov, 1966). Also, even greater variations can occur in the extinction coefficients of samples obtained from the same soil extract. These variations are a function of the molecular weight of the humic acid fractions (Swift *et al.*, 1970, see Figure 3.11). Such differences are not unexpected since the extinction coefficient will obviously be related to factors such as the contents of aromatic and phenolic groups within the humic acid molecules, and these will vary in the fractions of a single extract as well as between extracts from different soils. Thus care should be taken to try to compare only components from different soils subjected to the same isolation and fractionation procedures.

In addition to differences in extinction coefficients humic substances display differences in colour, or hue, ranging from dull grey-brown, through brown itself to a rich red-brown. In terms of spectral characteristics these differences are represented by changes in the gradient of the recorded spectrum. This gradient is most commonly measured between 400 and 600 nm and is expressed as the ratio of the optical densities recorded at these wavelengths. This value is known either as the colour quotient, or the E_4/E_6 ratio, and it is higher for the red-brown materials than for the grey-brown materials. Several attempts have been made to use this ratio as an index of humification (see Schnitzer and Khan, 1972; Flaig *et al.*, 1975 for reviews), i.e. the lower the E_4/E_6 ratio the greater the degree of humification. Although such a hypothesis is not generally accepted it does seem to be true that the more highly condensed and/or the higher molecular weight fractions tend to have low ratio values, whereas the more highly oxidized and/or lower molecular weight fractions tend to give higher ratio values.

Infrared spectra

Infrared spectroscopy is used extensively in chemistry to identify and demonstrate the presence of certain functional groups and to characterize organic compounds. As such it is a useful adjunct to chemical determinations, and in addition it is non-destructive and requires only small amounts of sample. Again the application of this technique to humic substances is somewhat limited by the nature of the material because the presence of numerous infrared active groups leads to overlapping of absorption bands with a consequent lack of resolution. However, certain absorption bands do appear as discrete, or

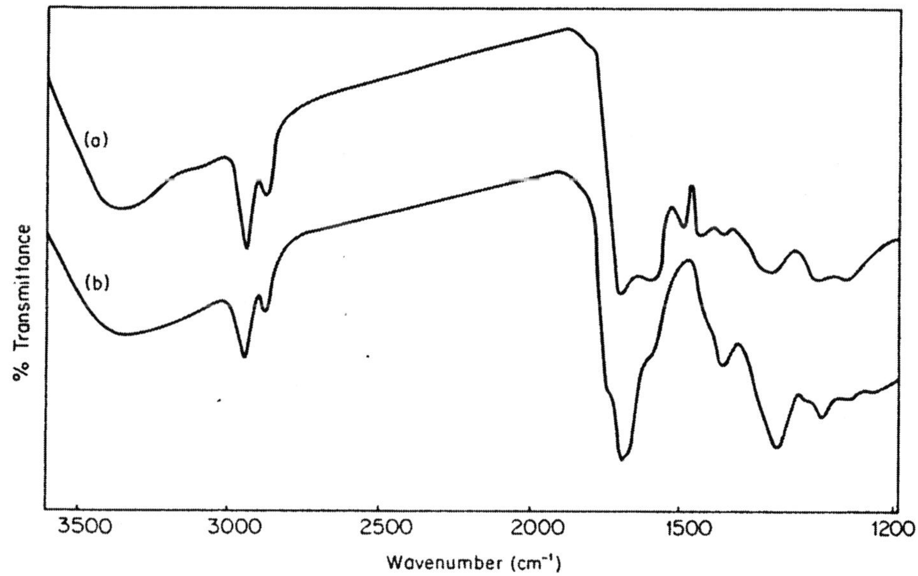

Figure 3.12 Infrared spectra of humic acids: extracted from an organic soil with 0.5N NaOH before (a) and after (b) autoxidation in alkaline solution. Note increased carbonyl (1720 cm^{-1}) and decreased C—H absorption in (b). (From Swift and Posner, 1972)

reasonably well defined peaks and these have proved to be useful for the chemical characterization of humic substances.

Most spectra are recorded by mixing finely ground humic substance with a suitable matrix material (usually *ca.* 1 mg sample with 0.5 g KBr) and pressing the mixture into a transparent disc. Care is necessary to exclude water during this process to prevent spurious results (Theng *et al.*, 1966). A typical humic acid spectrum is shown in Figure 3.12.

The main functional groups revealed by this technique are: hydrogen bonded —OH groups which give a broad absorption band with a maximum around 3400 cm^{-1}, aliphatic C—H giving sharper peaks at 2920 and 2850 cm^{-1}, carboxyl and carbonyl C═O around 1720 cm^{-1} and carboxylate ions ($—COO^-$) at 1610 and 1380 cm^{-1}, C═O conjugated to C═C gives a peak at approximately 1660 cm^{-1}, and a peak at 1610 cm^{-1} is generally thought to be due to aromatic $>C=C<$. It has been suggested that the peak at 1510 cm^{-1} has similar double-bond origins (Theng and Posner, 1967), although others attribute this peak to the presence of amino compounds (Butler and Ladd, 1969). Attempts have been made to assign other peaks which are sometimes observed, but the validity of such assignments is questionable when they are not supported by direct chemical evidence. Those peaks which are observed and have been assigned tend to support the structural information obtained chemically.

Apart from being used to characterize and compare humic materials obtained from various sources (e.g. Orlov *et al.*, 1962) infrared spectra have proved useful in observing and assessing changes produced by chemical reactions. The processes studied include methylation, acetylation, osazone formation, oxidation, reduction and prolonged exposure to acids and alkalis (Wagner and Stevenson, 1965; Theng and Posner, 1967; Schnitzer and Skinner, 1965a; Dubach *et al.*, 1964; Swift and Posner, 1972). Whilst yielding useful results these studies have also shown that even apparently simple, sharp peaks are usually compound peaks formed by the absorption of several closely related functional groups.

Fourier transform infrared offers a new technique in which the absorption characteristics of aqueous humic suspensions or solutions are digitally recorded over a number of scans (Nagar, 1977), and the spectrum for the water is subtracted out in the computerized set-up. The technique promises to give high resolution quickly over a wide range of the spectrum.

Nuclear magnetic resonance

It is necessary to have samples in solution for successful applications of (proton) ^1H n.m.r. because only protons of short relaxation times (*ca* 10^{-3} sec) can be detected by this technique. Thus the properties of the sample under investigation will be influenced by the solvent employed. Inevitably deutrated sodium hydroxide (NaOD) must be used in order to obtain the most comprehensive spectra.

Well differentiated n.m.r. spectra were obtained by Lüdemann *et al.* (1973) for lignins and synthetic and natural humic acids. They estimated the percentages of aromatic protons in the different spectra as: lignin, 30; brown coals, 26–35; synthetic humic acids (from pyrocatechol and hydroquinone), 64 and 78; chernozem and podzol soil humic acids 20 and 44, respectively. Later Lentz *et al.* (1977) used pulsed n.m.r. (100 MHz) and the Fourier transformation technique to compare the spectra for humic acids isolated at pH 4.5, 7 and 14 from the A and B horizons of an Erica podzol. Spectra (run in 0.1N NaOD) recorded the same percentage (34–35) of aromatic protons in the pH 4.5 extracts from the two horizons. Although the aromatic proton content of the B horizon was not influenced by the pH of extraction that of the A horizon was, and the evidence indicated that more aliphatic materials were extracted as the pH was increased in this instance. There was strong evidence for increased numbers of methoxyl groups (peak at 6.2 ppm) for the extracts from the more basic media.

Results by Sciacovelli *et al.* (1977) illustrate how elementary composition and n.m.r. chemical shift data of soil humic materials were influenced by the extractant used. Each of the ten fractions was subjected to infrared and 100 MHz n.m.r. analysis, and it was seen, for instance, that the n.m.r. responses of the dimethyl formamide and dimethylsulphoxide (dipolar aprotic solvents) extracts were very similar, and were less aliphatic than the materials soluble in benzene, dioxane, diethylether, etc.

Further advances in ^{1}H n.m.r. can be expected from applications of a new generation of 300 MHz instruments. However, ^{13}C-Fourier transform n.m.r., which is already widely used for studies on synthetic and bipolymers (such as polysaccharides, proteins, and nucleic acids) may well have a far greater impact on research on humic substances. In principle, for this technique nuclei are excited by applications of pulses of energy and their exponential decay back to the equilibrium state takes the form of a series of sine waves. A Fourier transform, for which an on-line computer is generally employed, relates this decay to a normal n.m.r. spectrum (Levy and Nelson, 1972; Williams and Fleming, 1973). ^{13}C n.m.r. has an additional advantage for studies with humic substances because it should be possible to use aqueous solvents.

Electron spin resonance

Steelink and Tollin (1967) have outlined the principles and applications of electron paramagnetic resonance or e.s.r. for detecting and estimating stable free radicals in humic substances. This spectroscopic technique permits transitions between different energy states to be observed when an external magnetic field induces splitting of the energy levels of a system of unpaired electrons. Spectra can be obtained for humic substances in the solid phase and in solution.

Humic materials provide sharp e.s.r. signals, but these are broadened when paramagnetic metal ions are present. Steelink (see Steelink and Tollin, 1967, for references to the original work) concluded that a polymer containing *ortho*- and *para*-quinhydrones might account for some of the observed properties of his e.s.r. spectra. These, on basification would give rise to semiquinone radical ions as outlined in reaction scheme (3.7).

Atherton *et al.* (1967) reached similar conclusions from their studies with acid boiled humic acids. Their data for the base-solubilized materials indicated that the signals were not from trapped radical species, and the disappearance of these signals on reduction and their reappearance on reexposure to air was considered to be suggestive of the presence of semiquinone-type species. Hyperfine structure or splitting was observed in the spectra.

Senesi *et al.* (1977) observed hyperfine splitting in the e.s.r. spectra of fulvic acids at pH 7.0 only after treatment with H_2O_2, and Senesi and Schnitzer (1977) have also investigated the effects of pH, reaction time, chemical reduction and irradiation on the spectra. The hyperfine splitting was rationalized on the basis of a two-step oxidation of the fulvic acid by H_2O_2; the first step was thought to involve the oxidation of a disubstituted hydroquinone structure to a semiquinone, and the second the oxidation of the semiquinone to the quinone.

Riffaldi and Schnitzer (1972a; 1972b) have quoted free radical contents of the order of 10^{17}–10^{18} spins g^{-1} of humic substances, and these decreased in the order humins > humic acids > fulvic acids. The spins g^{-1} in lignins are generally lower (Steelink and Tollin, 1967, p. 156). Hayes *et al.* (1975a) found that the relative amounts of the radicals in humic and fulvic acids depended on the solvent used for extraction, and their data (Table 3.1) suggested that e.s.r. might provide a useful means for observing artefact formation when extracting humic substances.

3.5.2 Molecular Weight

One of the fundamental physical properties of any chemical compound is its molecular weight. For a simple organic compound this is a single definable value which can be accurately determined, but with polymeric materials the situation is more complex. In the case of a protein, for instance, which is biosynthesized by a controlled pathway, there may be a single molecular weight value, or at least a series of molecular weights closely grouped around a single value. Many other naturally occurring polymers have relatively wide spreads of molecular weight since the processes involved in their synthesis often give rise to different numbers of monomer units in the molecules. Polymers exhibiting such spreads of molecular weight values are said to be polydisperse, and the degree of the spread, or the polydispersity observed within the system, can vary greatly depending upon the distribution of components of different molecular weights.

When techniques used for determining the molecular weights of well defined compounds are applied to polydisperse systems the results obtained can be used to give average molecular weight values. However, these are often meaningless. Of more interest is the molecular weight distribution, i.e. the range of molecular weights over which the polymer spreads, and the proportion of the whole sample found in any given molecular weight range. A diagrammatic representation of the molecular weight distribution of a polydisperse system is shown in Figure 3.13.

Early attempts to measure the molecular weights of humic acids using simple techniques based on colligative properties gave very low values and confusing

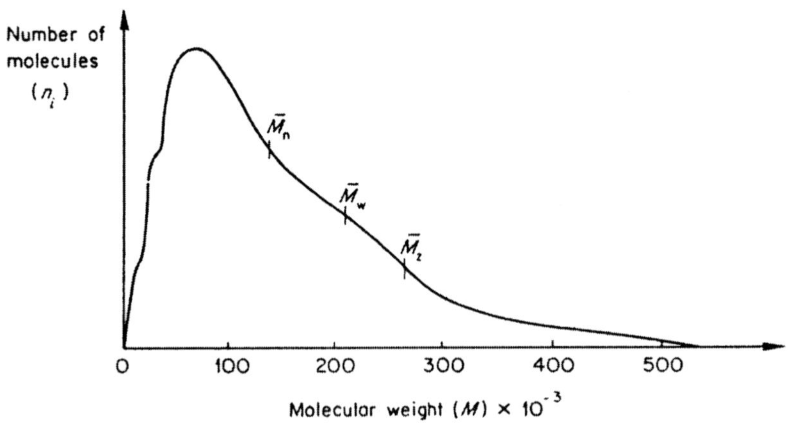

Figure 3.13 Diagrammatic representation of a possible molecular weight distribution, showing M_n, $\bar{M}_w$ and $\bar{M}_z$ values for a humic acid sample after removal of fulvic acid and low molecular weight contaminants

data arising from the effects of polydispersity and of contamination by low molecular weight impurities. Nowadays techniques and mathematical treatments are available to deal with such problems and average molecular weight values can be derived from one or more of the following expressions:

(a) The number-average molecular weight ($\overline{M}_n$), where

$$\overline{M}_n = \frac{\sum n_i M_i}{\sum n_i} \tag{3.32}$$

and the relevant information is obtained from measurements of colligative properties, which depend on the number of solute molecules present;

(b) The weight-average molecular weight ($\overline{M}_w$), given by

$$\overline{M}_w = \frac{\sum n_i M_i^2}{\sum n_i M_i} \tag{3.33}$$

and is influenced more by the weight of individual molecules, rather than their number;

(c) The *z*-average ($\overline{M}_z$), where

$$\overline{M}_z = \frac{\sum n_i M_i^3}{\sum n_i M_i^2} \tag{3.34}$$

has no simple physical analogy and arises through the mathematical treatment of data from certain experimental techniques, such as sedimentation.

In each case n_i is the number of molecules with a molecular weight M_i. For mono-disperse systems, in which all species have the same molecular weight, $\overline{M}_n = \overline{M}_w = \overline{M}_z$, but for a polydisperse system $\overline{M}_z > \overline{M}_w > \overline{M}_n$. The ratio $\overline{M}_w : \overline{M}_n$ is a useful measure of the polydispersity of a system; the higher the ratio the greater will be the polydispersity or the spread of molecular weights.

The number-average molecular weight ($\overline{M}_n$) is affected equally by all molecules. It is the type of mean we obtain, for instance, when computing the average height of a group of people. In molecular terms it gives a true average. However, in practice the result can be biased towards low molecular weight components (e.g. monomers) which may not be significant by weight but are by number. The weight-average values, on the other hand, tend to emphasize the contributions made by heavier molecules in the system and decrease the effects of the smaller components. Hence this average is probably the most useful for work with humic substances since it helps to overcome the problems associated with the presence of lower molecular weight components or contaminants. An awareness of the different averages in use is important to help explain some of the large discrepancies in the literature values quoted for these materials.

Most of the methods available for molecular weight determination of polymers have been applied to humic substances. However, osmometry, vapour pressure osmometry, gel chromatography, ultracentrifugation, viscometry, and light scattering have given the most useful results, and we shall limit our consideration to these techniques.

Osmometry

In practical terms osmometry is the only effective technique which can be used to determine the number-average molecular weights of high polymer systems. It requires the use of membranes which permit the passage of solvent but not of solute molecules. The membrane is used to separate a solution (solvent plus polymer) from pure solvent and the resultant difference in chemical potential causes solvent to pass into the solution. In osmometers this flow of solvent is used to create a hydrostatic head whose pressure, when equilibrium is achieved, is equal and opposite to the osmotic pressure across the membrane. Thus the osmotic pressure (π) can be directly measured, and its relationship to molecular weight, derived from van't Hoff's limiting law of osmotic pressure, is given by

$$\operatorname*{Lt}_{c \to 0} \frac{\pi}{c} = \frac{RT}{\overline{M}_n} \qquad (3.35)$$

where c is the concentration of solute, R is the gas constant, and T is the temperature (K). The molecular weight $\overline{M}_n$ is determined by extrapolating to zero concentration the values of π determined experimentally at known concentrations.

There are, however, a number of practical problems associated with the technique. Ideally membranes should allow only solvent to pass through, but they tend to be 'leaky' and to allow through some of the lower molecular weight solute species. The older membrane materials, such as cellophane, could exclude only molecules whose molecular weight exceeded 10,000 or even 20,000, but recent advances in membrane technology have produced materials with exclusion limits of 500–1000. Problems of non-ideal membranes can be minimized if equilibration times are rapid, or by extrapolating results back to zero time.

The method also loses sensitivity for high molecular weight materials. Equation (3.35) shows that when $\overline{M}_n$ is very large, and the concentration c is small as is necessary, π will be small. Thus it becomes difficult to measure π accurately for molecular weight values of approximately 200,000 and higher.

A further complication is introduced by the fact that humic substances are polyelectrolytes, and are macro-ionic in nature. Because of this the polymer solution on one side of the membrane will contain simple counterions as well as polymer molecules. Since humic acid polymers are highly charged the number of counterions in the solution will be orders of magnitude greater than the number of polymer molecules. Therefore, because osmotic pressure is a colligative property dependent solely on the number of species in solution, each counterion will be as effective as a polymer molecule in producing an osmotic effect. Any molecular weight value measured under such conditions will be meaningless.

This effect can be countered by adding to the system a simple electrolyte which is able to pass freely through the membrane. However, this introduces further problems arising from Donnan effects (see Chapter 5). Because of these the total concentration of simple electrolyte ions will be greater on the polyelectrolyte side of the membrane and will again lead to erroneous values. These effects can be minimized to negligible proportions if sufficient electrolyte is added to the system (Tanford, 1961, p. 225).

Where the required attention has been given to experimental technique and operating conditions osmotic pressure measurements have yielded useful results. Wright *et al.* (1958) using a solution of dialysed humic acid extract obtained from a podzol B horizon reported a mean molecular weight value of 50,800. Wake and Posner (1967) applied osmometry to measure the molecular weights of humic acid extracts fractionated by ultrafiltration through newly developed semipermeable membranes of different pore sizes. The values obtained for isolated fractions varied from 2000 up to approximately 34,000. They also showed that although the mean molecular weights for two whole extracts were very similar, the molecular weight distributions were considerably different. By use of a similar fractionation technique on a different humic acid extract Swift *et al.* (1970) determined $\overline{M}_n$ values ranging from 1500 to 65,000. When these fractions were passed down a gel chromatography column they were eluted at the positions predicted by the osmotic pressure molecular weight measurements.

The difference in chemical potential between a solvent and solution, resulting from the presence of a dissolved polymer, will also give rise to a difference in vapour pressures above the two liquids (in addition to an osmostic pressure between them if they are separated by a semipermeable membrane). In general, because this vapour pressure is much less than the osmotic pressure, measurement of the latter is preferred. However, improvements in design, and the commercial availability of vapour pressure osmometers has led to a renewed interest in the application of this rapid and relatively simple technique.

When drops of solvent and of solution are placed side by side in a closed container, maintained at constant temperature and saturated with solvent vapour, the solvent vapour will condense in the solution because the latter has the lower vapour pressure. As a result latent heat is discharged by the solvent condensing into the solution and the temperature of the solution will rise to a quasi-steady state value. If the instrument is accurately calibrated and care is taken to standardise variables such as drop size and time of measurement it can be shown, by a combination of the Clausius–Clapeyron equation and Raoult's law, that

$$\frac{W_2}{M_2}\frac{W_1}{M_1} = \text{constant} \times \Delta T \qquad (3.36)$$

where W_1 = weight of solvent in the solution, M_1 = molecular weight of solvent, W_2 = weight of solute in the solution, M_2 = molecular weight of solute

and ΔT = temperature increase of solution over solute.

A major improvement in instrumentation has been brought about by the use of thermistors whose resistance (R) changes with the change in temperature. Thus in practice, ΔT is measured as ΔR, i.e. the difference in resistance between sample and reference thermistors.

This technique is most effective in non-aqueous solvents or in aqueous systems where electrolyte is absent. However, Hansen and Schnitzer (1969c) have applied it to studies of acid soluble fulvic materials from a podzol B_h horizon where difficulties were experienced due to polyelectrolyte properties. They attempted to correct the molecular weight measurements by allowing for the dissociation of acidic functional groups. Although molecular weight values up to 3600 were recorded, most of the extract had values of the order of 1000 and less, which is rather low for polymeric systems. However, this particular humic fraction is thought to consist largely of low molecular weight components which would greatly influence the colligative properties and give low values for $\overline{M}_n$. Only values of $\overline{M}_n$ can be obtained in practice by this technique, and apart from their intrinsic merits such data are necessary to estimate the degree of polydispersity of samples from $\overline{M}_w/\overline{M}_n$ values.

Gel chromatography

Gel chromatography is a simple, relatively inexpensive, and very effective technique for studying polymers. It can be used as a method for separation, purification and fractionation as well as for determinations of molecular weights and molecular weight distributions of polymer systems. A review of the principles and applications of the technique is provided by Fischer (1969).

Although inorganic materials, such as porous glass beads, have been used, the gels most commonly employed consist of cross-linked polymers (e.g. polysaccharides, polystyrene, and polyamides) in the form of small granules. The gel structure is perfused by a system of pores and the size of these is determined by the degree of cross-linking in the polymer. These pores enable the gel to act as a chromatographic medium giving separations based on differences in molecular sizes. In practice the gel beads are swollen in an appropriate solvent (usually an aqueous salt solution for polymers of biological origin) and are packed into a column to form a stationary phase. When a solution, containing a mixture of molecules of varying sizes, is applied to the top of the column and eluted with solvent those molecules which cannot enter the pores in the beads will pass between them and will be eluted first from the column. Molecules smaller than the pore sizes of the gel will enter the pores and their passage through the column will be retarded. The extent to which this occurs will depend on the actual size and shape of the molecule, but the net result is that the solute molecules are eluted from the column in order of decreasing molecular size and, for a given polymer, decreasing molecular weight. This process is shown diagramatically in Figure 3.14.

When small samples are applied to the columns, solutes with discrete mole-

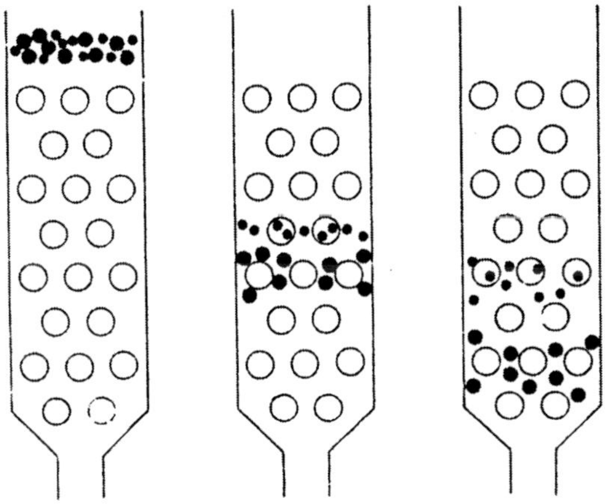

Figure 3.14 Diagrammatic representation of the gel chromatography process showing partial retention (b and c) of smaller molecules by the gel

cular weights are eluted as well defined peaks. The position at which the peak of a given solute is eluted can be characterized by calculating the partition coefficients denoted by K_{av} or K_d values where

$$K_{av} = \frac{V_e - V_0}{V_t - V_0} \tag{3.37}$$

Here V_0 is the void volume, or the volume of solvent outside the gel beads forming the bed, and it is also the volume in which a totally excluded solute will be eluted. V_e is the volume required to elute the solute being examined, and V_t is the total volume of the column. The value K_d is more accurate because it eliminates the volume of gel which is included in V_t but is not strictly a part of the inner phase; thus

$$K_d = \frac{V_e - V_0}{V_t - V_0 - V_g} \tag{3.38}$$

where V_g is the volume of the gel. For highly swelling gels the difference between K_{av} and K_d is not significant. However, the values will be different when the proportion of the total volume occupied by the actual gel matrix increases, for instance with glass beads or with only slightly swelling gels.

Ackers (1964) interpreted gel chromatography on the basis of steric and frictional resistance to the diffusion of solute in the gel pores, and derived a relationship between K_d and the Stokes' radius a of the solute where consideration was given to r, the effective radius of the gel pores (assumed to be cylindrical). A value for r can be obtained for a given gel by measuring the K_d for a solute of known Stokes radius. When K_d values are plotted against the logarithms of corresponding values of r sigmoid curves with extended linear regions are obtained which can be expressed as

$$K_d = -K_1 \log\left(\frac{a}{r}\right) + K_2 \tag{3.39}$$

where K_1 and K_2 are constants.

For any solute a is proportional to some fractional power x of the molecular weight M. This fractional power depends on the shape of the molecule but it will be the same for any series of similar molecules. Assuming that $a = M^x$ for a particular type of molecule, then for any series of molecules of this type

$$K_d = -k_1 \log M^x/r + k_2 \tag{3.40}$$

which simplifies to

$$K_d = -b \log M + c \tag{3.41}$$

where b and c are constants. Thus, it can be seen that there is a relationship between the molecular weight and the position of elution of a component, and that once calibrated for a particular group of polymers, gel chromatography provides a rapid and facile method for estimating molecular weights. Gels are available which operate over different molecular weight ranges and have different exclusion limits. By utilizing a range of gels it is possible to determine molecular weight values ranging from several thousands to millions.

This technique has been extensively and successfully applied to studies of humic substances. A number of problems are, however, encountered which, if they are not overcome, can invalidate the results. For instance, the gel material should be inert to the solute molecules so that there are no chemical or physical interactions between gel and solute. When any adsorption of the applied polymer molecules by the gel takes place the observed retention by the column is not solely caused by penetration into the pores and the resulting separation cannot be entirely attributed to molecular weight differences. Because of their chemical composition humic substances tend to be readily adsorbed.

A further problem arises as the result of charge interactions between residual charged groups on the gel and those on the humic substances leading to attractive or repulsive forces between the gel and charged humic substances. If not suppressed these charges interfere with the separation which again would not take place solely on the basis of molecular size differences.

Swift and Posner (1971) showed that such problems can be largely overcome by careful selection of the gel matrix and by the use of appropriate buffer solutions. Use of a buffer containing a large organic cation such as tris[2-amino-2(hydroxymethyl)propane-1,3-diol], or similar compounds is recommended. Even when such procedures are used there is some indication that a small amount of interaction between gel and solute can still take place, particularly in the cases of the very high molecular weight, less soluble humic acid fractions.

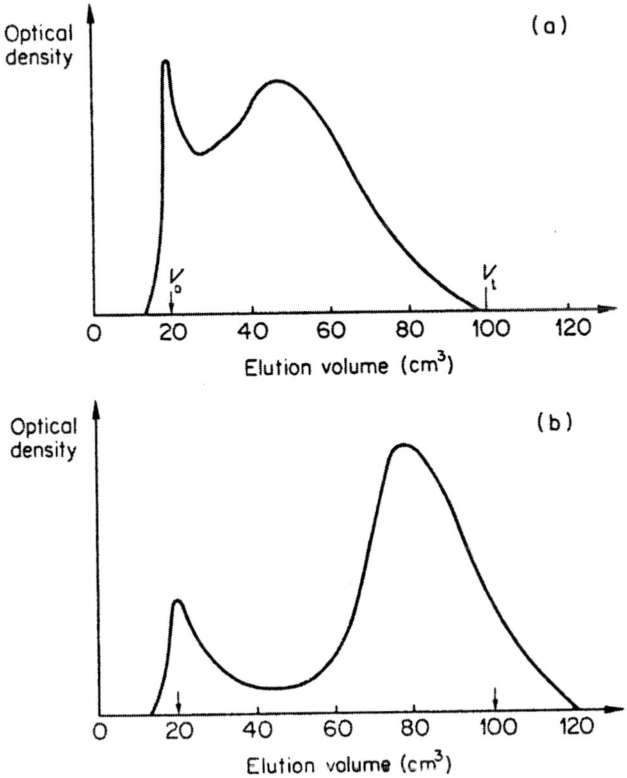

Figure 3.15 Fractionation of a humic acid by gel chromatography on Biogel P-150 (a) and Biogel P-300 (b). (These gels have nominal molecular weight exclusion limits of 150×10^3 and 300×10^3, respectively)

Despite these handicaps gel chromatography has proved to be a particularly useful technique for the purification, fractionation, and determination of the molecular weight of humic substances.

Typical elution patterns obtained by means of the gel chromatography of humic acid extracts are shown in Figure 3.15. For the system illustrated a peak consisting of excluded material appears at the void volume V_0, and the remainder is spread as a diffuse peak over the working range of the column. The proportions found in each of these peaks will depend upon the molecular weight distribution of extracts and the exclusion limit of the gel. Thus if a gel with a higher exclusion limit is used the size of the retained peak will be increased at the expense of the V_0 peak. Elution patterns such as these (e.g. Dubach *et al.*, 1964; Swift and Posner, 1971) confirm the polydisperse nature of humic substances and show that they cover a wide range of molecular weight values. Cameron *et al.* (1972b) separated by gel chromatography humic acid fractions ranging in molecular weight from 2000 to 1,500,000, and showed that the most abundant portion of the molecular weight distribution for a sodium hydroxide extracted humic acid was around 100,000.

When selecting gels for determinations of molecular weight values and distributions, consideration should be given to the molecular weight ranges over which the gels operate and in particular to their upper exclusion limit. The manufacturers of gels supply figures for these properties, and such values have been quoted extensively when reporting values for the gel chromatography of humic substances. However, the manufacturers values are obtained by calibrating with proteins or polysaccharides of known molecular weights. It is unlikely that humic acid polymers would have the same configurations or dimensions in solution as polysaccharides or proteins of similar molecular weights, and hence would behave differently in gel chromatography. Thus manufacturer calibrations are not likely to be valid for humic substances. This problem can be overcome by calibrating the gel materials with humic acid fractions of known molecular weights determined by osmometry or by ultracentrifugation on well fractionated samples. Such fractions have been used (Swift *et al.*, 1970; Cameron *et al.*, 1972a) to calibrate a series of gel chromatography materials and these calibrations were compared with those obtained using proteins or polysaccharides. In general, the calibrations from humic acid fractions differed significantly from those based on proteins which tend to have tightly-coiled, globular molecular configurations. However, there was reasonable agreement with some calibrations obtained using polysaccharides which tend to have less compact, randomly-coiled molecular configurations.

The technique of gel filtration is contributing significantly to the problem of isolating humic fractions of low polydispersity, and it has considerable potential for estimations of molecular weight values of the less polydisperse isolates. Such molecular weight determinations might be simplified, as suggested by Cameron *et al.* (1972a), by establishing a relationship between the molecular weights of humic substances and a selection of proteins or polysaccharides. This would permit the uses of these latter materials as secondary standards.

Ultracentrifugation

Sedimentation of finely divided solids from suspensions is achieved by normal centrifugation techniques. However, in order to sediment molecules from solution it is necessary to use the ultracentrifuge which, by its high rotational speeds, generates very strong force fields.

The usual design of analytical ultracentrifuge consists of a solid rotor which rotates about a central axis and contains a small sample cell, shaped like a segment of the rotor. When the cell is positioned approximately 65 mm from the axis or rotation and spun at 60,000 r.p.m. the contents experience a force of 250,000 times that of gravity.

If a dilute solution of macromolecules is spun in the cell the solute molecules will gradually sediment to the bottom of the cell. This process is shown dia-

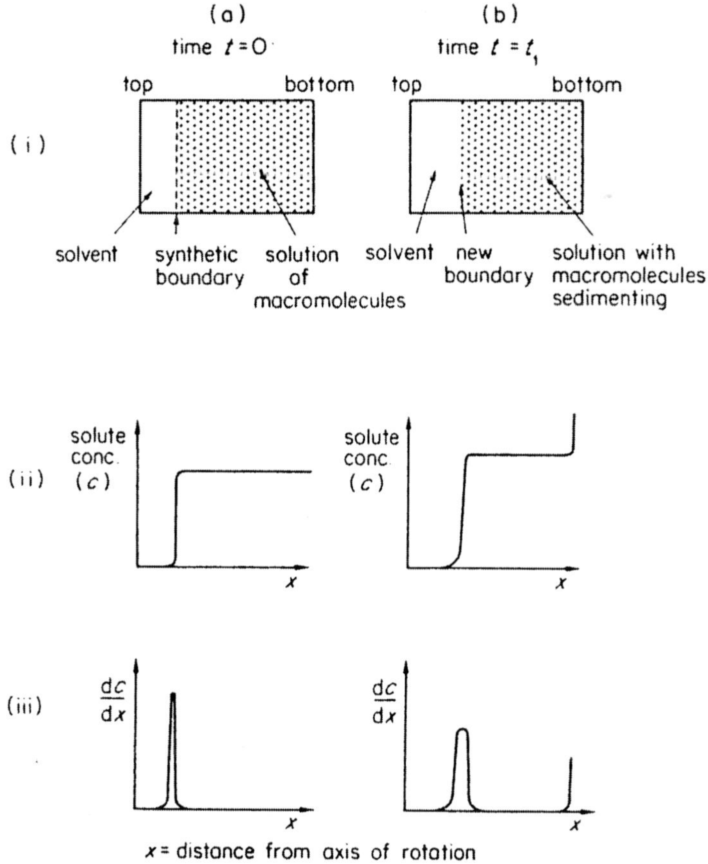

Figure 3.16 Diagrammatic representation of the ultracentrifugation process (using Schlieren optics) at the commencement (a) and during a run (b) showing (i) the distribution of cell contents, (ii) the plot of solute concentrations in the cell, and (iii) the peaks given by Schlieren optics as the derivatives of (ii)

grammatically in Figure 3.16. Several optical systems can be used to measure the changes in concentration that occur as the solute moves down through the cell. Many instruments are fitted with a version of the Schlieren optical system which measures differences in refractive index arising from concentration changes in the cell. For this system the change in concentration is given in a differential form so that the boundary between the pure solvent and the solution, formed by the sedimentation of the solute, appears as a peak. The rate of sedimentation of the solute can be determined by following the movement of this peak during an experimental run.

Measurements of optical densities provide the simplest technique for observing changes in concentration in coloured or ultraviolet-absorbing solutions.

For these the cell is scanned by a beam and the data can be presented as concentration, as a function of distance along the cell, or as the appropriate derivatives. The method is capable of a high degree of refinement, and the sensitivity of ultraviolet or visible absorption techniques allows very low solute concentrations to be used. Higher solution concentrations, of the order of one per cent or more of polymer, are generally required for the Schlieren system. Since molecular weight values are obtained by extrapolating measurements back to zero concentration it can be an advantage to use the absorption systems in which data for low concentration levels can be obtained.

Humic materials have a very intense coloration, and absorb strongly in the visible and ultraviolet regions of the spectrum (see Figure 3.11). This causes some problems in the use of traditional optical systems in the ultracentrifuge. The Schlieren system operates most effectively at higher concentrations, but the transmission of light through humic acid solutions at the desired concentrations is very low. Thus it is necessary to lower the concentration in order to transmit sufficient light for photographic recordings to be made. However, if the concentration is greatly reduced the Schlieren peak becomes too small and ill-defined for accurate measurements. Cameron *et al.* (1972b), using fast film, reported an upper concentration limit of 2.4 g dm^{-3} and a lower limit of 0.6 g dm^{-3}. The upper limit could probably be increased by use of a higher intensity light source of longer exposure times; indeed Piret *et al.* (1960) reported measurements of solutions containing 7.3 g dm^{-3} of humic acid.

The strong light absorption properties of humic substances would suggest that they could be effectively analysed by absorption optics. But because the optical extinction coefficients of humic acids vary as a function of molecular weight (Swift *et al.*, 1970), the optical density is not directly related to concentration for the highly polydisperse material. This difficulty can be overcome by applying suitable corrections where unfractionated extracts are used, as for instance in equilibrium ultracentrifugation experiments. However, when sedimentation coefficients are required, only well fractionated samples should be used in the ultracentrifuge. In this way variations in optical extinction coefficients will be small and may not require correction.

An optical system based on the Rayleigh interferometer technique can also be used to overcome the problem of variations in optical extinction coefficient in samples which exhibit a large molecular weight range. This system reponds to changes in refractive index which occur as solute concentration varies. It can be used at low solute concentrations, and since the refractive index increment per unit weight is not likely to vary with molecular weight among chemically closely related species, it can be used equally effectively on polydisperse or fractionated samples. Posner and Creeth (1972) have successfully used interferometry for their studies on unfractionated humic acids by equilibrium ultracentrifugation.

There are two principal methods which can be used to calculate molecular weight values from data obtained by observing concentration changes in the cell during an experimental run. The most common involves the use of values for sedimentation and diffusion coefficients. The sedimentation coefficient, s, represents the velocity of solute molecules divided by the centrifugal field so that

$$s = \frac{1}{x\omega^2} \cdot \frac{\mathrm{d}x}{\mathrm{d}t} = \frac{\mathrm{d}\log_e x}{\omega^2\,\mathrm{d}t} \tag{3.42}$$

where x is the distance of the solute/solution boundary (e.g. Schlieren peak) from the axis of rotation, ω the angular velocity and t the time. Thus a plot of $\log_e x$ against time should give a straight line and s can be obtained from its slope. As long as the density of the solute exceeds that of the solvent, which is generally the case, the solute will move to the bottom of the cell under the influence of the centrifugal force. This movement will be opposed by diffusion processes which will tend to redistribute the solute evenly through the cell. Thus the diffusion coefficient, D, of the solute must be determined to allow calculation of the molecular weight. This can be done in the ultracentrifuge, or preferably by an independent method. When s and D are used in the same calculation it is essential that they be determined at similar temperatures. It is also usual to extrapolate the results obtained to zero concentration of solute, and to correct to standard conditions; i.e. for water as solvent at 20 °C. These values are often written as $s^{\circ}_{20}w$, and $D^{\circ}_{20}w$. We shall assume here that the symbols s and D are the appropriate corrected values.

The molecular weight can be calculated from the Svedberg equation

$$M = \frac{RTs}{(1-\bar{v}\rho)D} \tag{3.43}$$

where M is the molecular weight, R the gas constant, T the temperature (K), $\bar{v}$ the partial specific volume of the solute, and ρ the solution density. The values of $\bar{v}$ and ρ can be determined by standard techniques using pycnometers.

A second method, which does not require measurement of the diffusion coefficient, calculates molecular weight values by using equilibrium ultracentrifugation. In this technique the rotor is run for a long time at relatively low speeds until sedimentation is equal and opposite to diffusion, and no net movement of solute occurs. It can then be shown that

$$M = \frac{2RT\,\mathrm{d}\log_e c}{(1-\bar{v}\rho)\omega^2\,\mathrm{d}x^2} = \frac{RT}{(1-\bar{v}\rho)} \cdot \frac{\mathrm{d}c/\mathrm{d}x}{\omega^2 xc} \tag{3.44}$$

where c is the solute concentration at distance x from the axis of rotation.

The problems caused by the polyelectrolyte nature and the polydispersity of humic substances must again be considered before meaningful results can be obtained from ultracentrifugation experiments. Humic acid molecules in solution carry a high negative charge which is usually balanced by suitable cations and these will be dissociated to give a diffuse, positively charged atmosphere around the negatively charged polymer molecule. The system will then behave as predicted by diffuse double layer theory (Chapter 5), and the polymer molecules will tend to repel each other. This behaviour will have a pronounced effect on ultracentrifugation experiments. The intermolecular repulsion will give anomalously high diffusion values, and in addition, during centrifugation the large macromolecules will tend to sediment faster than their counterions causing electrostatic drag and a lower value for the measured sedimentation coefficient. From equation (3.43) it can be seen that high D and low s values will result in anomalously low molecular weight values. These effects can be readily overcome by using an excess of neutral electrolyte; 0.1 to 0.2M additions of salt are usually sufficient. The added salt can itself produce secondary charge effects due to the difference in sedimentation behaviour between anion and cation, but this can be avoided by using an electrolyte, such as potassium chloride, where there is negligible difference in the behaviour of the two ions.

The importance of suppressing charge effects is evident from results obtained from ultracentrifugation studies on humic substances carried out by Flaig and Beutelspacher (1968). When sedimentation and diffusion coefficients were measured in water, a mean molecular weight value of 2050 was obtained. Addition of 0.2M sodium chloride suppressed the charge effects and the same humic material then gave a more realistic mean molecular weight value of 77,000.

The problem presented by the polydispersity of humic substances is more difficult to overcome. It must be remembered that in highly polydisperse systems numerous molecules of different sizes and different diffusion coefficients will be sedimenting at different velocities. This makes it impossible to obtain a well-defined sedimenting boundary. Even when such boundaries are formed artificially, as in a synthetic-boundary cell (see Figure 3.16), they will spread rapidly and disappear. Although short-lived Schlieren peaks can be obtained in this way these are highly asymmetric, and the values derived from them lead to erroneous results. The most effective way to deal with the problem is to decrease the polydispersity, by fractionation procedures, to a level where meaningful results can be obtained. Suitable gels and membranes are now available which allow fractionation of humic substances to acceptable levels of polydispersity.

Prior to the availability of the modern fractionation procedures some interesting applications of the sedimentation velocity technique were used in studies on humic substances. Stevenson *et al.* (1953) used a dialysed but unfractionated sodium pyrophosphate extract and obtained a weight-average molecular weight value of 53,000. The movement of the sedimentation boundary was followed by using a photodensitometer in conjunction with direct photographs of the cell contents; essentially an optical absorption technique. Later Piret *et al.* (1960) calculated a mean molecular weight value of 25,000 for a sodium hydroxide extract. In this study the diffusion coefficient was not measured directly but calculated from frictional data obtained from viscometric work. Charge effects were suppressed, but the spreading of the Schlieren peak due to polydispersity was considerable. Even though this peak remained stable long enough for measurements to be made, the asymmetry effects, referred to above, may have introduced considerable error into the molecular weight calculation.

Particular attention was given to decreasing polydispersity by Cameron *et al.* (1972b). They investigated molecular weight values by the ultracentrifugation of humic materials which had been extensively fractionated by means of gel chromatography and pressure filtration techniques. The processes involved were time consuming but led to the isolation of materials with relatively narrow limits of polydispersity. Fractions were taken at intervals over much of the molecular weight range of the parent material, and each was individually examined in the ultracentrifuge by the sedimentation velocity technique. Diffusion coefficients were measured independently and the molecular weight values calculated ranged from 2.3×10^3 to 1.4×10^6. This work indicates that more reliable and informative data can be obtained by working with fractions covering only narrow molecular weight ranges and that the averages obtained are more truly representative of the material being examined. For instance, a mean molecular weight value of the order of 5×10^4 for unfractionated material might be composed of molecules ranging from say 2×10^3 to 1.5×10^6, but in the case of the fractionated material the same value might refer to material with a molecular weight spread of the order of 4.5×10^4 to 5.5×10^4.

Useful results can be obtained for polydisperse materials if equilibrium ultracentrifugation is used. The technique requires a very high standard of instrument performance, sophisticated optics, and a prolonged running time. Posner and Creeth (1972) obtained $\overline{M}_n$ and $\overline{M}_w$ values for unfractionated humic acids and were able to estimate polydispersity from the $\overline{M}_n/\overline{M}_w$ ratio. When high molecular weight components sediment to the bottom of the cell at the rotor speed used they are not included in the calculations and the results obtained will not represent the true mean value. Thus even with this technique some prior fractionation is desirable.

Ultracentrifugation can also be used to determine molecular weight distributions by using a density gradient technique. For this a density gradient is established in the cell by use of suitable solutions so the buoyancy of the solution medium increases along the length of the cell. A point is reached for each solute molecule when the term $(1-\bar{v}\rho)$, in equation (3.43), equals zero and it ceases to sediment. When all of the solute species have reached equilibrium the distribution of solute within the cell can be determined (e.g. colorimetrically) and the molecular weight distribution can be calculated. The technique has the added advantage that less expensive, preparative ultracentrifuges with swing-out heads can be used.

Gradients are usually formed by using solutions of inorganic salts, such as rubidium chloride or caesium chloride, or of sucrose. In applying the technique to humic acid studies Cameron and Posner (1974) used a D_2O/H_2O density gradient to determine $\overline{M}_n$, $\overline{M}_w$, and $\overline{M}_z$ values for a number of extracts. They reported the following molecular-weight values for an unfractionated sodium pyrophosphate extracted humic acid from a Fen peat: $\overline{M}_n = 13.5 \times 10^3$, $\overline{M}_w = 216 \times 10^3$ and $\overline{M}_z = 1100 \times 10^3$. These results provide striking evidence for the extreme polydispersity of humic substances and of the usefulness of the density gradient procedure when working with such materials.

Viscometry

Viscosity is a measure of the frictional resistance that a flowing liquid offers to an applied shearing force. When fluid flows past a stationary phase the successive 'layers' extending from the stationary surface will move with increasing velocities. Velocity is described as Newtonian when the shearing force, τ, per unit between two parallel planes of liquid in motion relative to each other is proportional to the velocity gradient, dv/dx, between the planes and

$$\tau = \eta \, dv/dx \tag{3.45}$$

where η is the coefficient of viscosity. Newtonian viscosity is constant and independent of the rate of shear.

We will be interested here in deviations arising from the relative viscosity η_{rel} sometimes called the viscosity ratio as recommended by the International Union of Pure and Applied Chemistry (IUPAC), which is defined as the solution viscosity, η, divided by that of the pure solvent, η_0. These are conveniently measured by means of capillary viscometers such as the Ostwald–Fenske and the Ubbelohde instruments described in physical chemistry texts. These instruments are simple to operate and have almost invariably been used in viscometric studies on soil humic substances. Their main disadvantage is that the rate of shear varies from zero at the centre of the capillary to a maximum at the wall. Thus it should be remembered that anisotropic particles would assume various states of orientation over the cross-section of the viscometer, and although reproducible results would be obtained these would have little theoretical significance. For such systems rotational methods, which use the concentric cylinder, and the cone and plate techniques are particularly useful. The reader is referred to colloid chemistry texts for descriptions of the design and the use of such instruments, and only brief reference will be made here to the first of these methods. In this a thin film of liquid is placed between the concentric cylinders, the outer cylinder is rotated at a constant rate, and the shear stress is measured by considerations of the deflection of the inner cylinder which is suspended by a torsion wire.

Where an Ostwald-type viscometer is used the pressure at any instant which drives the liquid through the capillary will be proportional to its density. Thus

$$\eta = k\rho t \tag{3.46}$$

where k is the viscometer constant, ρ is the density of the liquid, and t is the time of flow. For two different liquids

$$\eta_1/\eta_2 = \frac{\rho_1 t_1}{\rho_2 t_2} \tag{3.47}$$

From this relationship it can be seen that at infinite dilution, where the densities of solvent and of solution are essentially the same, the viscosity ratios will be equal to those of the times of flow.

The specific viscosity, η_{sp}, is a measure of the relative increase in the viscosity of the solution over that of the solvent alone. Thus

$$\eta_{sp} = \frac{\eta_c - \eta_0}{\eta_0} \tag{3.48}$$

where η_0 is the viscosity of the solvent and η_c is that of the solution of concentration c. The specific viscosity divided by concentration η_{sp}/c, is known as the reduced viscosity (or viscosity number in IUPAC nomenclature), and the intrinsic viscosity, $[\eta]$, or limiting viscosity number (IUPAC), is defined as

$$[\eta] = (\eta_{sp}/c)_{c \to 0} \tag{3.49}$$

Values for $[\eta]$ can be obtained from extrapolations to zero concentrations of the plots of η_{sp}/c versus c as is implicit in the relationship

$$\eta_{sp}/c = [\eta] + k'[\eta]^2 c \tag{3.50}$$

or by the alternative relationship

$$\log \eta_r/c - [\eta] - k''[\eta]^2 c \tag{3.51}$$

where k' and k'' are constants, and c is expressed in g dl^{-1}.

The intrinsic viscosity is related to molecular weight by

$$[\eta] = KM^\alpha \tag{3.52}$$

in which the two empirical coefficients K and α are constants for a given polymer in a given solvent over a range of molecular weights. Equation (3.52) transforms to

$$\log [\eta] = \log K + \alpha \log M \tag{3.53}$$

which provides values for K and α from the plot of log $[\eta]$ versus log M. Obviously in order to use this method it is necessary to obtain at least two molecular weight values by independent methods for carefully fractionated monodisperse fractions of the polmer.

Both K and α are functions of the solvent as well as of the polymer. This empirical relation between viscosity and molecular weight must, however, not be used indiscriminately as it is most suitable for studies with linear polymers.

Viscosity data can be used to find viscosity-average molecular weight values ($\overline{M}_v$). Rafikov *et al.* (1964) give details of the mathematical manipulations required.

Viscosity is markedly influenced by changes in polymer conformations. We have repeatedly referred to the fact that when a polyelectrolyte solution is diluted increased ionization takes place and polymer expansion is enhanced as the result of repulsion by similarly charged functional groups along the polymer chain. When very high extensions are reached the effect is often reversed, and a characteristic convex curve is obtained for the η_{sp} vs. c plot when measurements are extended to very low concentrations. When dealing with gel filtration and ultracentrifugation, the addition of salt depresses the ionization effects in the

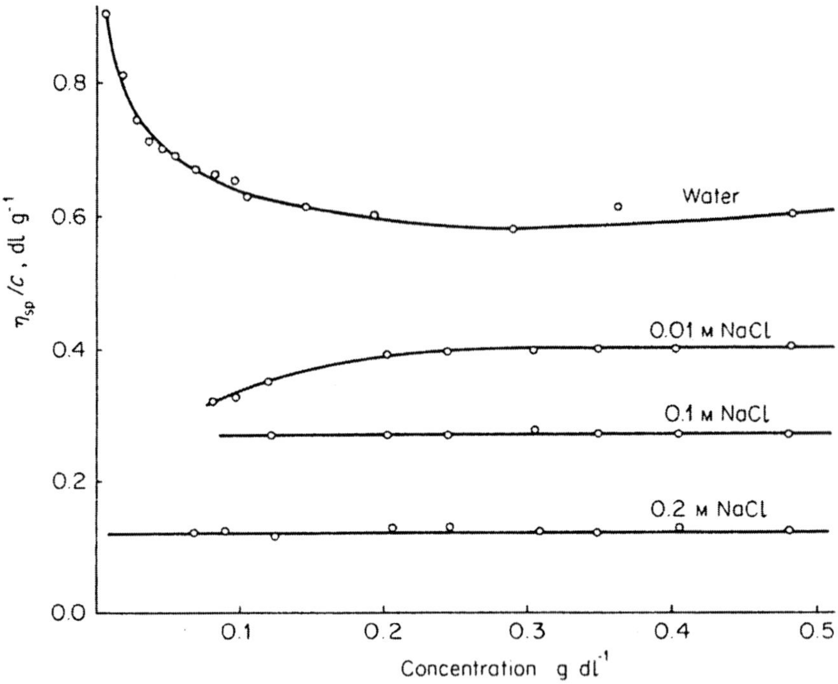

Figure 3.17 Plot of reduced viscosity (η_{sp}/c) versus concentration for a sodium humate showing the effects of added salts. (After Smith and Lorimer, 1964.) Reproduced by permission of the Agricultural Institute of Canada

polyelectrolyte. Thus it has been possible to obtain meanigful results from viscosity measurements in many instances where the addition of an appropriate amount of salt caused the polyelectrolytes to behave like neutral polymer molecules.

The influence of electrolytes on the viscometric properties of humic substances is well recognized (Flaig *et al.*, 1975). Mukherjee and Lahiri (1956) observed the characteristic increases in the η_{sp}/c values of dilute aqueous solutions of K^+-humate extracted from an air-oxidized low rank coal. These values were depressed in the presence of KCl at concentrations of $4–8 \times 10^{-3}$ g dl^{-1}. Data from Smith and Lorimer (1964) presented in Figure 3.17, show how the addition of salt influenced viscosity values in the case of humic acid materials extracted from a sphagnum peat. Piret *et al.* (1960) also observed a concave upwards relationship for the η_{sp}/c vs. c plot at low concentrations of a peat Na^+-humate in water. Although their viscosity data extended to very low concentrations of solute there was no evidence for diminished viscosity at the lowest dilutions measured. However, they observed a straight line relationship with a positive slope when measurements were carried out in the presence of 0.26M NaCl. When the NaCl concentration was only 0.1M a small convex upwards 'hump' was observed with a maximum at a polymer concentration of *ca.* 0.2 g dl^{-1}. This indicates that 0.1M NaCl was insufficient to completely depress the ionization of humic substances at very low concentrations.

Molecular weight data for humic substances, which have been reported from viscosity measurements up to the present time, can only be regarded as relative and not as absolute values because the materials used were invariably highly polydisperse. We suggest that it might be possible to derive more meaningful data from viscosity measurements by working with well fractionated samples and by careful additions of salt in order to derive $[\eta]$ values from plots of η_{sp}/c vs. c (equation 3.50). It will be necessary to observe carefully whether or not the values of K and α change from sample to sample. If they do not the viscosimetric method could be used as an inexpensive method for determining the molecular weights of samples and also for measurements of polydispersity.

The average molecular weight data, from viscometric measurements on humic acid by Piret *et al.* (1960) and Visser (1964) were calculated from equations designed for synthetic polymers of known structures and conformations. Thus the values of 5×10^4 and 36×10^3 quoted by the authors were only regarded as order of magnitude approximations by them because of the constraints implicit in the equations used.

Plots of η_{sp}/c vs. c gave Chen and Schnitzer (1976) straight line relationships for humic acid at pH 7.0, and for fulvic acid at pH values of 1.0 and 1.5 only. Salts were not added. It is surprising that the η_{sp}/c value did not change as the pH 7.0 humic acid sample was diluted. Considerable dissociation of the polymer carboxyls would be expected at this pH value to give increasing polymer expansion (and viscosities, as was observed at pH 8.5) on dilution. On the other hand it is equally difficult to explain a change in slope which was observed for a similar plot for fulvic acid at pH 2.0. This change might indicate some ionization, and only strong acids would be dissociated at such low pH values. Unionized fulvic and humic acids would be expected to behave like neutral polymers in viscosity measurements, even in the absence of additions of salt.

Light scattering

When light passes through a dilute polymer solution some will be scattered and from the intensity of the scattered light the molecular weight and shape of the polymer molecule can be calculated. However, it is important that the scattering of light be due only to the polymer and not to dust or other impurities in the system. Tanford (1961) and Billmeyer (1971) give more details about the principles of light scattering.

The fluorescence of humic substances gives rise to problems where light scattering studies are employed because it can cause considerable amounts of light to be detected as if it were scattered light. This problem can be overcome by judicious use of filters or monochromators. But, as yet, there is no evidence to indicate that such precautions have been taken when measurements were made on humic substances and hence the interpretations given for the data may not be highly accurate. Orlov and Gorskova (1965) report weight-average molecular weights by this technique of the order of 65×10^3 for humic acid extracts from chernozem and sod-podzolic soils but they warn that these values might be influenced by the presence of clay minerals in the preparations.

3.5.3 Molecular Shape and Size

Several of the methods used for determining the molecular weight of polymers can also be used to determine their shape and size, particularly if they involve considerations of frictional forces acting on solute macromolecules as they move through solvent media. The magnitude of these frictional forces is related to the shape and sizes of the macromolecules. Ultracentrifuge and viscosity measurements are particularly effective techniques for such studies, especially when they are used in conjunction with one another.

Before becoming involved in more detailed discussions about shapes and sizes of humic substances we must point out that the data which can be obtained in the laboratory do not necessarily provide accurate representations of the shapes and dimensions of these macromolecules in the soil. The laboratory experiments are carried out in solution whereas humic substances in the soil are largely in the solid or gel phase. Thus although the solution shapes and dimensions are important for an understanding of the system *per se* they need not necessarily be accurate representations of those pertaining in the soil. Many of the criticisms made earlier concerning the experimental conditions under which measurements were made in ultracentrifugation studies also apply to data recorded in this section. Thus proposals with regard to molecular configurations have not always been based on sufficiently sound experimental observations.

Information from frictional ratios

Interpretations about molecular shapes are often derived from a consideration of values obtained for the frictional ratio f/f_0, where f is the frictional coefficient of the molecule in question and f_0 the frictional coefficient for a condensed sphere (containing no solvent) occupying the same volume. Thus the value of f/f_0 will be a measure of the deviation of the molecule from a spherical shape and/or condensed form. Tightly coiled globular proteins, as would be expected, give f/f_0 values of 1.0 or only slightly greater. However, the frictional ratios for molecules which are highly solvated or are more extensive in one direction than another often give rise to values of approximately 2 to 3, and for extreme cases, such as the rod-shaped nucleic acids, even higher values are obtained. Flaig and Beutelspacher (1968) suggested that the humic acid molecule adopted a roughly globular conformation based on frictional ratio values of approximately 1.1, but these values were obtained from ultracentrifugation work in which charge effects were not suppressed. When salt was added their results gave the higher frictional ratio values of 1.2 and 1.5. This would indicate that the molecules were considerably distorted to ellipsoidal rather than spherical shapes, and with the longer axis several times greater than the shorter axis. Piret *et al.* (1960) used a combination of ultracentrifuge and viscosity measurements to obtain frictional ratios of 1.6 to 1.86. Their results indicated to them that the configurations of the molecule approximated to a prolate ellipsoid with axial ratio of about 7:1, and also that the structure was likely to be hydrated rather than condensed. Stevenson *et al.* (1953) reported a still higher frictional ratio indicating an even greater degree of asymmetry and/or hydration.

Once again the determinations of frictional ratio values require the averaging to a single figure of the large variations in the molecular parameters of the sample being measured. As we have already seen from the consideration of molecular-weight values this can result in gross distortion in the interpretation of the data. It is probable that advances in protein chemistry, in which rigid shapes such as globular spheres and oblate or prolate ellipsoids were found to be good approximations of the shapes of the molecules, have had an undue influence on the thinking of workers involved in determining the shapes of humic molecules. Before discussing the validity of such comparisons attention should be given to differences in the origins, chemical compositions, and the structures of the different molecules, as well as to comparisons of properties such as hydration and ion-exchange behaviour. From considerations of the regularity of protein structure, the nature of the bonds and cross-linkages which hold the protein molecules together, and the low hydration of spherical globular proteins, it would be difficult to visualize how humic substances, on the basis of what we know about them, should have similar shapes or tertiary structures. Certainly evidence which suggests similarities of shapes should be subjected to rigorous appraisal.

The ultracentrifuge study of well-separated humic acid fractions of low polydispersities by Cameron *et al.* (1972b) have given some reliable frictional ratio values. These fractions covered a wide range of molecular weight values and they showed that the values for the frictional ratio varied as a function of molecular weight. Now it can be shown (Tanford, 1961) that for shapes other than the hard sphere, represented by some globular proteins, the frictional ratio f/f_0 will vary as a function of molecular weight. It can be further shown that this function will be dependent upon the configuration of the polymer, e.g. spherical, oblate and prolate ellipsoids (all rigid), and flexible, random coils.

Comparisons of the experimental frictional data with the theoretical relationships derived from a consideration of these models indicated that only the oblate ellipsoid (rigid discus shape), and the flexible, random coil need be considered. As discussed above humic materials are unlikely to possess the degree of ordered structure necessary to maintain a rigid configuration. Their mode of formation, heterogeneous nature, and particularly their polyelectrolytic properties in solution are more likely to result in a less rigidly structured molecule whose polar and charged functional groups will be hydrated. Also the dissociation of charges, as discussed already, will lead to intramolecular repulsion causing molecular expansion. These are among the factors which, combined with additional experimental evidence, caused Cameron *et al.* (1972b) to propose the random coil model for humic acids. This model satisfied the observed experimental behaviour of the humates better than any other, and it is consistent with what might be predicted from considerations of the formation, environment, and chemical and physical behaviour of humic substances.

If it is assumed that this model is correct, or at least provides a good working hypothesis, it is worthwhile considering the properties it predicts and to compare them with experimental observations. A randomly-coiled humic acid molecule in solution can be visualized as a strand, with charges distributed along its length, which coils randomly with respect to both time and space. This results in a molecule which is roughly enclosed by a spherical shape within which the distribution of molecular mass is Gaussian, i.e. the mass density is greatest at the centre and decreases to zero at the outer limits of the sphere. The shape of the Gaussian mass distribution will vary according to whether the molecule is tightly or loosely coiled, and this will in turn depend upon factors such as the extent of solvent penetration, charge density, degree of dissociation, the nature of the counterion etc. Chain branching, which is thought to occur in humic substances, gives rise to a higher mass density within the sphere and results in a molecule that is more compact than a single linear chain of the same molecular weight.

In the study under discussion (Cameron *et al.*, 1972b) chain branching was

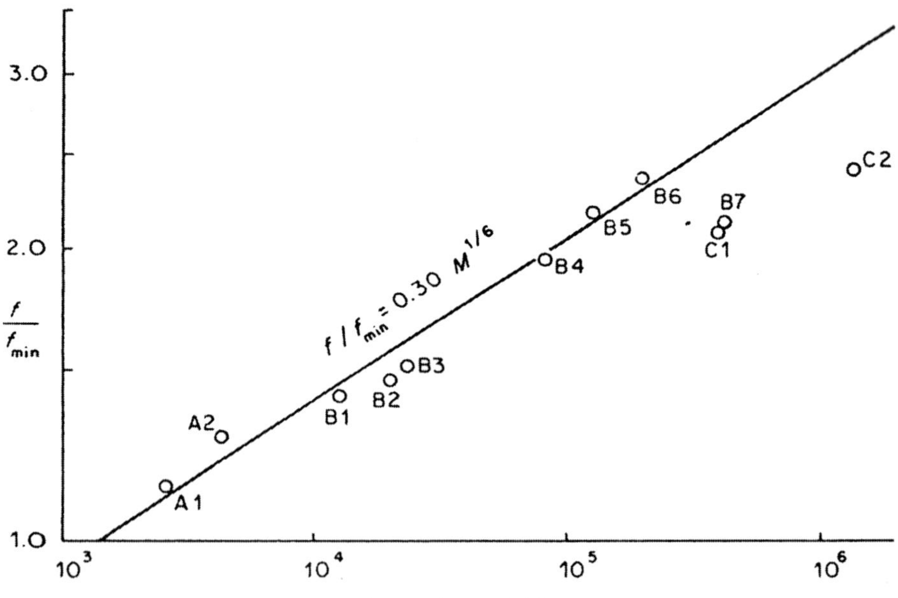

Figure 3.18 Experimental relationship between the frictional ratio and molecular weight of humic acid compared to a theoretical curve for linear random coils under theta solvent conditions. (After Cameron *et al.*, 1972b)

Table 3.5 Molecular parameters of humic acid fractions (A–C) separated by gel chromatography and subjected to ultracentrifugation experiments (Cameron *et al.*, 1972b). Reproduced by permission of Oxford University Press

Fraction	Extractant	Sedimentation[a] Conc. g dm^{-3}	Sedimentation[a] $s_{20} \times 10^{13}$ sec.	$D_{20} \times 10^7$ $cm^2\ sec^{-1}$	$M \times 10^{-3}$	$[f] \times 10^5$ dyne cm^{-1} sec $poise^{-1}$	f/f_{min}	$\frac{R_G}{M^{1/2}} \times 10^9$ cm	R_G nm
A1	Na pyro-phosphate	1.2	0.82	21.4	2.6	0.19	1.14	2.9	1.5
A2	Na pyro-phosphate	2.4	1.02	16.0	4.4	0.25	1.28	3.0	2.0
B1	20 °C NaOH	1.2	1.88	10.2	12.8	0.40	1.41	2.8	3.2
B2	20 °C NaOH	1.2	2.47	8.4	20.4	0.48	1.46	2.7	3.8
B3	20 °C NaOH	1.2	2.64	7.7	23.8	0.52	1.52	2.7	4.2
B4	20 °C NaOH	1.2	4.7	3.95	83	1.02	1.96	2.8	8.2
B5	20 °C NaOH	1.2	5.6	3.07	127	1.32	2.18	2.9	10.5
B6	20 °C NaOH	1.6 0.8 0.8	6.7 7.2 7.2	2.45	199	1.65	2.35	2.9	13.2
B7	20 °C NaOH	1.2	12.6	2.13	412	1.90	2.12	2.4	15.3
C1	60 °C NaOH	1.2	12.6	2.15	408	1.88	2.11	2.4	15.0
C2	60 °C NaOH	2.4	24.6	1.26	1360	3.21	2.41	2.2	25.5

[a] Sedimentation coefficients obtained by a least squares linear fit had typical standard errors of 4 per cent for lower molecular weight fractions, decreasing to 2 per cent at higher molecular weights.

Mean diffusion coefficients, obtained from 4 to 8 replicate runs, had standard errors between 1 to 5 per cent. Solution concentrations were approximately 0.1 g dm^{-3}.

thought to be particularly prevalent at high molecular weight values and the resulting more compact molecular dimensions caused f/f_0 to be lower than the theoretical prediction (Figure 3.18). The molecular coil in solution is perfused with solvent which can exchange with more solvent outside the sphere of influence of the macromolecule. However, because of the higher density of polymer chains in the central regions the rate of solvent exchange is relatively slow, and in the more compact structures the solvent is essentially trapped. This trapping effect is relatively small for the low molecular weight polymers but it increases markedly at higher molecular weights. Thus for the denser structures we can consider that the solvent is able to flow freely through the periphery of the molecule but that this freedom of movement is gradually restricted as the centre of the molecule is approached, and that finally the solvent is effectively trapped in the central regions. Thus the shape of the molecule in solution is essentially spheroidal but not condensed or rigid.

The values obtained for the spherical radii of the molecules are shown in Table 3.5 and range from 1.5 nm for a molecular weight of 2500 to 25.5 nm for a molecular weight of 1.4×10^6. These values agree well with some of those reported by Flaig and Beutelspacher (1968). Their data were also obtained from ultracentrifugation experiments in which charge repulsion effects were repressed, and they derived a radius figure of 6.9 nm for a molecular weight of 77×10^3 (*cf.* Table 3.5). When charge effects were not suppressed much lower values were obtained.

Other techniques have not given such well defined data. For instance Flaig and Beutelspacher (1951) report diameters of up to 10.0 nm using electron microscopy. This value is somewhat smaller than the maximum figures obtained from ultracentrifugation but perhaps not unreasonable because of the shrinking which must take place when the sample is dried for the microscopy. Wershaw *et al.* (1967) report maximum radii of 11.0 nm measured by X-ray scattering for sodium humate in solution. Their frictional coefficients were found to be between 1.3 and 2.5 showing that the molecules were not condensed and exhibited a degree of deviation from a true sphere. This technique would appear to be very promising, and although some fractionation of the samples used was achieved by gel chromatography it is probable that significant refinements in the data could be achieved by using well fractionated samples.

Influence of charge on the shape of humic substances

As we have seen from our earlier considerations of the properties of humic substances these materials contain considerable amounts of acidic and other functional groups which give rise to many charged sites in the polymer. Such sites will greatly affect the properties and behaviour of the molecules, both in solution and in the solid state, and we will now consider what some of the effects might be.

If, for the sake of argument only, we were to consider the molecular configuration of humic substances to be that of condensed, rigid structures rather than more open random coils, and further that this configuration is retained in solution then we would have to take into account a number of criteria. It would be especially necessary that all of the charged sites which are dissociated be located on or near the exterior of the molecules. Charge sites within the matrix of the polymer could not dissociate for two reasons; firstly, solvent which is required for dissociation is absent from the inner parts of a condensed structure, and secondly if dissociation of internal charged sites did occur then the resultant charge repulsion would lead to molecular expansion to destroy the condensed configuration.

None of these requirements are observed in humic substances. For instance they manifestly lack the ordered structure which would be necessary to ensure that all of the numerous charged sites are located at the surface. Indeed, all available evidence indicates that the charged sites are distributed along the length of the polymer and thus they can occur at any point in the polymer matrix. Observations of swelling properties when solvated, and of ion-exchange behaviour when in solution show that the molecules are readily penetrated by solvent and that most charged sites are dissociated and accessible for ion exchange. Such properties are incompatible with a condensed, rigid molecule but are compatible with the proposed flexible, random-coil model which is capable of molecular expansion.

3.5.4 The Solid or Gel Phase

As already emphasized (Section 3.5.3), the experimental determination of molecular shape is carried out in solution and the resulting description applies strictly only to the molecule when it is in this state. However, we can use the solution to predict what might occur when the molecule is precipitated by the addition of suitable counterions, or is dehydrated. In this way we can get some idea of what the molecular configuration might be when the humic substances are in their more common form in the soil i.e. in the solid or gel state. In order to do this it is probably more informative to consider first the forces and effects which cause a polymer such as humic acid to dissolve and remain in solution. Firstly, there are solvation effects which have been discussed in Section 3.2.1. Under normal aqueous conditions the solvent is water which will perfuse the molecular structure generally, but will solvate certain portions more strongly. For instance the polar functional groups, usually containing oxygen, but some with nitrogen or sulphur, and the charged sites (Section 3.2.2) plus their counterions will be strongly hydrated. However, solvation processes alone are not sufficient to carry any except the lowest molecular weight humic substances into solution, and reference has been made (Section 3.2.2) to the necessity for a degree of charge dissociation which allows some counterions to move into bulk solution and gives rise to electrostatic repulsion by the residual charges in the polymer, and eventually gives rise to the solution of the macromolecule. Although it is obvious that the process of molecular expansion due to the mutual repulsion of dissociated charge sites will have a very profound effect, there is a limit to the extent to which this expansion occurs. Thus, even when dissociation is complete a large proportion of the counterions, although separated from their charged site, are still found in the solvent enclosed by the polymer coils. These therefore remain within the domain of the macromolecule which resists further expansion. The foregoing considerations apply when the counterions are readily dissociated cations, such as sodium, potassium, and ammonium.

The situation is different when H^+, Ca^{2+}, and polyvalent cations such as Fe^{3+} are the counterions and these give rise to precipitation of the humic substances. These ions are either undissociated or only very weakly dissociated in the presence of humic substances and thus one of the major solubilizing forces is lost. When the polymer is H^+-saturated hydrogen bonding between the molecules shrinks the structure. While the introduction of counterions such as H^+, Ca^{2+}, and Fe^{3+} will not affect the solvation of functional groups or charged sites it will greatly decrease the amount of solvent within the polymer matrix. Thus the absence of charge repulsion will lead to molecular shrinkage and in the case of multivalent cations the simultaneous linking across two or more charged sites will cause further molecular contraction. Hence, in the processes of charge neutralization with non-dissociating ions, solvent expulsion and molecular shrinkage occur simultaneously, and result in precipitation. Thus, molecules of humic substances in the gel or solid state will be more compact, more tightly coiled and much less solvated compared with those in the solution state.

Even so, the molecule will still be far removed from a condensed form since there will still be a considerable degree of solvation of functional groups and charged sites. Further shrinkage and condensation can be induced by removal of water through drying, and under such conditions the molecules may well move into a highly condensed state. In this condition strong intramolecular bonds will be set up particularly between hydrophobic portions of the molecule. Such interactions probably account for the observed difficulty in rewetting organic soils and humic materials which have been exhaustively dried.

3.5.5 Investigations of Polyelectrolyte Properties of Humic Substances

We have repeatedly referred to the polyelectrolyte nature of humic materials because they exhibit many of the properties commonly associated with charged macromolecules, including polybasicity and ion exchange. Several techniques have been developed for examining the polyelectrolyte properties of biopolymers in particular, and many of these can be applied to investigations of humic substances.

Electrophoresis

This is one of the procedures used to describe the movement of ions in an electric field. When a particle of charge q is placed in an electric field of strength E it will be subject to a force Eq and will achieve a steady-state velocity μ under this force, where

$$\mu = Eq/f \tag{3.54}$$

and f is the frictional coefficient. For a charged macromolecule of radius R and carrying Z unit charges,

$$q = Z\varepsilon \tag{3.55}$$

where ε is a unit charge, i.e. 4.3×10^{-10} e.s.u. The frictional coefficient is given by

$$f = 6\pi\eta R \tag{3.56}$$

where η is the viscosity of the medium through which the particle is travelling and R is the radius of the particle. Thus the steady-state velocity of the molecule becomes

$$\mu = \frac{EZ\varepsilon}{6\pi\eta R} \tag{3.57}$$

where μ is the electrophoretic velocity. It is more usual, however, to determine a parameter which is not dependent on the strength of the electric field but only on the parameters of the charged molecule. This is the electrophoretic mobility U, defined as

$$U = \mu/E \tag{3.58}$$

Thus the technique would appear to offer an easy method for determining and comparing molecular sizes and charge densities.

Whilst this theoretical approach is straightforward the actual study of polyelectrolytes by electrophoresis is complicated by a number of factors. Macromolecular ions do not exist in isolation in solution but are accompanied by sufficient counterions to give an electrically neutral system. These counterions will not be randomly distributed through the solution but will tend to form localized clouds around the macromolecules. Under the influence of the electric field the counterions will tend to move in the opposite direction to the macroions, and in addition they will carry relatively large amounts of solvent with them. Both of these effects exert a drag on the movement of the macromolecules to give a resultant electrophoretic mobility below that predicted, and by an amount that is difficult to calculate. Another complication arises from the variable pK_a values which are exhibited for the acidic functional groups of polyelectrolytes in general and for humic substances in particular. At any given pH value it is difficult to estimate correctly the number of dissociated sites. Even if all sites are dissociated not all of the counterions will be outside the polymer matrix and so it will be very difficult to estimate the charge on the macromolecule. In any case it is considered that the charge or potential at the surface of the molecule should be replaced by the zeta-potential, which is the potential at the surface of shear of the charged particle. This relates to the true surface of the moving particle and includes solvation layers and attendant counterions. The zeta-potential is then somewhat lower than the potential at the surface of the molecule. In practice, various salts or buffers are added to the system to suppress deviations from ideality and to create and maintain an electrically neutral, pH-buffered system. Since added positive and negative ions will usually have different electrophoretic velocities the theoretical problem of estimating the final electrophoretic mobility of the macro-ion is further complicated and very difficult to determine.

It is obvious from this discussion that electrophoresis can be used to give semi-quantitative information about molecular parameters such as charge, size and shape, or to provide a fractionation based on molecular variations related to these properties. However, it is only of limited value for giving accurate measurements of actual molecular parameters, and its use for fractionations has been superceded by gel chromatography used in conjunction with ion-exchange chromatography techniques. Various types of electrophoresis procedures have been used to fractionate humic substances, and these are reviewed by Flaig *et al.* (1975).

Paper electrophoresis in alkaline buffers is the most common form of the technique used to fractionate humic materials. Other support media such as columns of glass or gel beads have also been used. If, however, the electrophoretic mobility is to be measured then it is best to employ the free boundary technique (Stevenson *et al.*, 1952). As with other fractionation procedures discrete fractions are not obtained but rather there is a gradation of properties with one fraction merging into another. In general, however, gray humic acids migrate very slowly, brown humic acids migrate more quickly and fulvic acids migrate more rapidly still. This observed electrophoretic behaviour supports the evidence and the view that the observed sequence is composed of molecules of increasing charge densities and decreasing molecular weights.

Many workers have observed fluorescent areas or fractions during electrophoresis experiments (e.g. Waldron and Mortensen, 1961), and this behaviour is usually associated with the more mobile materials. It is not clear whether the fluorescence is due to separate components or is characteristic of particular fractions of humic substances. It is possible that all fractions do in fact fluoresce, but that this is masked by the intense absorption of the grey-brown components.

Electrofocusing

This technique is similar to electrophoresis, and was designed to separate protein mixtures by utilizing their different isoelectric points. A column of gel is prepared having a pH gradient from top to bottom, and the gel material is instrumental in stabilizing the pH gradient. The mixture is applied to the top of the column and a potential difference is then applied between the top and the bottom. Under the influence of this field the charged molecules move until they reach a point in the column where the pH corresponds to their isoelectric point. At this point there is no net charge on the molecule and it ceases to move despite the electric field.

Although humic substances do not have an isoelectric point they can be fractionated by electrofocusing procedures (e.g. Cacco *et al.*, 1974). When there is an appropriate range of pH values within the gradient the humic substances will move until a sufficiently acidic pH value is reached where all the negatively charged sites are protonated and thus the molecule will cease to move because they are not dissociated. In this sense the process is more analogous to fractional precipitation than to isoelectric focusing, but nevertheless it does provide a potentially useful fractionation technique. Electrofocusing procedures provide a useful method for characterizing humic extracts from different soil types.

Ion exchange

One of the most important properties of soil humic substances is their ability to act as ion exchangers. This capacity to retain and exchange cations plays an important role in the supplying of nutrients to plants and for the general maintenance of soil fertility. The exchange capacity of humic materials is very high compared with most soil mineral components, and although they are generally less abundant than the inorganic colloids they often make a major contribution to the overall cation-exchange capacity of the soil. Humic materials are also highly efficient at combining strongly with a large number of heavy metal cations, some of which are essential plant micronutrients. Thus the behaviour of humic materials as an ion exchanger and complexing agent is of prime importance in the retention, mobilization, and cycling of nutrients.

Ion-exchange processes and the complexing of heavy metal ions are dealt with more extensively in the companion volume, and only aspects which relate to the chemical or molecular structure of humic substances will be considered here. As discussed earlier (Section 3.5.2) the molecular structure will profoundly influence accessibility to the ion-exchange sites and also the kinetics and mechanisms of exchange.

The first effect to consider is the Donnan membrane effect, the theory of which is dealt with in more detail in Chapter 5. Consider a solution containing macro-ions and counterions separated by a membrane from a solution containing only simple small ions. The small ions can pass through the membrane but the macro-ions cannot. When equilibrium is reached the distribution of diffusible, small ions is not uniform through the system. Counterions, which have charges opposite in sign to those on the macro-ions, will be present at higher concentration on the macro-ion side of the membrane and co-ions, which carry the same charge sign as the macro-ion, will be found in higher concentration on the other side of the membrane.

Whilst this theory has been derived and shown to be valid for the system described it has also been found to apply to single, charged, colloidal particles or macromolecules. In this case, instead of an actual membrane an imaginary boundary can be drawn around the particle or molecule (usually at a short distance from the particular or molecular surface) encompassing a small amount of solvent. Solvent and small ions can pass freely across this imaginary boundary, but the macro-ion obviously cannot, and the boundary thus becomes in effect analogous to a semipermeable membrane. The importance of this, as far as polyelectrolyte molecules are concerned, is that counterions can pass freely through the imaginary boundary and perfuse through the matrix of the molecule, whereas co-ions carrying the same charge as the macro-ion will be kept at a low concentration within the boundary and will be effectively excluded from the volume contained within the imaginary boundary. This phenomenon is often referred to as ion-exclusion, and the effect can be very large, particularly at low electrolyte concentrations, but it can, nonetheless be reduced by the addition of the electrolyte.

Since co-ions are excluded from the macro-ion the mechanism of ion exchange is now more apparent. Counterions from the solution enter the matrix of the exchanger and, in order to retain electrical neutrality, an equivalent number of other counterions must be displaced and move out of the matrix. Ion exchange is thus a simple stoichiometric process. The actual process of replacing one ion by another at an exchange site is very rapid and the rate at which ion-exchange processes occur is determined by a slower process, namely the rate at which ions diffuse to or away from the exchange sites. Since two different types of ions are usually involved this process is often referred to as interdiffusion.

In practice two separate, potentially rate determining processes can be identified. The first concerns the movement of ions within the matrix of the exchange medium and is referred to as *particle diffusion*. This process can become the rate limiting step when movement of ions through the particle to the exchange sites is limited. Such conditions can occur if the polymer is densely coiled and cross-linked. This leads to large frictional forces and tortuous pathways, and is accentuated if the diffusing ions are large and/or exhibit specific interactions with the polymer matrix. The second process concerns the rate of diffusion of ions across the adherent film of strongly held solvent adjacent to the polymer surface. The interdiffusion of ions across this zone is referred to as *film diffusion*. If the rate of particle diffusion is high and exchange within the particle is fast then film diffusion becomes the rate controlling step.

It has been shown (Helferich, 1962) that the nature of the rate controlling step can be represented mathematically using the expression

$$\frac{X\bar{D}S}{CDr_0}(5+2\alpha_{\mathrm{B}}^{\mathrm{A}}) \equiv 1 \qquad (3.59)$$

where X is the concentration of fixed ionic sites, C the concentration of the counterion in solution, $\bar{D}$ the interdiffusion coefficient in the ion exchanger, D the interdiffusion coefficient in the surrounding film, S the film thickness, r_0 the radius of exchanger, and $\alpha_{\mathrm{B}}^{\mathrm{A}}$ is a separation factor or selectivity coefficient. When this expression is equal to one, the half-times for film and particle diffusion are equal; if it is much greater than one then diffusion within the particle is the rate limiting step, and if it much less than one then diffusion through the surrounding film of liquid is the rate limiting step.

There is only limited awareness about the mechanisms and kinetics of the ion-exchange process in humic substances although much is known about the relative binding strength of the polymer for a wide variety of metal and other cations. The evidence which is available indicates that particle diffusion is the rate controlling step. However, the attainment of equilibrium for humic acid is found to be comparable with moderately cross-linked ion-exchange resins but more rapid than for highly cross-linked resins. In addition exchange appears to be complete when equilibrium is reached. These observations indicate that the humic substances have a structure sufficiently complex to cause some restriction to the rate at which exchange occurs, but that the structure is sufficiently open to allow access to all of the exchange sites. This concurs once more with the earlier structure, proposed in Section 3.5.3, of a moderately cross-linked random coil, but it is not compatible with any of the condensed structures which have been proposed.

Whilst structural organic cross-linkages are not thought to be numerous, other bridging linkages can be formed when the negative charge of the molecule is balanced by di- and trivalent cations. Thus when salts such as calcium-, iron-, or aluminium-humates are formed the effects produced are somewhat analogous to cross-linking, and molecular shrinking is observed. In this situation tortuosity and general frictional forces within the molecule will be greatly increased and the passage of solvent and of ions through the macromolecule will be restricted. As a result the ion-exchange process may well be slowed down and equilibrium achieved more slowly. Processes such as these, involving the more strongly held cations, are of great importance in the soil, but there is little evidence about the way in which they effect ion-exchange processes. There is considerable scope for more detailed studies on the ion-exchange processes which take place in humic substances, particularly in the area where these studies relate more closely to the actual conditions and exchange processes which occur in the soil.

3.6 SOIL POLYSACCHARIDES

Soil polysaccharides are composed of hexose (six-carbon), and pentose (five-carbon) sugar units and of some hexose derivatives. On average they make up about 10 per cent of the soil organic matter. Interest in these colloidal materials began during the mid 1940s when Martin (1945, 1946) established that slimy bacterial products, earlier shown (Waksman and Martin, 1939) to aggregate sand–clay mixtures, were polysaccharides. At about the same time (Haworth *et al.*, 1946) and Geoghegan and Brian (1946, 1948) were investigating the uses of bacterial dextran and levan polysaccharide slimes for the improvement of soil structure. Since then sporadic research interest has been directed to investigations of the genesis, isolation, purification, structural determinations, and the roles of polysaccharides in the soil environment.

During the 1950s statistical analysis (e.g. Rennie *et al.*, 1954; Chesters *et al.*, 1957) and in the early 1960s chemical treatments such as periodate degradations (e.g. Greenland *et al.*, 1961, 1962; Clapp and Emerson, 1965) further established the role of polysaccharides in the aggregation of soil particles. In the meantime Mehta *et al.* (1960) had shaken confidence in the aggregating influences of soil polysaccharides by showing that the crumb-structure of a Swiss braunerde resisted degradation in a periodate medium, but the later work by Greenland *et al.* (1961) suggested that these resistant crumbs of high organic matter contents were held together by fungal hyphae and myceliae, and other non-polysaccharide materials.

There are a number of reviews which deal with various aspects of the isolation, compositions, and the roles in the soil of sugars and polysaccharides (Mehta *et al.*, 1961; Gupta, 1967; Swincer *et al.*, 1969; Finch *et al.*, 1971; Greenland and Oades, 1975). Here we will deal briefly with the isolation, fractionation and purification, composition, and in a general way with structural determinations. Some consideration will also be given to polysaccharide clay interactions since these are likely to be important in soil aggregation processes.

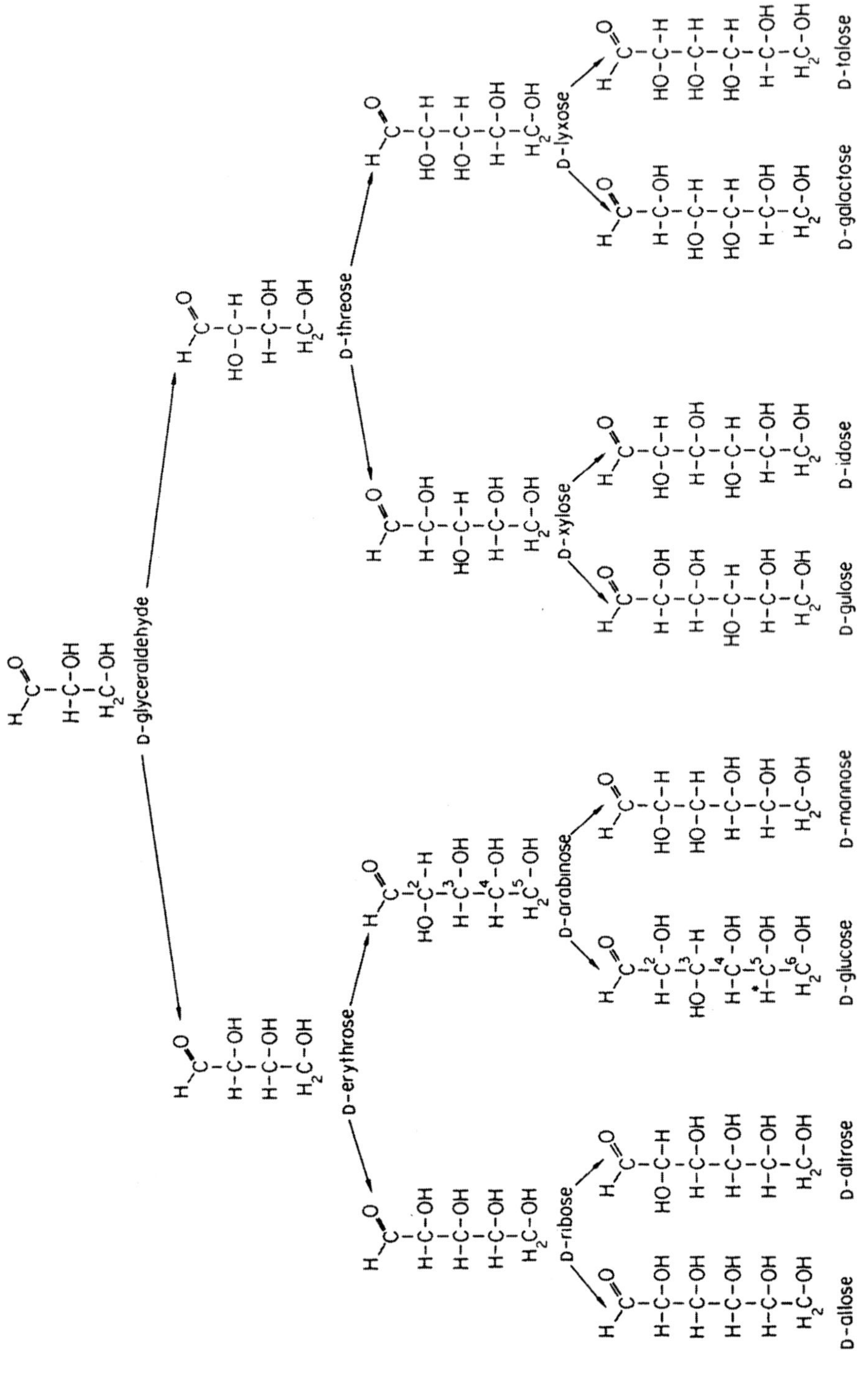

Figure 3.19 Open chain structures of aldoses in the D-configuration

3.6.1 The Composition and Structures of Monosaccharides

Before the composition, possible structures and properties of polysaccharides can be appreciated it is necessary to have an understanding of the structures of the sugar units which compose them. This treatment will provide an elementary knowledge of sugar structures. Texts on the chemistry of saccharides, such as those by Coffey (1967), Davidson (1967), Guthrie and Honeyman (1968), and Pigman and Horton (1970) give detailed treatises on the compositions and structures of monosaccharides (single sugar units), oligosaccharides (containing up to ten sugar units linked together), and polysaccharides. We will be concerned only with aldoses (which can contain an aldehyde functional group) although ketoses, especially fructose, have been found in soils.

It is easy to derive the aldopentose and aldohexose structures by starting from glyceraldehyde. All of the sugars listed in Figure 3.19 have the D-configuration which refers to the highest numbered asymmetric carbon atom (e.g. carbon-4 in D-arabinose, carbon-5 in D-glucose). The sugars are arranged systematically. By following the arrows it can be seen that two daughter sugars are built from each parent molecule, and the hydroxyls on the carbons α- to the aldehyde groups are first placed on the right hand side of that carbon and then on the left hand side. When all of the structures are drawn the names of the pentose and hexose structures can conveniently be remembered by the use of mnemonics. Thus *RAXL* refers to ribose, arabinose, xylose, and lyxose and the favoured mnemonic for the hexoses '*all altruists gladly make gum in gallon tanks*', has some resemblance to the names *allose*, *altrose*, *glucose*, *mannose*, *gulose*, *idose*, *galactose*, and *talose*. Note also that the arrangement of the hydroxyl groups follow a pattern. Thus in all cases the hydroxyl on carbon number 5 (C-5) are on the right, the first four of C-4 are on the right, and the second four on the left and so on. The L-sugars, which are the mirror images of the D-structures are less common in nature.

A solution equilibrium exists between the open chain structures and cyclic forms of pentose and hexose sugars. However, more than 90 % are in the cyclic form at any one time. Furanose (five-membered) and pyranose (six-membered) structures result from cyclization, through the formation of hemiacetal functional groups between C-4 and C-1 and between C-5 and C-1, respectively, as illustrated for the arabinose and glucose structures in Figure 3.20. Cyclization generates a new asymmetric centre at C-1 and thus gives rise to α- and β-anomers of the D- and L-sugars. These anomers are shown for the α-D- and for β-D-arabinose structures in Figure 3.20.

Five-carbon ring structures are almost planar whereas the cyclic six-carbon structures are puckered and assume conformations, or positions in space ranging from the '*boat*' to the '*chair*' extremes. The presence of one heterocyclic oxygen in the ring does not appreciably alter the conformations. Six-membered cyclic structures will preferentially assume the chair conformation where non-bonded interactions involving the bulky substituents (—OH and —CH_2OH

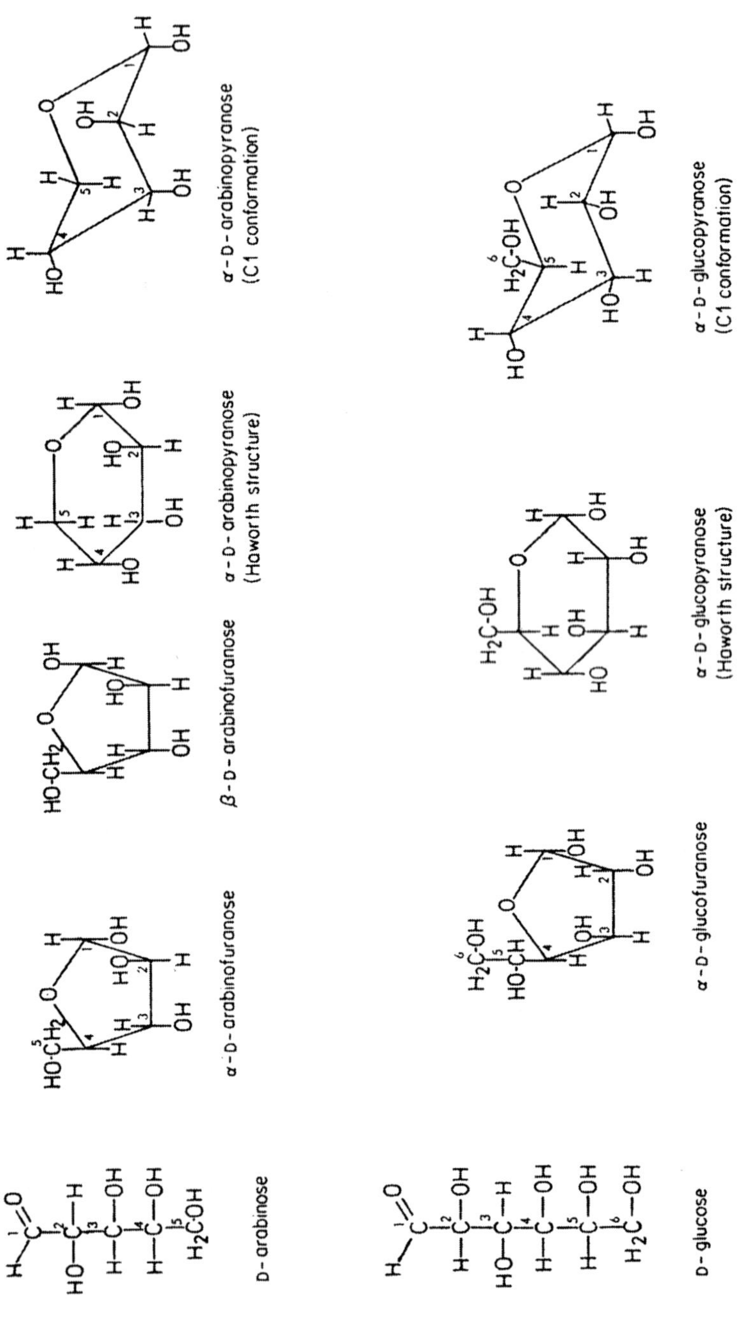

Figure 3.20 Open chain and ring structures of D-arabinose and D-glucose

groups in Figure 3.20) on the carbon atoms are at a minimum. However, these six-membered structures are still often represented by Haworth formulas as shown. From examination of the structures in Figure 3.20 it will be seen that assignment of the hydroxyl groups from the open-chains to furanose and Haworth structures is straightforward. It is also easy to convert the Haworth formulas to the C1 chair conformations shown. The sequence of transformations for β-D-glucopyranose, shown in scheme (3.60), illustrates how substituents or carbons of Haworth structures are assigned in the C1 conformations. The 1C conformation is obtained by 'flipping' the model for the C1.

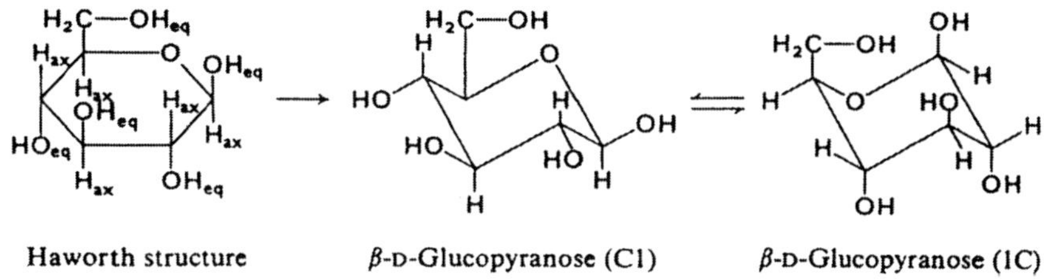

(3.60)

The C1 and 1C conformations are interconvertible in solution. However, the preferred and most stable conformation will be that in which the non-bonded interactions are least, and this is achieved when the bulky substituents are all equatorial as seen for the β-D-glucopyranose (C1) structure. By use of atomic models it is easy to realize how non-bonded interactions are greatest when the bulky groups are axial, in positions 1, 3, 5 and in 2, 4, because these two groups of positions are on the same side of the cyclic structure. Stacey and Barker (1960) have suggested that the favourable axial–equatorial arrangements of the bulky substituents could partially explain the abundance of D-glucose, D-galactose, and D-mannose in nature. This does not, however, explain the absence of D-allose and D-lyxose from polysaccharide structures.

Six derivatives of hexose sugars are found in soil polysaccharides. These are fucose, or 6-deoxy-L-galactose, rhamnose, or 6-deoxy-L-mannose (in which the —CH_2OH groups on C-6 are replaced by —CH_3), glucosamine, or 2-amino-2-deoxy-D-glucose, galactosamine, or 2-amino-2-deoxy-D-galactose (in which the —OH groups on C-2 are replaced by —NH_2), glucuronic and galacturonic acids (in which the —CH_2OH groups on C-6 are replaced by —CO_2H). The amino sugars are likely to be present in the polysaccharides as the acetyl (—$NHCOCH_3$) derivatives, but the acetyl group is lost during hydrolysis.

3.6.2 Structures of Polysaccharides

Polysaccharides are polymers in which acetal linkages are formed by elimination of water through condensation of the anomeric or glycosidic hydroxyl (on C-1) of one sugar with any non-glycosidic hydroxyl of another sugar molecule. The sugar residues released by the hydrolysis of a polysaccharide compose its

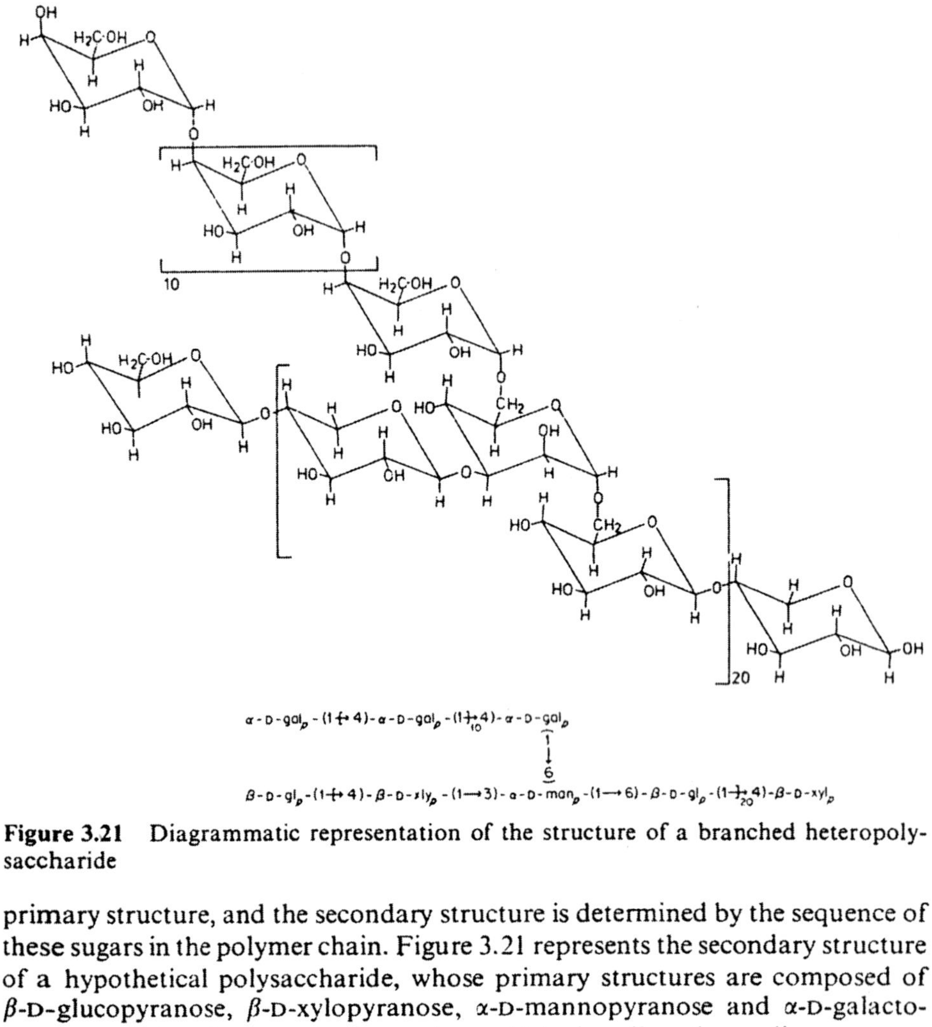

Figure 3.21 Diagrammatic representation of the structure of a branched heteropolysaccharide

primary structure, and the secondary structure is determined by the sequence of these sugars in the polymer chain. Figure 3.21 represents the secondary structure of a hypothetical polysaccharide, whose primary structures are composed of β-D-glucopyranose, β-D-xylopyranose, α-D-mannopyranose and α-D-galactopyranose units. A recognized notation which describes the cyclic monosaccharide components (and where *p* refers to the pyranose structure), and the types of linkages are also given. The repeating groups are enclosed by the large square brackets and the numbers 20 and 10 indicate that these groups are repeated 20 and 10 times in the 'backbone' and side chain, or branch, respectively.

Polysaccharides, sometimes referred to as glycans, can be broadly classified as homopolysaccharides (composed of one repeating monosaccharide) and heteropolysaccharides (composed of two or more monosaccharides). These structures are further subdivided into straight-chain and branched structures, and the polysaccharide in Figure 3.21 can be classed as a branched heteropolysaccharide. In this case the branching takes place through the α-D-galactopyranose(1 → 6)-α-D-mannopyranose linkage.

Bouveng and Lindberg (1961), Whistler (1965), Björndal *et al.* (1973) and Aspinall (1976) give details of the principles and techniques for determining polysaccharide structures. We present here only the rudiments of some procedures which can be used for structural determinations.

In order to ascribe detailed structure to any polysaccharide it is necessary: (a) to know what monosaccharides compose it; (b) to determine the relative abundances of these sugars; (c) to establish which hydroxyl groups are involved in the linkages in each case, and whether the glycosidic linkages are α or β; (d) to know the average chain lengths of the 'backbone' and of the branches; and (e) to know the extent of branching and the branching positions.

Determination of primary structure. Hydrolysis in acidic media is generally used to release the monosaccharide components from polysaccharides. This technique and the mechanisms involved have been reviewed by BeMiller (1967). It must be realized that losses occur during acid hydrolysis. These occur most frequently through the formation of furfuraldehyde derivatives (from the pentose and hexose residues released) which form brown polymeric structures. The extent of polymer formation can be decreased by carrying out the hydrolysis *in vacuo* or in an atmosphere of nitrogen. The hexopyranoside (glucose, galactose and mannose) linkages in the polysaccharide are more resistant to hydrolysis than the xylopyranose structure and advantage can be taken of this in order to achieve a partial hydrolysis.

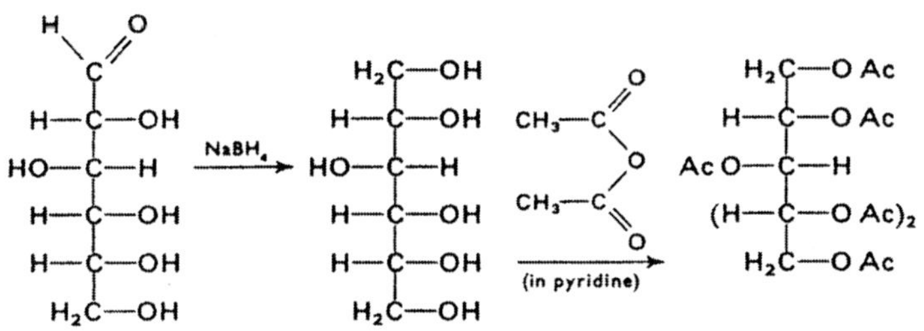

(3.61)

The sugars released by hydrolysis can be identified by paper chromatography and quantitatively determined by eluting the spots followed by colorimetric determination. More accurate identifications and determinations use gas–liquid chromatographic separations of the alditol acetate derivatives, formed from the alditol (or sugar alcohols) derivatives, as outlined for glucose in reaction scheme (3.61). Earlier we stated that less than 1 % of glucose is in the open chain form. However, as the glucitol is formed by reduction more of the straight chain structure becomes available through the equilibrium process until the reaction reaches completion. Quantitative determinations of the sugars, or of their derivatives, not only provide the primary structure of the polysaccharide but also give the ratios of the different sugar units present.

Determination of secondary structure. When polysaccharides are methylated, methyl ether ($—OCH_3$) derivatives are formed from all of the free hydroxyl groups. Older methylation procedures used dimethylsuphate in basic solution, or Purdie methylation which employs iodomethane and silver oxide catalyst. More recently procedures have been developed which require only milligram quantities of substrate. For instance, Björndal and Lindberg (1969) successfully methylated 5 mg of polysaccharide, in solution in methylsulfoxide containing methyl sulphinyl cation in an atmosphere of nitrogen, by adding iodomethane. Then the solution was hydrolysed in 0.25M H_2SO_4 and after neutralization the alditol acetates were formed and identified by g.l.c. and m.s. techniques. (See also Lindberg *et al.*, 1973; Dutton, 1974.)

The ether linkages resist hydrolysis in dilute acid, but the glycosidic linkages are, of course, labile. In the case of the polysaccharide structure illustrated in Figure 3.21 the following methylated sugars would be released in the ratios enclosed by brackets: 2, 3, 4, 6-tetra-*O*-methylglucose, (1); 2,3-di-*O*-methylxylose (21); 2,4-di-*O*-methylmannose, (20); 2,3,4-tri-*O*-methylglucose, (20); 2,3,4,6-tetra-*O*-methylgalactose, (1); 2,3,6-tri-*O*-methylgalactose (11). In the hydrolysis process both anomers are formed and it is thus not possible to predict from such data whether the residues were involved in α- or in β-linkages. Clearly the 2,3,4,6-tetra-*O*-methylglucose and -galactose can be assigned to non-reducing end groups (i.e. having the glycosidic hydroxyls involved in linkages). The amounts of 2,3,4-tri-*O*-methylglucose shows that this structure must have been derived from glucose in the chain and that the 6-hydroxyl, in addition to the glycosidic hydroxyl was involved in linkages. The same reasoning indicates that mannose was present in the chain and involved in branching through the 6-hydroxyl. The amounts of 2,3,6-tri-*O*-methylgalactose is consistent with a structure in which this sugar composed the branch and was thus involved in the linkage to the mannopyranose in the structural 'backbone'.

The fact that xylose is present in the 'backbone' structure and linked through its 1- and 4-OH groups is indicated by the amounts of 2,3-di-*O*-methyl xylose isolated; but xylose also composes the reducing-end group. The ring structure of the reducing end can open to free the aldehyde functional group, and this can form a phenylhydrazone derivative, or be reduced by sodium borohydride. Such derivatives can be isolated when the polymer is hydrolized to provide proof of the terminal residue containing the free glycosidic structure.

Two problems still remain to be solved: (1) the assignment of the sequence of glucopyranose, xylopyranose, and mannopyranose structures in the backbone, and (2) the determination of the configurations (α or β) of the glycosidic hydroxyls involved in the linkages. By means of partial hydrolysis, using decreasing concentrations of acid, it will be found that the disaccharide composed of the methylated xylose and mannose residues will be present in the greatest abundances as the concentration of acid in the hydrolytic medium is decreased, and the sequence can then be deduced. Enzymatic hydrolysis (Marshall, 1974; Snaith and Levvy, 1973) provides a method for establishing the configurations of the glycosidic linkages. This requires the addition of specific enzymes in order to cleave the appropriate α- or β-linkage.

Periodate oxidation is another widely used technique in polysaccharide structural studies. However, although meaningful methylation studies can be carried out on purified soil polysaccharides, we do not think that materials of sufficient purity are yet available to carry out periodate studies. Goldstein *et al.* (1965) and the standard texts referenced already give details of the technique.

3.6.3 Origins, Isolation, Purification, and Compositions of Soil Polysaccharides

Finch *et al.* (1971) and Greenland and Oades (1975) have summarized features of saccharides and polysaccharides of plants, animals, insects, and microorganisms, all of which, in one way or another, find their way to the soil environment. There they are subjected to microbial attack and a variety of new polysaccharides are biosynthesized. Cheshire (1977) has summarized studies which have shown that soil microorganisms can synthesize from ^{14}C-labelled glucose (Chahal, 1968; Cheshire *et al.*, 1969; Oades, 1974) or alfalfa (Keefer and Mortensen, 1963) the neutral sugars which are found in soil polysaccharides. However, Cheshire *et al.* (1969) showed that the polysaccharide material formed in this way was deficient in arabinose and xylose compared with natural soil polysaccharides, although the relative proportions of glucose, galactose, mannose, rhamnose and fucose were quite similar.

Cheshire and his associates (see Cheshire, 1977, and references therein) have done much to help our understanding of the origins of soil polysaccharides. Their long-term incubation studies, using ^{14}C-labelled preparations of glucose, glucose plus acetate, xylose, hemicellulose material, and cereal rye straw, indicated that:

1. part of the carbon chain of glucose is incorporated intact in the hexose and deohyhexose residues of newly biosynthesized polysaccharides;
2. the pattern of labelling of the newly synthesized sugars from a xylose substrate is similar to that from glucose, suggesting the possible conversion of xylose to glucose during the biosynthesis processes;
3. at low temperatures, in conditions which depress bacterial but not yeast numbers, the xylose content of polysaccharide from the labelled glucose substrate was similar to that of galactose and mannose, and was presumed to result from synthesis by the yeasts because a high proportion of yeast isolates synthesized xylose;
4. no synthesis from the labelled substrates or arabinose, comparable with that of xylose or the hexoses, was achieved, even when the conditions of light and temperature favoured algal growth (the carbohydrate of algae, like that of plants, contains xylose and arabinose);
5. the xylose in hemicellulose was almost completely degraded after 16 months incubation with soil, and the amounts of labelled glucose and

xylose which remained were influenced by the fineness of grinding of the rye straw.

Much work still remains to be done before the origins of soil polysaccharides can be assigned with confidence. At the present time there is no doubt that many of these polysaccharides, and in particular those containing hexoses and deoxyhexoses, have microbial origins, but the possibility remains that some are plant derived. The arabinose-containing structures, and some of those with xylose, might predominantly come from plants.

Isolation and purification of soil polysaccharides

Before the efficiency of any extractant for soil polysaccharides can be determined it is necessary to know the carbohydrate content of the soil. This can be done by estimating the amounts of sugars in a soil hydrolysate. Of the various hydrolytic procedures, those used by Cheshire and Mundie (1966) and Oades (1967a), based on the studies of Ivarson and Sowden (1962) and Gupta and Sowden (1965), appear to give the best results. The procedure involves pretreatment of the sample with 72 % H_2SO_4 followed by dilution to 1N acid and refluxing for a prescribed period. It is advisable to work out the optimum pretreatment and hydrolysis times, because these can vary from sample to sample. Where colorimetric procedures are used (e.g. anthrone), it is desirable to decolourize the hydrolysate products. Accurate quantitative analysis of the different sugars present can be made by the g.l.c. separation (Oades, 1976b) as outlined in Section 3.6.1. However, hexuronic acids (where C-6 is a carboxyl group) are degraded by the hydrolysis process and these should be separately determined (e.g. by the method of Bitter and Muir, 1962).

Methods for the extraction of polysaccharides from soil have been reviewed by Mehta *et al.* (1961), Swincer *et al.* (1969) and by Greenland and Oades (1975). Appreciable yields of polysaccharide (3.5–11 % of the soil organic matter) have been obtained by refluxing soils and compost with 98 % methanoic (formic) acid (Parsons and Tinsley, 1961), although this procedure might be expected to lead to decarboxylation of uronic acids. However, 0.5N NaOH is probably the best of the aqueous solvents where maximum yield of polysaccharide is desired. Swincer *et al.* (1968) found that pretreatment of soil with 1N HCl or HF (but not with 0.1N H_2SO_4) significantly increased polysaccharide yield in the basic extract. When the residue was acetylated, by means of acetic anhydride and concentrated H_2SO_4 at 60 °C for 2 hours, and extracted in chloroform, the yield of carbohydrate material was increased by 23 %. The amounts of carbohydrate extracted by the sequence 1N HCl, 0.5N NaOH, and the acetylated material ranged from 57 to 74 % of the sugar contents of the soils.

Soil polysaccharides can be contaminated by proteins or peptides, and amino acid residues are often reported in hydrolysate digests (e.g. Finch *et al.*, 1968; Swincer *et al.*, 1968). Deproteination can be achieved by the Sevag method

(Staub, 1965) as applied by Bernier (1958) and Mortensen (1960) or by trifluorotrichloroethane (Markowitz and Lange, 1965).

Removal of brown humic materials is a major problem in the purification of soil polysaccharides. There is evidence to indicate that these are bound to each other by hydrogen bonding, and by covalent bonds such as the phenolic glycoside linkage. The best methods so far applied for the removal of humic substances are gel chromatography (Hayes *et al.*, 1975b; Barker *et al.*, 1965, 1967), ion-exchange techniques and adsorption on polyclar-AT (a polyvinyl pyrrolidone product) or nylon-coated celite (Swincer *et al.*, 1968). Salts and low molecular weight materials can also be removed by gel chromatography by using a gel with a very low exclusion limit (see Section 3.5.2).

Fractionation of soil polysaccharides. Polysaccharides isolated from soil will contain a mixture of components of varying molecular weight, charge density, and monosaccharide composition. The use of paper and column electrophoresis techniques (Clapp, 1957) for fractionation of polysaccharides has been superceded by gel chromatography (Granath, 1965; Lee and Montgomery, 1965; Churms, 1970) for fractionations on the basis of molecular weight, and anion-exchange chromatography (Neukom and Kuendig, 1965) for separations on the basis of charge density. For gel column-chromatography dilute salt solutions should be used as eluents in order to suppress possible repulsion effects between the polysaccharide and residual charges on the gel. Buffered salt gradients are generally used as eluents in ion-exchange chromatography experiments. This allows the neutral components to emerge first, and the increasingly charged materials are eluted as the salt concentration in the gradient is increased (Barker *et al.*, 1967; Finch *et al.*, 1967, 1968; Hayes *et al.*, 1975b).

Sodium tetraborate complexes to varying extents with sugars (depending on the *cis–trans* relationships of their hydroxyls) to confer different charge characteristics to them. Clapp (1957) made use of this principle to isolate a neutral polysaccharide and, by column electrophoresis, to isolate and identify the monosaccharides in the hydrolysate digest. (Later Clapp and Davis, 1970, quantitatively estimated the sugars in a *Rhizobium* polysaccharide hydrolysate by this technique.) Similar types of complexes can be formed with the component sugars in polysaccharides, and these confer different charge characteristics to otherwise neutral polymer structures. Finch *et al.* (1967) have made use of this principle to check the homogeneity of neutral polysaccharides isolated by gel filtration and anion-exchange chromatography techniques. One polysaccharide, isolated from material precipitated by sodium bicarbonate, had a molecular weight (ultracentrifuge) of *ca.* 50,000, and moved as a single component during free boundary electrophoresis in borate (pH 9.1), phosphate (pH 7.0), and barbiturate (pH 9.0) buffers. The other neutral component stayed in solution in the bicarbonate medium, and moved as a single component during electrophoresis in phosphate and barbiturate buffers. In borate buffer, however, it

Table 3.6 Sugar components of soil polysaccharides and of soil polysaccharide fractions.

Reference	OM (%)	Fraction and MW data	Sugar composition (%)										
			Gl	Gal	Man	Xyl	Arab	Rib	Fuc	Rham	UA	AS	Others
Parsons and Tinsley (1961)													
Sandy loam	3.5	—	33.6	19.1	18.0	7.9	7.9	—	—	13.5	15.8	5.2	—
Meadow		—	37.7	20.1	18.2	8.2	7.6	—	—	8.2	17.9	5.0	—
Peat		—	26.8	20.8	15.8	8.9	7.9	—	—	18.8	16.0	3.3	—
Podzol A_o		—	33.6	23.1	17.9	7.5	6.7	—	—	11.2	16.3	3.1	—
Podzol B_h		—	31.8	18.2	20.5	9.1	6.8	—	—	13.6	13.5	4.3	—
Compost 1[a]		—	42.3	15.4	15.4	15.4	11.5	—	—	—	8.4	5.0	—
Compost 2[a]		—	47.2	16.7	13.9	13.9	8.3	—	—	—	6.7	4.3	—
Swincer *et al.* (1968)													
Old pasture	4.1	(HCl) 4×10^3–10^5 MW	28.1	21.0	19.4	5.1	7.7	0.4	7.8	10.5			
		(HCl) $>10^5$	31.5	15.5	20.8	6.5	7.1	0.5	5.5	12.5			
		(NaOH) $<4\times10^3$	27.4	11.1	26.8	11.0	12.2	4.7	7.5	9.1			
		(NaOH) 4×10^3–10^5	26.8	13.4	19.3	15.9	10.7	2.8	3.8	5.4			
		(NaOH) 10^5	32.9	15.3	17.2	11.0	6.5	0.4	7.5	9.1			
Fallow-wheat rotation	2.0	(HCl) $<4\times10^3$	63.1	9.7	7.3	3.1	12.6	1.4	1.4	1.4			
		(HCl) 4×10^3–10^5	44.4	22.1	21.7	3.0	1.7	0.1	3.3	3.7			
		(HCl) $>10^5$	30.2	10.3	22.9	10.6	5.2	0.5	8.7	11.7			
		(NaOH) $<4\times10^3$	36.6	10.1	24.7	8.3	9.0	3.4	3.4	4.1			
		4×10^3–10^5	36.5	15.2	20.9	12.6	9.2	1.1	1.7	2.8			
		$>10^5$	36.0	13.6	23.0	8.8	4.5	0.5	4.4	9.2			
Finch *et al.* (1968)													
		1	36.0	10.0	17.0	7.9	5.7	—	6.5	14.0	2.1	—	—
	80	2	60.0	6.0	11.0	6.9	2.9	—	10.2	0	3.0	—	—
		3	21.0	13.0	19.0	8.4	1.8	—	13.0	18.0	3.0	—	—
		4	22.0	17.0	6.4	4.2	13.0	—	13.0	11.0	14.0	—	—
Hayes *et al.* (1975b)													
		5	68.9	9.0	9.0	3.6	2.6	—	1.9	5.3	—	—	1.8
	80	6	32.3	21.6	8.7	18.8	10.3	—	3.3	4.4	—	—	1.0
		7	70.8	5.9	5.5	8.1	5.0	—	1.9	2.3	—	—	1.0
		8	19.5	31.8	17.3	7.9	4.0	1.5	3.5	12.7	—	—	1.5
		9	18.1	33.2	16.7	8.9	4.7	0.7	5.2	11.5	—	—	0.9
		10	21.3	24.3	16.9	11.2	8.3	0.6	7.5	9.1	—	—	1.0

Gl = glucose; Gal = galactose; Man = mannose; Xyl = xylose; Arab = arabinose; Rib = ribose; Fuc = fucose; Rham = rhamnose; UA = uronic acids; AS = amino sugars; HCl and NaOH refer to soil extractants; MW = molecular weight values estimated from gel filtration data; Fractions 1–10 were obtained by gel filtration and anion-exchange chromatography.

[a] Compost samples were taken after 11 and 48 months, respectively.

separated into four negatively charged peaks, and this was conclusive evidence for inhomogeneity.

The isolation of a pure soil polysaccharide is still a long and tedious process despite the advantages of modern laboratory technology. However, the task must not in any way be regarded as impossible. In fact some of the polymers already isolated could be regarded as sufficiently pure to justify the initiation of preliminary structural studies on them. Further purification improvements might be made by use of additional techniques employed in polysaccharide work, such as fractional precipitation, using quaternary ammonium salts (Scott, 1965) or barium hydroxide (Meier, 1965), prior to applications of ion-exchange and gel-chromatography. Also the use of borate buffers for anion-exchange chromatography would appear to offer a valuable technique for refined final stage fractionations. Considerations might also be given to the applications of immunological procedures whereby pure polysaccharides can be isolated from antigen–antibody precipitin-reaction products.

The composition of soil polysaccharides

Cheshire (1977), when providing data for the contents of sugars in the hydrolysate of a typical arable Scottish soil (5.6 % C), listed glucose, galactose, mannose, xylose, arabinose, fucose, rhamnose, glucuronic acid, galacturonic acid, glucosamine, and galactosamine as the digest monosaccharides. In addition to these, ribose is sometimes a hydrolysis product, although it invariably appears to be present in the lowest abundance. It would appear, from research reports for widely different soil types from different parts of the world, that the same sugars are present in all soils which contain organic matter.

Table 3.6 lists sugar contents of polysaccharide materials isolated from some British and Australian soils. The data are not complete in any instance. However those in the paper by Swincer *et al.* (1968), for the two Urrbrae fine sandy loam samples, indicate that amino acids and amino sugars amounted to 20–100 % and 10–15 % of the sum of the weights of the neutral sugars, respectively. The uronic acid contents for samples with molecular weights greater than 4000 were 30–40 % as large as the neutral sugar contents. These polysaccharides were extracted with HCl (1N) or NaOH (0.5N) solutions, and the data are quoted for fractions isolated by gel chromatography techniques.

Finch *et al.* (1968) estimated that amino acid contamination amounted only to 1–15 % of the sugar content of their polysaccharides isolated from the precipitate formed when the 0.6N H_2SO_4 soil extract was neutralized with sodium bicarbonate. The four polysaccharides in Table 3.6 were isolated by the anion-exchange chromatography of materials which were eluted as a reasonably discrete band during gel chromatography.

The six fractions described by Hayes *et al.* (1975b) were also obtained by anion exchange chromatography followed by treatment with Sephadex of samples which had previously been fractionated by gel chromatography on

Biogel. Extraction was with ethylenediamine and the polysaccharides were contained in the fulvic acid materials.

Parsons and Tinsley (1961) extracted their polysaccharides with anhydrous formic acid containing LiBr, and the polysaccharides were precipitated with cetavalon (a quaternary cetyl ammonium salt). This work was carried out before anion-exchange and gel chromatography techniques were used in soil polysaccharide studies. All of the Parsons and Tinsley polysaccharide materials were treated in the same way, and it can be seen that the sugar contents of the different soil extracts were quite similar. The plant origins of the composts are indicated by the relatively high glucose, xylose, and arabinose contents.

The polysaccharides of Swincer *et al.* (1968) were fractionated on the basis of molecular weight differences. It is clear from their data that the composition of the polysaccharides were influenced by cultivation practices. Differences between the compositions of the different molecular weight fractions and the influences of the extractants were most striking in the cases of samples from the cultivated soil. By contrast the various samples from the old pasture soil were very similar.

The data of Finch *et al.* (1968) and of Hayes *et al.* (1975b) as shown in Table 3.6, highlight the fact that the similarity in the molecular weights of soil polysaccharide extracts cannot be regarded as a criterion for homogeneity. Some of the polysaccharides of similar molecular sizes, which were fractionated by ion-exchange chromatography, were distinctly different from each other, although there were similarities between others. The high glucose contents of fractions 1, 6, and 10 might suggest plant origins, but this was not substantiated by the relative abundances of xylose and arabinose.

Where only small amounts of neutral sugars are present in the digests of extensively fractionated soil polysaccharides it is tempting to infer that these arise from contaminations by other polysaccharides. However, it will be seen from an examination of the relative abundances of the sugars in all of the hydrolysates in Table 3.6 that it is unlikely that there are fewer than four sugar residues in the majority of soil polysaccharides.

3.6.4 Persistence of Polysaccharides in Soil

Reese (1968) and Finch *et al.* (1971) have discussed some of the features of enzymatic transformations which may be involved in the degradation of soil polysaccharides. We will only briefly refer here to some of the concepts which are involved in these degradations.

Polysaccharide biosynthesis and degradation is unified by the glycosyl transfer reaction which can be expressed by

$$\underset{\text{(donor)}}{\text{G—OR}} + \underset{\text{(acceptor)}}{\text{R'O—H}} \longrightarrow \text{G—OR'} + \text{RO—H} \qquad (3.62)$$

In the biosynthesis reaction the acceptor is the growing polysaccharide chain, R′OH, and in degradation the acceptor is a small molecule such as water. The glycosyl group G is transferred in each case. Thus the degradation reaction may be represented as

$$G_1—O—G_2 + HOH \longrightarrow G_1OH + HOG_2 \tag{3.63}$$

Cleavage can take place as indicated by the broken lines in (a) and (b) in scheme (3.64), but almost invariably glycone–oxygen fission (a) occurs.

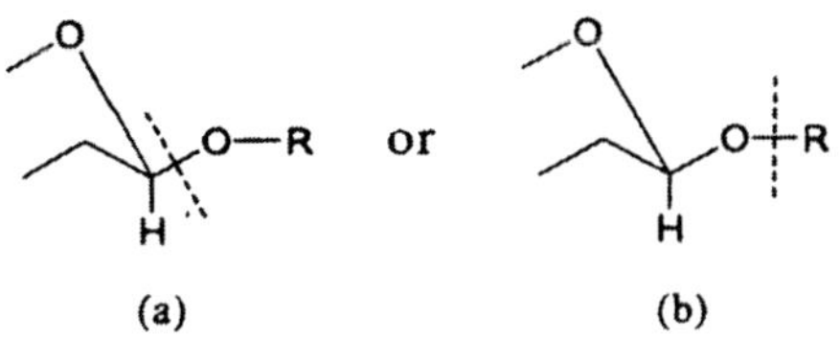

(3.64)

Two classes of degradative enzymes exist; one class (exoenzymes) catalyses the cleavage of the non-reducing terminal residues, and the other (endo-enzymes) that of glycosidic bonds throughout the polymer chains. It is easy to distinguish between these two classes because molecular weight is more readily decreased by the action of exoenzymes. However, their action can be blocked by the presence of bulky substituents, such as *O*-acetyl groups on the sugar hydroxyls, which inhibit the approach of the enzymes to the glycosidic bonds. In many cases minor modifications in the donor glycone, such as inversion of configuration at one carbon atom leads to greatly reduced activity. However, the influence of the different types of linkages is most important in so far as donor aglycone specificity in hydrolysis is concerned. Thus branching, and changes from 1 → 4 to 1 → 6 linkages, as illustrated in Figure 3.20 will give rise to different stereochemical arrangements in the vicinity of the glycosidic linkages, and will require different enzyme systems for cleavage. It should also be remembered that all glycosidases are specific for the α or β configuration at the anomeric carbon atom.

We concluded in Section 3.6.2 that it is highly likely that most of the polysaccharides synthesized in soil contain four or more sugar residues. The greater the variety of residues and of linkages the greater should be the resistance of the polysaccharide to enzymatic attack. Unfortunately lack of structural information at this time does not allow predictions to be made on the basis of these criteria. However, it is highly likely that a variety of enzymes are required in order to achieve extensive depolymerization. Cheshire and Anderson (1975) have shown that single purified enzymes have little degradative influence on soil polysaccharides and this is understandable in the light of the foregoing considerations.

Finch *et al.* (1971) have referred to unpublished results which indicated that polysaccharide mixtures from an organic soil were less susceptible to decomposition (as measured by CO_2 evaluation) than glucose, sucrose, cellulose, and ryegrass when each of these different materials was supplied as the sole source of carbon for microorganisms in aerated liquid culture media. Some of the polysaccharides decompose quite readily, as was shown by Cheshire *et al.* (1974) when a large proportion of their alkali-soluble soil polysaccharide was decomposed in 4–8 weeks during incubations with fresh soil suspensions.

It would appear, however, that soil polysaccharides are protected to some extent in their native environments. Among the reasons offered for this are (Cheshire, 1977): (i) adsorption by clays resulting in a degree of inaccessibility to enzymes and microorganisms; (ii) a preservative effect from tannins; and (iii) the formation of complexes with metals.

Lynch and Cotnoir (1956) have provided evidence to show that adsorption on montmorillonite slows microbial degradation of polysaccharides, and Finch *et al.* (1967) and others have observed interlamellar adsorption of soil polysaccharide by Na^+-montmorillonite. Such intimate associations with the clays would alter the conformations of the molecules and inhibit the approach of the hydrolytic enzymes to the glycosidic linkages. Chemical degradation is also slowed as the result of adsorption, and Finch *et al.* (1967) have shown that uptake of periodate by a kaolinite adsorbed soil polysaccharide was ten times slower than that by the polysaccharide in solution. Arguments along these lines might also apply for polysaccharides adsorbed to humic substances or bound to them by Ca^{2+} or Mg^{2+} bridging (Barker *et al.*, 1967), or entrapped within coiled humic acid molecules.

The preservative effects of tannins, as reviewed by Cheshire (1977) might be regarded as an extension of this argument. Tannins have some of the structural features of humic substances and could form covalent (e.g. phenolic glycosides), ionic, and hydrogen bonded linkages with the polysaccharide.

Martin *et al.* (1966) proved that the resistance to decomposition of some polysaccharides in the soil is increased by complexing with Fe, Cu, and Zn. This might be related to inhibition of hydrolytic enzymes and/or to changes in polymer conformations resulting from bridging by the metals of the carboxyl groups of the uronic acid residues.

3.6.5 Soil Polysaccharide–Clay Interactions

The beneficial effects of polysaccharides on soil structure can almost certainly be attributed to their interactions with clays. Some emphasis is given to the mechanisms of polymer adsorption by clays in the chapter on 'Adsorption' by Burchill, Greenland, and Hayes in the companion volume of this book, and more detailed attention is given there to clay–polysaccharide interactions. Here we will be concerned in a more general way with the adsorption of polysaccharides of known structures, and of soil polysaccharides with reasonably well-defined clay preparations.

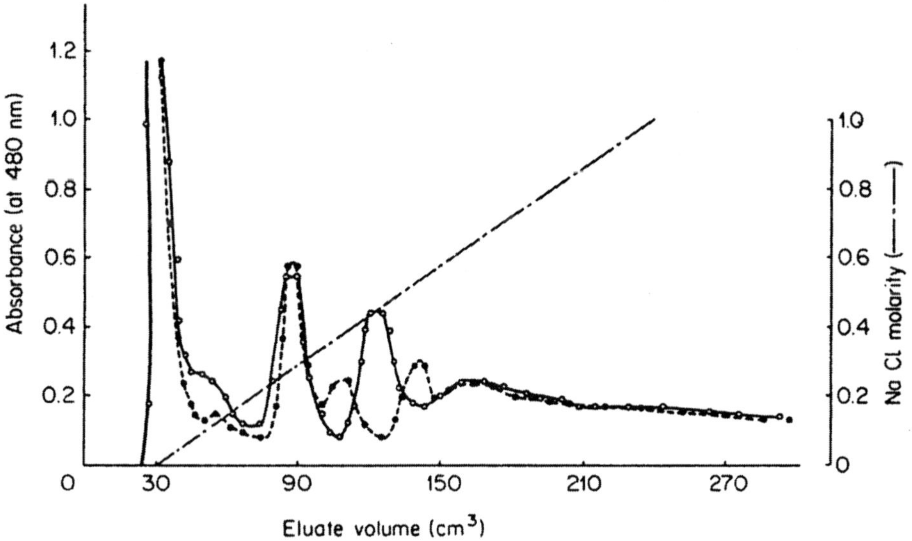

Figure 3.22 Anion-exchange chromatography on DEAE-A50 Sephadex in *p*H 6.0 phosphate buffer of a soil polysaccharide mixture before (—o—o—) and after (—•—•—) treatment with H^+-montmorillonite. (After Finch *et al.*, 1967)

Clapp and his associates (Clapp *et al.*, 1968; Clapp and Emerson, 1972; Olness and Clapp, 1973, 1975) have carried out a number of studies on the adsorption by montmorillonite of a variety of microbial polysaccharides whose structures are understood. These included an extracellular rhizobial polysaccharide and two dextrans, B-512F, and a commercial product 'Polytran' with molecular weights of the order of 2×10^6. The dextrans differed in the nature of the linkages of the glucose units; B-512F (signifying the NRRL code for the synthesizing strain of *Leuconastoc mesenteroides*) and 'Polytran' were linked 95 % α-1 → 6, 5 % α-1 → 3 and 75 % β-1 → 3, 25 % β-1 → 6, respectively.

Olness and Clapp (1973, 1975) have clearly shown that the nature of these linkages significantly affected the amounts (and hence the conformations) of polymers adsorbed on the clay surfaces. In the case of B-512F almost all of the adsorbed dextran could lie on the Na^+-montmorillonite surface. Polytran, however adsorbed significantly in excess of the amount necessary to cover the adsorbent surface and this would suggest the presence of loop and train conformations. The authors concluded that the availability of the primary alcohol groups (because of the lesser involvement of the C-6 hydroxyl in linkages) in Polytran enabled this polymer to form a stable complex with the clay from a shorter length of polymer chain than for B-512F.

It will not be possible to make comparable predictions about the mechanisms of soil polysaccharide–clay interactions until more is known about the polymer structures. However, there is ample evidence from different laboratories to prove that at least some soil polysaccharides are adsorbed by clays. Data from Finch *et al.* (1967), summarized in Figure 3.22, show that all of the polysaccharides which they isolated were not adsorbed to the same extents by a resin treated H^+-montmorillonite clay preparation. Comparison of the elution patterns of polymer from DEAE-A50 anion-exchange Sephadex before and after reaction with the clay shows that most of the component eluted at *ca.* 120 cm^3 (in a sodium chloride gradient in pH 6.0 phosphate buffer) was adsorbed by the clay. Other work by Barker *et al.* (1967) suggested that this component contained one uronic acid unit per 6 to 7 sugar residues. More highly charged components (eluted at 130–250 cm^3) were not adsorbed. Only a small proportion of the charge on these components was, however, contributed by uronic acids. Very little adsorption was observed for the essentially neutral (eluted at 25–40 cm^3) and lesser charged (75–100 cm^3) polysaccharide components.

The study by Lynch *et al.* (1956) of the adsorption of cellulose preparations, starch dextrins, glycogen, and corn polysaccharides by H^+- and Ca^{2+}-clay preparations showed that these polymers were adsorbed to different extents. Carboxymethylcellulose ($\alpha\,\beta$-1 $\rightarrow$ 4 linked polyglucose structure in which the —OH substituent on C-6 of glucose is substituted by —COOH) was the least adsorbed of the cellulose derivatives by H^+-montmorillonite. Also only small amounts of alginates and pectinates (charged polysaccharides) were adsorbed by H^+- and Na^+-montmorillonite. At the low pH values of the H^+-clay medium dissociation of the uronic acid carboxyls would not be expected, and some hydrogen bonding (as suggested from infrared evidence by Lynch *et al.*, 1956) between the clays and acidic polysaccharides might be predicted. But the failure of the highly acidic polysaccharides to be extensively adsorbed would suggest that hydrogen bonding was inhibited in these instances.

It is not necessary, however, to invoke specific mechanisms such as hydrogen bonding for polymer adsorptions by clays. As Greenland (1965) and Parfitt and Greenland (1970) have stated, multiple dispersion forces, arising from the numerous points of contact per polymer segment, as well as entropy factors may be most important. The increased entropy which would favour the reaction could be provided by displacement of a number of small molecules, such as water, by a single macromolecule.

Therefore it would appear that the ability of any (soluble) polysaccharide to adsorb on clay may be governed by the stereochemistry of the polymer because this can determine the extent to which the macromolecules can make contacts with the surfaces. Also the conformations of the polymers in the adsorbed states may well determine whether or not a particular soil polysaccharide will stabilize aggregates. For instance part of the contribution to aggregate stabilisation might arise from polysaccharide loops from one adsorbing surface forming trains on another (see Chapter by Burchill, Greenland and Hayes in the companion volume). However, such proposals must be regarded as speculative until results from further adsorption studies are available.

REFERENCES

Achard, F. (1786), Chemische Untersuchung des Torfs. *Grell's Chem. Ann.*, **2**, 391.

Ackers, G. K. (1964), Molecular exclusion and restricted diffusion processes in molecular-sieve chromatography, *Biochemistry*, **3**, 723–730.

Alexandrova, L. N. (1960), The use of sodium pyrophosphate for isolating free humic substances and their organic mineral compounds from the soil, *Soviet Soil Sci.*, **2**, 190–197.

Aspinall, G. O. (1976), Polysaccharide methodology, in *Organic Chemistry* **Ser 3, Vol. 7**, *Carbohydrates*, Butterworths, London, Boston, pp. 201–221.

Atherton, N. M., Cranwell, P. A., Floyd, A. J. and Haworth, R. D. (1967), Humic acid—I. ESR spectra of humic acids, *Tetrahedron*, **23**, 1653–1667.

Bailey, N. T., Briggs, G. C., Lawson, G. J., Scruton, J. H. and Ward, S. G. (1965), Observation on the structure of humic acid, *VIth Intern. Kohlenwiss. Tagung*, (Münster), Beitrag Nr. 3.

Barker, S. A., Finch, P., Hayes, M. H. B., Simmonds, R. G. and Stacey, M. (1965), Isolation and preliminary characterization of soil polysaccharides, *Nature* (London), **205**, 68–69.

Barker, S. A., Hayes, M. H. B., Simmonds, R. G. and Stacey, M. (1967), Studies on soil polysaccharides, I, *Carbohyd. Res.*, **5**, 13–24.

Barton, D. H. R. and Schnitzer, M. (1963), A new experimental approach to the humic acid problem, *Nature* (London), **198**, 217–218.

Bazilevich, N. I. (1974), Soil-forming role of substance and energy exchange in the soil–plant system, *Trans. 10th Intern. Congr. Soil Sci.* (Moscow), **6**, 17–27.

Behrman, E. J. and Edwards, J. O. (1967), in A. Streitwieser and R. W. Toft (eds.), *Progress in Physical Organic Chemistry*, p. 93, Interscience, New York.

Belcher, R. (1948), The anodic oxidation of coal. Parts I–V, *J. Soc. Chem. Ind.*, **67**, 213–221 and 265–270.

BeMiller, J. N. (1967), Acid-catalyzed hydrolysis of glycosides, *Advan. Carbohyd. Chem.*, **22**, 25–108.

Bentley, R. and Campbell, I. M. (1968), Secondary metabolism of fungi, in E. H. Stotz (ed.), *Comprehensive Biochemistry*, Vol. 20, pp. 415–483, Elsevier, Amsterdam.

Berger, St. and Rieker, A. (1974), Identification and determination of quinones, in S. Patai (ed.), *The Chemistry of the Quinonoid Compounds*, **Part 1**, pp. 163–229, Wiley, New York and London.

Bernier, B. (1958), Characterisation of polysaccharides isolated from forest soils, *Biochem. J.*, **70**, 590–598.

Billmeyer, Jr., F. W. (1971), *A Textbook of Polymer Science*, Wiley, New York and Toppan, Tokyo.

Bitter, T. and Muir, M. H. (1962), A modified uronic acid carbazole reaction, *Anal. Biochem.*, **4**, 330–334.

Björndal, H. and Lindberg, B. (1969), Polysaccharides elaborated by *Polyporus fomentarius* (Fr.) and *Polyporius igniarius* (Fr.) I. Water soluble neutral polysaccharides from the fruit bodies, *Carbohyd. Res.*, **10**, 79–85.

Björndal, H., Lindberg, B., Lönngren, J., Méssáros, M. and Thompson, J. L. (1973), Structural studies of the capsular polysaccharide of *Klebsiella* Type 52, *Carbohyd. Res.*, **31**, 93–100.

Blackall, E. L., Hughes, E. D. and Ingold, C. K. (1952), Kinetics and mechanism of aromatic nitration. IX. Nitrosating agents in nitration catalysed by HNO_2, *J. Chem. Soc.*, 28–32.

Blom, L., Edelhausen, L. and van Krevelen, D. W. (1957), Chemical studies on the properties of coal. XVIII—Oxygen groups in coal and related products, *Fuel*, **36**, 135–153.

Bondietti, E., Martin, J. P. and Haider, K. (1971), Influence of nitrogen source and clay on growth and phenolic polymer production by *Stachybotrys* species, *Hendersonula toruloidea* and *Aspergillus sydowi*, *Soil Sci. Soc. Amer. Proc.*, **35**, 917–922.

Bone, W. A. and Himus, G. W. (1936), *Coal, its Constitution and Uses*, Longman Green, London.

Bone, W. A., Parsons, L. G. B., Sapiro, R. H. and Groocock, C. M. (1935), Researches on the chemistry of coal: VIII. The development of benzenoid constitution in the lignin-peat-coal series, *Proc. Roy. Soc.* (London), **A148**, 492–522.

Bone, W. A., Horton, L. and Ward, S. G. (1930), Researches on the chemistry of coal. VI. Its benzenoid constitution as shown by its oxidation with alkaline permanganate, *Proc. Roy. Soc.* (London), **A127**, 480–510.

Boos, R. N. (1964), Rapid thermal conductivity microanalytical method for combined oxygen determination, *Microchem. J.*, **8**, 389–94.

Bouveng, H. O. and Lindberg, B. (1961), Methods in structural polysaccharide chemistry, *Advan. Carbohyd. Chem.*, **15**, 53–89.

Bracewell, J. M. and Robertson, G. W. (1973), Humus type discrimination using pattern recognition of the mass spectra of volatile pyrolysis products, *J. Soil Sci.*, **24**, 421–428.

Bracewell, J. M. and Robertson, G. W. (1976), A pyrolysis–gas chromatography method for discrimination of soil humus types, *J. Soil Sci.*, **27**, 196–205.

Braun, D., Cherdron, H. and Kern, W. (1971), *Techniques of Polymer Synthesis and Characterisation*, Wiley-Interscience, New York.

Brauns, F. E. (1960), *The Chemistry of Lignin*, Supplement. Volume, Academic Press, New York.

Bremner, J. M. (1950), Some observations on the oxidation of soil organic matter in the presence of alkali, *J. Soil Sci.*, **1**, 198–204.

Bremner, J. M. (1967), Nitrogeneous compounds, in A. D. McLaren and G. H. Peterson (eds.), *Soil Biochemistry*, **Vol. 1**, pp. 19–66, Marcel Dekker, New York.

Bremner, J. M. and Lees, H. (1949), Studies on soil organic matter: II. The extraction of organic matter from soil by neutral reagents, *J. Agric. Sci.* **39**, 274–279.

Briggs, G. C. and Lawson, G. J. (1970), Chemical constitution of coal: 16. Methylation studies on humic acid, *Fuel*, **49**, 39–48.

Brooks, J. D. and Sternhell, S. (1957), Chemistry of brown coals. I. Oxygen-containing functional groups in Victorian brown coals, *Aust. J. Appl. Sci.*, **8**, 206–221.

Brooks, J. D., Durie, R. A. and Sternhell, S. (1957), Chemistry of brown coals. I. Oxygen-containing functional groups in Victorian brown coals, *Aust. J. Appl. Sci.*, **8**, 206–221.

Brooks, J. D., Durie, R. A. and Sternhell, S. (1958), Chemistry of brown coals. III. Pyrolytic reactions. *Aust. J. Appl. Sci.*, **9**, 303–320.

Brown, H. C. (1961), New selective reducing reagents, *J. Chem. Education*, **38**, 173–179.

Brown, J. K. and Wyss, W. F. (1955), Oxygen groups in bright coals, *Chem. Ind.* (London), 1118.

Bühler, C. and Simon, W. (1970). Curie point pyrolysis gas chromatography, *J. Chromatogr. Sci.*, **8**, 323–329.

Burdon, J., Craggs, J. D., Hayes, M. H. B. and Stacey, M. (1974), Reactions of sodium sulphide: I. With compounds containing hydroxyl groups, *Tetrahedron*, **30**, 2729–2733.

Burges, N. A., Hurst, H. M. and Walkden, B. (1964), The phenolic constituents of humic acid and their relation to the lignin of plant cover, *Geochim. Cosmochim. Acta.*, **28**, 1547–1554.

Butler, J. H. A. and Ladd, J. N. (1969), Effect of extractant and molecular size on the optical and chemical properties of soil humic acids, *Aust. J. Soil Res.*, **7**, 229–239.

Cacco, G., Maggioni, A. and Ferrari, G. (1974), Electrofocusing: A new method for characterization of soil humic matter, *Soil Biol. Biochem.*, **6**, 145–148.

Cameron, A. E. (1938), Oxidation-reduction potentials of unstable organic system, *J. Phys. Chem.*, **42**, 1217–27.

Cameron, R. S. and Posner, A. M. (1974), Molecular weight distribution of humic acid from density gradient ultracentrifugation profiles corrected for diffusion, *Trans. 10th Intern. Congr. Soil Sci.* (Moscow), **2**, 325–331.

Cameron, R. S., Swift, R. S., Thornton, B. K. and Posner, A. M. (1972a), Calibration of gel permeation chromatography materials for use with humic acid, *J. Soil Sci.*, **23**, 342–349.

Cameron, R. S., Thornton, B. K., Swift, R. S. and Posner, A. M. (1972b), Molecular weight and shape of humic acid from sedimentation and diffusion measurements on fractionated extracts, *J. Soil Sci.*, **23**, 394–408.

Chahal, K. S. (1968), Biosynthesis and characterization of soil polysaccharides, in *Isotopes and Radiation in Soil Organic Matter Studies*, IAEA, Vienna, pp. 207–218.

Chang, H.-M. and Allan, G. G. (1971), Oxidation, in K. V. Sarkanen and C. H. Ludwig (eds.), *Lignins*, pp. 433–485, Wiley, New York and Chichester.

Chen, Y. and Schnitzer, M. (1976), Viscosity measurements on soil humic substances, *Soil Sci. Soc. Am. J.*, **40**, 866–875.

Cheshire, M. V. (1977), Origins and stability of soil polysaccharide, *J. Soil Sci.*, **28**, 1–10.

Cheshire, M. V. and Anderson, G. (1975), Soil polysaccharides and carbohydrate phosphates, *Soil Sci.*, **119**, 356–362.

Cheshire, M. V., Cranwell, P. A., Falshaw, C. P., Floyd, A. J. and Haworth, R. D. (1967), Humic acid—II. Structure of humic acids, *Tetrahedron*, **23**, 1669–1682.

Cheshire, M. V., Cranwell, P. A. and Haworth, R. D. (1968), Humic Acid—III. *Tetrahedron*, **24**, 5155–5167.

Cheshire, M. V. and Mundie, C. M. (1966), The hydrolytic extraction of carbohydrates from soil by sulphuric acid, *J. Soil Sci.*, **17**, 372–381.

Cheshire, M. V., Mundie, C. M. and Shepherd, H. (1969), Transformation of ^{14}C glucose and starch in soil, *Soil Biol. Biochem.*, **1**, 117–130.

Cheshire, M. V., Greaves, M. P. and Mundie, C. M. (1974), Decomposition of soil polysaccharide, *J. Soil Sci.*, **25**, 483–498.

Chesters, G., Attoe, O. J. and Allen, O. N. (1957), Soil aggregation in relation to various soil constituents, *Soil Sci. Soc. Amer. Proc.*, **21**, 272–277.

Choudri, M. B. and Stevenson, F. J. (1957), Chemical and physicochemical properties of soil colloids. III. Extraction of organic matter from soils, *Soil Sci. Soc. Amer. Proc.*, **21**, 508–513.

Churms, S. C. (1970), Gel chromatography of carbohydrates, *Advan. Carbohyd. Res.*, **25**, 13–51.

Clapp, C. E. (1957), *High molecular weight water-soluble muck. Isolation and determination of constituent sugars of a borate complex forming polysaccharide, employing electrophoretic techniques*, Ph.D. Thesis, Cornell University.

Clapp, C. E. and Davis, R. J. (1970), Properties of extracellular polysaccharides from *Rhizobium*, *Soil Biol. Biochem.*, **2**, 109–117.

Clapp, C. E. and Emerson, W. W. (1965), The effect of periodate oxidation on the strength of soil crumbs. I. Qualitative studies. II. Quantative studies, *Soil Sci. Soc. Amer. Proc.*, **29**, 127–134.

Clapp, C. E. and Emerson, W. W. (1972), Reactions between Ca-montmorillonite and polysaccharides, *Soil Sci.*, **114**, 210–216.

Clapp, C. E., Olness, A. E. and Hoffmann, D. J. (1968), Adsorption studies of a dextran on montmorillonite, *Trans. 9th Intern. Congr. Soil Sci.* (Adelaide), **1**, 627–637.

Coffey, S. (ed.), (1967), *Rodd's Chemistry of Carbon Compounds, Vol. 1, Aliphatic Compounds, Part F*, Elsevier, Amsterdam.

Craggs, J. D. (1972), *Sodium sulphide reactions with humic acid and model compounds*, Ph.D. Thesis, University of Birmingham.

Craggs, J. D., Hayes, M. H. B. and Stacey, M. (1974), Sodium sulphide reactions with humic acid and model compounds, *Trans. 10th Intern. Congr. Soil Sci.* (Moscow), **2**, 318–324.

Davidson, E. A. (1967), *Carbohydrate Chemistry*, Holt, New York.

Dence, C. W. (1971), Halogenation and nitration, in K. V. Sarkanen and C. H. Ludwig (eds.), *Lignins*, pp. 373–432, Wiley, New York and Chichester.

Deuel, H. and Dubach, P. (1958a), Decarboxylierung der organischen Substanz des Bodens. II. Nachweis von Uronsäuren, *Z. Pflanzenernähr. Düng. Bodenk.*, **82**, 97–106.

Deuel, H. and Dubach, P. (1958b), Decarboxylierung der organischen Substanz des Bodens. III. Extraktion and Fraktionierung decarboxylierbar Humusstoffe, *Helv. Chim. Acta*, **41**, 1310–1321.

Dormaar, J. F. (1969), Reductive cleavage of humic acids of chernozem soils, *Plant Soil*, **31**, 182–184.

Drozdova, T. V. (1959), Chitin and its transformation in natural processes. Melanoiden formation, *Uspekhi Souremennoi Biol.*, **47**, 277–296.

Dryden, J. G. C. (1952), Solvent power for coals at room temperature. Chemical and physical factors. *Chem. Ind.* (London), 502–508.

Dubach, P. and Mehta, N. C. (1963), The chemistry of soil humic substances, *Soils Fertil.*, **26**, 293–300.

Dubach, P., Mehta, N. C., Jakab, T., Martin, F. and Roulet, N. (1964), Chemical investigations on soil humic substances, *Geochim. Cosmochim. Acta*, **28**, 1567–1578.

Dutton, G. G. S. (1974), Applications of gas–liquid chromatography to carbohydrates, II. *Advan. Carbohyd. Chem. Biochem.*, **30**, 9–110.

Eller, W. (1925), Zur Frage der Konstitution der naturlichen Huminsäuren, *Brennstoff-Chem.*, **6**, 52–54.

Ellis, G. P. (1959), The Maillard reaction, *Advan. Carbohyd. Chem.*, **14**. 63–134.

Enders, C. (1943a), Wie ensteht der Humus in der Natur, *Die Chemie*, **56**, 281–292.

Enders, C. (1943b), Über den Chemismus der Huminsäurebildung unter physiologischen Bedingungen. IV. Mitt. Die Rolee der Mikroorganismen bei den Humifizierungs-vorgangen, *Biochem. Z.*, **315**, 259–292.

Enders, C. and Fries, G. (1936), Zur Analogie von Melanoidinen und Huminsauren, *Kolloid Z.*, **76**, 289–291.

Enders, C. and Sigurdsson, S. (1948), Über den Chemismus der Huminsäurebildung unter physiologischen Bedingungen. VII Mitt. *Biochem. Z.*, **318**, 44–46.

Enders, C., Tschapek, M. and Glane, R. (1948), Vergleichende Untersuchungen einiger kolloider Eigenschaften von natürlichen Huminsäuren und synthetischen Melanoidinen, *Kolloid Z.*, **110**, 240–244.

Farmer, V. C. and Morrison, R. I. (1960), Chemical and infrared studies on phragmites peat and its humic acid, *Sci. Proc. Roy. Dublin Soc.*, **Ser A1**, 85–104.

Felbeck, G. T. (1965a), Studies on high pressure hydrogenolysis of organic matter from a muck soil, *Soil Sci. Soc. Amer. Proc.*, **29**, 48–55.

Felbeck, G. T. (1965b), Chemistry of humic substances, *Advan. Agron.*, **17**, 327–368.

Felbeck, G. T. (1967), Normal alkanes in muck soil organic-matter hydrogenolysis products, *Int. Soil Sci. Soc., Trans II and IV Comm.* 1966 (Aberdeen), pp. 11–17.

Finch, P., Hayes, M. H. B. and Stacey, M. (1967), Studies on soil polysaccharides and their interactions with clay preparations, *Int. Soil Sci. Soc., Trans. II and IV Comm.*, 1966 (Aberdeen), pp. 19–32.

Finch, P., Hayes, M. H. B. and Stacey, M. (1968), Studies on the polysaccharide constituents of an acid extract of a Fenland muck soil, *Trans. 9th. Int. Congr. Soil Sci.*, **3**, 193–201.

Finch, P., Hayes, M. H. B. and Stacey, M. (1971), The biochemistry of soil polysaccharides, in A. D. McLaren and J. J. Skujins (eds.), *Soil Biochemistry, Vol. 2*, pp. 254–319, Marcel Dekker, New York.

Finley, K. T. (1974), The addition and substitution chemistry of quinones, in S. Patai (ed.), *The Chemistry of the Quinonoid Compounds, Part 2*, pp. 879–1144, Wiley, New York and Chichester.

Fischer, L. (1969), An introduction to gel chromatography, in T. S. Work and E. Work (eds.), *Laboratory Techniques in Biochemistry and Molecular Biology*, **Vol 1**, Part II, pp. 157–390, North Holland Publishing Co. Amsterdam.

Flaig, W. (1950), Zur Kenntnis der Huminsäuren. 1. Zur chemischen Konstitution der Huminsäuren, *Z. Pflanzenernahr. Dung. Bodenk.*, **51**, 193–212.

Flaig, W. (1960), Comparative chemical investigations on natural humic compounds and their model substances, *Sci. Proc. Roy. Dubl. Soc.*, **Ser. A1**, 149–162.

Flaig, W. (1966), Chemistry of humic substances in relation to coalification, *Advan. in Chem. Ser.*, **55**, 58–68.

Flaig, W. and Beutelspacher, H. (1951), Zur Kenntnis der Haminsäuren. 2. Elektronenmikroskopische Untersuchungen an natürlichen und synthetischen Huminsäuren, *Z. Pflanzenernahr Dung. Bodenk*, **52**, 1–21.

Flaig, W. and Beutelspacher (1954), Physikalische Chemie der Huminsäuren, *Landbouwk. Tijdschr.*, **66**, 306–336.

Flaig, W. and Beutelspacher, H. (1968), Investigations of humic acids with the analytical ultracentrifuge, in *Isotopes and Radiation in Soil Organic Matter Studies*, IAEA Symposium (Vienna), pp. 23–30.

Flaig, W., Beutelspacher, H., Riemer, H. and Kalke, E. (1969), Effect of substituents on the redox potential of substituted 1,4-benzoquinones, *Ann. Chem.*, **719**, 95–111.

Flaig, W., Beutelspacher, H. and Rietz, E. (1975), Chemical composition and physical properties of humic substances, in J. E. Gieseking (ed.), *Soil Components*, Vol. 1, pp. 1–219, Springer-Verlag, Berlin, New York.

Flaig, W. and Salfeld, J. Chr. (1958), UV-Spektren und Konstitution von *p*-Benzochinonen, *Liebigs. Ann. Chem.*, **618**, 117–139.

Flaig, W. and Salfeld, J. Chr. (1960a), Zwischenstufen dei der Bildung von Huminsäuren aus Phenolen, *Trans. 7th. Intern. Congr. Soil Sci.* (Madison), **2**, 648–656.

Flaig, W. and Salfeld, J. Chr. (1960b), Nachweis der Bildung von Hydroxy-*p*-Benzochinon als Zwischenprodukt bei der Autoxydation von Hydrochinon in schwach alkalischer Losung, *Naturwissenschaften*, **47**, 516.

Flaig, W., Scheffer, F. and Klamroth, B. (1955), Zur Kenntnis der Huminsäuren. 8. Zur Charakterisierung der Huminsäuren des Bodens, *Z. Pflanzenernähr Düng. Bodenk.*, **7**, 33–37.

Flaig, W. and Schulze, H. (1952), Uber den Bildungsmechanismus der Synthesehuminsäuren, *Z. Pflanzenernähr Düng. Bodenk.*, **58**, 59–67.

Forsyth, W. G. C. (1947), Studies on the more soluble complexes of soil organic matter. 1. A method of fractionation, *Biochem. J.*, **41**, 176–181.

Foster, R. and Foreman, M. I. (1974), Quinone complexes, in S. Patai (ed.), *The Chemistry of the Quinonoid Compounds*, Part 1, pp. 257–333, Wiley, New York and Chichester.

Freudenberg, K., Lautsch, W. and Engler, K. (1940), Lignin, XXXIV. Formation of vanillin from spruce lignin, *Chem. Ber.*, **73B**, 167–171.

Fujii, S. (1963), Infra-red spectra of coal: The absorption band at 1600 cm^{-1}, *Fuel*, **42**, 17–23.

Ganz, E. (1944), Contribution a l'étude de la structure des acides humique naturels, *Ann. Chim.*, **19**, 202–216.

Gascho, G. J. and Stevenson, F. J. (1968), An improved method for extracting organic matter from soil, *Soil Sci. Soc. Amer. Proc.*, **32**, 117–119.

Geoghegan, M. J. and Brian, R. C. (1946), Influence of bacterial polysaccharides on aggregate formation in soils, *Nature* (London), **158**, 837.

Geoghegan, M. J. and Brian, R. C. (1948), Aggregate formation in soil. 2. Influence of various carbohydrates and proteins in aggregation of soil particles, *Biochem. J.*, **43**, 14.

Gierer, J. and Norén, I. (1962), Reactions of lignin on sulfate digestion. II. Model experiments on the cleavage of aryl alkyl ethers by alkali, *Acta Chem. Scand.*, **16**, 1713–29.

Goldstein, I. J., Hay, G. W., Lewis, B. A. and Smith, F. (1965), Controlled degradation of polysaccharides by periodate, reduction and hydrolysis, in R. L. Whistler (ed.), *Methods of Carbohydrate Chemistry*, **5**, 361–370.

Gomez Aranda, V., Osacar Flaquer, J. and Revuelta Blanco, G. (1972), Gas-chromatography study of the flash pyrolysis products of humic material, *An. Quim.*, **68**, 1407–1410 (*Chem. Abs.*, **78**, 149594q).

Gottlieb, S. and Hendricks, S. B. (1945), Soil organic matter as related to newer concepts of lignin chemistry, *Soil Sci. Soc. Amer. Proc.*, **10**, 117–125.

Granath, K. A. (1965), Gel filtration: Fractionation of dextran, in R. L. Whistler (ed.), *Methods in Carbohydrate Chemistry*, **5**, 20–28.

Green, J. W. (1963), Determination of carbonyl groups, in R. L. Whistler (ed.), *Methods in Carbohydrate Chemistry*, **3**, 49–54.

Greene, G. and Steelink, G. (1962), Structure of soil humic acid. II. Some copper oxide oxidation products, *J. Org. Chem.*, **27**, 170–174.

Greenland, D. J. (1965), Interaction between clays and organic compounds in soils, II. *Soils Fertil.* (Harpenden, U.K.), **28**, 521–532.

Greenland, D. J., Lindstrom, G. R. and Quirk, J. P. (1961), Role of polysaccharides in stabilisation of natural soil aggregates, *Nature* (London), **191**, 1283–1284.

Greenland, D. J., Lindstrom, G. R. and Quirk, J. P. (1962), Organic materials which stabilize natural soil aggregates, *Soil Sci. Soc. Amer. Proc.*, **26**, 366–371.

Greenland, D. J. and Oades, J. M. (1975), Saccharides, in J. E. Gieseking (ed.), *Soil Components*, Vol. 1, pp. 213–261, Springer Verlag, Berlin.

Griffith, S. M. and Schnitzer, M. (1976), The alkaline cupric oxide oxidation of humic and fulvic acids extracted from tropical volcanic soils, *Soil Sci.*, **122**, 191–201.

Gupta, U. C. (1967), Carbohydrates, in A. D. McLaren and G. H. Peterson (eds.), *Soil Biochemistry*, Vol. 1, pp. 91–118, Marcel Dekker, New York.

Gupta, U. C. and Sowden, F. J. (1965), Studies on methods for the determination of sugars and uronic acids in soils, *Can. J. Soil Sci.*, **45**, 237–240.

Guthrie, R. D. and Honeyman, J. (1968), *An Introduction to the Chemistry of Carbohydrates*, 3rd edition, Clarendon Press, Oxford.

Haider, K. and Martin, J. P. (1967), Synthesis and transformation of phenolic compounds by *Epicoccum nigrum* in relation to humic acid formation, *Soil Sci. Soc. Amer. Proc.*, **31**, 766–772.

Haider, K. and Martin, J. P. (1975), Decomposition of specifically carbon-14-labeled benzoic and cinnamic acid derivatives in soil, *Soil Sci. Soc. Amer. Proc.*, **39**, 657–662.

Haider, K., Nagar, B. R., Saiz, C., Meuzelaar, H. L. C. and Martin, J. P. (1977a), Studies on soil humic compounds, fungal melanins, and model polymers by pyrolysis mass spectrometry, *Proc. IAEA-FAO-Agrochemica Symposium* (Braunschweig, 1976), **2**, 213–219.

Haider, K., Martin, J. P. and Rietz, E. (1977b), Decomposition in soil of ^{14}C-labeled coumaryl alcohols; Free and linked into dehydropolymer and plant lignins and model humic acids, *Soil Sci. Soc. Amer. J.*, **41**, 556–562.

Hamaker, J. W. and Thompson, J. M. (1972), Adsorption, in C. A. I. Goring and J. W. Hamaker (eds.), *Organic Chemicals in the Soil Environment*. Part 1, pp. 49–143, Marcel Dekker, New York.

Hansen, E. H. and Schnitzer, M. (1966), The alkaline permanganate oxidation of Danish illuvial organic matter, *Soil Sci. Soc. Amer. Proc.*, **30**, 745–748.

Hansen, E. H. and Schnitzer, M. (1967), Nitric acid oxidation of Danish illuvial organic matter, *Soil Sci. Soc. Amer. Proc.*, **31**, 79–85.

Hansen, E. H. and Schnitzer, M. (1969a), Zinc-dust distillation of soil humic compounds, *Fuel*, **48**, 41–46.

Hansen, E. H. and Schnitzer, M. (1969b), Zinc-dust distillation and fusion of a soil humic acid, *Soil Sci. Soc. Amer. Proc.*, **33**, 29–36.

Hansen, E. H. and Schnitzer, M. (1969c), Molecular weight measurements of polycarboxylic acids in waters by vapour pressure osmometry, *Anal. Chim. Acta.*, **46**, 247–254.

Hästbacka, K. (1962), A kinetic study of the reactions of vanillyl alcohol, *Soc. Sci. Fenn. Comm. Phys.-Math.*, **26**, 1–51.

Haworth, R. D. (1971), The chemical nature of humic acid, *Soil Sci.*, **111**, 71–79.

Haworth, W. N., Pinkard, F. W. and Stacey, M. (1946), Function of bacterial polysaccharides in soil, *Nature* (London), **158**, 836–837.

Hayashi, T. and Nagai, T. (1961), On the components of soil humic acids (Part 8). The oxidative decomposition by alkaline potassium permanganate and nitric acid: *Soil and Plant Food*, **6**, 170–175.

Hayes, M. H. B. (1960), *Subsidence and humification in peats*, Ph.D. Dissertation, Ohio State University.

Hayes, M. H. B. (1970), Adsorption of triazine herbicides on soil organic matter, including a short review on soil organic matter chemistry, *Residue Reviews*, **32**, 131–174.

Hayes, M. H. B., Stacey, M. and Swift, R. S. (1972), Degradation of humic acid in a sodium sulphide solution, *Fuel*, **51**, 211–213.

Hayes, M. H. B., Swift, R. S., Wardle, R. E. and Brown, J. K. (1975a), Humic materials from an organic soil: A comparison of extractants and of properties of extracts, *Geoderma*, **13**, 231–245.

Hayes, M. H. B., Stacey, M. and Swift, R. S. (1975b), Techniques for fractionating soil polysaccharides, *Trans. 10th Intern. Congr. Soil Sci.* (Moscow), **Suppl. Vol.**, 75–81.

Helferich, F. (1962), *Ion Exchange*, McGraw-Hill, New York.

Herédy, L. A. and Neuworth, M. B. (1962), Low temperature depolymerization of bituminous coal, *Fuel*, **41**, 221–231.

Hildebrand, J. H. and Scott, R. L. (1962), *Regular Solutions*, Prentice-Hall, Inc, Englewood Cliffs, New Jersey.

Hodge, J. E. (1953), Chemistry of browning reactions in model systems, *J. Agric. Food Chem.*, **1**, 928–943.

Hubacher, M. H. (1949), Determination of carboxy groups in aromatic acids, *Anal. Chem.* **21**, 945–947.

Hurst, H. M. and Burges, N. A. (1967), Lignin and humic acids, in A. D. McLaren and G. H. Peterson (eds.), *Soil Biochemistry*, Vol. 1, pp. 260–286, Marcel Dekker, New York.

Imuta, K. and Ouchi, K. (1968), Separation of depolymerization products of Tempoku coal, *Nenroyo Kyokai-shi* (Japan), **47**, 888–93, (*Chem. Abs.*, **71**, 72716b, 1969).

Imuta, K. and Ouchi, K. (1969), Analysis of depolymerization products of Tempoku coal. II. *Nenroyo Kyokai-shi.* (Japan), **48**, 900–904 (*Chem. Abs.*, **72**, 134970, 1970).

Ivarson, K. C. and Sowden, F. J. (1962), Methods for the analysis of carbohydrate material in soil, *Soil Sci.*, **94**, 245–250.

Jackson, M. P., Swift, R. S., Posner, A. M. and Knox, J. R. (1972), Phenolic degradation of humic acid, *Soil Sci.*, **114**, 75–78.

Jakab, T., Dubach, P., Mehta, N. C. and Deuel, H. (1962), Abbau von Huminstoffen. 1. Hydrolyse mit Wasser und Mineralsäuren, *Z. Pflanzenernähr. Düng. Bodenk.*, **96**, 213–217.

Jakab, T., Dubach, P., Mehta, N. C. and Deuel, H. (1963), Abbau von Huminstoffen. 2. Abbau mit Alkali, *Z. Pflanzenernähr. Düng. Bodenk.*, **102**, 8–17.

Keefer, R. F. and Mortensen, J. L. (1963), Biosynthesis of soil polysaccharides. I.—Glucose and alfalfa tissue substrates, *Soil Sci. Soc. Amer. Proc.*, **27**, 156–160.

Kenyon, J. and Symons, M. C. R. (1953), The oxidation of carboxylic acids containing a tertiary carbon atom, *J. Chem. Soc.*, 2129–2132.

Khan, S. U. and Schnitzer, M. (1971). Further investigations of the chemistry of fulvic acid, a soil humic fraction. *Can. J. Chem.*, **49**, 2302–2309.

Khan, S. U. and Schnitzer, M. (1972). Permanganate oxidation of humic acids extracted from a gray wooded soil under different cropping systems and fertilizer treatments, *Geoderma*, **7**, 113–120.

Kimber, R. W. L. and Searle, P. L. (1970), Pyrolysis gas chromatography of soil organic matter. II. The effects of extractants and soil history on the yields of products from pyrolysis of humic acids. *Geoderma*, **4**, 57–71.

Kononova, M. M. (1961), *Soil Organic Matter. Its nature, its role in soil formation and in soil fertility*, Pergamon Press, Oxford.

Kononova, M. M. (1966), *Soil Organic Matter. Its nature, its role in soil formation and in soil fertility*, 2nd edition, Pergamon Press, Oxford.

Kononova, M. M. (1975), Humus of virgin and cultivated soils, in J. E. Gieseking (ed.), *Soil Components*, Vol. 1, pp. 475–526, Springer Verlag, Berlin, New York.

Kononova, M. M. and Bel'chikova, N. P. (1961), Rapid methods of determining the humus composition of mineral soils, *Soviet Soil Sci.* **10**, 75–87.

Kukharenko, T. A. and Savel'ev, A. S. (1951), Hydrogenation of humic acids on a nickel catalyst, *Dokl. Akad. Nauk. SSSR*, **76**, 77–80 (*Chem. Abs.*, **45**, 8451*h*, 1951).

Kukharenko, T. A. and Savel'ev, A. S. (1952), Neutral products from the hydrogenations of humic acids of various origins, *Dokl. Akad. Nauk. SSSR*, **86**, 729–732, (*Chem. Abs.* **47**, 8037*c*, 1953).

Kukharenko, T. A. and Kredenskaja, T. E. (1956), Exhaustive cleavage of humic acids with sodium in liquid ammonia, *Khim. i Tekhnol. Topliva*, **6**, 25–24 (*Chem. Abs.*, **50**, 16005, 1956).

Lai, Y. Z. and Sarkanen, K. V. (1971), Isolation and structural studies, in K. V. Sarkanen and C. H. Ludwig (eds.), *Lignins*, pp. 165–240. Wiley, New York and Chichester.

Lawson, G. J. and Purdie, J. W. (1966). Chemical constitution of coal XI—Estimation of sub-humic acids produced by ozonization of humic acid. *Fuel*, **45**, 115–130.

Lee, D. G. (1969), Hydrocarbon oxidation using transition metal compounds. In R. L. Augustine (ed.), *Oxidation* **Vol. 1**, pp. 1–51, Marcel Dekker, New York.

Lee, Y. C. and Montgomery, R. (1965), Separations with molecular sieves, in R. L. Whistler (ed.), *Methods in Carbohydrate Chemistry*, **5**, 28–34.

Lentz, H., Lüdemann, H.-D. and Ziechmann, W. (1977), Proton resonance spectra of humic acids from the solum of a podzol, *Geoderma*, **18**, 325–328.

Leopold, B. (1952), Lignin. XI. The alkaline nitrobenzene oxidation of lignin and lignin models, *Svensk Kem. Tid.*, **64**, 18–26.

Levy, G. C. and Nelson, G. L. (1972), *Carbon-13 Nuclear Magnetic Resonance for Organic Chemists*, Wiley, New York and Chichester.

Lewis, S. N. (1969), Peracid and peroxide oxidations, in R. L. Augustine (ed.), *Oxidation*, Vol. 1, Marcel Dekker, New York.

Lindsey, A. S. (1974), Polymeric quinones, in S. Patai (ed.), *The Chemistry of the Quinonoid Compounds*, Part 2, pp. 793–855, Wiley, New York and Chichester.

Lindberg, B., Lönngren, J. and Thompson, J. L. (1973), Degradation of polysaccharides containing uronic acid residues, *Carbohyd. Res.*, **28**, 351–357.

Lönngren, J. and Svensson, S. (1974), Mass spectrometry in structural analysis of natural carbohydrates, *Adv. Carbohyd. Chem. Biochem.*, **29**, 41–106.

Lobo, P. F. S., Flexor, J. M. Rapaire, J. L. and Sieffermann, G. (1974), Determination of humic fraction retention times of two ferrallitic soils utilizing natural and thermonuclear radiocarbon, *Cah. ORSTOM, Ser. Pedol.*, **12**, 115–123 (in French) (*Chem. Abs.*, **82**, 19623v, 1975).

Lynch, D. L. and Cotnoir, L. J. (1956), The influence of clay minerals in the breakdown of certain organic substances, *Soil Sci. Soc. Amer. Proc.*, **20**, 367–370.

Lynch, D. L., Wright, L. M. and Cotnoir, L. J. (1956), The adsorption of carbohydrates and related compounds on clay minerals, *Soil Sci. Soc. Amer. Proc.*, **20**, 6–9.

Lüdemann, H. D., Lentz, H. and Ziechmann, W. (1973), Protonenresonanspektroskopie von Ligninen und Huminsäuren bei 1000 Megahertz, *Erdöl Kohl-Erdg.-Petrochem. Verein. Brennstoff-Chem.*, **26**, 506–509.

MacDonald, A. M. G. (1965), The oxygen flask method, in C. N. Reilley (ed.), *Advan. in Anal. Chem. and Instrumentation*, **4**, 75–116.

Maillard, L. C. (1912), Action des acides amines sur les sucre; formation des melanoidines par voie methodique, *C.R. Acad. Sci.*, Paris, **154**, 66–68.

Maillard, L. C. (1916), Synthese des materies humiques par action des acides amines sur les sucres reductours, *Ann. Chim.*, **5**, 258–317.

Maillard, L. C. (1917), Identite des matieres humiques de synthese avec les matieres humiques naturelles, *Ann. Chim.*, **7**, 113–152.

Markowitz, A. S. and Lange, C. F. (1965), Removal of proteins with trifluoro-trichloroethane: Deproteinizing action of trifluorotrichloroethane and recovery of carbohydrate-rich materials from the aqueous phase, in R. L. Whistler (ed.), *Methods in Carbohydrate Chemistry*, **5**, 6–8.

Marshall, J. J. (1974), Application of enzymic methods to the structural analysis of polysaccharides: I. *Advan. Carbohyd. Chem. Biochem.*, **30**, 257–370.

Martin, F. (1975), Pyrolysis gas chromatography of humic substances from different origin, *Z. Pflanzenern. Bodenk.*, **4/5**, 407–416.

Martin, F. (1976), Effects of extractants on analytical characteristics and pyrolysis gas chromatography of podzol fulvic acids, *Geoderma*, **15**, 253–265.

Martin, F. E., Dubach, P., Mehta, N. C. and Deuel, H. (1963), Bestimmung fur funktionellen Gruppen von Huminstoffen, *Z. Pflanzenernähr. Düng. Bodenk.*, **103**, 29–39.

Martin, J. P. (1945), Microorganisms and soil aggregation. I. Origin and nature of some of the aggregating substances, *Soil Sci.*, **59**, 163–174.

Martin, J. P. (1946), Microorganisms and soil aggregation. II. Influence of bacterial polysaccharides on soil structure, *Soil Sci.*, **61**, 157–166.

Martin, J. P. and Haider, K. (1969), Phenolic polymers of *Stachybotrys atra*, *Stachybotrys chartarum* and *Epicoccum nigrum* in relation to humic acid formation, *Soil Sci.*, **107**, 260–270.

Martin, J. P. and Haider, K. (1971). Microbial activity in relation to soil humus formation, *Soil Sci.*, **111**, 54–63.

Martin, J. P. and Haider, K. (1976), Decomposition of specifically carbon-14-labeled ferulic acid: Free and linked into model humic-acid type polymers, *Soil Sci. Soc. Amer. J.*, **40**, 377–380.

Martin, J. P. and Haider, K. (1977), Decomposition in soil of specifically ^{14}C-labeled

DHP and corn stalk lignins, model humic acid-type polymers, and coniferyl alcohols, *Proc. IAEA-FAO-Agrochemica Symposium* (Braunschweig, 1976), **2**, 23–32.

Martin, J. P., Haider, K. and Saiz-Jimenez, C. (1974), Sodium amalgam reductive degradation of fungal and model phenolic polymers, soil humic acids and simple phenolic compounds, *Soil Sci. Soc. Amer. Proc.*, **38**, 760–765.

Martin, J. P., Haider, K. and Wolf, D. (1972), Synthesis of phenols and phenolic polymers by *Hendersonula toruloidea* in relation to humic acid formation, *Soil Sci. Soc. Amer. Proc.*, **36**, 311–315.

Martin, J. P., Richards, S. J. and Haider, K. (1967), Properties and decomposition and binding action in soil of 'humic acid' synthesized by *Epicoccum nigrum*, *Soil Sci. Soc. Amer. Proc.*, **31**, 657–662.

Martin, J. P., Ervin, J. O. and Shepherd, R. A. (1966), Decomposition of the iron, aluminium, zinc and copper salts or complexes of some microbial and plant polysaccharides in soil, *Soil Sci. Soc. Amer. Proc.*, **30**, 196–200.

Martin, F., Saiz-Jimenez, C. and Cert, A. (1977), Pyrolysis-gas chromatography-mass spectrometry of soil humic fractions: I. The low boiling point compounds, *Soil Sci. Soc. Am. J.*, **41**, 1114–1118.

Marton, J. (1971), Reactions in alkaline pulping, in K. V. Sarkanen and C. H. Ludwig (eds.), *Lignins*, pp. 639–694, Wiley, New York and Chichester.

Matsuda, K. and Schnitzer, M. (1972). The permanganate oxidation of humic acids extracted from acid soils, *Soil Sci.*, **114**, 185–193.

Maximov, O. B. and Krasoskaya, N. P. (1977), Action of metallic sodium on humic acids in liquid ammonia, *Geoderma*, **18**, 227–228.

Maximov, O. B., Shvets, T. V. and Elkin, Yu. N. (1977), On permanganate oxidation of humic acids, *Geoderma*, **19**, 63–78.

Mayaudon, J. and Sarkar, J. M. (1974), Etude des diphenol oxydases extraites d'un litiere de foret, *Soil Biol. Biochem.*, **6**, 269–274.

Mehta, N. C., Streuli, H., Müller, M. and Deuel, H. (1960), Role of polysaccharides in soil aggregation, *J. Sci. Food Agric.*, **11**, 40–47.

Mehta, N. C., Dubach, P. and Deuel, H. (1961), Carbohydrates in the soil, *Advan. Carbohyd. Chem.*, **16**, 335–355.

Meier, H. (1965), Fractionation by precipitation with barium hydroxide, in R. L. Whistler (ed.), *Methods in Carbohydrate Chemistry*, **5**, 45–46.

Mendez, J. and Stevenson, F. J. (1966), Detection of phenolic carboxylic acids by gas-liquid chromatography, *J. Gas Chromatog.*, **4**, 483–485.

Meneghel, R., Petit-Sanlotte, C. and Block, J. M. (1972), Sur la caractérization en l'isolement des produits de dégradation d'un acide humique aprés oxydation peracétique, *Bull. Soc. Chim. Fr.*, **7**, 2997–3001.

Meuzelaar, H. L. C., Haider, K., Nagar, B. R. and Martin, J. P. (1977), Comparative studies of pyrolysis-mass spectra of melanins, model phenolic polymers, and humic acids, *Geoderma*, **17**, 239–252.

Meuzelaar, H. L. C., Kistemaker, P. G. and Posthumus, M. A. (1974), Recent advances in pyrolysis mass spectrometry of complex biological materials, *Biomedial Mass Spectrom.*, **1**, 312–319.

Morita, H. (1962), Composition of peat humus and its derivatives. Oxidative conversion to lignin model compounds. *J. Org. Chem.* **27**, 1079–1080.

Morrison, R. I. (1958), The alkaline nitrobenzene oxidation of soil organic matter, *J. Soil Sci.*, **9**, 130–147.

Morrison, R. I. (1963), Products of the alkaline nitrobenzene oxidation of soil organic matter, *J. Soil Sci.*, **14**, 201–216.

Mortensen, J. L. (1960), Physico-chemical properties of a soil polysaccharide, *Trans. 7th Intern. Congr. Soil Sci.*, **2**, 98–104.

Mortensen, J. L. and Schwendinger, R. B. (1963), Electrophoretic and spectroscopic characterization of high molecular weight components of soil organic matter, *Geochim. Cosmochim. Acta*, **27**, 201–208.

Moschopedis, S. E. (1962), Studies in humic acid chemistry. III—The reaction of humic acids with diazonium salts, *Fuel*, **41**, 425–435.

Mukherjee, P. N. and Lahiri, A. (1956), Studies on the rheological properties of humic acid from coal, *J. Colloid Sci.*, **11**, 240–243.

Müller, M., Mehta, N. C. and Deuel, H. (1960), Chromatographische Fraktionierung von Bodenpolysacchariden au Cellulose-Anionenaustauchern, *Z. Pflanzenernähr. Düng Bodenk.*, **90**, 139–145.

Murphy, D. and Moore, A. W. (1960), A possible structural basis of natural humic acid, *Sci. Proc. Roy. Dublin Soc.*, **Ser A1**, 141–195.

Murzakov, B. G. (1972), Use of gas chromatography to study soil organic matter, *Soviet Soil Sci.*, **4**, 474–478.

Nagar, B. R. (1963), Examination of the structure of soil humic acids by pyrolysis gas chromatography, *Nature* (London), **199**, 1213–1214.

Nagar, B. R. (1977), Applications of computer techniques in humus research, *Proc. IAEA-FAO-Agrochemica* Symposium (Braunschweig, 1976), **2**, 171–175

Nagar, B. R., Waight, E. S., Neuzelaar, H. L. C. and Kistemaker, P. G. (1975), Studies on the structure and origin of soil humic acids by Curie point pyrolysis in direct combination with low voltage ionization mass spectrometry, *Plant Soil*, **43**, 681–685.

Neukom, H. and Kuendig, W. (1965), Fractionation on diethylaminoethylcellulose columns: Fractionation of neutral and acidic polysaccharides by ion-exchange column chromatography on diethylaminoethyl (DEAE)-cellulose, in R. L. Whistler (ed.), *Methods in Carbohydrate Chem.*, **5**, 14–17.

Neyroud, J. A. and Schnitzer, M. (1974a), The chemistry of high molecular weight fulvic acid fraction, *Can. J. Chem.*, **52**, 4123–4132.

Neyroud, J. A. and Schnitzer, M. (1974b), The exhaustive alkaline cupric oxide oxidation of humic acid and fulvic acid, *Soil Sci. Soc. Amer. Proc.*, **38**, 907–913.

Neyroud, J. A. and Schnitzer, M. (1975), The alkaline hydrolysis of humic substances, *Geoderma*, **13**, 171–188.

Nigh, W. G. (1973), Oxidation by cupric ion, in W. J. Trahanovsky (ed.), *Oxidation in Organic Chemistry*, pp. 1–96, Academic Press, New York.

O'Callaghan, M. (1977), *Some studies in soil chemistry*, M.Sc. thesis, Chemistry Department, University of Birmingham.

Oades, J. M. (1967a), Carbohydrates in some Australian soils, *Aust. J. Soil Res.*, **5**, 103–115.

Oades, J. M. (1967b), Gas liquid chromatography of alditol acetates and its application to the analysis of sugars in complex hydrolysates, *J. Chromatogr.* **28**, 246–252.

Oades, J. M. (1974), Synthesis of polysaccharides in soil by microorganisms, *Trans. 10th Intern. Congr. Soil Sci.* (Moscow), **3**, 93–99.

Ogner, G. and Schnitzer, M. (1971), Chemistry of fulvic acid, a soil humic fraction and its relation to lignin, *Can. J. Chem.*, **49**, 1053–1063.

Olness, A. and Clapp, C. E. (1973), Occurrence of collapsed and expanded crystals in montmorillonite-dextran complexes, *Clays and Clay Minerals*, **21**, 289–293.

Olness, A. and Clapp, C. E. (1975), Influence of polysaccharide structure on dextran adsorption by montmorillinote, *Soil Biol. Biochem.*, **7**, 113–118.

Orlov, D. S. (1966), Spectrophotometric analysis of humics substances, *Pochvovedenie*, **11**, 84–95 (*Chem. Abs.*, **66**, 28016w, 1967).

Orlov, D. S. and Gorskova, E. I. (1965), Size and shape of humic acid particles from black earth and sod-podzolic soils, *Nauchr. Dokl. Vyssei. Shkoly. Biol. Nauki.*, 207–212 (*Chem. Abs.*, **73**, 4586, 1966).

Orlov, D. S., Rozanova, O. H. and Matyukhina, S. G. (1962), Infrared absorption spectra of humic acids. *Pochvovedenie*, **1**, 17–25.

Otsuka, H. (1975), Volcanic soil at Ohnobaru, Tarumi, Kagoshima Prefecture. Accumulated state of humus in the soil profile, and sugar, uronic acid, and amino acid contents and amino acid composition in fulvic acid, *Nippon Dojo-Hiryogaku Zasshi*, **46**, 180–184 (*Chem. Abs.*, **83**, 177210*h*).

Ouchi, K., Imuta, K. and Yamashita, Y. (1965), Catalytic depolymerisation of coals. I—Depolymerization of Yūbari coal by *p*-toluene sulphonic acid as catalyst, *Fuel*, **44**, 29–38.

Ouchi, K. and Brooks, J. D. (1967), The isolation of certain compounds from depolymerized brown coal, *Fuel*, **46**, 367–377.

Packter, N. M. (1973), *Biosynthesis of Acetate-Derived Compounds*, Wiley, London.

Parfitt, R. L. and Greenland, D. J. (1970), Adsorption of polysaccharides by montmorillonite, *Soil Sci. Soc. Amer. Proc.*, **34**, 862–866.

Parsons, J. W. and Tinsley, J. (1961), Chemical studies of polysaccharide material in soils and composts based on extraction with anhydrous formic acid, *Soil Sci.*, **92**, 46–53.

Parsons, J. W. and Tinsley, J. (1975), Nitrogeneous substances, in J. E. Gieseking (ed.), *Soil Components*, **Vol. 1**, pp. 263–304, Springer-Verlag, Berlin, New York.

Pigman, W. and Horton, D. (eds.), (1970), *The Carbohydrates*, Academic Press, New York.

Piper, T. J. and Posner, A. M. (1972), Sodium amalgam reduction of humic acid—I. Evaluation of the method; II. Application of the method, *Soil Biol. Biochem.*, **4**, 513–531.

Piper, T. J. and Posner, A. M. (1975), Phenolic degradation of humic acid: Model compound studies, *Soil Sci.*, **119**, 132–135.

Piret, E. L., Hein, R. F., Besser, E. D. and White, R. G. (1957), Oxidation of peat to organic acids. *Ind. Eng. Chem.*, **49**, 737–741.

Piret, E. L., White, R. G., Walther, Jr., H. C. and Madden, Jr., A. J. (1960), Some physico-chemical properties of peat humic acids, *Sci. Proc. Roy. Dubl. Soc.* **A1**, 69–79.

Posner, A. M. and Creeth, J. M. (1972), A study of humic acid by equilibrium ultracentrifugation, *J. Soil Sci.*, **23**, 50–57.

Rafikov, S. R., Pavlova, S. A. and Iverdokhlebova, J. (1964), *Determination of Molecular Weights and Polydispersity of High Polymers*, Israel Program for Scientific Translations, Jerusalem.

Reese, E. T. (1968), Microbial transformation of soil polysaccharides, in *Organic Matter and Soil Fertility* (Pontif. Acad. Sci. Scripta varia 32), North Holland Publishing Co; and Wiley, New York.

Rennie, D. A., Truog, E. and Allen, O. N. (1954), Soil aggregation as influenced by microbial gums, level of fertility, and kind of crop. *Soil Sci. Soc. Amer. Proc.*, **18**, 399–403.

Riffaldi, R. and Schnitzer, M. (1972a), Electron spin resonance spectrometry of humic substances, *Soil Sci. Soc. Amer. Proc.*, **36**, 301–305.

Riffaldi, R. and Schnitzer, M. (1972b), Effects of diverse experimental conditions on ESR spectra of humic substances, *Geoderma*, **8**, 1–10.

Riffaldi, R. and Schnitzer, M. (1973), Effects of 6N HCl hydrolysis on the analytical characteristics and chemical structure of humic acids, *Soil Sci.*, **115**, 349–356.

Roberts, J. D. and Caserio, M. C. (1965), *Basic Principles of Organic Chemistry*, W. A. Benjamin, Inc., New York, Amsterdam.

Ross, S. D., Finkelstein, M. and Rudd, E. J. (1975), *Anodic Oxidation*, Academic Press, New York.

Rybicka, S. M. (1959), The solvent extraction of low rank vitrain, *Fuel*, **38**, 45–54.

Saiz-Jimenez, C., Haider, K. and Martin, J. P. (1975), Anthraquinones and phenols as intermediates in the formation of dark-coloured, humic acid-like pigments by *Eurotium echinulatum*, *Soil Sci. Soc. Amer. Proc.*, **39**, 649–653.

Scheffer, F. and Kickuth, R. (1961a), Chemische Abbauversuche an einer natürlichen Huminsäure: I, *Z. Pflanzenernähr. Düng. Bodenk.*, **94**, 180–188.

Scheffer, F. and Kickuth, R. (1961b), Chemische Abbauversuche an einer natürlichen Huminsäure: II, *Z. Pflanzenernähr. Düng. Bodenk.*, **94**, 189–198.

Schnitzer, M. (1974a), The methylation of humic substances, *Soil Sci.*, **117**, 94–102.

Schnitzer, M. (1974b), Alkaline cupric oxide oxidation of a methylated fulvic acid. *Soil Biol. Biochem.*, **6**, 1–6.

Schnitzer, M. and Desjardins, J. G. (1964), Further investigations on the alkaline permanganate oxidation of organic matter extracted from a podzol B_h horizon, *Can. J. Soil Sci.*, **44**, 272–279.

Schnitzer, M. and Desjardins, J. G. (1970), Alkaline permanganate oxidation of methylated and unmethylated fulvic acid, *Soil Science Soc. Amer. Proc.*, **34**, 77–79.

Schnitzer, M. and Gupta, U. C. (1965), Determination of acidity in soil organic matter, *Soil Sci. Soc. Amer. Proc.*, **29**, 274–277.

Schnitzer, M. and Khan, S. U. (1972), *Humic Substances in the Environment*, Marcel Dekker, New York.

Schnitzer, M. and Neyroud, J. A. (1975), Further investigations on the chemistry of fungal 'humic acids'. *Soil Biol. Biochem.*, **7**, 365–371.

Schnitzer, M. and Ortiz de Serra, M. I. (1973a), The chemical degradation of a humic acid, *Can. J. Chem.*, **51**, 1554–1566.,

Schnitzer, M. and Ortiz de Serra, M. I. (1973b), The sodium-amalgam reduction of soil and fungal humic substances, *Geoderma*, **9**, 119–128.

Schnitzer, M., Ortiz de Serra, M. I. and Ivarson, K. (1973), The chemistry of fungal humic acid-like polymers and of soil humic acids, *Soil Sci. Soc. Amer. Proc.*, **37**, 229–236.

Schnitzer, M. and Skinner, S. I. M. (1965a), The carboxyl group in soil organic matter preparations, *Soil Sci. Soc. Amer. Proc.*, **29**, 400–405.

Schnitzer, M. and Skinner, S. I. M. (1965b), Organo-metallic interactions in soils: 4. Carboxyl and hydroxyl groups in organic matter and metal retention, *Soil Sci.*, **99**, 273–284.

Schnitzer, M. and Skinner, S. I. M. (1966), A polarographic method for the determination of carbonyl groups in soil humic compounds, *Soil Sci.*, **101**, 120–124.

Schnitzer, M. and Skinner, S. I. M. (1968), Gel filtration of fulvic acid, a soil humic compound, in *Isotopes and Radiation in Soil Organic Matter Studies*, I.A.E.A., Vienna, pp. 41–55.

Schnitzer, M. and Skinner, S. I. M. (1974a), The peracetic acid oxidation of humic substances, *Soil Sci.*, **118**, 322–331.

Schnitzer, M. and Skinner, S. I. M. (1974b), The low temperature oxidation of humic substances, *Can. J. Chem.*, **52**, 1072–1080.

Schnitzer, M. and Wright, J. R. (1960a), Studies on the oxidation of the organic matter of the A_0 and B_h horizons of a podzol, *Trans. 7th Intern. Congr. Soil Sci.* (Madison) **2**, 112–119.

Schnitzer, M. and Wright, J. R. (1960b), Nitric acid oxidation of the organic matter of a podzol, *Soil Sci. Soc. Amer. Proc.*, **24**, 273–276.

Schöniger, W. (1969), The present status of organic elemental microanalysis, *Plenary*

Lectures of IUPAC Symposium on Analytical Chemistry (Birmingham), Butterworths, London, 1970, pp. 497–512.

Schuffelen, A. C. and Bolt, G. H. (1950), Some notes on the synthesis of humic compounds, *Landbouwk. tijdschr.*, **62**, ste, Jaargang No. 4/5.

Sciacovelli, O., Senesi, N., Solinas, V. and Testini, C. (1977), Spectroscopic studies on soil organic fractions I. IR and NMR spectra, *Soil Biol. Biochem.*, **9**, 287–293.

Scott, J. E. (1965), Fractionation by precipitation with quaternary ammonium salts, in R. L. Whistler (ed.), *Methods in Carbohydrate Chemistry*, **5**, 38–44.

Senesi, N., Chen, Y. and Schnitzer, M. (1977), Hyperfine splitting in electron spin resonance spectra of fulvic acid, *Soil Biol. Biochem.*, **9**, 371–372.

Senesi, N. and Schnitzer, M. (1977), Effects of pH, reaction time, chemical reduction and irradiation on ESR spectra of fulvic acid, *Soil Sci.*, **123**, 224–234.

Seymour, R. B. (1971), *Introduction to Polymer Chemistry*, McGraw-Hill, New York.

Shemyakin, M. M. and Shchukina, L. A. (1956), Oxidative–hydrolytic splitting of carbon–carbon bonds of organic molecules, *Quart. Rev.* (London), **10**, 261–282.

Simmonds, P. G. (1970), Whole microorganisms studied by pyrolysis–gas chromatography–mass spectrometry: significance for extraterrestrial life detection experiments, *Appl. Microbiol.*, **20**, 567–572.

Simmonds, P. G., Shulman, G. P. and Stembridge, C. H. (1969), Organic analysis by pyrolysis–gas chromatography–mass spectrometry. A candidate experiment for the biological exploration of Mars, *J. Chromatogr. Sci.*, **7**, 36–41.

Smith, D. G. and Lorimer, J. W. (1964), An examination of the humic acids of sphagnum peat, *Can. J. Soil Sci.*, **44**, 76–87.

Snaith, S. M. and Levvy, G. A. (1973), α-D-Mannosidase, *Advan. Carbohyd. Chem. Biochem.*, **28**, 401–445.

Stacey, M. and Barker, S. A. (1960), *Polysaccharides of Microorganisms*, Oxford University Press, London.

Staub, A. M. (1965), Removal of proteins: Sevag method, in R. L. Whistler (ed.), *Methods in Carbohydrate Chemistry*, **5**, 5–6.

Steelink, C., Reid, T. and Tollin, G. (1963), On the nature of the free-radical moiety in lignin, *J. Amer. Chem. Soc.*, **85**, 4048–4049.

Steelink, C. and Tollin, G. (1967), Free radicals, in A. D. McLaren and G. H. Peterson (eds.), *Soil Biochemistry*, **Vol. 1**, pp. 147–169, Marcel Dekker, New York.

Stevenson, F. J. and Butler, J. H. A. (1965), Chemistry of humic acids and related pigments, in G. Eglinton and M. T. J. Murphy (eds.), *Organic Geochemistry*, pp. 534–557, Springer Verlag, Berlin.

Stevenson, F. J., Marks, J. D., Varner, J. W. and Martin, W. P. (1952), Electrophoretic and chromatographic investigations of clay adsorbed organic colloids. 1. Preliminary investigations, *Soil Sci. Soc. Amer. Proc.*, **16**, 69–73.

Stevenson, F. J., Marks, J. D., Varner, J. W. and Martin, W. P. (1953), Physico-chemical investigations of clay adsorbed organic colloids, 2. *Soil Sci. Soc. Amer. Proc.*, **17**, 31–34.

Stevenson, F. J. and Mendez, J. (1967), Reductive cleavage products of soil humic acids, *Soil Sci.*, **103**, 383–388.

Stewart, R. (1964). *Oxidation Mechanisms*, W. A. Benjamin, Inc., New York.

Stewart, R. (1965), Oxidation by permanganate, in K. B. Wiberg (ed.), *Oxidation in Organic Chemistry*, **Part A**, pp. 1–68, Academic Press, New York.

Swift, R. S. (1968), *Physico-chemical studies on soil organic matter*, Ph.D. Thesis, University of Birmingham.

Swift, R. S. and Posner, A. M. (1971), Gel chromatography of humic acid, *J. Soil. Sci.*, **22**, 237–249.

Swift, R. S. and Posner, A. M. (1972), Autoxidation of humic acid under alkaline conditions, *J. Soil Sci.*, **23**, 381–393.

Swift, R. S., Thornton, B. K. and Posner, A. M. (1970), Spectral characteristics of a humic acid fractionated with respect to molecular weight using an agar gel, *Soil Sci.*, **110**, 93–99.

Swincer, G. D., Oades, J. M. and Greenland, D. J. (1968), Studies on soil polysaccharides. I. The isolation of polysaccharides from soil. II. The composition and properties of polysaccharides in soils under pasture and under fallow wheat rotation, *Aust. J. Soil Res.*, **6**, 211–239.

Swincer, G. D., Oades, J. M. and Greenland, D. J. (1969), Extraction, characterization and significance of soil polysaccharides, *Advan. Argon.*, **21**, 195–235.

Tanford, C. (1961), *Physical Chemistry of Macromolecules*, Wiley, New York.

Teryama, H. and Wall, F. T. (1955), Reduced viscosities in the presence of added salts, *J. Polymer Sci.*, **16**, 357–365.

Theng, B. K. G. and Posner, A. M. (1967), Nature of the carbonyl groups in soil humic acids, *Soil Sci.*, **101**, 191–201.

Theng, B. K. G., Wake, J. R. H. and Posner, A. M. (1966), Infrared spectrum of humic acids, *Soil Sci.*, **102**, 70–72.

Thomas, R. L., Mortensen, J. L. and Himes, F. L. (1967), Fractionation and characterization of a soil polysaccharide extract, *Soil Sci. Soc. Amer. Proc.*, **31**, 568–570.

Tobolsky, A. V. and Yannas, I. V. (1971), Thermodynamics of polymer solutions, in A. V. Tobolsky and H. F. Mark (eds.), *Polymer Science and Materials*, Wiley, New York, London.

Ubaldini, I. and Siniramed, C. (1933), Richerche sugli acidi uminici—Nota I. Dosaggio dei gruppi carbossilici e fenolici, *Ann. Chim. Applicata*, **23**, 585–597.

Van Krevelen, D. W. (1961), *Coal*, Elsevier, Amsterdam.

Vasilyevskaya, N. A., Glebko, L. I. and Maximov, O. B. (1971), Determination of humic acid quinoid groups, *Soviet Soil Science*, **4**, 63–68.

Verma, L. and Martin, J. P. (1976), Decomposition of algal cells and components and their stabilization through complexing with model humic acid-type phenolic polymers, *Soil Biol. Biochem.*, **8**, 85–90.

Visser, S. A. (1964), A physico-chemical study of the properties of humic acids and their changes during humification, *J. Soil Sci.*, **15**, 202–219.

Vogel, A. I. (1956), *Practical Organic Chemistry*, 3rd edition, Longman, London.

Wagner, G. A. and Stevenson, F. J. (1965), Structural arrangement of functional groups in soil humic acids as revealed by infrared analysis, *Soil Sci. Soc. Amer. Proc.*, **29**, 43–48.

Wake, J. R. H. and Posner, A. M. (1967), Membranes for measuring low molecular weights by osmotic pressure, *Nature* (London), **213**, 692–693.

Waksman, S. A. (1938), *Humus*, Williams and Wilkins, Baltimore.

Waksman, S. A. and Martin, J. P. (1939), The role of microorganisms in the conservation of the soil, *Science*, **90**, 304–305.

Waldron, A. C. and Mortensen, J. L. (1961), Soil nitrogen complexes: II. Electrophoretic separation of organic components, *Soil Sci. Soc. Amer. Proc.* **25**, 29–32.

Wallis, A. F. A. (1971), Solvolysis by acids and bases, in K. V. Sarkanen and C. H. Ludwig (eds.), *Lignins*, pp. 345–372, Wiley, New York and Chichester.

Ward, S. G. (1947), Coal—its constitution and utilisation as a chemical and as a raw material, *J. Inst. Fuel.*, **21**, 67–70.

Waters, W. A. (1964), *Mechanisms of Oxidation of Organic Compounds*, Wiley, New York.

Watson, J. R. (1971), Ultrasonic vibration as a method of soil dispersion, *Soils Fertil.*, **34**, 127–134.

Weedon, B. C. L. (1963), Alkali fusion and some related processes, in A. Weissberger (ed.), *Techniques of Organic Chemistry*, **Vol. 9**, part 2, chapter 12, Wiley, New York.

Wershaw, R. L., Burcar, P. J., Sutula, C. L. and Wiginton, B. J. (1967), Sodium humate solution studied with small angle X-ray scattering, *Science*, **157**, 1429–1431.

Wershaw, R. L. and Bohner, G. E. (1969), Pyrolysis of humic and fulvic acids, *Geochim. Cosmochim. Acta*, **33**, 757–762.

Whistler, R. L. (ed.), (1965), *Methods in Carbohydrate Chemistry*, **Vol. 5**, Academic Press, New York.

Whistler, R. L. and Kirby, K. W. (1956), Composition and behaviour of soil polysaccharides, *J. Amer. Chem. Soc.*, **78**, 1755–1759.

Whitehead, D. C. and Tinsley, J. (1964), Extraction of soil organic matter with dimethylformamide, *Soil Sci.*, **97**, 34–42.

Wildung, R. E., Chesters, G. and Behmer, D. E. (1970), Alkaline nitrobenzene oxidation of plant lignins and soil humic colloids, *Plant and Soil*, **32**, 221–237.

Williams, D. H. and Fleming, I. (1973), *Spectroscopic Methods in Organic Chemistry*, 2nd edition, Wiley, London, New York.

Wright, J. R., Schnitzer, M. and Levick, R. (1958), Some characteristics of the organic matter extract by dilute inorganic acids from a podzolic B horizon, *Can. J. Soil Sci.*, **38**, 14–22.

Wright, J. R. and Schnitzer, M. (1959a), Oxygen containing functional groups in the organic matter of a podzol soil, *Nature* (London), **184**, 1462–1463.

Wright, J. R. and Schnitzer, M. (1959b), Alkaline permanganate oxidation of the organic matter of the A_0 and B_{21} horizons of a podzol, *Can. J. Soil. Sci.*, **39**, 44–53.

Wright, J. R. and Schnitzer, M. (1960), Oxygen containing functional groups in the organic matter of the A_0 and B_h horizon of a podzol, *Trans. 7th Intern. Congr. Soil Sci.*, (Madison), **2**, 120–127.

Yokakawa, C., Watanabe, Y., Kajiama, S. and Takiegami, Y. (1962), Studies on the chemical structure of coal. I—Stepwise oxidative degradation of coal, *Fuel*, **41**, 209–219.

Young, D. K., Sprang, S. R. and Yen, T. F. (1977), Preliminary investigation of the precursors of the organic components in sediments—melanoidin formations, in T. F. Yen (ed.), *Chemistry of Marine Sediments*, pp. 101–110, Ann Arbor Science.

Zunino, H. and Martin, J. P. (1977), Metal-binding organic macromolecules in soil: 2. Characterization of the maximum binding ability of the macromolecules, *Soil Sci.*, **123**, 188–202.

CHAPTER 4

Surfaces of soil particles

D. J. Greenland and C. J. B. Mott
both of Department of Soil Science, University of Reading

4.1 INTRODUCTION

It has been suggested that the relationship of surface chemistry to soil science is similar to that of biochemistry to biology. Certainly the fact that surfaces of soil particles are extensive and reactive leads to a dominance of surface chemistry in soil behaviour, but it is not an exclusive relationship, because many precipitation reactions, oxidation and reduction processes, and biological reactions take place more or less independently of the surface processes. Nevertheless it is difficult to understand soil properties fully without an appreciation of the nature and extent of the surfaces of the soil constituents, and this chapter is devoted to a discussion of this topic.

The surfaces of clay particles often differ from the constitution which might be expected from their internal structures because of strongly adsorbed materials normally present on those surfaces. Thus interpretations based on comparisons with colloids with carefully 'cleaned' surfaces, such as are considered essential for most studies in surface chemistry, can be misleading. Soil colloids are invariably formed in a 'dirty' environment, and apart from materials newly formed in the soil, will have had the opportunity to adsorb any of several materials to which they are exposed. In most instances these include relatively simple ionic forms and hydrous oxides of aluminium, silicon and iron, and in surface soils a great variety of large and small organic molecules. Although various procedures have been devised to 'clean up' soil minerals, so that for instance better X-ray diffraction patterns and infrared spectra can be obtained, the reactive surfaces are those where the 'impurities' are present, and the understanding of soil properties requires an appreciation of these contaminated surfaces, as well as the constitution of the surfaces of the 'pure' clay minerals (cf. Plate 4.1). Although there is still much to be learnt, sufficient information has been obtained in recent years to provide at least an indication of the most common surface constitutions to be found in soil minerals, and the first part of this chapter is concerned with this.

The extent and accessibility of the colloid surfaces also influence their relation to soil behaviour. The surface areas of soil particles are most frequently measured by the various adsorption methods used for other finely divided solids, although because of the irreversible changes which can be induced in soils by predrying, careful consideration of the suitability of these techniques to specific problems is essential.

The accessibility of the surfaces of soil particles is a subject which has received a great deal of attention. It is extremely important in relation to ion exchange and retention, persistence and efficiency of pesticides, and in general to all adsorption and flow phenomena in soils. Flow in soils proceeds most rapidly in the larger pores, and only slowly in smaller pores. The arrangement of the soil particles, and of the spaces left between them—of the soil structure or fabric as it is usually termed—is therefore of great importance in relation to movement of gases and solutions in soils.

The latter part of this chapter is concerned with the measurement of the surface areas of soil particles, and the micro- and meso-pore size distribution in soils and clays and its determination.

4.2 SURFACES OF CLAY PARTICLES

4.2.1 The Major Types of Particle Surface

The surfaces of clay particles may be very broadly divided into two groups: (i) the siloxane type and (ii) the hydrous oxide type.

Siloxane surfaces

Siloxane-type surfaces are typically those of the basal surfaces of the micas and related 2:1 lattice type minerals (Chapter 2), including smectites and vermiculites and their interstratifications. The hydrous oxide type surfaces occur at all 'broken edges' of clay particles, on all surfaces of crystalline and amorphous oxides, oxy-hydroxides and hydroxides occurring in soils, on allophane and the related quasi-crystalline imogolite. Kaolinites and chlorites might be expected to have both hydrous oxide and siloxane basal surfaces, depending on the position of cleavage of the crystal, since both structures consist of alternate sheets of gibbsite (or brucite) and silicate type (Chapter 2). Siloxane surfaces are composed of a planar or near-planar layer of oxygen ions, underlain by silicon ions. The numbers of silicon ions may be sufficient to render the layer free of resultant charge, as in the minerals pyrophyllite and talc. The surface is then essentially hydrophobic (Chapter 6). More commonly the surface carries a resultant charge because of the isomorphous substitution of the silicon ions by others of lower valency, as in the micas. In the common micas, muscovite and biotite, every fourth silicon is replaced by an aluminium ion, and the resultant charge deficit is balanced internally by potassium ions in the interlayer positions (Chapter 2). At the external surfaces however these potassium ions are readily exchangeable, and the charge deficit appears as a net charge on the surface,

Table 4.1 Density of negative charge on surfaces of clay minerals with predominantly constant charge surfaces

(a) Calculated charge densities on external surfaces of crystals of 2:1 phyllosilicates

Mineral	Basal area per unit cell, nm^2	Charge on basal area	Calculated charge density C/m^2	Calculated charge density $\mu eq\,m^{-2}$	Charges meq per 750 m^2	Area (nm^2) per charge
Pyrophyllite	0.459 × 2	0	0	0	0	
Muscovite mica	0.469 × 2	1 × 2	0.341	3.54	2.66	0.469
Margarite brittle mica	0.458 × 2	2 × 2	0.699	7.26	5.45	0.229

(b) Measured cation exchange capacities and surface areas and derived mean charge densities for internal plus external surfaces

Mineral type & origin	Surface area, $m^2\,g^{-1}$ External	Internal	Total	CEC meq/g	Derived charge density $C\,m^{-2}$	$\mu eq\,m^{-2}$	Charges meq per 750 m^2	Area (nm^2) per charge
Fithian illite	93	0	93	0.26	0.27	2.80	2.10	0.60
Willalooka illite	132	0	132	0.41	0.30	3.10	2.33	0.54
Sarospatak illite/smectite	43	109	152	0.16	0.101	1.05	0.79	1.58
Kenya vermiculite	1	780	780	1.24	0.152	1.58	1.18	1.04
Wyoming montmorillonite	47	753	800	0.98	0.118	1.22	0.92	1.36

balanced by the suite of exchangeable cations (Chapter 9). For the micas the area per charge is approximately 0.45 nm^2, corresponding to the half-basal area of the unit cell, i.e. the product of the *a* and *b* dimensions of the unit cell, for muscovite, 0.52×0.90 nm^2 (Table 4.1). The complexities of isomorphous substitution in the micas are discussed in Chapter 2. Some differences in the forces acting on ions at the surface arise when the charge deficit originates not in the tetrahedral layer of silicon ions next to the surface but in the underlying octahedral layer.

Alteration of micas in soils normally produces materials of relatively high aluminium content and lower potassium content, the so-called hydrous micas or illites. Measured surface charge densities of these materials are usually lower than those of the unaltered micas (Table 4.1). This may be related to the fact that the micas alter by development of interstratified layers of lower charge, and these layers are sites of cleavage, the newly exposed surfaces having the lower charge density typical of the vermiculite or smectite minerals.

The vermiculite and smectite minerals differ from the micas in that the charge deficit due to isomorphous substitutions is always less than in the micas. The electrostatic force between successive layers of unit cells is also less, and the interlayer cations balancing the charge can be hydrated by water and other liquids and give rise to exchange reactions. Little information is available about

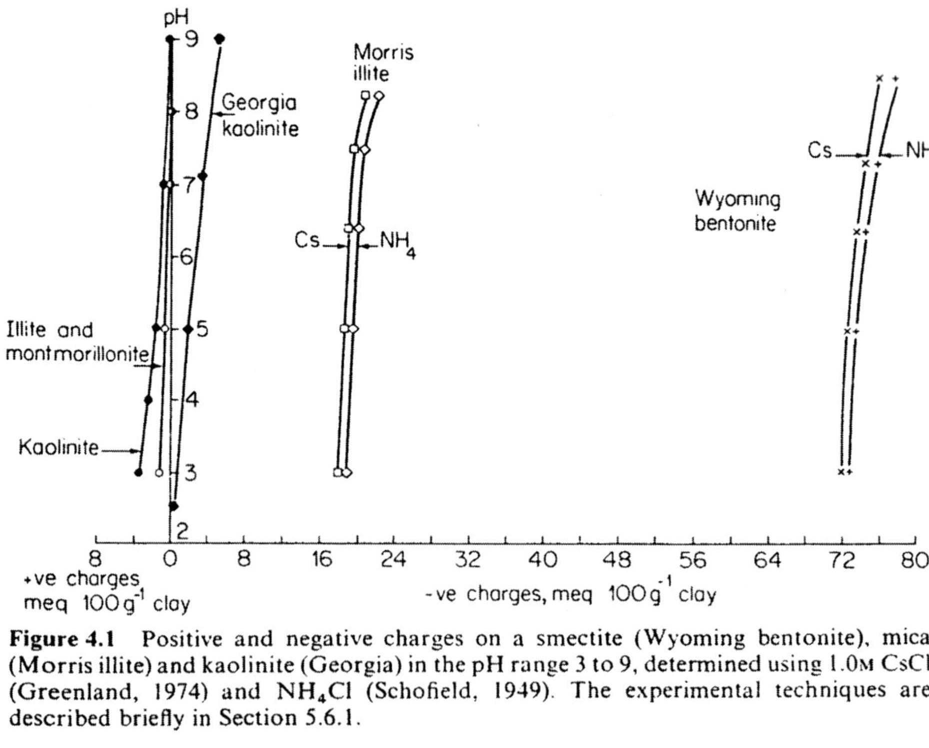

Figure 4.1 Positive and negative charges on a smectite (Wyoming bentonite), mica (Morris illite) and kaolinite (Georgia) in the pH range 3 to 9, determined using 1.0M CsCl (Greenland, 1974) and NH_4Cl (Schofield, 1949). The experimental techniques are described briefly in Section 5.6.1.

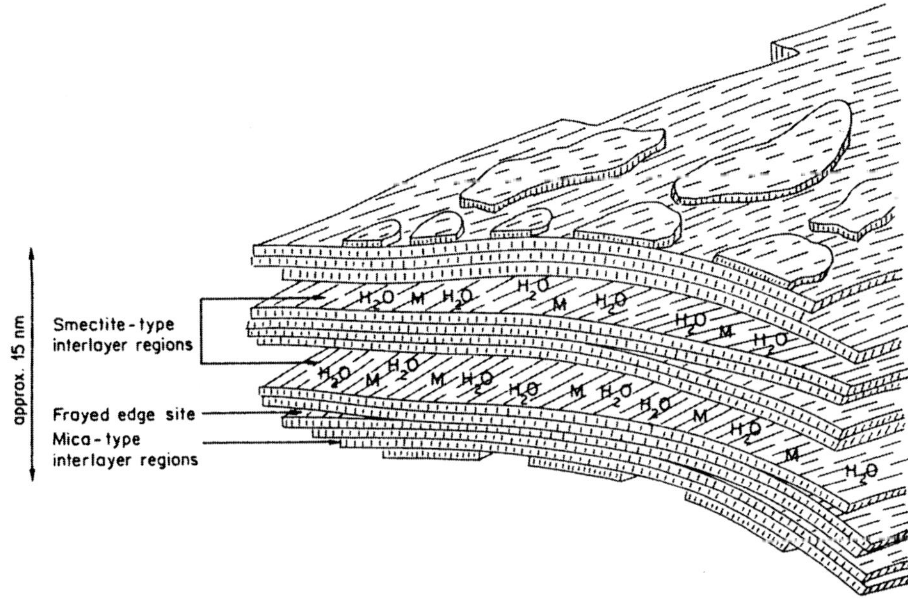

Figure 4.2 Schematic illustration of 'illite' particle in a soil clay, with two smectite-type layers interstratified with mica-type layers. The crystal is shown to be flexed deliberately, as the thin crystals of illites and smectites in soil clays are unlikely to be rigid

the uniformity of charge on the internal and external surfaces of smectite and vermiculite crystals. It is usually assumed that they are in fact similar. Typical values for such materials are included in Table 4.1. The charge due to isomorphous substitution is independent of pH, cation, electrolyte concentration (Figure 4.1 and Chapter 5) but minor effects arise from hydroxyl groups at particle edges.

Mica-type minerals in soils are commonly interstratified with smectite- or vermiculite-type layers (Section 2.3.5). Thus these minerals also have internal surfaces with low charge densities. Sarospatak illite/smectite is an example. The relative extent of internal and external surfaces will be discussed later.

As well as differences in charge density, there are also differences in the distribution of oxygen ions on the siloxane surface. This depends on the precise ionic composition of the adjacent tetrahedral and octahedral layers (Chapter 2). Although of less importance than the differences in charge density, it may influence properties such as the selectivity of the surface for specific ions. In addition to exchange sites on external and internal basal surfaces, partial interlayer cleavage at edges of mica particles (Figure 4.2) can provide surfaces of rather special properties. Strong bonding of potassium has been postulated to occur at such sites.

The structure of water layers at clay surfaces has also been thought to be related to the structure of the oxygen surface. An early proposal was that a regular hexagonal array of water molecules was directly hydrogen bonded to the underlying hexagonally packed oxygens of the siloxane surface of montmorillonite. Unfortunately this structure left no place for the exchangeable cations, which even on a low charged montmorillonite surface are only about 1.2 nm apart, corresponding to the charge density of 1 charge per 1.4 nm^2 (Table 4.1). It is in fact now abundantly clear that the structure of the water layers adjacent to siloxane surfaces is determined by hydration of the exchangeable cations, and the surface structure plays a minor role, with only weak hydrogen bonds between water molecules and surface oxygens (Chapter 6).

Hydrous oxide surfaces

The term 'hydrous oxide' is used here for any oxide, oxyhydroxide or hydroxide. The surfaces of the various hydrous oxides commonly found in soils are similar in so far as the charge is dependent on pH and electrolyte concentration in the surrounding medium. The charge arises from association or dissociation of protons, and may be represented by

$$\underset{\text{more acid}}{\mathrm{M{-}OH_2^+} \overset{\mathrm{H^+}}{\rightleftharpoons}} \mathrm{M{-}OH} \underset{\text{more alkaline}}{\overset{\mathrm{OH^-}}{\rightleftharpoons} \mathrm{M{-}O^- + H_2O}} \qquad (4.1)$$

where M represents a coordinated metal or silicon ion in the clay surface. The extent of association or dissociation of protons is determined by the concentration of protons near the surface, and hence by pH and electrolyte concentration of the soil solution (Chapter 5). The pH at which the surface is uncharged is known as the point of zero charge (p.z.c.). It depends primarily on the affinity of the ion M^{n+} for electrons. Thus there is an approximate relationship between cation valency and p.z.c. of its oxide (Parks, 1965):

Monovalent ions, giving oxides M_2O, p.z.c. $<$pH 11.5
Divalent ions, giving oxides MO, $8.5<$p.z.c.<12.5
Trivalent ions, giving oxides M_2O_3, $6.5<$p.z.c.<10.4
Quadrivalent ions, giving oxides MO_2, $O<$p.z.c.<7.5
Other ions, giving oxides M_2O_5, p.z.c. <0.5
and MO_3.

In accordance with this, the zero points of charge of the hydrous oxides most important in soils are as follows: SiO_2, close to pH 2; TiO_2, close to 4.5; MnO_2, close to 4; Fe_2O_3, between 6.5 and 8; Al_2O_3 between 7.5 and 9.5 (Parks, 1965).

Fairly wide ranges have been reported for the zero points of charge of these oxides. These arise because of differences in hydration state of the surface, the effects of impurities, and differences associated with different coordination numbers for the cation, and 'specific adsorption' effects which are important with certain ions (Chapter 5).

In general it appears that the mineralogical form of a particular oxide, oxyhydroxide or hydroxide is of relatively little importance in determining the p.z.c. This is because all oxide surfaces tend to hydrate rather readily, so that in the wet condition which invariably prevails within the soil, all oxide surfaces are hydrated, and quartz (SiO_2), hematite (α-Fe_2O_3) and corundum (α-Al_2O_3) have surfaces essentially the same as those of silica gel ($SiO_2 \cdot nH_2O$), goethite (α-$FeO \cdot OH$) and gibbsite ($Al(OH)_3$) respectively. The best evidence of the ease with which the surfaces of the oxides are hydrated comes from infrared studies, which show that it is extremely difficult to remove all the surface hydroxyl groups and that these are re-established very rapidly on exposure to water vapour at elevated temperatures, and in hours or days at room temperature (Blyholder and Richardson, 1962; Peri, 1965, 1966). The surface hydroxyls are directly bound to the silicon, iron or aluminium ion, and are readily distinguished from adsorbed water, which becomes hydrogen bonded to them. This direct association of surface hydroxyls with the adsorbed water is in marked distinction to the inertness of the oxygens of a siloxane surface.

The hydration of the surface may in fact penetrate to some depth. For instance Breeuwsma (1973) considered that the water lost on heating non-porous haematites was 5 to 15 times the amount of water required to hydroxylate the surface. Similar conclusions have been reached for other oxides (Lyklema, 1968). It is not known whether the hydrated surfaces are similar to amorphous hydrated gels, or to regular hydroxides or oxy-hydroxides.

Heating a silanol surface ($\rangle Si{-}OH$) of silica sufficiently strongly transforms it to a siloxane ($\rangle Si{-}O{-}Si\langle$) surface of very low or zero charge density (Laskowski and Kitchener, 1969). This surface is hydrophobic. In contrast to the negatively charged siloxane surfaces of 2:1 minerals, it will react readily with water vapour to form a silanol surface and regain its hydrophilicity. The reason why the silanol and siloxane surfaces differ so markedly in their behaviour towards water is that the surface–water interaction is the sum of the dispersion forces, the hydration energy of non-polar sites and the ionization forces. Whenever the last two terms are small, as they will be on a siloxane surface of low charge, the work of adhesion between water and surface will be much less than the very large cohesive energy of water. The strong polarity of the exchange ions on more charged surfaces is sufficient to overcome this large cohesive energy.

The charge on the surface arises from the association of the surface hydroxyls with a proton below the p.z.c., giving a positively charged surface, or loss of a proton above the p.z.c., giving a negatively charged surface (Figure 4.3). The theoretical maximum charge on the surface is determined by the density of hydroxyl groups, which is in the region of one per 0.20 nm^2, (see for SiO_2 e.g. Fripiat and Uytterhoeven, 1962 and Peri and Hensley, 1968) but this charge is

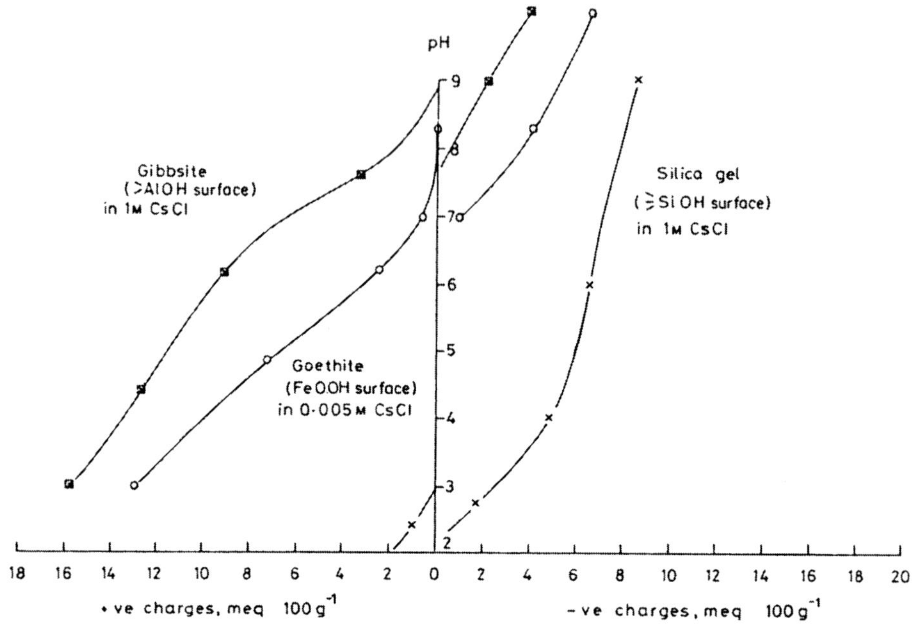

Figure 4.3 Charges on gibbsite ($Al(OH)_3$), goethite ($FeO \cdot OH$) and silica ($SiO_2 \cdot nH_2O$) surfaces in the pH range 3 to 9. The measured charges depend on the electrolyte and concentration used to determine the charge (see Section 5.6); those used are shown on the figure (Mashali, 1976)

only attainable at pHs well removed from the point of zero charge and in strong electrolyte, so that in the normal pH range encountered in soils (pH 5 to 8) the charge density is much less (Table 4.2) and of course the net charge is zero at the p.z.c.

A further important characteristic of the hydrous oxides of iron and aluminium is that they participate very readily in ligand exchange reactions with anions such as silicate, phosphate, molybdate, borate and selenite. The chemisorbed anions modify the surface properties substantially. Thus adsorption of phosphate by alumina will produce a surface with a p.z.c. several pH units lower than that of the original aluminium hydroxide (Al—OH) surface (Parks, 1965). As the soil solution in most soils is saturated with respect to silica, it is to be expected that iron and aluminium hydroxides in soils will normally have reacted with silicate, and so the p.c.z. will often tend towards that of silica. The very low p.z.c.s found for naturally occurring iron hydrous oxides (Schuylenborgh and Sanger, 1949) and the apparent negative charges found on the iron and aluminium oxides in soils (Deshpande, Greenland and Quirk, 1964; Tweneboah, Greenland and Oades, 1967) are in accord with this. Even soils containing very large proportions of hydrous oxides of iron and aluminium have been found to have low p.z.c.s, particularly surface soils (van Raij and Peech, 1972; Gallez, Juo and Herbillon, 1976) and others where organic anions

Table 4.2 Approximate negative and positive charge densities on hydrous oxides at pH 4 and pH 8 in 0.005M CsCl

Mineral	Surface area, $m^2\ g^{-1}$	– ve charge pH 4				– ve charge pH 8			
		Meq $100\ g^{-1}$	µeq m^{-2}	$C\ m^{-2}$	nm^2 $(charge)^{-1}$	Meq $100\ g^{-1}$	µeq m^{-2}	$C\ m^{-2}$	nm^2 $(charge)^{-1}$
Gibbsite $Al(OH)_3$	25.8	0	0	0		0.5	0.19	0.018	8.74
Goethite $FeO{\cdot}OH$	41.4	0	0	0		0.9	0.22	0.021	7.55
Silica $SiO_2{\cdot}4H_2O$	10.0	2.7	2.7	0.26	0.615	4.6	4.6	0.44	0.361
		+ ve charge pH 4				+ ve charge pH 8			
Gibbsite $Al(OH)_3$	25.8	9.2	3.57	0.34	0.465	0.9	0.349	0.034	4.76
Goethite $FeO{\cdot}OH$	41.4	11.0	2.66	0.26	0.624	0.1	0.024	0.002	69.2
Silica $SiO_2{\cdot}4H_2O$	10.0	0	0	0		0	0	0	

Most soils have a pH in the range 5 to 8, and the soil solution is dominated by Ca, so that in soils the charges on particle surfaces similar to those of the hydroxides are close to those measured with 0.005M CsCl in this pH range. However the surface areas used to determine the charge densities are those obtained by adsorption of N_2 after intensive degassing and therefore drying of the samples. These areas may be less than those which obtain when water is present, and consequently the charge densities shown may be too high.

and phosphate may have reacted with the iron and aluminium oxide surfaces, as well as silicate.

In addition to the effect that surface adsorbed impurities have on the p.z.c., structural impurities also cause a shift in the pH of the p.z.c., in the directions of the p.z.c. of the oxide of the impurity (Parks, 1965). Thus in the mixed hydrous oxide systems formed by silica and alumina the p.z.c. depends on the Si:Al ratio (Wada, 1967; Wada and Harada, 1969).

In soils, allophane (Chapter 2.6) mostly has a predominance of silica, and a correspondingly low p.z.c. This may be further reduced by reaction of aluminol (—Al—OH) sites with organic anions, which are bound very strongly by allophane. However in allophanes where the Si:Al ratio approaches 1:1, p.z.c.s as high as 6 may be found (Figure 4.4) whereas other allophane soils have p.z.c.s below 4.

Soils containing much allophane (Ando soils, or Andepts, of areas of relatively recent volcanic activity) have properties differing substantially from others, because allophane has a very large surface area consisting of hydrous oxide groups. The accessibility of all the hydroxyls in such materials to water is shown by their rapid deuteration at room temperature when they are in contact with D_2O vapour (Wada, 1966; Russell, McHardy and Fraser, 1969).

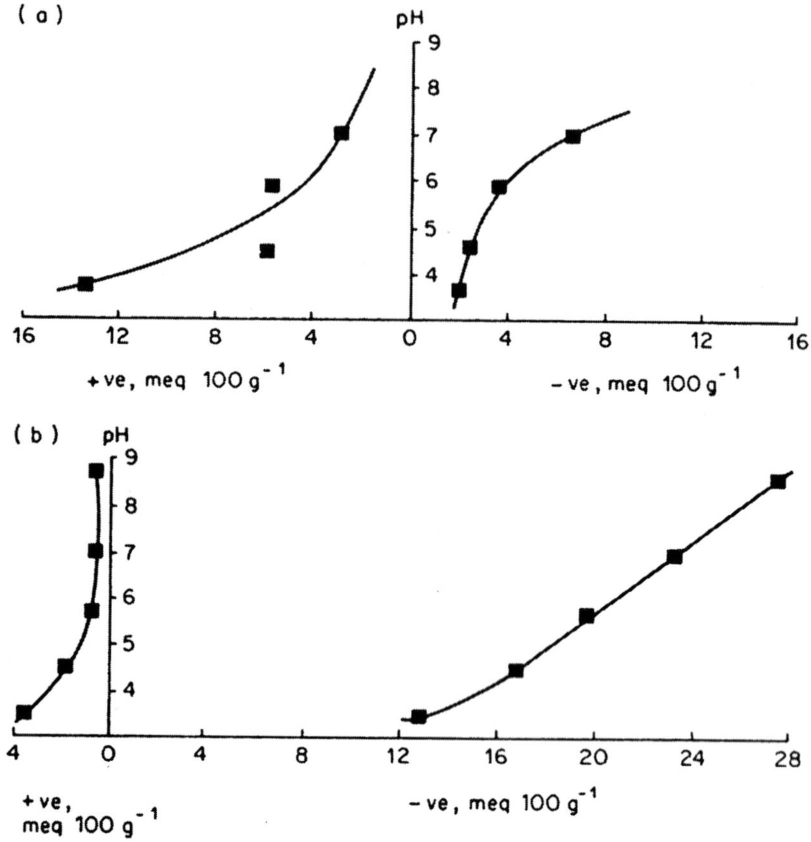

Figure 4.4 Charges on soils rich in allophane (Andepts) in the range pH 3 to pH 9, determined in 0.005M CsCl (a) a Japanese soil with a high Al/Si ratio and a dominance of imogolite in the clay fraction (Yoshinaga and Aomine, 1962) (b) a New Guinea soil with a lower Al/Si ratio and lower proportion of imogolite. (Greenland, Wada and Hamblin, 1969.) Charge data from Greenland, 1974.

Kaolinite and chlorite surfaces

In the discussion so far no mention has been made of the very important and common groups of clay minerals, the kaolinites and chlorites. The classical structure of kaolinite (Section 2.3.5) indicates that crystals of these minerals will expose a siloxane and a gibbsite face on the two opposite basal surfaces. The charge on the faces has been a matter of some controversy. The normal observation is that samples of 'pure' kaolinites have a cation exchange capacity of only 2 to 5 meq 100 g^{-1}. In much American work it has been assumed that these low cation exchange capacities can be attributed to the fact that the basal surfaces are uncharged, the observed charges arising at the broken edges of the crystals, where silanol groups would give rise to a pH-dependent negative

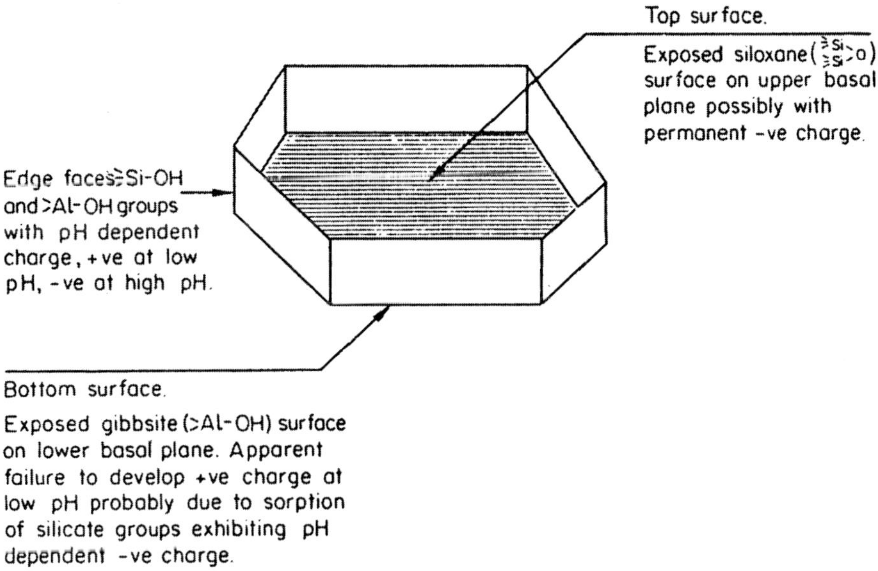

Figure 4.5 Schematic illustration of kaolinite particle (cf. Plate 4.1 c and d) indicating the nature of the basal and edge surfaces of the crystal

charge. Careful separation of pH-dependent charges which may occur at the edges, from permanent charge due to isomorphous substitution, suggested that both types of charges could be present in kaolinite samples (Figure 4.5). Schofield and Samson (1954) suggested that isomorphous substitution occurred in the external layers of the crystals, so that a permanent charge resided on the basal surfaces. The charge was small because of the low specific surface areas of kaolinites. However a question remains about this work because of the difficulty of recognizing the contribution which can arise from small amounts of impurity, one or two per cent of a smectite impurity being able to account for the small permanent charge measured. Such small amounts of smectite are very difficult to detect. The most important feature of Schofield and Samson's work was that they showed very clearly that kaolinite developed a small amount of positive charge at low pH. This was assumed to be at the edges of the crystal, and due to proton acceptance by an aluminol group. They depicted the amphoteric behaviour of the edges of the kaolinite layers as:

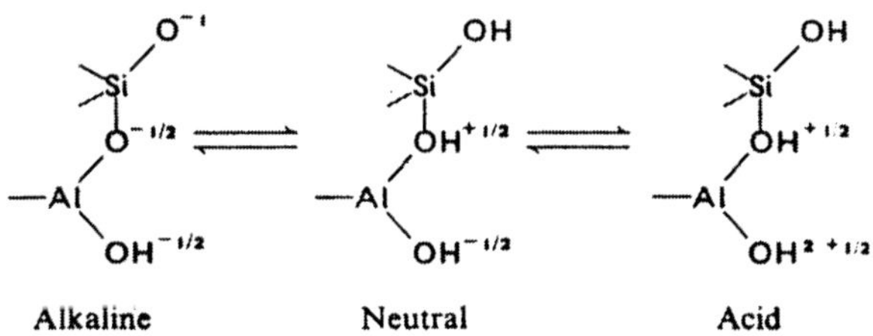

The negative charges on the basal surfaces, if due to isomorphous substitution, would persist independently of pH, so that in the region of pH 4 to 5 the basal faces were considered to be negatively charged, and the edges positively charged. In support of these ideas, 'self-flocculation' of kaolins was observed in this pH range, and considered to be due to interaction between oppositely charged edges and faces. Electron micrographs obtained by Thiessen some years earlier (van Olphen, 1963) showed negatively charged silver iodide particles aggregating at the edges of kaolinite crystals. Plate 4.2 shows a similar effect, iron hydroxide at pH 3, and so positively charged, adsorbed on the basal surface of a dickite crystal, but not adsorbing on the positively charged edges of the crystal.

The concept of variable charge at the edges of the crystal lattice is generally accepted. Two aspects of the work of Schofield and Samson have received little attention, the fact that the Al—OH groups of the gibbsite surface would be expected to behave in much the same manner as aluminol groups at the edges,

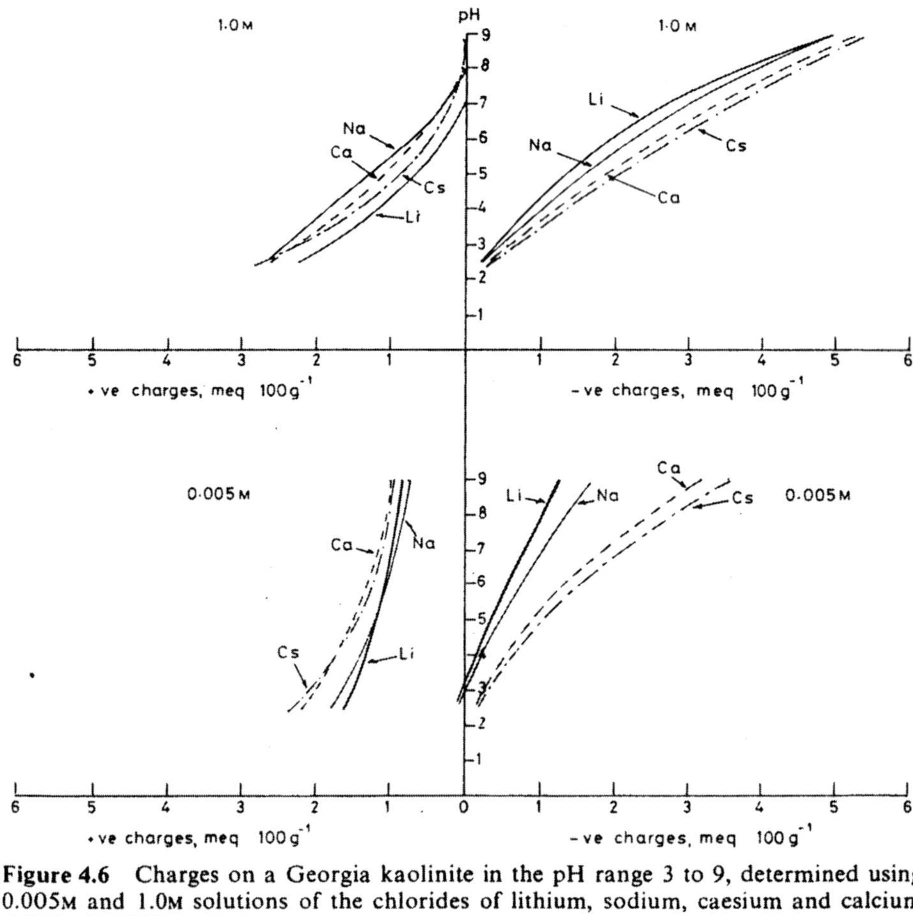

Figure 4.6 Charges on a Georgia kaolinite in the pH range 3 to 9, determined using 0.005M and 1.0M solutions of the chlorides of lithium, sodium, caesium and calcium (Mashali, 1977)

but do not apparently give rise to any positive charge, and the fact that the observed increase in negative charge as the pH increases above neutrality is considerably more than can be accommodated at the edges.

The reason may well be the high reactivity of aluminol surfaces towards silicate anions. Kaolinites are formed in silica-rich weathering environments, and hence the gibbsite face is exposed to silicate anions which sorb readily to it (Ferris and Jepson, 1975; Jepson, Jeffs and Ferris, 1976). Single crystals of kaolinite analyzed by electron scanning techniques (the EMMA IV) have a very slight excess of silica (Jepson and Rowse, 1975) corresponding to the excess that could arise from silicification of the gibbsite face. This would explain satisfactorily the pH-dependent negative charge in excess of that which can be accommodated at the edges, and the small positive charge (Mashali and Greenland, 1975). It is still not clear whether the one siloxane basal surface has a permanent charge. For the sample of Georgia kaolinite for which the charge characteristics are shown in Figure 4.6 it must be zero, but samples of halloysite (Figure 4.7) and some kaolinites (e.g. Greenbushes, described by Bolland *et al.*, 1976) appear to possess a permanent negative charge which cannot be attributed to impurities.

4.2.2 Hydrous Oxide Coatings on Clay Particles

Surface coatings of hydrous oxides are often reported to be present on the surfaces of clay minerals. Very little direct evidence of the presence of such coatings has been reported, although electron micrographs of carbon replicas of particles from a South Australian soil (Plate 4.1a) have shown these clays to be coated with a substantial siliceous crust, probably of silica-alumina similar to poorly ordered 'illite' (Wada and Greenland, 1970). Jones and Uehara (1973) have also published electron micrographs showing clay particles embedded in 'gel hulls', but these were particles in soils developed from volcanic ash and containing relatively large amounts of amorphous gel material (allophane).

The rather frequently made suggestion that the kaolinite particles, which form the main constituent of the clay fraction of many tropical soils, are coated with iron oxides to give the soils their red colour is probably not always correct. Electron micrographs show the iron oxide to be present in many such soils as small, discrete particles (probably ferrihydrite) (Plates 4.3 and 2.11d). Although coatings of iron hydroxide can be precipitated onto kaolinite particles under suitable (acid) conditions (Greenland, 1975 and Plate 4.2) similar features are not normally seen in soil clays, although in some hydromorphic soils where iron hydroxides may be seasonally dissolved under reducing conditions and re-precipitated under oxidizing conditions, precipitates of iron hydroxide may envelop clay particles (Habibullah, Greenland and Brammer, 1971, Plate 4.4).

Many chlorites found in soil clays contain aluminium rather than magnesium hydroxide as a partial or complete interlayer (Chapter 2). Aluminium-hydroxy

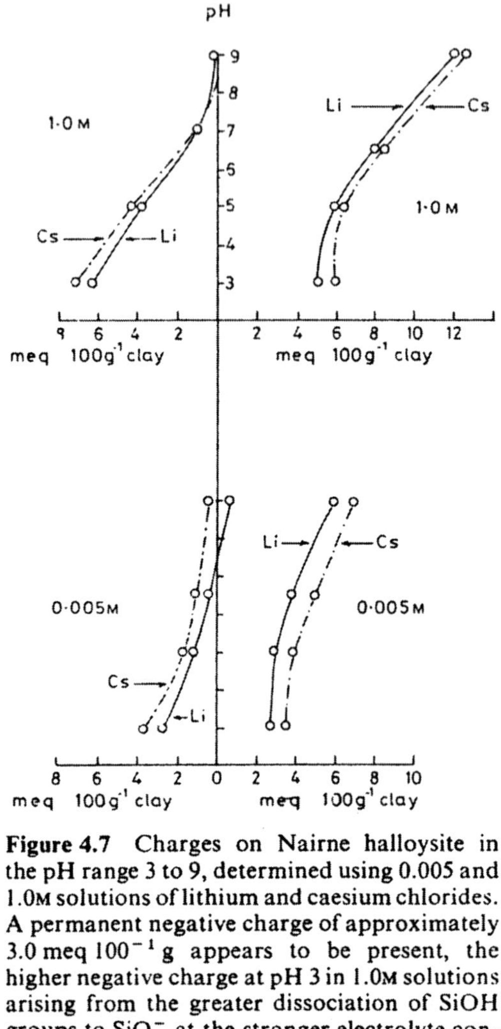

Figure 4.7 Charges on Nairne halloysite in the pH range 3 to 9, determined using 0.005 and 1.0M solutions of lithium and caesium chlorides. A permanent negative charge of approximately 3.0 meq 100^{-1} g appears to be present, the higher negative charge at pH 3 in 1.0M solutions arising from the greater dissociation of SiOH groups to SiO^- at the stronger electrolyte concentration (Mashali, 1977)

interlayers occur rather commonly in soil clays, and it could be that the conditions which lead to interlayer formation could produce 'aluminium outerlayers'. However there is no experimental evidence for this, and unless the soil solution to which they were exposed was considerably richer in aluminium than silicate, it would not be expected that the alumina surface would persist as such.

Several studies of the weathering of biotite have shown that iron is expelled from the mineral lattice during the alteration process, and appears as a hydrous oxide coating on the surfaces of the hydrobiotite or vermiculite particles formed (Roth, Jackson, de Villiers and Volk, 1967; Roth, Jackson and Syers, 1969;

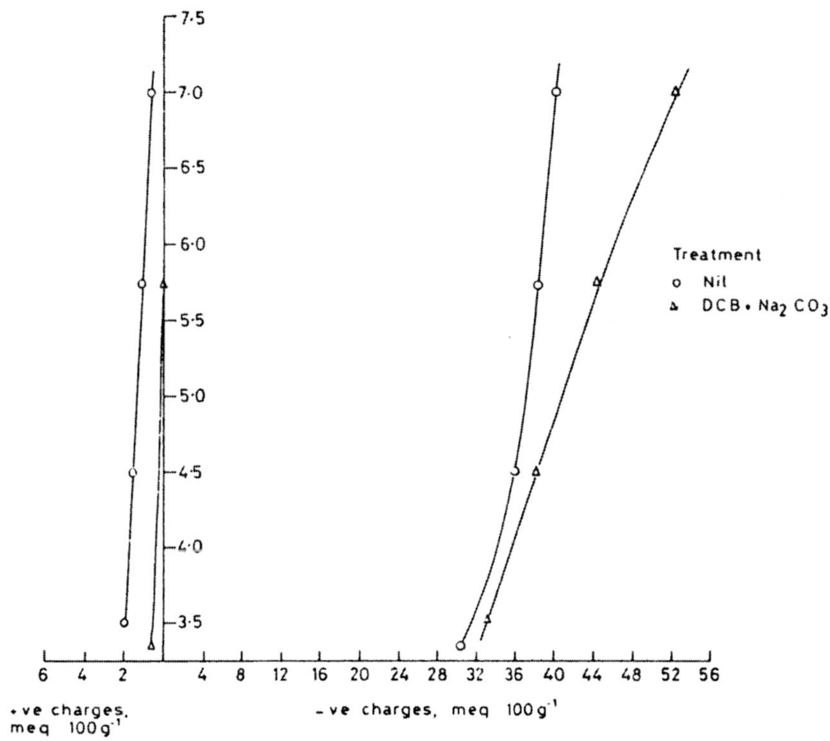

Figure 4.8 Charges on the clay fraction from the subsoil of the Urrbrae loam, Adelaide, South Australia, determined using 0.005M CsCl before (O) and after (△) treatment with sodium dithionite, citrate and bicarbonate (DCB) and 5 % sodium carbonate to remove hydrous oxides of iron and aluminium (cf. Plate 4.1 a and b)

Farmer *et al.*, 1971). There is some evidence that in soils containing vermiculite these hydrous iron oxides persist as interlayer or surface coatings, since treatment with dithionite to remove them increases the negative charge (Roth, Jackson and Syers, 1969). In several other soils, including a wide variety of red, tropical soils removal of iron oxides did not lead to an increase in negative charge (Deshpande, Greenland, and Quirk, 1964; Tweneboah, Greenland, and Oades, 1967). Further critical studies are needed to determine how commonly oxide coatings occur. The fact that cation exchange capacities of soils can usually be related to the clay mineralogy—high in soils containing vermiculite and montmorillonite, moderate to low in soils containing mica, low in soils containing kaolinite—suggests that they are more the exception than the rule. For the soil shown in Plate 4.1a removal of the surface coating did not make a very great change in the charge characteristics (Figure 4.8) indicating that the crust and underlying material are similar in composition.

4.2.3 Organic Coatings on Clay Particles

Clay minerals and hydrous oxides are strong adsorbents for many organic compounds. As soils contain both, a natural clay–organic complex develops in most soils. The mechanisms of association between clay and organic material are discussed in detail in relation to adsorption in the companion volume. Some organic polymers are known to be adsorbed to form a surface coating on clay particles (Greenland, 1965). Some uncharged soil polysaccharides may behave in this way, although studies of their adsorption by montmorillonite suggest that precipitation reactions may predominate over surface adsorption (Parfitt and Greenland, 1970).

What is known of the behaviour of humic and fulvic materials at clay surfaces suggests that they do not form a surface coating. The dominant mechanism of their association with clays appears to be through complexing with calcium, iron or, probably most importantly, aluminium ions at the clay surface (Greenland, 1971). They only become associated with siloxane surfaces by virtue of interaction with exchangeable ions or precipitated hydroxy polymers at the surface. They are not adsorbed in the interlayers of expanding lattice minerals, except for fulvic materials of exceptionally low molecular size which have been shown to be able to penetrate the interlayer space of aluminium montmorillonite in extremely acid (pH 2.5) conditions (Schnitzer, 1968). Interlamellar adsorption of organic compounds in soil clays seems to be very rare. Thus in soils surfaces of 2:1 phyllosilicate minerals are probably normally free of organic material, except where hydrous oxide coatings act as an intermediary. On the other hand hydrous oxides of iron and aluminium in soils are often associated with humic and fulvic materials. The adsorption mechanism is probably ligand exchange with surface hydroxyls. The strength of association suggests that there may be several carboxyl groups involved, making it difficult to displace the humic group.

However it also appears to be true that much of the humic group remains unattached to the oxide surface.

4.2.4 Calcium Carbonate Particles and Coatings

A very large number of soils contain calcite or amorphous calcium carbonate. Surprisingly little work has been done on the detailed constitution and surface properties of such materials. As with many other particles in soils it is probable that their properties are modified by foreign ions on their surface layers.

4.3 SURFACE AREAS OF SOIL PARTICLES

4.3.1 General

To develop quantitative relationships between surface properties of soil colloids, and the behaviour of soils, it is normally essential to measure the surface area of the soil particles. Unfortunately in any soil the particles normally consist of a very diverse assortment of shapes and sizes. The areas of finely divided solids of this type must usually be determined by adsorption of molecules of known dimensions, using some method to establish the amount required to form a monolayer. It is then simple to multiply the number of molecules in the monolayer by the projected area of each to obtain the surface area of the solid.

The surface area of clays is large. It is easy to calculate in the case of a 2:1 mineral cleaved into individual layers (the thinnest particles that can be obtained in any given instance), that the total surface area is approximately 760 $m^2\ g^{-1}$. This comes about from the product of the *a* and *b* dimensions of the unit cell (e.g. 0.515 nm × 0.892 nm for pyrophyllite) for each of Avogadro's number of such cells, divided by their unit cell atomic weight ($Si_8Al_4O_{20}(OH)_4$ = 728). This has to be multiplied by 2 (each unit cell has a top and a bottom) to give the result in m^2 per g. So for pyrophyllite (ignoring non-basal surfaces) the calculated surface area is

$$\frac{(0.515 \times 0.892 \times 10^{-18}) \times (6 \times 10^{23}) \times 2}{728} m^2\ g^{-1} = 757\ m^2\ g^{-1}$$

Although this might seem to be only a theoretical calculation, it is a highly significant result of practical value. For clays such as the smectites and vermiculites which exhibit interlayer swelling it is the actual area available to water molecules and exchange cations. The same calculation applies to both the pyrophyllite example and to montmorillonite because the unit cell dimensions and atomic weight of the two clays are so similar. This is true for all the 2:1 minerals, and the figure of 760 m^2 is often used as a 'marker' for this whole group. Consider for example an 'illite' or clay mica which has a measured area of 76 $m^2\ g^{-1}$ by the nitrogen method described below. On *average*, each crystal or stack must consist of 10 lamellae, so that the measured area is not the 'theoretical' one of 760 but the observed 76 $m^2\ g^{-1}$.

In a similar way, micas often have very low areas indeed, because the high surface charge density ensures that the stack size is large. Such a relationship is evident from Chapter 2. These low areas give rise to the view that such clays are of low exchange capacity, and low chemical activity. In terms of charge per unit area, this is mistaken; neglect of the extensive factor easily leads to the error.

It is not possible to do simple area calculations for other soil materials. The above, for example, has tacitly neglected edge areas, assuming them to be insignificant for platey 2:1 minerals. This is not true for 1:1 minerals like kaolinite which is platey but thick (Plate 2.1) and positively misleading for minerals of a more complex topography like halloysite. Crystal shape is not definable for the oxides in their various modifications; particle size and the degree to which they can be described as 'amorphous' is a factor with both hydrous oxides and organic matter.

This means that the surface area of a soil cannot be predicted from a knowledge, however detailed, of components as individuals. Structural organization is the determining factor. The use of surface area and associated porosity measurements is a key to this organization, as well as an aid to the understanding of chemical behaviour.

4.3.2 Surface Area Measurements

There are basically two different methods that can be used. The simpler is direct observation by the electron microscope which, whilst of real value for more regular solids such as synthetic catalysts, demands an immense labour if sufficient observations are to be made to cover the variability in particle size and shape in the average soil. It is however occasionally worth such labour for a few samples which can subsequently be used to calibrate results obtained by an indirect method. The second is a whole group of related methods depending upon an adsorption process of some kind, and based upon the simple idea that the amount of material adsorbed is related to the surface area available. Adsorption methods may be divided into vapour phase and solution phase techniques. The first of these is the more important. The text by Gregg and Sing (1967) contains both comprehensive practical details of laboratory determinations and an elegant account of the theory necessary to understand them.

Vapour phase sorption onto solids

The number of molecules of a gas which will sorb onto a given solid surface at equilibrium depends upon the partial pressure of the gas in question and the area of surface available. The variation in amount sorbed, x, with change in partial pressure, p/p_0, at any temperature T, is a function of the gas–solid interaction energy and the coverage of the surface; the relationship between x and p/p_0 is the adsorption isotherm, where p is the actual vapour pressure at temperature T, and p_0 is the saturated vapour pressure at that temperature. The vast majority of all known physisorption isotherms may be grouped into only six main types. Figure 4.9 is based on the classification of Brunauer, Deming, Deming and Teller (BDDT) (1940).

Of these types I, II and IV have so far been found to be relevant to soil systems. Type I is the Langmuir isotherm, and frequently describes adsorption from solution (Chapter 10) but seldom gas sorptions. The Langmuir isotherm is in fact a special case of the more general BET equation which is discussed below. Type II isotherms are reversible and obtained for the sorption of nitrogen on non-porous or macroporous solids. An approximation to it is given by adsorption of several gases onto a coarse grained kaolinite but most clays are too microporous to give this behaviour. Some oxides have been specially prepared which exhibit Type II isotherms, and these are of great importance as non-porous reference solids. However, Type IV is incomparably the most important

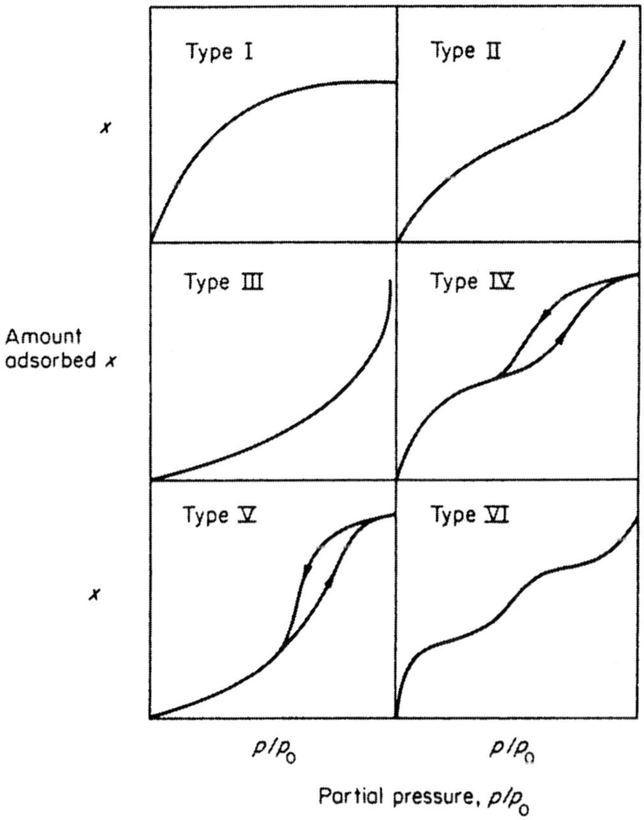

Figure 4.9 The six types of isotherms for the adsorption of gases and vapours by solids (after Brunauer, Deming, Deming and Teller, 1940)

so far as soils are concerned and is usually shown when non-polar gases, such as nitrogen are sorbed onto their surfaces. The shape of the isotherm indicates a strong interaction between gas molecules and the surface. The hysteresis loop on the desorption path is very common, and occurs over the capillary condensation region. To complete the picture, Types III and V are characteristic of weak interaction (convex slopes) and Type VI is a stepwise sorption never yet shown by a soil to an inert gas. These three types are difficult to analyze.

The BET equation

Although there is no one general theory available to cover all types of sorption behaviour, the best known description, the BET equation (Brunauer, Emmett and Teller, 1938) works satisfactorily for Type IV isotherms and has been applied widely to soils. Basically, BET is an extension of the Langmuir treatment to multilayer sorption. The important assumption is that the first layer of gas molecules is attracted to the surface with an energy greater than that of the second, third and subsequent layers. The energy of the first layer for a given gas is characteristic of the solid: the heat of subsequent sorption is simply the heat of condensation of the gas with respect to its own liquid phase. Such layers build up in an unrestricted fashion as p/p_0 increases. When it reaches unity and the pressure of the gas is the saturation vapour pressure, p_0, then the number of layers is infinite. Summation along the isotherm gives the BET equation

$$\frac{p}{x(p_0-p)} = \frac{1}{x_m C} + \frac{(C-1)p}{x_m C p_0} \tag{4.2}$$

where x_m is the amount of gas (measured either as a volume or directly as a weight) which forms the monolayer, and C is a constant which is related exponentially to the first layer heat of adsorption. Experimental points, therefore, are plotted as $p/[x(p_0-p)]$ against p/p_0. The relationship is found to be linear only over the range from about $p/p_0=0.03$ to around 0.2 for soils (sometimes a little higher for some other solids), and obviously the multilayer build up combined with the pore geometry of the solid is straining the simplified BET approach. However over this restricted range, the sum of the measured slope and intercept of the graph gives $1/x_m$. Then, assuming a close-packed arrangement of nitrogen molecules on the surface which gives a mean coverage of 0.162 nm^2 per N_2 molecule, the surface area being covered by the monolayer amount can be calculated. A typical BET plot for a soil clay is shown in Figure 4.10.

Experimental methods for gas sorption studies

A non-polar gas is normally used to measure the external surface areas of soils and clays. Because any sample has to be 'degassed' of any sorbed molecules before beginning a surface area determination, all adsorbed water, including interlamellar water, is lost. This causes any expanding clay lattice to collapse and the non-polar gas cannot subsequently enter interlayer areas. Nitrogen is almost universally used as the adsorbent. However it has the disadvantage of a specific field gradient—quadrupole contribution at the surface, and this enhances the C value. Luckily most soils so far investigated have high C values, and the effect is relatively very small. Krypton is sometimes used for very low area solids. It requires the use of special apparatus since the saturated vapour pressure at 77 K is only 2.35×10^2 Pa.

It is convenient to conduct inert gas sorptions at 77 K as this is the boiling point of the easily obtainable liquid nitrogen. Because sorption involves a phase change from gas to liquid it is best done at or below the boiling point of the liquid.

Soil aggregates can be used for gas sorption without any pretreatment if the

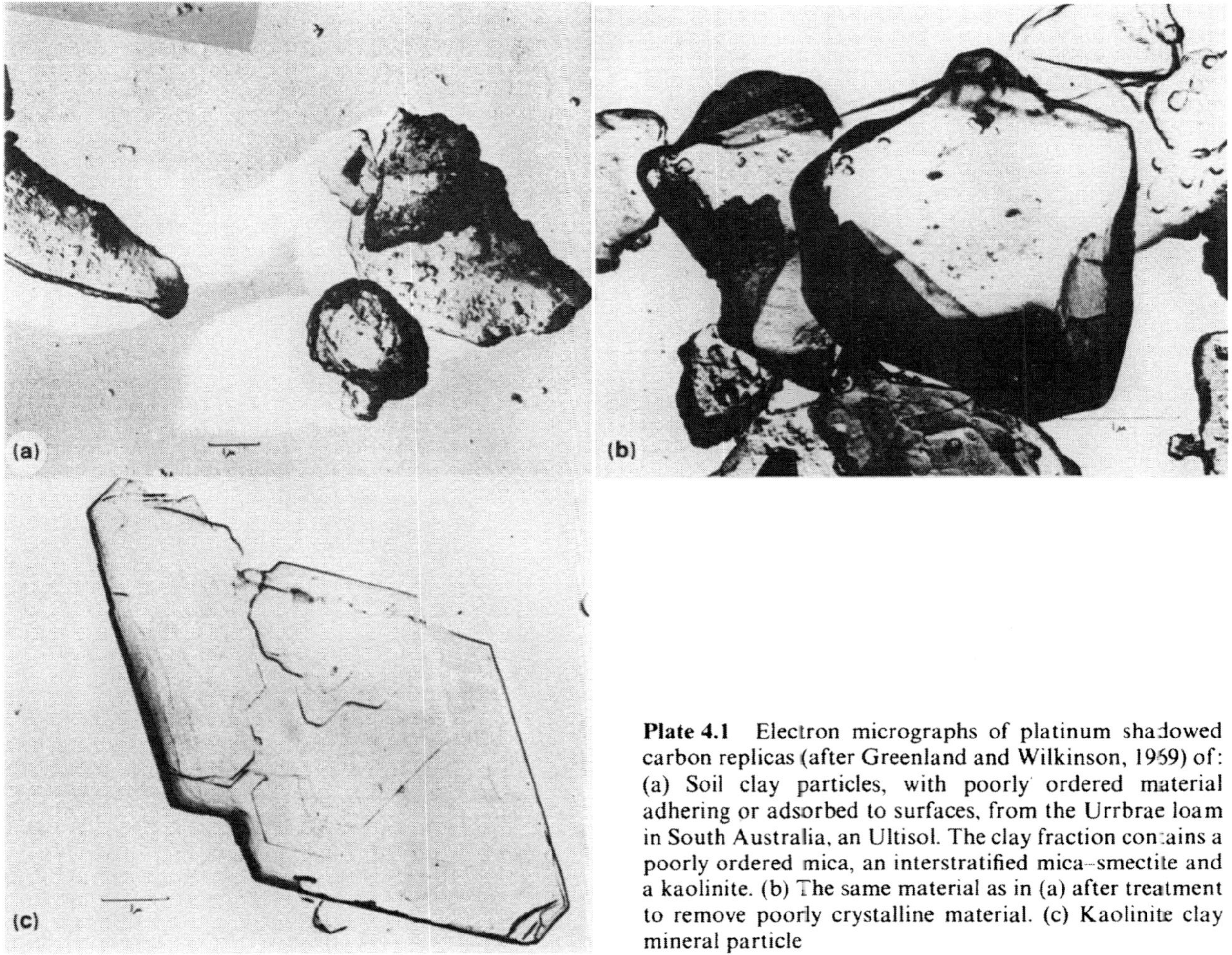

Plate 4.1 Electron micrographs of platinum shadowed carbon replicas (after Greenland and Wilkinson, 1959) of: (a) Soil clay particles, with poorly ordered material adhering or adsorbed to surfaces, from the Urrbrae loam in South Australia, an Ultisol. The clay fraction contains a poorly ordered mica, an interstratified mica–smectite and a kaolinite. (b) The same material as in (a) after treatment to remove poorly crystalline material. (c) Kaolinite clay mineral particle

[*Facing page 340*

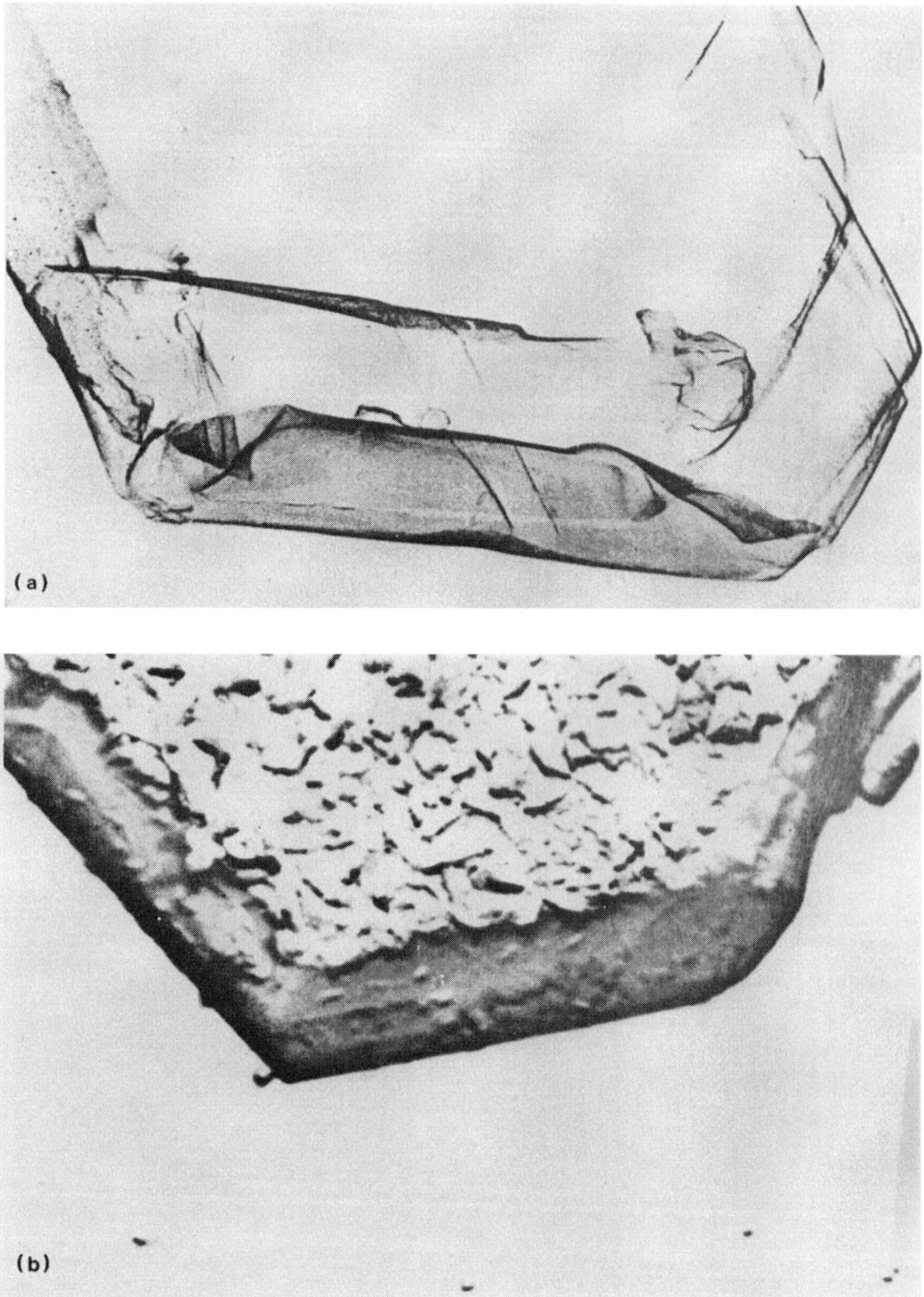

Plate 4.2 Dickite particle, (a) before and (b) after precipitation of ferric hydroxide at pH 3. Note: the edge faces which develop a predominantly positive charge at pH 3 remain essentially free of the similarly charged iron hydroxide, whereas the basal surfaces, which are either uncharged or negatively charged, accept a surface coating of the hydroxide. (From Greenland, 1975.) Both $\times$ 18,000

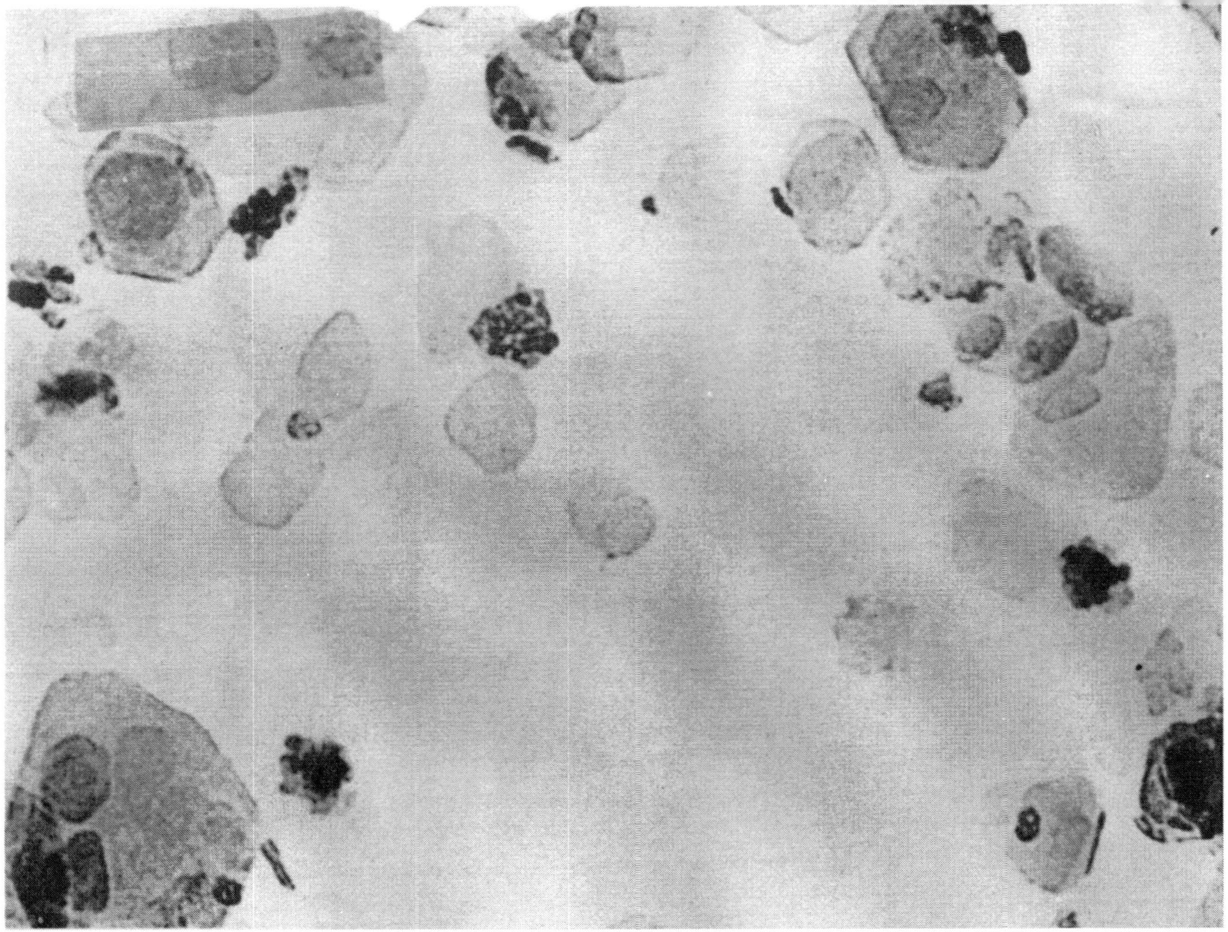

Plate 4.3 Electron micrograph of clay particles from a tropical red earth at Katherine, northern Australia. The small, electron dense particle ('blackberries') are probably hematite or ferrihydrite. The translucency of the small hexagons suggests they are free of surface iron. Dithionite dissolves iron and removes the electron dense material, leaving only kaolinite (Greenland, Oades and Sherwin, 1968). × 140,000

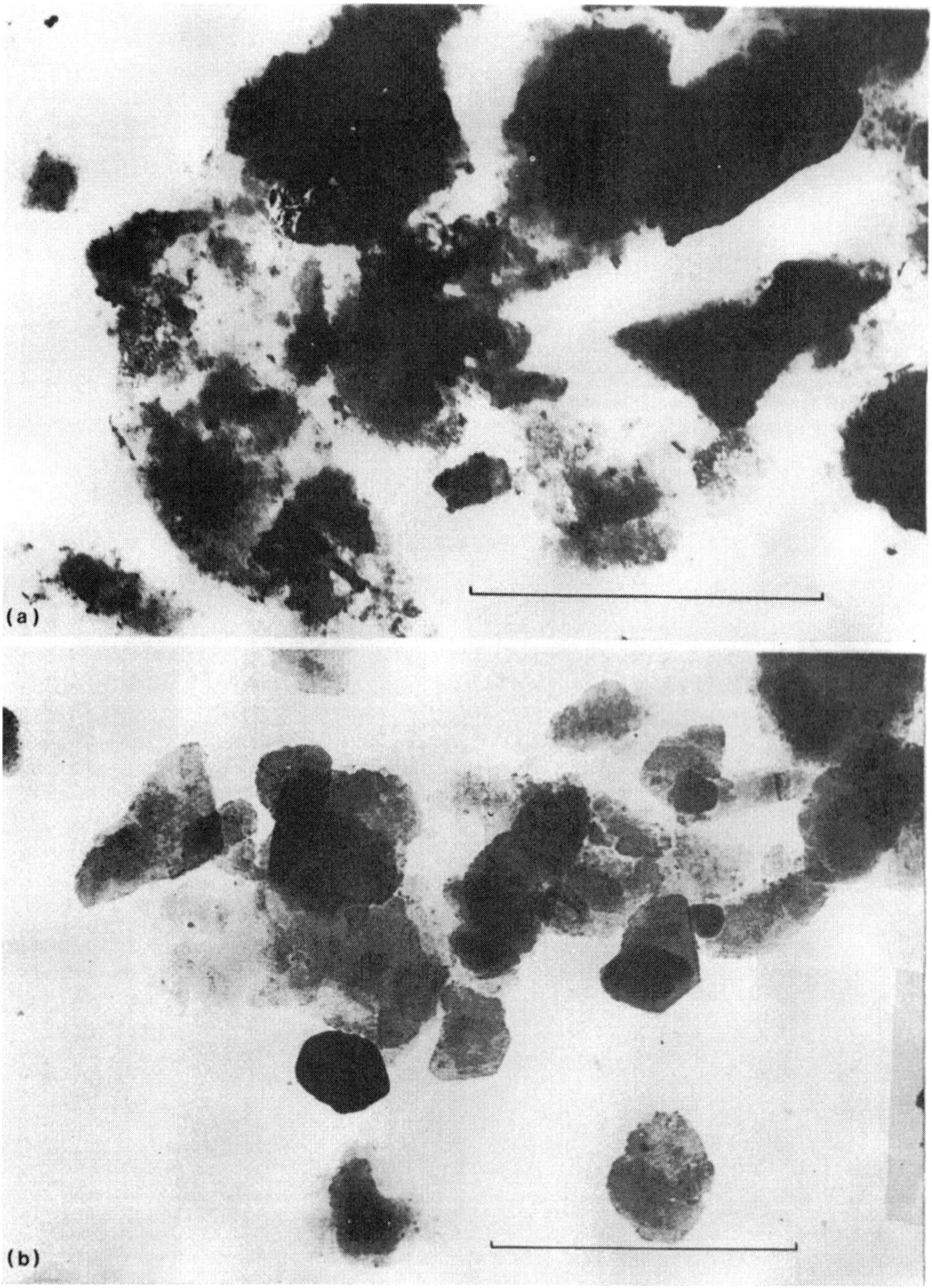

Plate 4.4 Electron micrograph of clay particles from a seasonally waterlogged soil from Bangladesh. (a) before and (b) after dissolution of iron-rich material by treatment with dithionite (Habibullah, Greenland and Brummer, 1971). Markers represent 1 μm

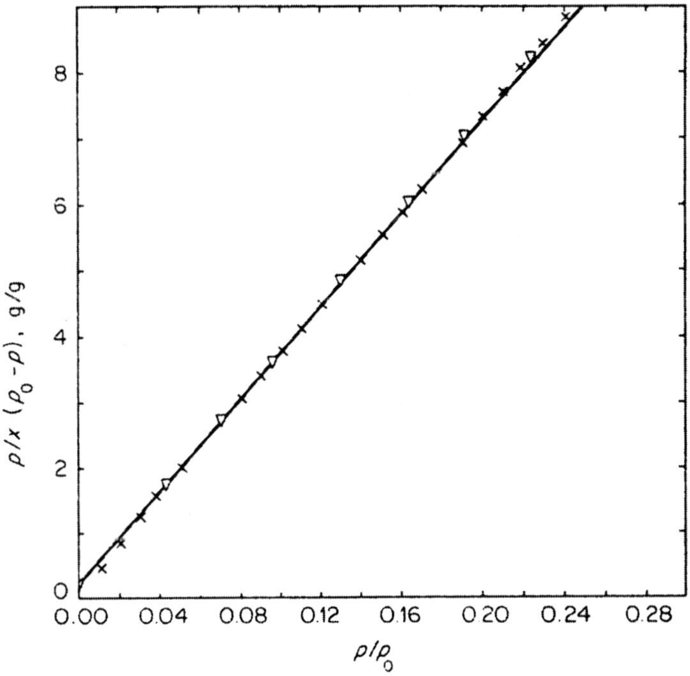

Figure 4.10 BET plot for the adsorption of N_2 at 77 K by clay fraction from the Denchworth clay (a Mollisol). Slope = 35.3 g/g, intercept = 0.23 g/g. Monolayer capacity, x_m = 0.0282 g/g, C = 157. Derived 'BET' area = 98 $m^2 g^{-1}$

measurement of the area of the whole soil is required. Very frequently a clay fraction is extracted and used, since in most cases this fraction contributes most to the result. If the sample is in the form of a fine powder and is to have its area measured by the gravimetric adsorption method it should be lightly pelleted. The necessary degassing under vacuum (10^{-2} Pa) at an elevated temperature for several hours should be at temperatures not beyond 70 °C, because changes due to dehydroxylation of hydrous oxides and carbonization of organic matter occur above this temperature. Organic matter has a strong influence on surface area determinations, and even when present in small amounts has a major effect on determined areas (Burford, Deshpande, Greenland and Quirk, 1964). The BET method yields easily reproducible values for the external surface area of soils and clays. The figures obtained are not necessarily correct in an absolute sense, but valuable for comparative work.

4.3.3 The Micro- and Mesoporosity of Soils

A feature of the BET analysis is the unrestricted build up of molecular layers of sorbed gas. It takes no account of porosity. Such pores will of course have to be very small (less than a few nm) by soil standards in order to affect the sorption. The terminology in use in this field is based upon the IUPAC (1972) classification of pore sizes:

Micropores <2.0 nm equivalent cylindrical diameter (e.c.d.)
Mesopores 2.0 to 50 nm e.c.d.
Macropores >50 nm e.c.d.

(These names are often misleading if used without clear definition in the soils literature; they would have a different significance, for example, to an optical microscopist, who often terms all pores not visible in the optical microscope, i.e. all those less than about 1000 nm e.c.d., micropores.)

Micropores are of a size in which four or fewer N_2 molecular layers can form, so that condensation is inhibited. Also, the potential fields from opposite walls of the pore overlap and this causes an apparent increase in attractive force. The C constant increases, the isotherm becomes more convex, and the 'knee' sharper. As a result the isotherm is distorted towards increased adsorption. The surfaces contained in micropores are therefore thought to become covered by a process of pore filling rather than by a 'BET' monolayer build up. Unfortunately in a solid of unknown microporosity a simple analysis cannot sort out the contributions of surface adsorption and micropore condensation. Thus the BET area will not be an absolutely correct figure. A considerable amount of work has been done in the determination of sorption in the low pressure (Henry's Law) region, where surface coverage is less than that of a monolayer. Only well characterized solids have been studied, and there are no reports of results obtained with soils. A promising approach is that of Gregg and Langford (1969). They treated a carbon sample at 77 K with n-nonane vapour, and this filled the micropores and covered any external surfaces. On raising the temperature it was found that the nonane could not have been evaporated from the micropores because a nitrogen isotherm subsequently measured on the 'outgassed' sample gave a BET area which agreed closely with that determined by electron microscopy. The n-nonane thus, in this example, acted as a selectively strongly sorbed molecule for the micropore volume.

So far as soils are concerned there generally seems to be a significant micropore volume, especially with clay soils. It should be emphasized once more that this is not interlayer volume but the pores formed between the irregularly aligned clay crystals.

Pore size distribution measured by gas adsorption

Pore volumes may be derived by analyses of gas adsorption isotherms. The equation which relates the relative vapour pressure of a liquid held in a capillary with its radius (r) is the Kelvin equation

$$\ln p/p_0 = -\frac{2V}{rRT}\gamma \cos \theta \tag{4.3}$$

where γ is the surface tension, V the molar volume of the liquid, and θ the contact angle between the liquid and solid. Cylindrical capillaries are of course highly unlikely in soil aggregates and the shape of the Type IV isotherm which soils so frequently exhibit indicates that the dominant pore shape is that which occurs between parallel plates. Since the Kelvin equation given above has as its more general form

$$\frac{dv}{ds} = -\frac{V\gamma \cos\theta}{RT \ln (p/p_0)} \tag{4.4}$$

where v is the volume of the capillaries and s the surface area of their walls, we can replace r, the radius for capillaries, by d, the distance apart of the plates.

If now the simplifying assumption is made that the contact angle is zero, and that, for nitrogen, the value of V is 34.6 cm^3 $mole^{-1}$ and γ is 8.85 dyne cm^{-1}, the equation reduces to

$$d = \frac{4.14}{\log (p/p_0)} \tag{4.5}$$

This equation governs the emptying or filling of parallel plate capillaries in the mesopore range, and its effect is superimposed on BET multilayer formation. When a given partial pressure is reached on a desorption isotherm and the pores of the corresponding plate separation d empty, a film of nitrogen of thickness t corresponding to the amount adsorbed at that pressure, remains on the walls. The true size of the pore which empties therefore at this partial pressure is $(d+2t)$. The value of t has been determined experimentally for many non porous solids, and a number of theoretical expressions have been given. Table 4.3 gives some values involved in the mesopore region.

In brief, the calculation of pore size distribution begins with the total porosity at the top of the desorption branch of the isotherm (Figure 4.11, total nitrogen

Table 4.3 Calculated parallel plate separations $(d+2t)$ nm corresponding to the partial pressure (p/p_0) of sorbed nitrogen at 77 K. t is the thickness of the layer of nitrogen retained when the capillary empties according to the Kelvin Equation (4.5)

p/p_0	t, nm	d, nm	$(d+2t)$, nm
0.98	2.21	48.1	52.52
0.95	1.63	18.5	21.76
0.90	1.27	9.06	11.60
0.80	0.98	4.27	6.23
0.70	0.85	2.67	4.37
0.60	0.75	1.87	3.37
0.50	0.68	1.38	2.74
0.40	0.62	1.04	2.28
0.30	0.56	0.79	1.91

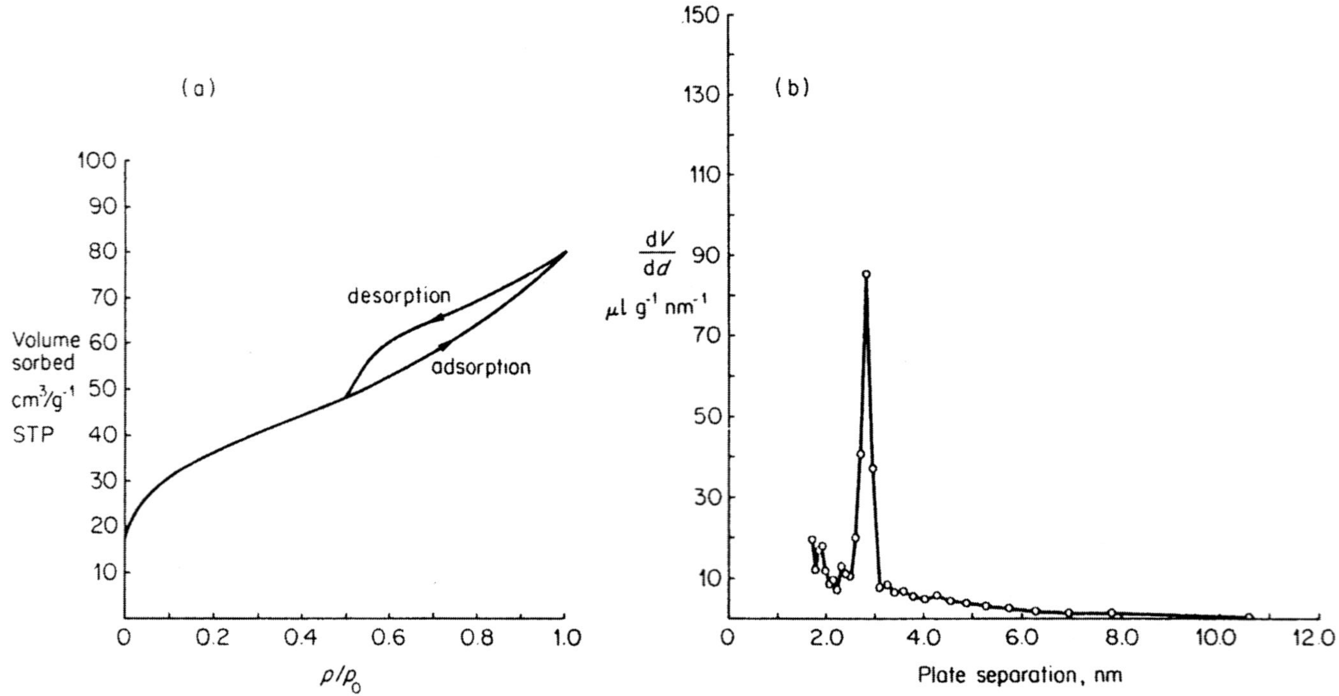

Figure 4.11 (a) Adsorption isotherm for nitrogen on a soil clay, and (b) the pore size distribution derived from the isotherm in (a) for the range 1.7 to 10.0 nm. In (a) an hysteresis loop, typical of Type IV isotherms, is seen. It is usually attributed to irregularly shaped capillaries forming necks which tend to retain sorbate during the desorption, but not during the adsorption, process

uptake, calculated as a liquid volume, at high p/p_0), and applies the Kelvin equation to the remainder of the curve, in small decrements, to $c. p/p_0 = 0.2$. The t correction is made and the area of surface corresponding to this decrement calculated. This procedure is carried on down the isotherm in small steps. The method was originally described by Innes (1957). It is a calculation ideally suited to a computer.

There are perhaps two conclusions of general interest from this analysis:

(i) Most soils exhibit considerable mesoporosity at around 3 nm and this seems to be due to pores found between the finest clay particles present.
(ii) The cumulative surface area calculated from the successive decrements of the isotherm has a lower value than the BET area, and this indicates a significant number of micropores to be present. However the combination of pore filling (in micropores) and capillary condensation (in mesopores) makes analysis uncertain. Much more work with selective adsorbate or 'probe' molecules is required with soil systems.

A typical isotherm, with its derived pore size distribution is shown in Figure 4.11.

Structural alteration on drying

The vacuum degassing procedure produces high evaporation forces within the soil aggregates and one result of this is an enhancement of the observed mesoporosity. Other, slower, methods of drying, for example, vapour exchange with methanol, followed by ether, then evaporation) or critical point drying (vapour exchange with methanol, followed by liquid CO_2 subsequently evaporated at its critical point in order to avoid a liquid meniscus) give rise to smaller values for the mesoporosity than direct drying. The surface area, however, is little affected, and the labour involved in slow drying seems not to be rewarded in this respect.

4.3.4 Use of Polar Molecules for Surface Area Determination

Adsorption isotherms for polar molecules such as water on soils and clays can be determined using either volumetric or gravimetric techniques. Given the various uncertainties involved it is usual to make the determination of the isotherm by relatively crude but simple methods. For instance, partial pressures of water (relative humidities) are given precisely at a fixed temperature by the saturated solutions of various salts. Thus the range can be covered by a series such as:

Saturated solution	*Relative humidity at 25 °C*
$CuSO_4$	98
$ZnSO_4$	90
NH_4Cl/KNO_3	72.6
$Mg(NO_3)_2$	55

continued on next page

Saturated solution	*Relative humidity at 25 °C*
$Zn(NO_3)_2$	42
$CaCl_2$	32.3
CH_3COOK	20
LiCl	15
$ZnCl_2$	10

Weighed, degassed samples are placed in a series of vacuum desiccators containing saturated salt solutions in a constant temperature room. After at least a week the samples are reweighed and an isotherm constructed.

If only comparative values of area are required and there is to be no attempt at isotherm analysis, use can be made of the fact that the water monolayer is completed for many soils and clays at approximately $p/p_0=0.2$ (Quirk, 1955). It is therefore possible to obtain an estimate of the BET water area by a one point determination at this relative humidity. Large errors may result, though, if the exchange cations are very different from one sample to another (see below).

The BET water area

The disadvantages of using a polar molecule is that it is more strongly adsorbed on specific sites on the surface, and so not uniformly distributed. Water molecules are attracted to the bare exchange cations and cluster round them. This means that the monolayer and multilayer processes overlap, and so make the BET approach of doubtful validity, as the meaning of the monolayer coverage is quite arbitrary. In addition, the geometry of clustering depends upon the exchange ion. For instance, the calcium ion attracts water strongly, but potassium is far less ready to hydrate. The result is that the apparent area of a calcium saturated montmorillonite is much greater than that of the same clay when potassium is the exchangeable cation.

The advantage of using polar ions is that they explore the internal area of swelling lattice clays as well as the external, and this provides a direct indication of the extent of interlayer sites available for sorption and ion exchange in a soil. The nitrogen and water areas for kaolinite clays are closely similar but the water area will be very much higher than the nitrogen area for a soil containing a smectite or clay in which vermiculite or smectite layers are interstratified with others.

A further complication arises in relation to the influence of the exchange cation when interlayer adsorption of water occurs. The ion influences not only the clustering of the water molecules, but also the forces between the opposing faces of the clay lamellae. When the charge density of the lamellae is high, and the hydration energy of the cation low, the attractive forces may exceed the hydration energy and the interlayer space will close. Thus when potassium is the exchange cation on vermiculite, the basal spacing normally falls to 1.0 nm corresponding to the thickness of the aluminosilicate layer, whereas with calcium or magnesium present it will remain at 1.4 nm, with two layers of water molecules in the interlayer region. Of course the loss of water from the interlayer region depends on the external as well as internal conditions. The electrolyte concentration or the relative humidity of the surrounding medium compete with the hydration energy of the interlayer cations and the surfaces for water molecules. At very low external solution concentrations water is not drawn out of the interlayer space in vermiculite, even when caesium is the exchange cation, and thus the charge on vermiculite as well as of other clays can be measured with 0.005M caesium chloride (Greenland, 1974). With lithium or sodium as the exchangeable cation, the extent of swelling is linearly related to the square root of the electrolyte concentration of the surrounding medium (Figure 6.10).

Because of the ease with which the 'quasi-crystals' of a smectite can separate, the external surface area of any one sample of the material can differ substantially depending on its pretreatment. Thus Wyoming montmorillonite shows considerable differences in the surface area determined by nitrogen adsorption when saturated with different exchangeable cations (Table 4.3). High areas obtained when Cs is the exchange cation may arise from partial penetration of some interlayer regions (Aylmore, Sills and Quirk, 1970).

Where water is adsorbed in interlayer regions, the application of the BET equation and determination of the surface area assuming each H_2O molecule to cover 0.12 nm^2 gives an area well below that expected (760 $m^2 g^{-1}$ for Wyoming montmorillonite) because in the interlamellar regions there is only one molecular layer of water, shared by opposite surfaces. Thus the determined area should be doubled, or each adsorbed water molecule should be assumed to cover 2×0.12 nm^2. As mentioned above, where no interlayer sorption occurs, good agreement is obtained between N_2 and H_2O BET areas (Table 4.3).

Other polar molecules

Many other polar molecules have been used to determine specific surface areas of soils and clays, but usually from solution or liquid rather than gas phase. Examples are ethylene glycol, methylene blue, *o*-phenanthroline, and cetyl pyridinium bromide. It is worth mentioning the first and last of these in more detail.

Ethylene glycol sorption for surface area measurement

Although ethylene glycol has been widely used for relatively rapid measurements (Brindley, 1966), it suffers from the same theoretical disadvantages as water. A typical procedure is to soak the dried sample with the glycol, place it in a vacuum desiccator at room temperature and periodically remove and weigh it. The liquid in the pores evaporates steadily with time, but when the 'monolayer' condition is reached the weight remains sensibly constant from one weighing to another because of the strong attraction of the surface for the first layer of adsorbent molecules. In practice it is often difficult to identify the plateau in the

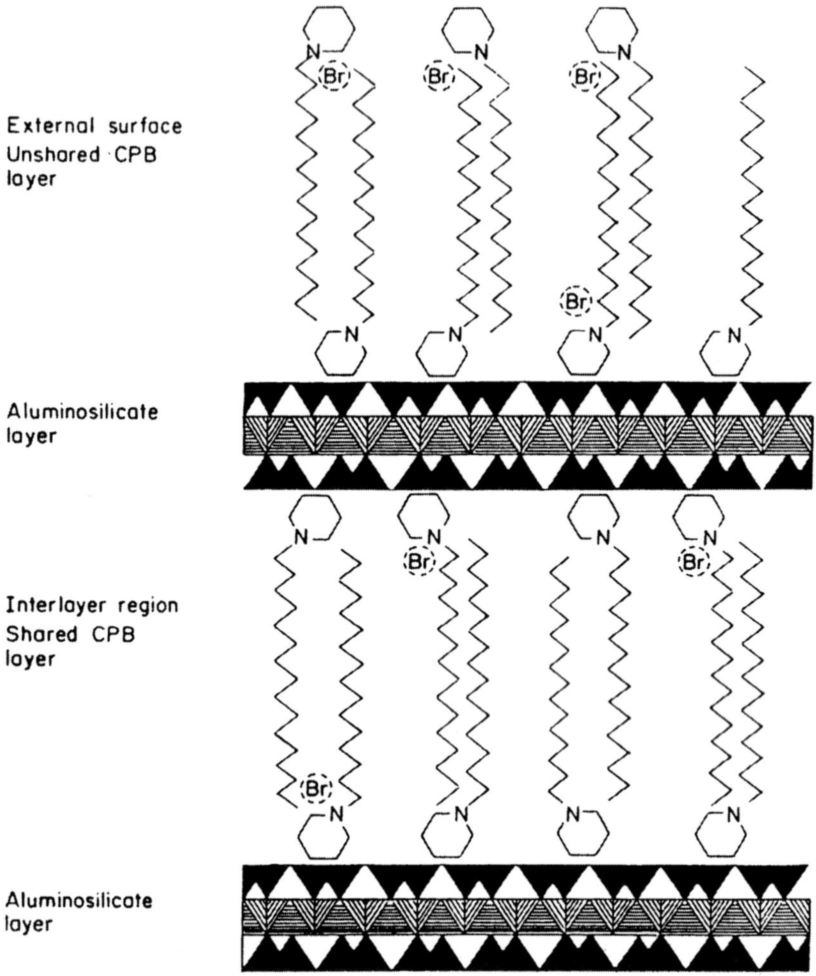

Figure 4.12 Cetyl pyridinium bromide (CPB) adsorption on the internal and external surfaces of a smectite clay at maximum adsorption (after Greenland and Quirk, 1963). The unshared layer on external surfaces leads to a mean surface coverage per cetyl pyridinium group of 0.27 nm^2, whereas in the interlayer region each adsorbed pyridinium group requires its projected area of 0.54 nm^2. The $d(001)$ value for the montmorillonite—CPB complex is 4.2 nm, which, allowing for the aluminosilicate layer thickness of 1.0 nm, gives 3.2 nm for the CPB layer corresponding closely with the known dimensions of CPB ions (figure not to scale).

drying curve. However, for some soils the method seems satisfactory, and of course many samples can be determined simultaneously. The mean area per adsorbed molecule is approximately 0.33 nm^2, but varies depending on the exchange cation, and between interlayer and external sites.

N-Cetyl pyridinium bromide (CPB) sorption for surface area measurement

This was developed as a suitable technique by Greenland and Quirk (1962, 1964). The molecule is $C_{16}H_{33}C_6H_5N^+ Br^-$, and X-ray work has shown that at the plateau of the adsorption isotherm it is sorbed with the pyridine ring on the surface and the hydrophobic cetyl chain projecting outwards at right angles (Figure 4.12). In this position, the projection of each alkylpyridinium covers 0.54 nm^2 on the surface. The ions however form a bilayer on external surfaces, because the mean coverage per molecule on a non-expanding clay lattice (area known from nitrogen sorption) is exactly half this—0.27 nm^2. On an internal surface the mode of sorption is the same, but because each interlayer region has a top and bottom surface, the effective area per molecule is 0.54 nm^2.

The cetyl pyridinium bromide is sorbed according to a Langmuir-type isotherm, Type I, and so the maximum uptake by the surface is given by the plateau of the isotherm. Concentrations in solution can be determined with good precision by ultraviolet spectrophotometry. If the external area of the sample can be determined using nitrogen, the amount of CPB which would be required to saturate this surface at 0.27 nm^2 per molecule may be calculated. All that remains must be sorbed on internal surface at 0.54 nm^2 per molecule, and so this area can be separately evaluated. The CPB method is therefore capable of differentiating internal from external area, quantitatively (Table 4.4).

Table 4.4 Specific surface areas of clays determined by N_2, H_2O and CPB adsorption (from Greenland and Quirk, 1962 and 1964)

Clay				Surface areas, $m^2 g^{-1}$		
				Total by		
Source	Mineral type	Exchange cation	External By N_2	H_2O	CPB	Internal, from N_2 and CPB
Wyoming	montmorillonite	Na	14	300	800	784
		Ca	40	340	600	560
		Cs	146	—	—	—
Kenya	vermiculite	Mg	<1	—	720	720
Morris Illinois	illite (interstratified clay mica)	Na	70	—	186	116
Sarospatak	illite (interstratified mica–smectite)	—	43	57	152	109
Fithian Illinois	illite (clay mica)	Na	93	—	96	3
Lewistown, Montana	kaolinite	—	14	13	11	—
Malone, Australia	kaolinite	—	17	21	19	—

The limitations of the method are first that CPB is unable to cover surfaces of a low surface density of charge (less than about 1 site/2 nm^2); second, that it cannot penetrate clays with rigid, microporous structures; thirdly that, to ensure uniform penetration of the lamellae by the CPB, pretreatment with Na^+ is often required; and finally that, since organic matter interferes with the sorption, soils with anything but low organic C contents must be pretreated either with hydrogen peroxide or sodium hypobromite. Both cause other changes in the soil materials which may affect surface area directly. Despite all these points, which apply to other adsorption methods also, the method works well for some soils.

It should perhaps be emphasized that the pyridinium cation is not only neutralizing the negative charges on the clay surface—if it were, it would become a method for CEC—but covering surface to give a statistical bilayer. This is because the Br^- ion accompanies cations adsorbed to complete the bilayer once the cation exchange capacity of the clay has been satisfied.

SUMMARY

Surfaces of soil particles can be characterized in terms of constitution, area and accessibility. Constitution differs markedly between the siloxane surfaces of 2:1 type clay minerals, dominated by negative charges arising from isomorphous substitution in the clay lattice, and whose hydration is determined by the interaction between the exchange cation and water molecules, and the amphoteric hydroxylic surfaces of the hydrous oxides, whose charge depends upon the ionic constitution of the surrounding solution, and where hydration arises from hydrogen bonding interactions between oxygens and hydroxyls of the solid surface and water molecules.

Surface areas can be determined by several adsorption methods. The application of the BET equation to the sorption of N_2 at 77 K over the partial pressure range to approximately 0.2 is the recognized method for determining the 'external' surface area of degassed soil particles. Analysis of the complete isotherm yields a mesopore size distribution (Kelvin equation) from approximately 2 to 20 nm, though an ideal geometry must be assumed. Because of the combination of the significant micropore volume and mesoporosity it is impossible to be completely certain that the BET monolayer amount is the true one and so there is some uncertainty about absolute areas. Despite this, there is no doubt that the method detects small changes in true area and is much the best means of comparing this between one soil and another.

Adsorption of polar molecules provides an experimentally simpler method for determining surface area, but the derived areas are subject to several uncertainties arising from the interaction of polar molecules with the surface and between themselves.

REFERENCES

Aylmore, L. A. G., Sills, J. D. and Quirk, J. P. (1970), Surface area of homoionic illite and montmorillonite clay minerals as measured by the sorption of nitrogen and carbon dioxide, *Clays Clay Min.*, **18**, 91–96.

Blyholder, G. and Richardson, E. A. (1962), Infra-red and volumetric data on the adsorption of ammonia, water and other gases on activated iron(III) oxide, *J. Phys. Chem.*, **66**, 2597–2602.

Bolland, D. A., Posner, A. M. and Quirk, J. P. (1976), Surface charge of kaolinites in aqueous suspension, *Aust. J. Soil Res.*, **14**, 197–216.

Breeuwsma, A. (1973), Adsorption of ions on hematite (α-Fe_2O_3); a colloid chemical study, *Meded. Landbouwhogeschool Wageningen*, **73**, 1–123.

Brindley, G. W. (1966), Ethylene glycol and glycol complexes of smectites and vermiculites, *Clay Minerals*, **6**, 237–259.

Brunauer, G., Emmett, P. H. and Teller, E. (1938), Adsorption of gases in multimolecular layers, *J. Amer. Chem. Soc.*, **60**, 309–319.

Brunauer, G., Deming, L. G., Deming, W. S. and Teller, E. (1940), A theory of the van der Waals adsorption of gases, *J. Amer. Chem. Soc.*, **62**, 1723–1732.

Burford, J. R., Deshpande, T. L., Greenland, D. J. and Quirk, J. P. (1964), Influence of organic materials on the determination of the specific surface areas of soils, *J. Soil Sci.*, **15**, 192–201.

Deshpande, T. L., Greenland, D. J. and Quirk, J. P. (1964), Charges on iron and aluminium oxides in soils, *Trans. 8th Int. Congr. Soil Sci.*, Bucharest, **3**, 1213–1225.

Farmer, V. C., Russell, J. D., McHardy, W. J., Newman, A. C. D., Ahlrichs, J. L. and Rimsaite, J. Y. H. (1971), Evidence for loss of protons and octahedral iron from oxidised biotites and vermiculites, *Min Mag.*, **38**, 121–137.

Ferris, A. P. and Jepson, W. B. (1975), The exchange capacities of kaolinite and the preparation of homoionic clays, *J. Colloid Interface Sci.*, **51**, 245–259.

Fripiat, J. J. and Uytterhoeven, J. (1962), Hydroxyl content in silica gel 'Aerosil', *J. Phys. Chem.*, **66**, 800–805.

Gallez, A., Juo, A. S. R. and Herbillon, A. (1976), Surface and charge characteristics of selected soils in the tropics. *Soil Sci. Soc. Amer. J.*, **40**, 601–608.

Greenland, D. J. (1965), Interaction between clays and organic compounds in soils. *Soils and Fertilizers*, **28**, 415–425 and 521–532.

Greenland, D. J. (1971), Interactions between humic and fulvic acids and clays, *Soil Science*, **111**, 34–41.

Greenland, D. J. (1974), Determination of pH dependent charges of clays using caesium chloride and X-ray fluorescence spectrography, *Trans. 10th Int. Congr. Soil Sci.*, Moscow, **2**, 278–285.

Greenland, D. J. (1975), Charge characteristics of some kaolinite–iron hydroxide complexes, *Clay Minerals*, **10**, 407–416.

Greenland, D. J., Oades, J. M. and Sherwin, T. W. (1968), Electron microscope observations of iron oxides in some red soils, *J. Soil Sci.*, **19**, 123–126.

Greenland, D. J. and Quirk, J. P. (1962), Surface areas of soil colloids, *Trans. Comm. IV and V, Int. Soc. Soil Sci.*, New Zealand, pp. 79–87.

Greenland, D. J. and Quirk, J. P. (1963), Determination of surface areas by adsorption of cetyl pyridinium bromide from aqueous solution, *J. Phys. Chem.*, **67**, 2886–2887.

Greenland, D. J. and Quirk, J. P. (1964), Determination of the total specific surface areas of soils by adsorption of cetyl pyridinium bromide, *J. Soil Sci.*, **15**, 178–191.

Greenland, D. J. and Wilkinson, G. K. (1969), Use of electron microscopy of carbon replicas and selective dissolution analysis in the study of the surface morphology of clay particles from soils, *Proc. Int. Clay Conference*, Tokyo, **1**, 861–870.

Greenland, D. J., Wada, K. and Hamblin, A. (1969), Imogolite in a volcanic ash soil from Papua, *Aust. J. Sci.*, **32**, 56–58.

Gregg, S. J. and Langford, J. F. (1969), Evaluation of microporosity, with special reference to carbon black, *Trans. Faraday Soc.*, **65**, 1394–1400.

Gregg, S. J. and Sing, K. S. W. (1967), *Adsorption, Surface Area and Porosity*, Academic Press, London.

Habibullah, A. K. M., Greenland, D. J. and Brammer, H. (1971), Clay Mineralogy of some seasonally flooded soils of E. Pakistan, *J. Soil Sci.*, **22**, 179–190.

Innes, W. B. (1957), Use of a parallel plate model in calculation of pore size distribution, *Anal. Chem.* **29**, 1069–1073.

IUPAC (1972), Manual of symbols and terminology for physicochemical quantities and units, *Pure Appl. Chem.*, **31**, 577–638.

Jepson, W. B. and Rowse, J. B. (1975), The composition of kaolinite—an electron-microscope microprobe study. *Clays Clay Min.*, **23**, 310–317.

Jepson, W. B., Jeffs, D. S. and Ferris, A. P. (1976), The adsorption of silica on gibbsite and its relevance to the kaolinite surface, *J. Colloid Interface Sci.*, **55**, 454–461.

Jones, R. C. and Uehara, G. (1973), Amorphous coatings on mineral surfaces, *Soil Sci. Soc. Amer. Proc.*, **37**, 792–798.

Laskowski, J. and Kitchener, J. A. (1969), The hydrophilic–hydrophobic transition on silica, *J. Colloid Interface Sci.*, **29**, 670–679.

Lyklema, J. (1968), Structure of the electrical double layer on porous surfaces, *J. Electroanal. Chem. Interfacial Electrochem.*, **18**, 341–348.

Mashali, A. M. (1977), The charge characteristics of clays, oxides and hydrous oxide minerals, Ph.D. thesis, University of Reading.

Mashali, A. M. and Greenland, D. J. (1975), Dependence of charge characteristics of kaolinites on pH and electrolyte concentration, *Proc. Int. Clay Conf.*, Mexico, **1**, 240–241.

Parfitt, R. L. and Greenland, D. J. (1970), Adsorption of polysaccharides by montmorillonite, *Soil Sci. Soc. Amer. Proc.*, **35**, 862–866.

Parks, G. A. (1965), The isoelectric points of solid oxides, solid hydroxides and aqueous hydroxo complex systems, *Chem. Rev.* (London), **65**, 177–198.

Peri, J. B. (1965), Infra-red gravimetric study of the surface hydration of γ-alumina, *J. Phys. Chem.*, **69**, 211–219, 220–230.

Peri, J. B. (1966), Infra-red study of OH and NH_2 groups on the surface of a dry silica aerogel, *J. Phys. Chem.*, **70**, 2937–2945.

Peri, J. B. and Hemsley, A. L. (1968), The surface structure of silica gel, *J. Phys. Chem.*, **72**, 2926–2933.

Quirk, J. P. (1955), Significance of surface areas calculated from water vapour sorption isotherms by use of the B.E.T. equation, *Soil Sci.*, **80**, 423–430.

Roth, C. B., Jackson, M. L. and Syers, J. K. (1969), Deferration effect on structural ferrous-ferric iron ratio and c.e.c. of vermiculites and soils, *Clays Clay Min.*, **17**, 253–264.

Roth, C. B., Jackson, M. L., de Villiers, J. M. and Volk, V. V. (1967), Surface colloids on micaceous vermiculite, *Trans. Comm. II and IV, Int. Soc. Soil Sci.*, Aberdeen, pp. 217–221.

Russell, J. D., McHardy, W. J. and Fraser, A. R. (1969), Imogolite: a unique aluminosilicate, *Clay Minerals*, **8**, 87–99.

Schnitzer, M. (1968), Reactions between organic matter and inorganic soil constituents, *Trans. 9th Int. Congr. Soil Science*, Adelaide, **1**, 635–642.

Schofield, R. K. (1949), Effect of pH on electric charges carried by clay particles, *J. Soil Sci.*, **1**, 1–8.

Schofield, R. K. and Samson, H. R. (1954), Flocculation of kaolinite due to the attraction of oppositely charges faces, *Discuss. Faraday Soc.*, **18**, 135–145.

Schuylenborgh, J. and Sanger, A. M. H. (1949), The electrokinetic behaviour of iron and aluminium hydroxides and oxides, *Rec. Trav. Chim.*, **68**, 999–1010.

Tweneboah, C. K., Greenland, D. J. and Oades, J. M. (1967), Changes in charge characteristics of soils after treatment with 0.5M calcium chloride at pH 1.5, *Aust. J. Soil Res.*, **5**, 247–261.

van Olphen, H. (1963), *An Introduction to Clay Colloid Chemistry*. Wiley–Interscience, New York.

van Raij, B. and Peech, M. (1972), Electrochemical properties of some Oxisols and Alfisols in the tropics, *Soil Sci. Soc. Amer. Proc.*, **36**. 587–593.

Wada, K. (1966), Deuterium exchange of hydroxyl groups in allophane, *Soil Sci. Plant Nutrition*, (Tokyo), **12**, 176–182.

Wada, K. (1967), A structural scheme of soil allophane, *Amer. Min.*, **52**, 690–708.

Wada, K. and Greenland, D. J. (1970), Selective dissolution and differential infra-red spectroscopy for characterisation of amorphous constituents in soil clays, *Clay Minerals*, **8**, 241–254.

Wada, K. and Harada, Y. (1969), Effects of salt concentration and cation species on the measured cation exchange capacity of soils and clays. *Int. Clay Conf.*, Tokyo, *Proceedings*, **1**, 561–571.

Yoshinaga, N. and Aomine, S. (1962), Allophane in some Ando soils. *Soil Sci. Plant Nutrition*, (Tokyo), **8**, (2), 6–13.

CHAPTER 5

Surface–electrolyte interactions

P. W. Arnold
Department of Soil Science, University of Newcastle-upon-Tyne

5.1 INTRODUCTION: THE PHYSICAL CHEMISTRY OF COLLOIDAL SUSPENSIONS

This chapter is concerned with understanding the factors which influence the partition of ions between solid surfaces and aqueous solutions. There are many aspects of soil science such as some of those concerned with soil formation, transfer of nutrient ions from soil particles to plant roots, soil pollutant interaction and soil management which are critically dependent on what happens in the space of the order of 1 nm wide between solid and equilibrium solution. A broad survey of the development of ideas on soil–electrolyte interaction shows that there have been two main avenues of approach. One views the soil as an electrically charged surface with its adsorbed hydrated ions as one phase in equilibrium with solution, which is seen as the second phase. The other approach focuses on the electrical double layer extending from the charged solid surface out into the solution. In the latter approach the solid surface, the outer equilibrium solution and the make up of the transition zone between these two phases must all be involved in any detailed consideration of the micro structure of the system. For a full and proper understanding of soil–electrolyte interaction anything less than a detailed analysis of the micro-structure is likely to be inadequate and in this it must be remembered that the interaction between electrical double layers which overlap with one another is likely to be important.

The study of soil–electrolyte interactions is, of course, central to understanding some of the complex problems of soil–plant relationships. The way in which plants respond to changes in soil composition through changes in the soil solution has interested scientists for many years. Although it has long been recognized that ion uptake by roots is concentration dependent the study of plant nutrition via the soil solution has had a chequered history. As well as being dilute, soil solution can vary in composition from day to day and it is most sensitive to changes in soil moisture content. Dilution of most soil–soil solution systems containing mono- and divalent cations displaces the equilibrium such that the adsorption of divalent ions increases and the adsorption of monovalent ions decreases. Changes in the soluble anion content, for example of nitrate, also alter the amounts and proportions of cationic constituents in the soil solution and for many years no satisfactory means of interpreting analytical data on solution composition could be found. Liebig encountered such difficulties as early as the middle of the last century, which led him to conclude that as a means of understanding plant nutrition, soil solution studies could not be profitably pursued. Despite further work over many years there was, especially after the middle 1920s, a notable absence of soil solution studies at least until some fundamental progress had been made in developing ideas on soil–electrolyte interaction. The fact that some of the ideas were, in retrospect, quite simple cannot detract from the impact they have had, and in this connection the significance of Schofield's (1947) Ratio Law in understanding cation exchange phenomena cannot be over estimated. In its generalized form the Ratio Law

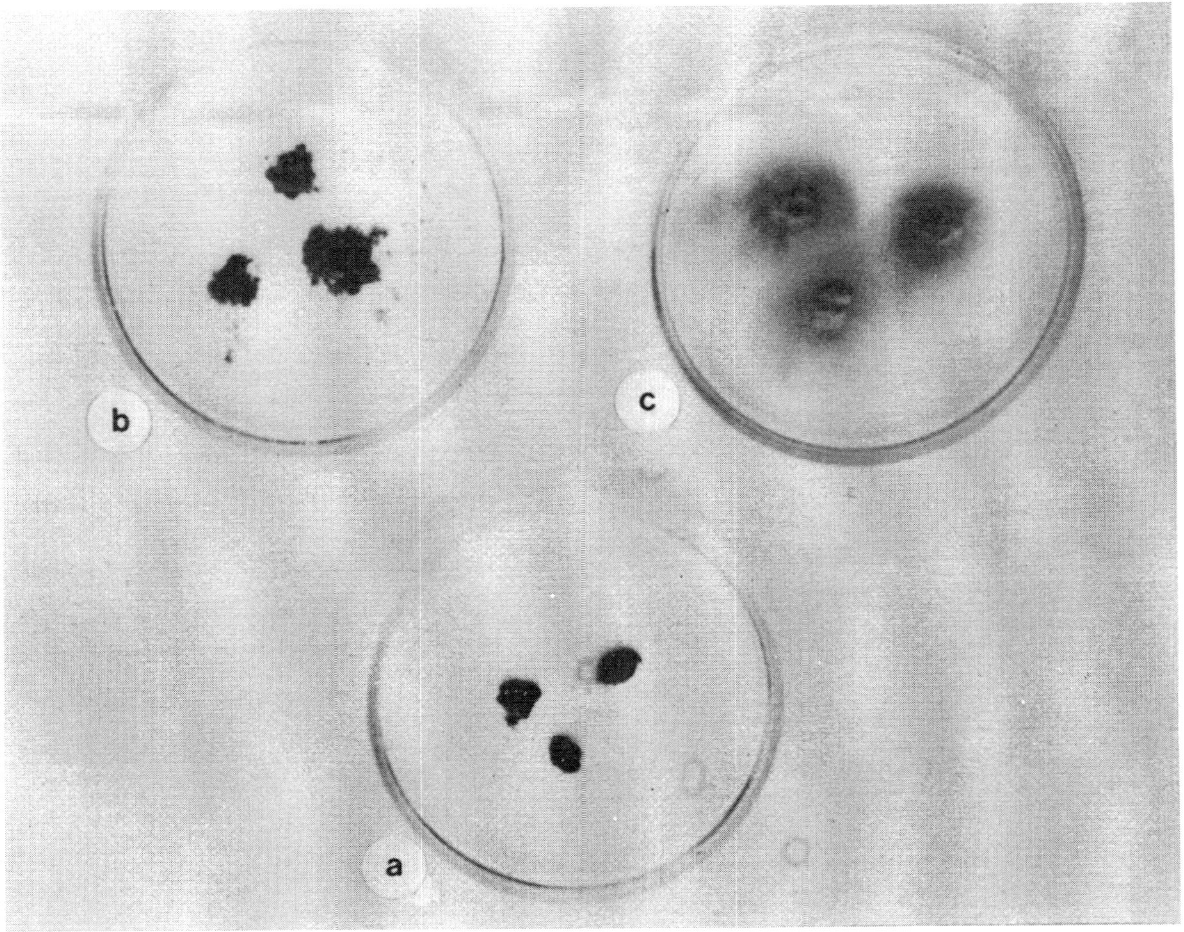

Plate 5.1 The contrasting behaviour of aggregates from different soils in water: (a) from a structurally stable soil, (b) from a soil where the aggregates slake but the clay does not disperse, and (c) from a soil which disperses. Dispersed soils are usually very impermeable to water and plant roots and are extremely difficult to cultivate.

[*Facing page 356*

states that 'When cations in a dilute solution are in equilibrium with a larger number of exchangeable ions, a change in the concentration of the solution will not disturb the equilibrium if the activities of all the monovalent cations are changed in one ratio, those of all the divalent cations in the square of that ratio and those of all the trivalent cations in the cube of that ratio'. This can easily be checked by taking, for example, $a_{M^+} = 0.004$ and $a_{M^{2+}} = 0.01$ both in mol dm^{-3}, which gives a value of

$$\frac{a_{M^+}}{(a_{M^{2+}})^{1/2}} = 0.04$$

If the activity of a_{M^+} is reduced by some factor, say 0.5, then the activity of M^{2+} would have to be reduced by the same factor squared to maintain the value 0.04. Thus

$$\frac{0.004 \times 0.5}{\{0.01 \times (0.5)^2\}^{1/2}} = 0.04$$

Expressed in even simpler terms the Ratio Law states that if a soil holds both monovalent and divalent cations in readily exchangeable form, the ratio of the amounts held is dependent on the ratio of the activity of the monovalent cation in solution to the square root of the divalent cation activity in solution. Comparable statements involving the cube root of trivalent cation activities can be made in relation to either mono- or divalent cations. As in other branches of dilute solution chemistry, for many purposes it is quite adequate and acceptable to use ion concentrations in place of ion activities. The conditions under which the Ratio Law is expected to hold are discussed on page 390.

The soil solution usually contains appreciable Ca, Mg, Na and K along with bicarbonate, carbonate, chloride, sulphate, nitrate, phosphate and silicic acid (H_4SiO_4). In very acid soils the concentrations of Al, Fe and Mn are usually appreciable in the soil solution but it is difficult to generalize about other ions. In fact, the total inorganic phosphate concentration can be very low in many soil solutions, as can the concentration of several other constituents involved in plant nutrition such as Cu, Zn, molybdate and boric acid (H_3BO_3). Although there will be few direct references to plant nutrition in this Chapter some of the physico-chemical problems concerning soil–ion interaction which are discussed have a direct bearing on the understanding of many problems in plant nutrition (Fried and Broeshart, 1967; Epstein, 1972; Nye and Tinker, 1977).

Most natural soil solutions are dilute with the ions in dynamic equilibrium with the labile counter ions associated with the soil colloid surfaces. Taking, as an example, a soil with predominantly negatively charged particles we find an accumulation of cations and a deficit of anions in the vicinity of the solid surfaces relative to the equilibrium solution (Figure 5.1). The thermal motion of the ions counteracts the electrostatic interaction. Thus, cations being attracted

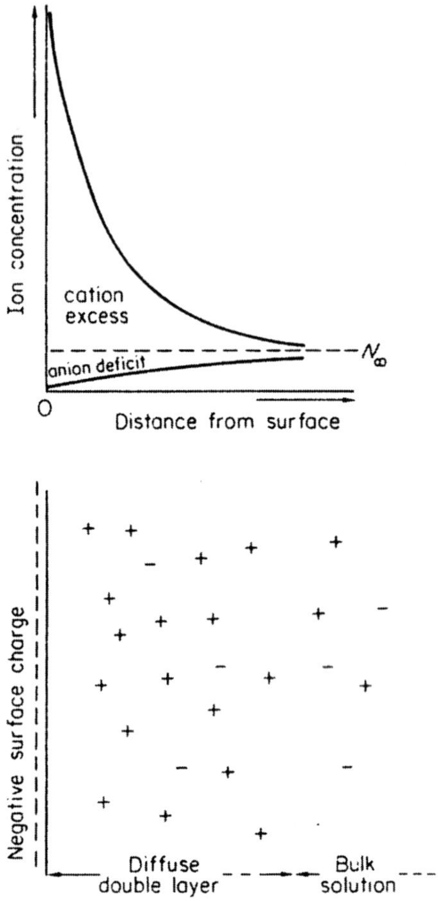

Figure 5.1 Distribution of ions at clay particle surface. Electrical neutrality maintained in diffuse double layer by an excess of cations and a deficit of anions, relative to the outer (or bulk) solution of concentration N_∞

and anions repelled, the cation concentration increases as the surface is approached while the anion concentration decreases. The zone of changing electrical potential between the solid with its surface charge and the equilibrium solution is known as the electrochemical double layer. As will be seen later, many ions show special affinity for soil surfaces which means that forces other than coulombic ones are often involved. However, in the absence of such special affinity, the lower the overall electrolyte concentration, and the lower the valency of the counter ion, the thicker the diffuse double layer will be. Because of the importance of the effects of double layer interaction between adjacent

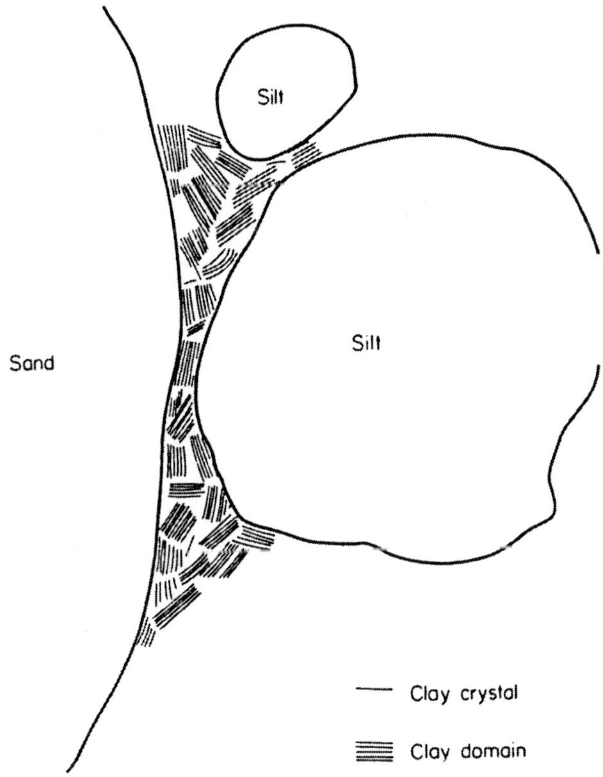

Figure 5.2 Oriented clay crystals (including interstratified quasi-crystals) packed in domains between larger mineral grains

particles, an indication of the approximate thickness of diffuse double layers is useful. For monovalent ions held by coulombic forces, as the concentration in the outer solution increases from 1×10^{-5} mol dm^{-3} through 1×10^{-3} to 1×10^{-1} the double layer diminishes in thickness from about 100 to 10 to 1 nm. At the same molar concentrations, adsorbed divalent ions give double layers which extend about 50, 5 and 0.5 nm out from the solid surface.

The colloidal particles in the majority of soils in their natural state are flocculated, and clustered together in aggregates. At the fine end of the particle size range, clays are normally held together in packages (domains) by a variety of agencies, and these domains are bonded to form aggregates (Figure 5.2 and Plate 5.1). Historically, there was much interest in repulsive forces between soil particles, particularly in the clay fraction. This interest was understandable because if repulsive forces are dominant soils become dispersed and are virtually unmanageable in the agronomic sense. However, it is the balance of attractive and repulsive forces which determines whether a soil is flocculated or dispersed, and by the mid 1950s a swing towards studying the attractive forces between soil particles was well under way and certain salient soil features began to be understood more widely. In the present context it may suffice to concentrate on comparing the behaviour of sodium and calcium dominated soils, with an emphasis on how expansible clays behave towards these two ions.

Calcium dominated soils and clays are normally flocculated, but any appreciable proportion of exchangeable sodium (15 % of the exchange capacity is often considered critical) leads to considerable swelling, and as more water is introduced, dispersion. Double layer theory is remarkably successful in explaining this type of behaviour, as well as the fact that increasing electrolyte concentration in the external solution will prevent dispersion even of sodium clays.

5.2 INTERACTION BETWEEN SOIL PARTICLES: FLOCCULATION AND DISPERSION*

5.2.1 Interparticle Forces

The attractive or van der Waals forces which exist between molecules are of three types. The most widespread attractive forces are those first explained by London (1930) and known as 'dispersion' forces. The London forces arise from characteristic vibrations of a molecule with a frequency which is the same as occurs in the equation for the dispersion of light by the molecule, hence the use of the term dispersion force. The force is due to the polarization of one molecule caused by the fluctuations in the charge distribution of another, the overall effect being a net attraction between molecules. The attractive interaction energy is, to a first approximation, inversely related to the sixth power of the distance between atoms or molecules, which is the same as in molecules with permanent dipoles or dipole induced dipoles, which constitute the other causes of mutual attraction: the attractive force is obtained by differentiation with respect to account for practically all the van der Waals attraction. Most of the quantitative relationships encountered in the sector are concerned with potential energy of attration; the attractive force is obtained by differentiation with respect to distance, so that the 'dispersion' force is inversely proportional to the seventh power of the distance between molecules. For a collection of molecules the 'dispersion' forces are approximately additive so that the attractive energy between two particles is calculated by summing the attraction between all molecular pairs. Such summations show that attractive energy between particles decays very much less rapidly than between individual atoms or molecules. For example, for parallel plates separated by a distance of $2d$ (up to about 20 nm) the approximate attractive potential energy per unit area is given by

$$V_A = -\frac{A}{48\pi d^2}$$

where A is a constant which depends on the nature, including the polarizability, of the particles and the medium separating them. The main point to note is

* Dispersion is used in two different senses, to refer to London dispersion forces and as here synonymously with deflocculation.

that the attractive energy for parallel plates decreases approximately as the inverse square of the distance between particles. This shows very clearly the much longer range of attractive energy between collections of molecules as compared with isolated atoms and molecules. Soil colloid particles with their atmospheres of counterions are, as a whole, electrically neutral so that adjacent particles in a dispersed system will show no interaction until they approach sufficiently closely for their electrical double layers to interpenetrate. Thus the repulsive energy between particles arises from the interaction, or interpenetration, of electrical double layers around particles which, as will be seen, decrease almost exponentially in potential with distance. Basically, the problem was to decide how to deal with changes in the potential between two interacting particles. As with other electrical double layer problems, the model involving parallel plates is easier to deal with than the model for spheres.

DLVO theory

The quantitative theory to evaluate the balance of repulsive and attractive forces when particles approach each other was worked out by Derjaguin and Landau (1941) and independently by Verwey and Overbeek (1948) on the basis of interacting Gouy–Chapman type electrical double layers (see Section 5.4). The theory is often referred to as the DLVO theory. For charged parallel plates separated by a distance of $2d$ such that the diffuse parts of the double layers overlap as shown in Figure 5.3a, the potential energy of repulsion is given by

$$V_R = \frac{64\, N_\infty kT}{K} \gamma^2 \cdot \exp(-2Kd)$$

where N_∞ is the electrolyte concentration, k the Boltzmann constant, T the absolute temperature, K is the reciprocal length parameter in the Debye–Hückel theory and

$$\gamma = \left[\frac{\exp(ze\psi_0/2kT) - 1}{\exp(ze\psi_0/2kT) + 1}\right].$$

In the latter expression ψ_0 is the surface potential relative to some point far away, usually taken as zero, and e is the charge on the electron and z the valency of the counter ion. The total energy of interaction is obtained by summation of the attractive and repulsive energies as shown in Figure 5.3b. Because of the way in which these energies decay with distances, V_A will predominate at small and large interparticle distances, while at intermediate distances V_R may predominate, often showing a maximum at a critical distance as shown in Figure 5.3b. This maximum can constitute an energy barrier to flocculation, which must be exceeded for flocculation to occur. Clearly, in theory, if the repulsive energy maximum is large compared to the thermal energy the dispersion can be stable. The repulsive force increases both with the thickness of

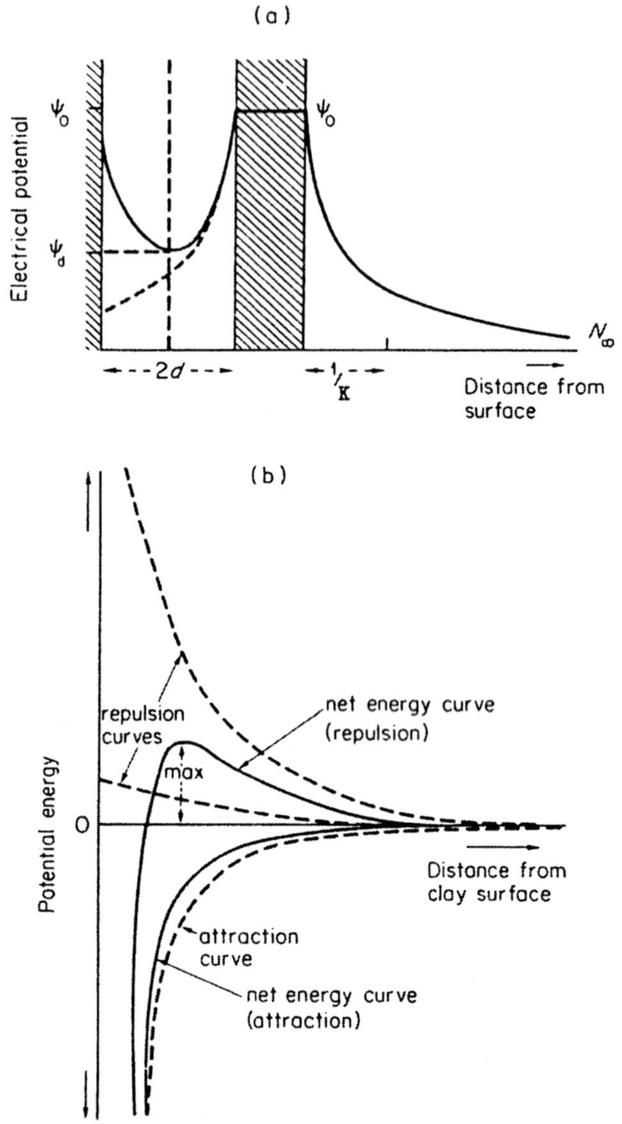

Figure 5.3 (a) Electrical potential of overlapping diffuse double layers between two charged clay plates separated by a distance $2d$; minimum potential ψ_d. Effective thickness of unrestricted diffuse double layer shown as $1/K$, the distance for the potential to fall to $1/e$ of its value, ψ_0, at the clay surface. (b) Potential energies of repulsion (two examples) and attraction (one example) and net (total) interaction energy curves (solid lines) between two clay plates. One net energy curve shows a maximum (repulsion) at a critical distance between the clay surfaces and may be conducive to stable dispersion. The other net energy curve shows attraction only and must favour rapid flocculation. (Reproduced by permission of Elsevier Scientific Publishing Co. from 'The Interaction Between Colloidal Particles' by J. Th. Overbeek, in *Colloid Science*, H. R. Kruyt, ed., 1952)

the diffuse double layer and the potential at the surface of the particle. In the parallel plate model it is the midway ion concentration between plates (assumed to be identical) that is of special interest because of its key importance in understanding the phenomenon of clay swelling.

The DLVO theory involving parallel plates should apply in principle to most clays with a platy habit, but behaviour will be modified by particle size and shape, surface charge density and particular affinities of the clay surfaces for individual ions (Norrish, 1954). Although a preponderance of face-to-face alignment of clay plates is expected in most flocculated systems, there are situations when face-to-edge flocculation can occur, namely, when the positive edges of clay sheets are attracted to the negative faces (Schofield and Samson, 1954), sometimes referred to as the house of cards effect. There is no doubt that certain cations occupying positions between some clay sheets can prevent their free expansion. Among the common soil cations, calcium is well known for exerting such an effect (see page 375).

The Schulze–Hardy rule

Prior to the DLVO theory the flocculating power of electrolyte solution was, according to the Schulze–Hardy rule, known to increase markedly with increase of the valency of the ion of opposite charge to that of the particle, as well as on the concentration of electrolyte. Other factors, such as the concentration of the colloid and the nature of the co-ions were of relatively small influence. From DLVO theory it emerged that in the absence of complicating factors, such as specific adsorption of ions into the inner part of the electrical double layer, the amounts of ion required to cause a certain rate of flocculation are in the ratio $1/(\text{valency})^6$. In other words, the critical concentrations of mono-, di- and trivalent cations required for rapid flocculation of a negatively charged soil would, in theory, decrease in the ratios

$$\frac{1}{1^6}:\frac{1}{2^6}:\frac{1}{3^6}, \quad \text{or} \quad 1:0.016:0.0014$$

5.2.2 Surface Charge and Potential

Inorganic surfaces

Most soils from temperate regions are predominantly negatively charged. However, they invariably possess some sites where the charge depends on pH, and this manifests itself as an increasing cation exchange capacity with increasing pH (Chapter 4). Soils, whose active surface constituents are dominated by oxides and hydrous oxides do, in fact, behave differently from those dominated by clays where charge arises from isomorphous lattice substitution. The majority of hydrophobic colloids acquire their surface charge by the adsorption of potential determining ions which by strict definition are ions common to the colloid and to the aqueous medium. By a slightly less strict definition such ions are those which leave the solution, cross over the real solid solution boundary, and become part of the solid surface. On this basis any ion which is associated with the Si or metal ions as a ligand may be capable of altering the surface potential and hence fall into the category of potential determining. For oxides and hydrous oxides in aqueous media the potential determining ions are frequently H^+ and OH^-. When reference is made to constant surface potential colloids it has to be understood that the surface electrical potential is only constant so long as the activity of the potential determining ions remains constant. Thus the electrical charge properties of soil colloids cover a spectrum of behaviour. For one end member the charge on the surface is fixed and remains independent of the solution composition, but the electric potential is sensitive to indifferent electrolyte concentration. This member is usually termed the constant charge surface. For the other end member, the constant potential surface, it is the charge which changes with changing concentration of indifferent electrolyte.

Organic surfaces

Turning from clays and hydrous oxides, with their contrasted surface properties, humified organic matter acquires its charge through the ionization of carboxylic, phenolic and possibly other groups. With complex molecules it is often difficult to specify single acidic dissociation constants because loss of protons tends to become increasingly difficult as more functional groups become negatively charged. Thus in the dissociation

$$\underset{(1-f)}{\mathrm{R{-}COOH}} \rightleftharpoons \underset{f}{\mathrm{R{-}COO}} + \mathrm{H^+}$$

the negative logarithm of the dissociation constant, normally written as

$$\mathrm{p}K_a = \mathrm{pH} - \log\frac{f}{1-f}.$$

where f is the fraction of carboxylic groups which has lost protons, is expected to increase with increasing f. In many surface soils it is usual to find that between 10 and 90 per cent of the total cation exchange capacity stems from the functional groups of the organic matter. In physico-chemical studies on soil organic matter it is often convenient to use so-called humic acid or fulvic acid preparations as the starting materials. Because of the methods used for obtaining such preparations, and because of the ways in which well humified organic matter interacts with clay (see Chapters 3 and 7), there are often doubts about whether the information obtained about the humic functional groups and properties dependent on functional groups can be applied to the original soil. As described in Chapter 3, the results of functional group analysis can be difficult to interpret, partly because the particular preparations have not been properly described, and partly because the constitution of the colloids is so complex that the normal methods of functional group analysis do not give such clear cut separations as they do for simple organic compounds.

Posner (1964) examined the potentiometric behaviour of soil organic matter in the form of humic acid and, recognizing difficulties about the choice of the end point and the apparent irreversibility of titrations against acid and base, concluded that humic acid is not a typical polyelectrolyte in which ionization is influenced by the charge on the molecule as it is neutralized. He concluded that with variations both within and between molecules, the shape of a titration curve conforms to a Gaussian distribution of acid pK values with a deviation of ± 1.7 units about a mean value. The evidence showed the functional groups are carboxylic, phenolic and with some other group or groups probably covering the overlap region (pH 7–8) between the main carboxylic and phenolic ranges. Titration curves carried out with alkali metals in the presence of their chlorides were identical, whereas using calcium or barium in place of alkali metal indicated that there were groups capable of complexing with these ions. Cation exchange capacities in the acid region were in the range of 330–340 meq per 100 g (0.0033–0.0034 mol charges g^{-1}) of humic acid with a tendency for divalent cations to give the higher values. Because of the great tendency for soil organic matter to chelate transition metals, the potentiometric behaviour of organic matter is likely to be dependent on the type of soil in which the organic matter formed. This must be recognized as a serious complication in making any generalized statements about the potentiometric behaviour of soil organic matter. Borggaard (1974) examined the effect of altering the speed of titrations of EDTA-extracted humic acids. He found that provided autoxidation and other complications were prevented, protolytic equilibrium is established instantly, at least for NaOH additions; furthermore the titrations were reversible. It therefore appears possible that some of the supposed complexity of the humic acid titration results from artefacts induced by the procedure used to extract the humic acid and the conditions under which titrations were conducted.

van Dyk (1971) discussed the terms complex and chelate in relation to di- and polyvalent metal–humic acid interaction and stated that according to IUPAC nomenclature, such slightly dissociated salts are polynuclear complexes. Although this does not stipulate the nature of the bond, it emphasizes that a number of metal ions are attached to an equal number of bi- or polydentate ligands on one humic acid molecule. The metal–humic substance bonds, especially for some transition metals can be so strong that some functional groups have to be regarded as permanently blocked, and ash free humic and fulvic acid preparations have probably never been made, though Posner's (1964) and Borggaard's (1974) humic acids had very low ash contents.

5.3 CHEMICAL POTENTIALS OF IONS IN SOIL–SOIL SOLUTION SYSTEMS

The condition for thermodynamic equilibrium between soil and its solution requires that the chemical potential of each diffusible molecular species must be the same at all points in the system. If we take MA as an unionized molecular species then the chemical potential is defined by the standard isotherm

$$\mu_{MA} = \mu^{\circ}_{MA} + RT \ln a_{MA}$$

where μ°_{MA} is the chemical potential in the standard state. Because the latter is a constant, $\mu_{MA} = RT \ln a_{MA} + \text{constant}$. For unchanged molecules we need not go beyond thinking about chemical potentials at equilibrium. Thus,

$$(\mu_{MA})_{soil} = (\mu_{MA})_{solution}$$

For a charged ionic species, say M^+, associated with a negatively charged colloid, there is, however, a complication arising from changes in the electrical potential with distance from the colloid surface. Certainly, for a particular ionic species it must be recognized that the electrical potential changes as we approach or move away from the charged surface, but when equilibrium is established it can be stated that the electrochemical potential of each diffusible ionic species must be the same at all accessible points in the system. Thus,

$$\bar{\mu}_{M^+} = \mu_{M^+} + z_{M^+} F\psi$$

in which

$$\mu_{M^+} = \mu^{\circ}_{M^+} + RT \ln a_{M^+}$$

where z_{M^+} is the valency of the ion, F the Faraday constant and ψ is the electric potential of the phase to which $\bar{\mu}_{M^+}$ and μ_{M^+} refer.

As discussed by Marshall (1956) for points I, II and III at different distances from a colloid surface where the electrical potentials are ψ^{I}, ψ^{II} and ψ^{III} the electrochemical potential is the same in all three locations. This is formally recognized by

$$\mu^{I}_{M^+} + z_{M^+} F\psi^{I} = \mu^{II}_{M^+} + z_{M^+} F\psi^{II} = \mu^{III}_{M^+} + z_{M^+} F\psi^{III}$$

and because ψ changes, the ion activity must change in order to keep the electrochemical potential constant. It is not thermodynamically possible to define the activity of an ion in an aqueous soil or similar system in the same way as it is for an unchanged molecule, without referring to some other quantity. It is, however, not necessary to refer the activity of an ion to the electrical potential because, as will be shown, the products and ratios of the activities of certain pairs of ions are the same at every accessible point in the system when equilibrium is established. In a soil-dilute electrolyte system we are usually mainly concerned with ions but because the chemical potential of a whole compound e.g. MA, may be written as the sum of those of its constituents, we can write

$$(\mu_{M^+})_{soil} + (\mu_{A^-})_{soil} = RT \ln a_{M^+} + RT \ln a_{A^-} + \text{constant terms}$$

where M^+ and A^- are the cation and an ion, respectively. Similarly, if a divalent cation is involved, along with a monovalent anion, we may write

$$(\mu_{M^{2+}})_{soil} + (2\mu_{A^-})_{soil} = RT \ln a_{M^{2+}} + 2RT \ln a_{A^-} + \text{constant terms}$$

or

$$(\tfrac{1}{2}\mu_{M^{2+}})_{soil} + (\mu_{A^-})_{soil} = \tfrac{1}{2}RT \ln a_{M^{2+}} + RT \ln a_{A^-} + \text{constant terms}$$

where the activities 'a' are those in the equilibrium solution. When two cations are present together in the equilibrium system, a_{A^-} can be taken as the same for both components. It follows, therefore, that

$$(\mu_{M^+})_{soil} - (\tfrac{1}{2}\mu_{M^{2+}})_{soil} = RT \ln a_{M^+} - \tfrac{1}{2}RT \ln a_{M^{2+}} + \text{constant terms}$$

and hence

$$(\mu_{M^+} - \tfrac{1}{2}\mu_{M^{2+}})_{soil} - \text{constant terms} = RT \ln \frac{a_{M^+}}{(a_{M^{2+}})^{1/2}}$$

The latter relationship involves the concept of referring the activity of one ion to another and overcomes the difficulty of changing electrical potential from point to point; the relationship requires no experimental proof because it is merely a restatement of the conditions for thermodynamic equilibrium. In his work on cations in aqueous soil systems, Schofield found that the relative chemical potentials of pairs of soil cations are not affected by changes in the equilibrium soil solution, provided the requirements of the Ratio Law are met. As stressed by Beckett (1964a) this is not self-evident but, within the limits described by Schofield and Taylor (1955a), holds for predominantly negatively charged soils.

5.4 ELECTRICAL DOUBLE LAYERS AT SURFACES: THE GOUY–CHAPMAN DIFFUSE DOUBLE LAYER

5.4.1 Ion Distribution at a Charged Surface

Although it is possible to visualize a situation where counterions are packed compactly on a charged surface so that the counterion charge balances the charge on the colloid surface, such a simple picture is not possible if the electrical double layer has appreciable thickness or, if because of their thermal energy, the counterions diffuse away from the surface so that a diffuse double layer is formed. Any model for a negatively charged colloid possessing a diffuse double layer involving the attraction of cations, often referred to as counterions, must also involve at least some repulsion of anions, often referred to as co-ions, which possess the same sign as the surface charge. The total surface charge per unit area on a negative surface is thus balanced by a surface excess of cations Γ_+ and a deficit of anions Γ_-, both taken relative to the external solution as depicted in Figure 5.1. For a positively charged surface the balance would be provided by an excess of anions over cations in the double layer. Gouy (1910) and Chapman (1913) independently proposed the diffuse double layer theory as applied to flat interfaces. In the present treatment only the bare essentials are considered, so that the reader may add details at a later stage. The main theory involves the solid surface regarded as an infinite plane carrying a non-discrete or smeared surface charge, which may be negative or positive. The overall effect is that the surface carries a certain charge density and as a result of this, ions in the solution phase distribute themselves so that electrical neutrality is maintained and the tendency for the ions to diffuse away is counteracted by coulombic forces of attraction. In the Gouy–Chapman theory it is only the coulombic forces that are considered. In the model there is no such thing as a specifically adsorbed ion and hence the electrolyte which is present, or its constituent ions, are said to be indifferent, meaning that the ions show no interaction with the surface other than that arising from electrostatic (coulombic) attraction or repulsion.

The Gouy–Chapman theory of the electrical double layer is based on simplified assumptions and hence is not always quantitatively correct in its predictions. It does, however, serve as the basis from which most other models of the electrical double layer are derived and judged on this basis it assumes considerable importance.

5.4.2 Mathematical treatment of the Gouy–Chapman model

Taking the simplest Gouy–Chapman model where the ions are regarded as charged points and the dielectric constant of the solvent is everywhere the same, the distribution of ions in solution in the direction normal to a charged plane surface will follow the pattern laid down by the Boltzmann equation

$$N_{ix} = N_{i\infty} \exp\left(-\frac{z_i e \psi_x}{kT}\right) \tag{5.1}$$

where $N_{i\infty}$ is the number of ions per unit volume in the solution outside the influence of the surface, N_{ix} the number of ions at distance x from the surface and ψ_x is the electrical potential at the same position. z, e, k and T are the valency of the ion (sign included), the charge of a proton, the Boltzmann constant and the absolute temperature, respectively. Clearly, $N^+ > N^-$ when ψ is negative and $N^- > N^+$ when ψ is positive. In other words, at places of negative potential, the positive ions are concentrated and the negative ions are repelled, whereas for places of positive potential the reverse is true. In formulating equation (5.1) it is customary to assume that the potential far out in the solution is zero. Because each ion carries a charge $z_i e$ the net charge density, ρ, in any small volume of solution is given by the algebraic sum of the ionic charges in the small volume.

Thus

$$\rho_x = \sum_i z_i e N_i x \tag{5.2}$$

which quantifies the space or volume charge.

The general relationship between the electrical potential and the space density is given by the Poisson equation

$$\nabla^2 \psi = -\frac{\rho}{\varepsilon_r \varepsilon_0} \tag{5.3}$$

where ∇^2 is the Laplacian operator of potential ψ, and ε_r and ε_0 are the relative permittivities of the medium and of a vacuum, respectively. Combining equations (5.1), (5.2) and (5.3) gives the Poisson–Boltzmann equation.

$$\nabla^2 \psi = -\frac{1}{\varepsilon_r \varepsilon_0} \sum_i z_i e N_{i\infty} \exp\left(-\frac{z_i e \psi_x}{kT}\right)$$

Because we are concerned only with changes in potential in a direction normal to the surface, $\nabla^2 \psi$ becomes $d^2 \psi_x / dx^2$ giving

$$\frac{d^2 \psi_x}{dx^2} = -\frac{1}{\varepsilon_r \varepsilon_0} \sum_i z_i e N_{i\infty} \exp\left(-\frac{z_i e \psi_x}{kT}\right) \tag{5.4}$$

At this stage it can be mentioned that according to equations (5.1) and (5.2) there is an exponential relationship between charge density ρ and the potential. Remembering that

$$\exp(-a) = 1 - a + \frac{a^2}{2!} - \frac{a^3}{3!} + \cdots$$

the ion summation term in equation (5.4) becomes

$$\sum z_i e N_{i\infty}\left[1 - \frac{z_i e \psi_x}{kT} + \left(\frac{z_i e \psi_x}{kT}\right)^2 \frac{1}{2!} - \left(\frac{z_i e \psi_x}{kT}\right)^3 \frac{1}{3!} + \cdots\right]$$

giving

$$\sum z_i e N_{i\infty} - \sum z_i e N_{i\infty} \frac{z_i e \psi_x}{kT} + \sum z_i e N_{i\infty} \frac{1}{2!}\left(\frac{z_i e \psi_x}{kT}\right)^2 - \cdots$$

In this expanded expression the first term is zero because the system is electrically neutral. If $z_i e \psi_x$ is much less than kT, $(z_i e \psi_x / kT \ll 1)$ i.e. the electrostatic energy much less than thermal energy which is generally acceptable as true at least when dealing with 1 : 1 electrolytes, only the linear term involving ψ is numerically significant, hence

$$\nabla^2 \psi = \frac{d^2 \psi_x}{dx^2} = -\frac{1}{\varepsilon_r \varepsilon_0} \sum \frac{z_i^2 e^2 N_{i\infty} \psi_x}{kT} \tag{5.5}$$

and for a single symmetric binary electrolyte, the following approximate differential is obtained

$$\frac{d^2\psi_x}{dx^2} = \frac{2}{\varepsilon_r \varepsilon_0} \frac{z^2 e^2 N_\infty \psi_x}{kT}$$

$$= K^2 \psi_x \tag{5.6}$$

This simple solution involving K, which is identical to the reciprocal length parameter in the Debye–Hückel theory of electrolyte interaction, is not suitable for dealing with many colloid problems because valencies of the positive and negative ions occur in it in symmetrical manner and we do not always wish to restrict ourselves to a single binary electrolyte. In the Gouy–Chapman context $1/K$ represents the distance from the plane surface at which the electrical potential has decayed to $1/e$ of its value ψ_0 at the surface. It is usual to refer to $1/K$ as the 'thickness' of the diffuse double layer (Verwey and Overbeek, 1948) and the decay of the potential with distance from the surface is given by

$$\psi_x = \psi_0 \exp(-Kx) \tag{5.7}$$

which is purely exponential.

If the restrictions implicit in equation (5.6) are not acceptable an alternative approach involving a first integration of (5.5) can easily be done using the relation

$$2\frac{d^2\psi}{dx^2} = \frac{d}{dx}\left(\frac{d\psi}{dx}\right)^2$$

Thus, if each side of equation (5.4) is multiplied by $2d\psi_x/dx$ we obtain

$$\frac{d\psi_x}{dx} \cdot \frac{d}{dx}\left(\frac{d\psi_x}{dx}\right)^2 = -\frac{2}{\varepsilon_r \varepsilon_0} \sum z_i e N_{i\infty} \exp\left(-\frac{z_i e \psi_x}{kT}\right) \frac{d\psi_x}{dx}$$

which can be integrated for the boundary condition as $x \to \infty$, $\psi_x \to \psi_x$ and $d\psi_x/dx \to 0$ to give

$$\left(\frac{d\psi_x}{dx}\right)^2 = \frac{2kT}{\varepsilon_r \varepsilon_0} \sum_i N_{i\infty} \left\{\exp\left[-\frac{z_i e \psi_x}{kT}\right] - 1\right\} \tag{5.8}$$

Turning now to the double layer system as an entity, the charge on the surface σ must be balanced by the charge in the solution again considered in the x direction only giving,

$$\sigma = -\int_0^\infty \rho_x \, dx \tag{5.9}$$

From equations (5.3) and (5.9)

$$\sigma = \int_0^\infty \varepsilon_r \varepsilon_0 \frac{d^2\psi_x}{dx^2} dx$$

and, for the boundary conditions already stated, the above equation can be integrated to give

$$\sigma = -\varepsilon_r \varepsilon_0 \left(\frac{d\psi_x}{dx}\right)_{x=0} \tag{5.10}$$

Combining (5.8) and (5.10) we obtain the commonly referred to Gouy–Chapman equation namely

$$\sigma = -\varepsilon_r \varepsilon_0 - \left\{\frac{2kT}{\varepsilon_r \varepsilon_0} \sum N_{i\infty} \exp\left(-\frac{z_i e \psi_0}{kT}\right) - 1\right\}^{1/2}$$

giving

$$\sigma = \left\{2\varepsilon_r \varepsilon_0 kT \sum N_{i\infty} \exp\left(-\frac{z_i e \psi_0}{kT}\right) - 1\right\}^{1/2} \tag{5.11}$$

The equations for $d\psi_x/dx$, given as its square in (5.8), and σ, given by (5.11), are difficult to use, but become considerably simpler if the system contains only one symmetrical electrolyte of valency z, which was of course, one of the restrictions associated with equation (5.6). In fact, such a simplification is of relatively small consequence because the valency of the ion with the same charge as the surface is usually unimportant and so, for the sake of simplifying the mathematics, can be chosen as equal to the valency of the counter ion. Thus from equation (5.8)

$$\frac{d\psi_x}{dx} = -\left(\frac{2kTN_{i\infty}}{\varepsilon_r \varepsilon_0}\right)^{1/2} \left\{\exp\left(\frac{ze\psi_x}{2kT}\right) - \exp\left(-\frac{ze\psi_x}{2kT}\right)\right\} \tag{5.12}$$

and remembering that the hyperbolic sine of a, that is $\sinh a$ is equal to $\frac{1}{2}(e^{+a} - e^{-a})$

$$\frac{d\psi_x}{dx} = -\left(\frac{8kTN_\infty}{\varepsilon_r \varepsilon_0}\right)^{1/2} \sinh \frac{ze\psi_x}{2kT}$$

From this is seen that the slope $d\psi/dx$ decreases in proportion to $(N_\infty)^{1/2}$; in this relationship the concentration applies to ions which do not affect the potential at the surface, ψ_0, and, clearly, with increasing N_∞ the potential decays more rapidly, as illustrated in Figure 5.4. With the same simplification already cited, namely, that the system contains only one symmetrical electrolyte of valency z, the equation (5.11) for σ reduces to

$$\sigma = (8N_\infty \varepsilon_r \varepsilon_0 kT)^{1/2} \sinh \frac{ze\psi_0}{2kT} \tag{5.13}$$

5.4.3 Electrolyte Concentration and Surface Potential

The surface charge density on the colloid in Figure 5.4 is assumed to be constant ($0.0158\ C\ m^{-2}$). Such a value is small compared with the negative charge density carried by a typical montmorillonite clay, which would be some seven or eight times larger. For example, if the montmorillonite possessed a cation exchange capacity of 1.0 meq g^{-1} and a surface area of 800 $m^2 g^{-1}$, then the charge density would be 0.12 C m^{-2}. The small value of 0.0158 C m^{-2} would correspond to the positive charge carried by a goethite at a pH a little above 7.0 in an indifferent uni–univalent electrolyte at a concentration of about 1×10^{-2} mol dm^{-3} (Chapter 4, Figure 4.3 and Table 4.2).

Equation (5.12) without any simplifications gives a relationship between the potential and the position within the double layer, thus

$$\frac{d\psi_x}{\exp\left(-\frac{ze\psi_x}{2kT}\right) - \exp\left(\frac{ze\psi_x}{2kT}\right)} = \left(\frac{2kTN_\infty}{\varepsilon_r \varepsilon_0}\right)^{1/2} dx$$

which on integration, with the potential equal to ψ_0 at $x = 0$, gives

$$\exp\frac{ze\psi_x}{2kT} = \frac{\exp\left(\frac{ze\psi_0}{2kT}\right) + 1 + \left\{\exp\left(\frac{ze\psi_0}{2kT}\right) - 1\right\} \exp - Kx}{\exp\left(\frac{ze\psi_0}{2kT}\right) + 1 - \left\{\exp\left(\frac{ze\psi_0}{2kT}\right) - 1\right\} \exp - Kx} \qquad (5.14)$$

in which $K = (2z^2e^2 N_\infty/\varepsilon_r \varepsilon_0 kT)^{1/2}$ (see equation (5.6)). The relation (5.14) gives the fall in the potential as a function of the distance from the surface at a particular surface potential and a particular electrolyte concentration. It approximates to exponential decay and can be simplified to give equation (5.7) which has already been mentioned as truly exponential. Since K is directly proportional to the square root of the ion concentration in the solution, the potential fall is increased by adding indifferent electrolyte.

Viewing the Gouy–Chapman theory in general terms, the surface potential, ψ_0 from equation (5.13) depends both on the surface charge density and, through K, on the concentration of electrolyte in the solution. As a mathematical relationship it can be seen that if the concentration of electrolyte is increased, then the surface charge density must increase or the surface potential must decrease or, as might happen with soils which are frequently regarded as possessing permanent charge and also responding to potential determining ions, both could change.

5.4.4 Swelling and Double Layer Theory

The application of Gouy–Chapman double layer theory to a plane surface of constant charge (see Figure 5.4) shows that the surface potential decreases in a complicated manner with increasing electrolyte concentration such that sinh $(ze\psi_0/2kT)$ is inversely proportional to the square root of the electrolyte concentration. This manifests itself for instance in the relation between clay

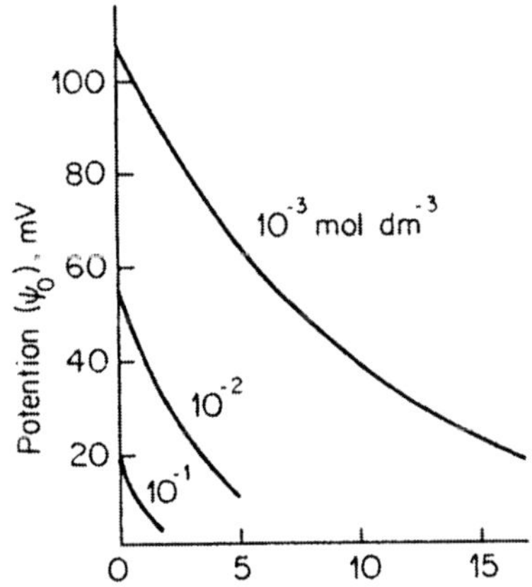

Figure 5.4 Relation between distance from surface and solution potential for different bulk uni–univalent electrolyte concentrations. The surface charge density assumed to be 0.0158 C m^{-2}. Reproduced by permission of Cambridge University Press from *An Introduction to the Principles of Surface Chemistry* by Aveyard and Haydon, 1973

swelling and electrolyte concentration. Thus, Norrish (1954) was able to follow the interlamellar swelling of Na-montmorillonite in salt solutions, and found that the interlamellar separation was proportional to $1/(N_\infty)^{1/2}$ (Figure 6.10). In contrast, for plane surfaces of fixed potential which, of course, requires a constant concentration of potential determining ions in the system, the surface charge increases with the square root of the electrolyte concentration as inferred again from equation (5.13). For colloids for which the surface charge density is a function of the surface potential, the relationship between σ and ψ_0 at one concentration of uni–univalent electrolyte (10^{-2} mol dm^{-3}) is shown in Figure 5.5.

Whatever other complications may arise, a choice has to be made between either a constant surface charge model or a constant surface potential model. For effects such as clay flocculation which often depend on reducing the thickness of the double layer, as mentioned at the beginning of this Chapter, an increase in the valency of the counterion has, other factors being equal, much more pronounced effect than an increase in concentration of indifferent electrolyte (van Olphen, 1963).

Swelling of clay upon wetting is a manifestation of the repulsive forces acting between clay particles. The osmotic force arising from ions in the diffuse double layer, which causes repulsion between charged colloidal surfaces, has been considered by Langmuir (1938). The excess concentration of ions at the mid-plane between two parallel clay plates with overlapping diffuse layers, compared with the concentration in the outer solution, can be obtained using

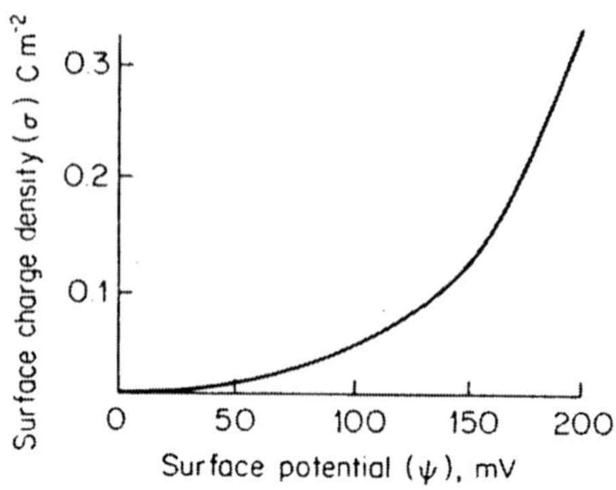

Figure 5.5 Surface charge density as a function of surface potential. The electrolyte is uni-univalent at 10^{-2} mol dm^{-3}. Adapted by permission of Cambridge University Press from *An Introduction to the Principles of Surface Chemistry* by Aveyard and Haydon, 1973

the Boltzmann distribution law. Thus, for negatively charged plates the positive ion concentration is $N_\infty e^\mu - N_\infty$ and, for negative ions, the concentration is $N_\infty e^{-\mu} - N_\infty$, where N_∞ is the outer solution concentration and $\mu = ze\psi_d/kT$ in which ψ_d is the mid-plane potential. The total ion excess at the mid-plane is

$$(N_\infty e^\mu - N_\infty) + (N_\infty e^{-\mu} - N_\infty) = N_\infty(e^\mu - 1) + N_\infty(e^{-\mu} - 1)$$
$$= 2N_\infty(\cosh\mu - 1)$$

The osmotic pressure difference π between the mid-plane arising directly from the excess of ions is given by

$$\pi = 2N_\infty kT(\cosh\mu - 1)$$

This van't Hoff osmotic pressure equation is valid for both constant charge and constant potential surfaces, with μ computed for the appropriate model (see Verwey and Overbeek (1948) for constant potential model, and van Olphen (1963) for constant charge model). The osmotic pressure must be opposed by a confining pressure acting triaxially or at right angles to the parallel plates and can be regarded as a repulsive or swelling pressure. In effect, the confining pressure is needed to keep the plates at a distance $2d$ apart.

There have been many studies on the interaction between clay particles and swelling behaviour, particularly involving the effects of divalent cations such as calcium. The behaviour of Ca clay is usually very different from that of sodium dominated clay, and shows that forces in addition to those normally found in diffuse double layers are operative. Russell (1934) was the first to suggest that clay particles could be bridged by divalent and more highly charged cations were instrumental in holding them together. Norrish (1954) stated that for a polyvalent cation to take up additional water when situated between clay sheets, the free energy loss of the water forming the hydration shell must be greater than the work done in increasing the interlamellar spacing to accommodate the water. The ways in which cations influence the hydration of clays is disucussed in Chapter 6. In connection with swelling pressure, Warkentin *et al.* (1957) found that, starting with loose floccs of Na- or Ca-montmorillonite, the pressures were greater than expected on the first compression with either type of cation present. Subsequent compressions and decompressions showed Na-montmorillonite in dilute NaCl behaved very much according to theory but Ca-montmorillonite behaved erratically. After the first compression the clay platelets were still largely randomly oriented with appreciable voids making the distance between many clay particles quite large. There is no doubt that many ions, particularly Ca of the common dominant exchangeable ions, have an orientating effect on clay platelets and, as the initially loose flocculated mass of clay moves towards an oriented structure, very much aided by Ca ions, the water relationships become more and more dominated by the development of tactoids. Relatively small differences in the treatment received by samples, e.g. variation in rates of compression or the length of time a sample is left in a wet state, can have marked effects on the behaviour of samples. The shrinking and swelling history of a Ca clay, in particular, determine the size of tactoids. When clays contain both Ca and Na exchangeable ions there is often a preferential positioning of calcium between clay plates, leaving the peripheral exchange sites on the tactoids to be occupied by sodium ions. The implications of this have been discussed by Quirk (1968), O'Connor and Kemper (1969) and McNeal (1970) and others and is, no doubt, responsible for some of the so called irreversible tendencies encountered in cation exchange equilibria in soils.

5.4.5 Effects of Na Ions on Soil Behaviour

Electrical double layer theory is often involved in the more detailed treatments of how soils behave when they contain enough exchangeable sodium to affect their structural stability and permeability. Both these latter problems involve an understanding of flocculation and structural stabilization, swelling and permeability, and dispersion and clay movement and are of special significance to irrigation in many arid regions. The mechanism by which soil permeability is reduced when soils disperse has been studied by many workers including Quirk and Schofield (1955) and Rowell, Payne and Ahmad (1969). In relating the quality of saline irrigation water to its effects on soils it is perhaps surprising that soils from many different arid regions tend to adsorb sodium relative to divalent cations in a more or less set and characteristic manner. In terms of double layer theory this implies that the soils possess similar surface charge densities (Eriksson, 1952; Bolt, 1955). Although there are uncertainties in the estimation of surface areas and hence surface charge densities it usually emerges that despite geographical diversity many soils of arid regions often have similar clay mineralogy and, being low in organic matter, respond in similar ways to increasing sodium levels and changing concentrations of salts. When soil mineralogy and organic matter contents differ amongst soils it seems that observed trends in experimental data (Pratt and Grover, 1964) conform reasonably well to electrical double layer theory, with the relative adsorption of monovalent to divalent ions decreasing with increasing surface charge density.

5.4.6 Limitations of the Gouy–Chapman Model

The Gouy–Chapman diffuse double layer theory can be applied to spherical particles as well as to flat plates, although the mathematical treatment of the former is more difficult (Verwey and Overbeek, 1948). Regardless of the geometry of the treatment, the Gouy–Chapman theory has some quite easily identifiable shortcomings, including the fact that the ions in solution are regarded as point charges and hence could reach impossibly high concentrations near the colloid surface, calculated as $>100\,\text{mol dm}^{-3}$ for quite reasonable estimates of surface electrical potential and electrolyte concentration in solution. Furthermore, the value of the dielectric constant is assumed to be independent of distance from the surface, which is almost certainly incorrect because the dielectric constant of a polar liquid like water is known to vary with the electric field strength and, in addition, the surface charge is assumed to be uniformly smeared over the colloid surface in a non-discrete manner. A further assumption of the theory is that ionic properties are solely dependent on their charges with ions possessing no individuality, which again is far too broad a generalization.

5.5 STERN-TYPE DOUBLE LAYERS

The first serious attempt to modify the Gouy–Chapman theory and make it more appropriate for application to real (non-ideal) systems was due to Stern (1924). He recognized that ions of finite size cannot approach a surface more closely than their effective radii allow and he recognized that ions may interact in more than a simple coulombic manner with charged surfaces. In the Stern model the double layer is divided into two parts with a compact layer adjacent to the surface in which the potential changes linearly from ψ_0 to ψ_δ, as in a Helmholtz classical molecular condenser-type double layer. The remainder of the model comprises a diffuse Gouy–Chapman layer as illustrated in Figure 5.6a in which the potential drops from ψ_δ to ψ_∞, the latter normally taken as zero. Usually the potential change in the compact 'Stern' layer increases with electrolyte concentration, and particularly when polyvalent counter ions are involved, it is possible for a change in sign in the potential at the outer edge of the Stern layer to occur as depicted in Figure 5.6b.

Because of the limitations of the Gouy–Chapman theory, most present double layer treatments are based on the Stern model or some suitable development

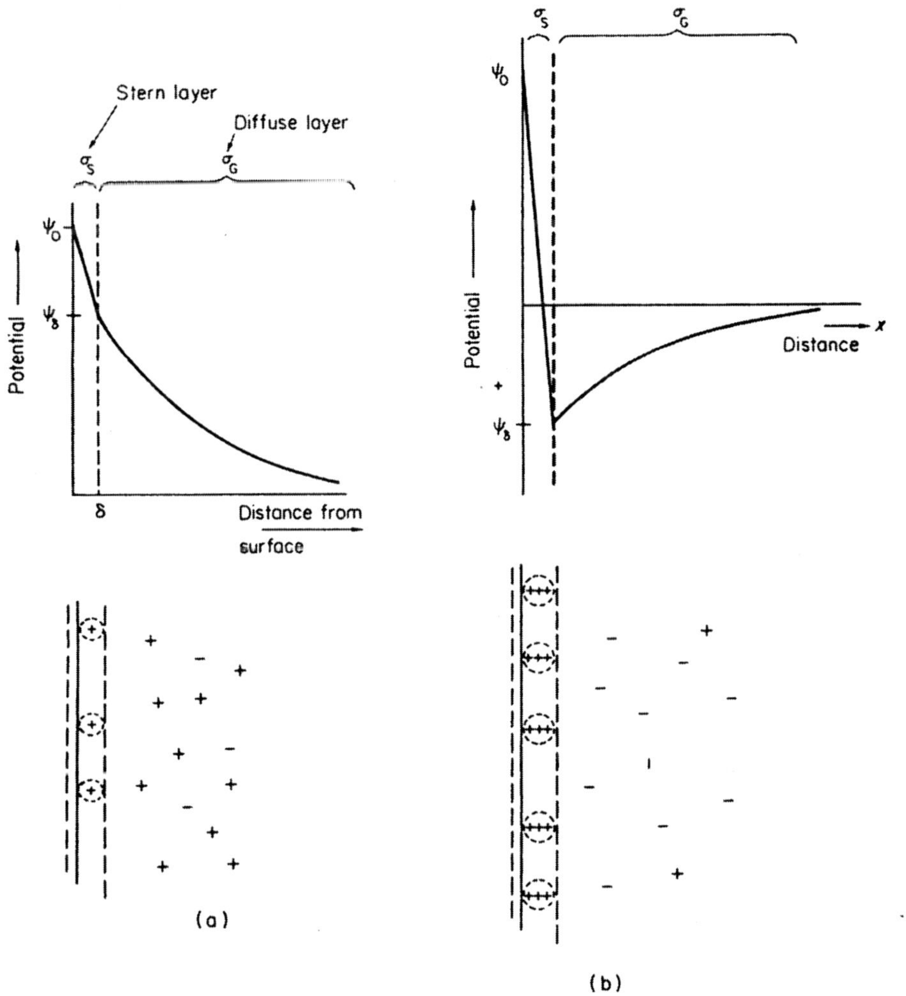

Figure 5.6 (a) Schematic representation of potential decay curve with distance from surface, showing Stern layer containing some specifically adsorbed cations, (+). Potential falls from ψ_0 to ψ_δ across Stern layer. (b) Stern layer with considerable specific adsorption of cations (+++) and showing charge reversal, ψ_0 negative to ψ_0 positive, across Stern layer

from it. For example, Grahame (1947) refined the Stern model by splitting the Stern layer into two to allow consideration of two types of strongly adsorbed ion or ions. Nearest the solid surface Grahame recognized an inner Helmholtz plane in which the adsorbed ions lose some of their water of hydration and an outer Helmholtz plane supposed to contain normally hydrated counterions close to the colloid surface.

In the Stern model ψ_δ is the potential at the boundary between the compact

'Stern' layer of thickness δ, and the diffuse Gouy–Chapman layer. The total surface charge, σ, can be regarded as split between that which is balanced by the Stern layer and that balanced by the diffuse component of the double layer. Thus,

$$\sigma = -(\sigma_S + \sigma_G)$$

where the subscripts S and G refer to the Stern and diffuse or Gouy–Chapman parts of the system, Figure 5.6b. Stern estimated the extent to which ions enter the compact layer using principles analogous to Langmuir (1918) adsorption theory at one type of site, namely, that only a single layer of adsorbate is formed, and adsorption at one site does not affect adsorption at another (the energy of adsorption is independent of surface coverage). Thus, neglecting ions of the same sign as the surface, Stern theory gives

$$\sigma_S = \frac{N_S ze}{1 + N_A/MN_\infty \exp -(ze\psi_\delta + \phi)/kT} \tag{5.15}$$

in which N_S is the number of adsorption sites per unit area, N_A the Avogadro number, M is the molecular weight of the liquid medium, ϕ the specific adsorption potential of the counter ions and the other symbols, such as N_∞ the number of ions per unit volume far removed from the influence of the colloid, have their usual significance. Because 1 in the denominator of equation 5.15 (actually taken as 2 in Stern's original work) is small compared to the exponential term in the denominator, the equation becomes

$$\sigma_S = \frac{N_S ze MN_\infty}{N_A} \exp \frac{ze\psi_\delta + \phi}{kT}$$

The value of σ_G is given by equation (5.16) (see (5.13)) using the appropriate value of ψ_δ, namely

$$\sigma_G = (8N_\infty \varepsilon_r \varepsilon_0 kT)^{1/2} \sinh \frac{ze\psi_\delta}{2kT} \tag{5.16}$$

The charge on the surface is also given by the expression for an electrical condenser

$$\sigma = -\frac{\varepsilon_r' \varepsilon_0}{\delta}(\psi_0 - \psi_\delta) \tag{5.17}$$

where ε_r' is the relative permittivity of the Stern layer. From these relationships it is possible to calculate values for σ_S, σ_G and ψ_0 for selected values of ψ_δ and given electrolyte concentrations. In the Stern treatment there are, however, a number of weaknesses such as the number of available positions in the liquid being taken as N_A/M, the uncertainty in the number of sites available for adsorption on the surface, and uncertainties about the values of δ and ε_r'. Thus despite the quantitative appearance of the Stern treatment it is probably only qualitatively correct. In practice, because of the various uncertainties it is necessary to use reasonable values of ε_r'/δ which can be adjusted in the light of experiment when calculating net surface charge from equation (5.17). There have been numerous attempts to improve on the Stern model, including that of Grahame (1947), already mentioned, and Wright and Hunter's (1973a, 1973b).

5.6 CONSTANT POTENTIAL (REVERSIBLE) HYDROXYLIC SURFACES

As already indicated, most progress is likely to be made by examining soils with simple rather than with complicated mineral compositions, and in recent years some of the highly weathered Fe and Al rich soils from various parts of the humid tropics have been profitably studied. It is, however, the studies of oxide surfaces which have contributed most towards a better understanding of soils dominated by constant potential surfaces, in contrast to most others which are dominated by constant charge surfaces. In aqueous systems surfaces of oxide minerals are hydroxylated (Chapter 4) and a reasonable model can be presented as

$$MOH_2^+ \overset{H^+}{\rightleftharpoons} MOH \overset{OH^-}{\rightleftharpoons} MO^- + H_2O$$

which indicates that H^+ and OH^- are surface potential determining ions and that the reaction $MOH + H^+ \rightarrow M^+ + H_2O$ is unlikely to take place.

5.6.1 The Point of Zero Charge (p.z.c.)

In any unit of surface the pH of the solution at which there is no net surface potential is called the point of zero potential (p.z.p.). If the surface potential determining ions are H^+ and OH^- then according to the Nernst equation the surface potential is given by

$$\psi_0 = \frac{RT}{zF} \ln \frac{a_{H^+}}{(a_{H^+})_{p.z.p.}}. \tag{5.18}$$

If, as seems inevitable, the p.z.p. is identified with the point of zero charge (p.z.c.) on the surface then equation (5.18) can be rewritten as

$$\psi_0 = -\frac{2.303RT}{zF}(\text{pH} - \text{pH}_{p.z.c.}) \tag{5.19}$$

where $\text{pH}_{p.z.c.}$ is the pH of the system at which the surface of the solid carries no net charge. Clearly, a gain of H^+ by the surface cannot be distinguished from a loss of OH^- and vice versa. For hydroxylated reversible surfaces the p.z.c. makes an excellent reference point and from equation (5.19) it should be possible to calculate the surface charge for any value of the surface potential at a particular symmetrical indifferent electrolyte concentration, if Gouy–Chapman principles apply. Alternatively, the net surface charge can be calculated from equation (5.17) if the model involves a Stern layer. As discussed in Chapter 4, there are two types of surface. When the surface is reversible, as for the oxides and allophane, the potential at any pH should be given by equation (5.19) and, provided the supporting electrolyte is indifferent, the potential should be independent of the electrolyte concentration. The surface charge density, however, would change with changing electrolyte concentration according to equation (5.13). The other type of interface is, of course, the one with constant surface charge, as on the main surfaces of the 2:1 minerals. For the latter there is no mechanism by which the surface charge can change, and so the surface potential must change in response to changing indifferent electrolyte concentration (equation (5.13)).

5.6.2 Determination of p.z.c.

Experimentally there are two obvious ways to obtain information about the surface of a hydrous oxide or hydrous oxide-rich soil. One is by an apparently simple potentiometric titration approach, in which, if H^+ and OH^- are the only potential determining ions, the surface charge density from measured surface area, is given by

$$\sigma_0 = F(\Gamma_{H^+} - \Gamma_{OH^-})$$

where F is the Faraday constant and Γ_{H^+} and Γ_{OH^-} are the adsorption densities of the potential determining ions (p.d.i.). Usually, titrations are performed in indifferent electrolyte at various concentrations and, when appropriately matched, the point where the σ_0 versus pH curves cross each other gives the p.z.c. of the interface. A typical example (Breeuswma and Lyklema, 1971) for a synthetic haematite (α-Fe_2O_3) is given in Figure 5.7). If an interface is initially uncharged the net adsorption is given by $\Gamma_{H^+} - \Gamma_{OH^-}$; an excess of H^+ ions must give the surface a positive charge, whereas an excess of OH^- ions must make the surface negative. On this basis, regardless of the initial state of the interface, there must be for amphoteric surfaces a particular pH at which the net charge on the solid surface is zero, usually referred to as the p.z.c. In the present context it has so far been assumed that only H^+ and OH^- are potential determining. However, for most soil surfaces there are many ions which can be chemisorbed and which change the surface potential. It is surprising that soil scientists and others interested in oxide and hydroxylated surfaces did not until recently exploit Verwey's (1935) statement that 'hydrogen ion and hydroxyl ion have to be considered as potential determining ions' for Fe_2O_3, Al_2O_3, Cr_2O_3 and similar materials. The important clue that H^+ and OH^- are p.d.i. for oxides and oxide-rich soils remained largely neglected for about a quarter of a century.

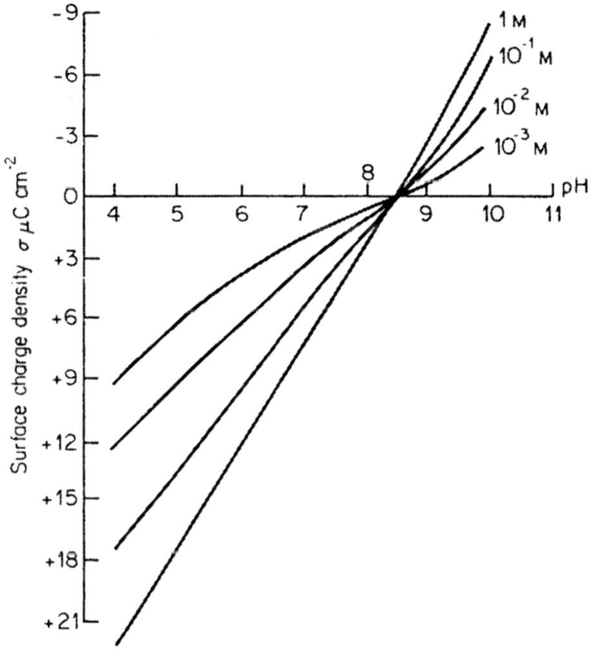

Figure 5.7 Surface charge of synthetic haematite (α-Fe_2O_3) as a function of pH in the presence of four different concentrations of KCl (Breeuwsma and Lyklema, 1971). Reproduced by permission of The Chemical Society

The other experimental approach to determining surface charge characteristics involves the estimation of the adsorption of both positive and negative ions from a suitable (indifferent) electrolyte as a function of pH and concentration. In principle, the method is that adopted by Schofield (1949). It has the advantage that it involves the direct determination of positive and negative surface charges as well as net charge. According to van Raij and Peech (1972) both positive and negative charges coexist in some soils, due presumably to the steric disposition of charges on the surface which makes impossible the mutual cancellation of all opposite charges. Schofield and Samson (1954) also noted the coexistence of positive and negative charges on kaolinite particles (Chapter 4). Using this method the p.z.c. on reversible soils should be the same as obtained from potentiometric titrations, provided the ions used are indifferent.

As discussed later in this Chapter any tendency for the cation to be specifically adsorbed into the inner part of the double layer will lower the p.z.c. whereas the specific adsorption of an anion will tend to shift the p.z.c. to a higher value (see Figure 5.12). If either of the ions were to be chemisorbed by the surface, that is become potential determining, then the p.z.c. would move in the opposite directions to those cited in the previous sentence.

5.6.3 Potential Determining Ions (p.d.i.)

Although according to Grahame's model (1947), specifically adsorbed ions close to the solid surface, in the inner Helmholtz plane, can lose some of their water of hydration, this leaves enough water to differentiate such ions from others which cross the solid–liquid interface and, joining the solid, become potential determining.

Insoluble oxides like goethite and hematite fit the constant potential model rather well and at least some types of soils or subsoils, particularly Oxisols, do the same. From what has been stated, for the analysis of any one system either the constant potential model, or the constant charge model has to be used; the models are mutually exclusive and, as far as is known, cannot be combined into one comprehensive or universal model.

In many different kinds of solid–aqueous solution interface phenomena the majority of hydrophobic colloids acquire their surface charge by the adsorption

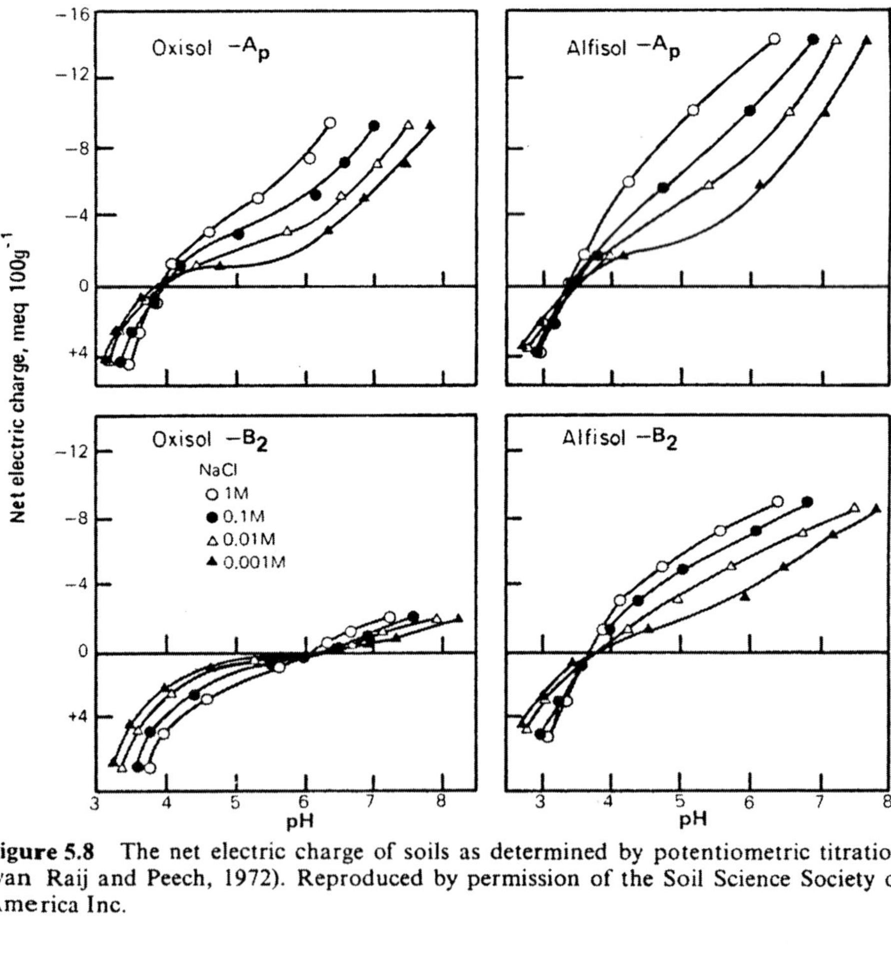

Figure 5.8 The net electric charge of soils as determined by potentiometric titration (van Raij and Peech, 1972). Reproduced by permission of the Soil Science Society of America Inc.

of potential determining ions which are common to the colloid and to the aqueous medium, for instance silver chloride in a silver nitrate or potassium chloride solution. On this basis clays with their permanent or fixed surface charges arising from isomorphous substitution in the crystal lattice are the exception and not the rule. Surface potential determining ions should not, after adsorption, be chemically distinguishable from ions already in the lattice, and so p.d.i.s should not alter the chemical potential of the surface of the solid. For many oxide surfaces H^+ and OH^- behave as potential determining ions although they are not always constituents of the lattice (in the sense that Ag^+ and Cl^- are constituents of AgCl). It is, therefore, debatable whether H^+ and OH^- can, without qualification, be termed p.d.i. on the hydroxylated surfaces of insoluble oxides, and on the edges of clay mineral lattices. Unfortunately, the variations in surface potential with pH cannot be tested experimentally on commonly occurring iron, aluminium and similar oxides and hydroxides because suitable electrodes cannot be made of these poor electrical conductors. This means that the validity of the Nernst equation (5.19) cannot be checked by direct measurement. However, for most practical purposes we can accept that H^+ and OH^- are p.d.i. and assume that equation (5.19) applies. Although some configurational changes on soil constituents may accompany H^+ and OH^- adsorption and desorption, such changes may not be large, particularly as OH^- is a common lattice constituent of the more reactive surfaces of soils.

Thus it can be concluded that for oxides in aqueous media the most important pair of p.d.i.s is H^+ and OH^- and the development of surface charge is measured by their adsorption. The presence of other p.d.i.s in an aqueous oxide system causes the p.z.c. to move to a lower pH in the case of anions and to a higher pH in the case of cations. Clearly, p.d.i.s are adsorbed into (or lost from) the surface, and hence fall into the category of chemisorbed ions.

5.6.4 The p.z.c. of Soils

With even the most oxidic soils there will normally be particles present with appreciable permanent negative charge. Van Raij and Peech (1972) noted that the p.z.c. derived from the point of intersection of titration curves on soils obtained with different concentrations of electrolyte shown in Figure 5.8 was not always the same as the p.z.c. determined by direct measurement of adsorbed ions over a suitable range of pH values. As a rule the p.z.c. obtained from direct measurement of ion adsorption (e.g. Na^+ and Cl^-) is lower than that obtained from potentiometric titration curves. This difference may be attributed to the fact that some of the negative charge on the original soil was balanced by strongly adsorbed aluminium ions, whereas using the direct ion adsorption technique such negative charge would be balanced only, or at least principally, by the particular selected index cation. On this basis the procedure involving the direct measurements of cation and anion adsorbed would be preferable to potentiometric titration because the pretreatment of the soil would remove much of the adsorbed aluminium. For soils carrying considerable negative charge the titration curves at different electrolyte concentrations do not intersect and hence the p.z.c. cannot be determined. To construct net electric charge–pH curves comparable to those shown in Figure 5.8, the amount of H^+ and OH^- adsorbed by the soil at any pH value was taken as equal to the amount of HCl or NaOH added to the suspension minus the amount of acid or base required to bring the same volume and the same concentration of NaCl solution, without soil, to the same pH. The net electric charge was calculated from the amount of H^+ and OH^- adsorbed relative to the p.z.c. In such titrimetric work questions such as the time allowed for equilibration and the concentration of the suspensions usually require careful thought.

As a rule the presence of oxides and hydroxides of Fe and Al and short range order constituents tend to place the p.z.c. at higher pH values whereas clays with their structural negative charges and soil organic matter will bring the p.z.c. of the soil to lower pH values. On this basis the p.z.c. should reflect the overall mineralogical and organic matter composition of the soil.

The net electric charges and the p.z.c. of the A_p and B_2 horizons from Oxisols and an Alfisol are shown in Figure 5.8, as determined at various electrolyte concentrations by van Raij and Peech (1972). The effects of lower organic matter contents are seen in the B_2 samples, the lower organic contents displacing the p.z.c. to higher pH values. They calculated variations in net surface charge densities with changing surface potential and electrolyte concentration, the former being directly related to pH and $pH_{p.z.c.}$ through Nernst equation (5.19) for reversible interfaces. Thus for the Gouy–Chapman model and a symmetric electrolyte in solution

$$\sigma_0 = (8N_\infty \varepsilon_r \varepsilon_0 kT)^{1/2} \sinh \frac{ze\psi_0}{2kT}$$

in which ψ_0 comes from equation (5.19). Calculation of the surface charge according to the Stern model is more complicated using equations (5.15), (5.16), and (5.17) and involving best estimates of N_s, ε_r' and δ. Following van Olphen (1963) the value of N_s was taken as 10^{15} cm^{-2} and the value of δ/ε_r' was adjusted so as to give the best fit of curves to the experimental results obtained at four different concentrations of NaCl, a value of 0.015 nm for the latter ratio being found to give the best fit of the curves to the experimental data for all the soil samples. The plot of surface charge against potential, labelled 'Stern' in Figure 5.9, agreed quite well with the experimental points at least down to 0.01 mol dm^{-3} NaCl for three soils at two depths. The experimental data did not agree with the calculations based on the Gouy–Chapman model as seen in Figure 5.9. They described how the p.z.c. can be obtained, using 0.2 mol dm^{-3} NaCl by plotting the net surface charge against the pH and interpolating the curves to zero charge. Their results showed that both positive and negative charges can coexist in soils, indicating that steric factors prevent the complete cancellation of opposite charges. The magnitude of both the positive and nega-

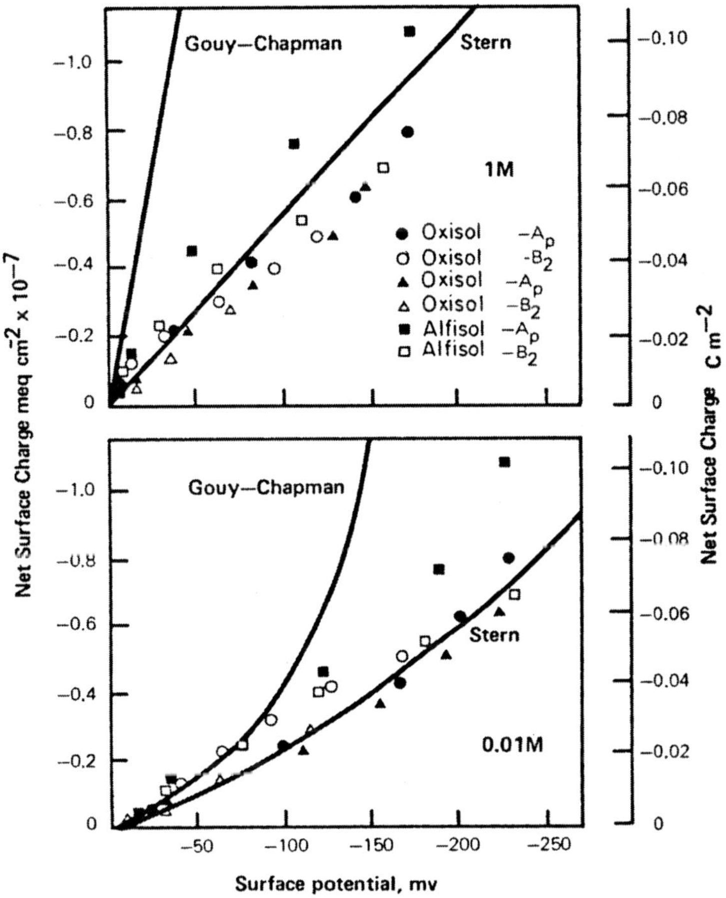

Figure 5.9 Comparison of the net negative surface charge of soils as determined by potentiometric titration with that calculated by the Gouy–Chapman and the Stern theory for reversible interfaces (van Raij and Peech, 1972). Reproduced by permission of the Soil Science Society of America Inc.

tive charges was found to decrease with decreasing electrolyte concentration. At constant pH such an effect agrees with double layer theory for reversible interfaces. Although soil types which carry a net positive charge are rare when viewed globally, an understanding of how they behave can be helpful when dealing with less extreme soil types.

5.6.5 ΔpH Values

The traditional method of deciding whether a soil is negatively or positively charged is to measure its pH in 1.0 mol dm^{-3} KCl and in water. On this basis the pH difference expressed as

$$\Delta pH = pH_{KCl} - pH_{H_2O}$$

gives the sign of the charge carried by the colloids. As is well known the vast majority of soil types give a lower pH in the KCl and are negatively charged. Obviously, soils which happen to be at the p.z.c. will give the same pH in water and in KCl. The above comments apply only if the electrolyte, in this case KCl, is indifferent. If either the anion or the cation is specifically adsorbed into the Stern layer (or IHP) then very different effects can be observed. For example, Mekaru and Uehara (1972) gave data to show that sulphate supplied as K_2SO_4 gave positive values for ΔpH even when the soil surface initially possessed a net negative charge. They stressed that it is imperative to use an indifferent electrolyte if one wishes to interpret ΔpH values for the purpose of establishing whether the surface charge is negative or positive. In this connection it is important to realize that soils do not have to be positively charged to adsorb anions specifically, nor do soil surfaces have to be negatively charged to adsorb cations specifically. A possible definition of a specifically adsorbed ion is one that is adsorbed at the p.z.c.

5.6.6 Specific Adsorption

So far in this chapter the use of the term specific adsorption has been used sparingly because it is far too general a term unless its use is restricted as recommended by some authors. It is now generally agreed that ions which are chemisorbed, as opposed to physisorbed (Everett, 1972) are potential determining ions or at least alter the surface potential. Following the convention normally adopted in colloid science the term specific adsorption should be used only for adsorption other than non-specific adsorption into the electrical double layer; such specifically adsorbed ions are in the Stern layer or Inner Helmholtz Plane (IHP), depending on the model adopted.

At this stage it is helpful to consider a possible working model which deals with the basic question of where and how does an adsorbed species 'sit'. For this purpose it does not matter whether we take positive, negative or neutral surfaces, or an anion or a cation as adsorbate. Once one particular combination has been considered, other combinations should follow. Consider, for example, a reversible surface which is negatively charged and a cation which is specifically adsorbed but, as already agreed by definition, does not become chemisorbed by the solid surface. In other words, the cation probably becomes partially dehydrated, retaining at least some of its inner hydration sheath and it occupies a zone near the surface in the Stern layer, as illustrated in Figure 5.12c. According to Lyklema (1971) what happens is that the positive charge carried by the specifically adsorbed cation, which ends up near the surface, promotes some desorption of H^+ or adsorption of OH^- and the surface becomes more negative and thus more H^+ would be needed in the solution to bring the surface to its p.z.c. The pH of the p.z.c. must, therefore, fall. Data which are available so far seem to support this model and explain related phenomena such as super-equivalent adsorption (q.v.).

Still keeping to the same example, the adsorption of the positive ions into the IHP or Stern layer makes the potential in the slipping plane, the zeta potential (see page 400), more positive provided the slipping plane more or less coincides with the boundary between the Stern and diffuse parts of the double layer. Any measurements based on zeta potential such as measurements of the isoelectric point (i.e.p., see page 401) before and after adsorption will emerge as effects in a direction opposite to that in which the p.z.c. moves. Thus, for a particular specific adsorption, Δp.z.c. and Δi.e.p. move in opposite directions. Because the occurrence of adsorption into the inner part of the electrical double layer is often judged by either Δp.z.c. or Δi.e.p. it is very important that a proper distinction is made between the two. It is evident that some authors do not make it clear whether they are concerned with Δp.z.c. or Δi.e.p. and readers are thus often left to resolve the ambiguities as best they can.

5.7 THE DONNAN PRINCIPLE AND ITS APPLICATION TO SOILS

5.7.1 Donnan Equilibria

As already mentioned, the soil solution may be considered as a system of distinct phases, one associated with the surface and the other the bulk solution. Donnan principles (Donnan, 1911) may then be applied. Reference to Donnan equilibrium principles is needed on several grounds but largely because of the general beneficial influence they had on the early development of ideas con-concerned with soil–electrolyte interaction. The classical Donnan system, made up of colloid plus salt solution in equilibrium across a semipermeable membrane with more salt solution assumes that ions are completely dissociated from the colloid and that there exists a uniform charge distribution in the non-diffusing part of the electrolyte solution as well as in the diffusible part.

A Donnan system for a negatively charged colloid and a uni–univalent electrolyte can be represented as

colloid M^+	¦	
$[Z]$	¦	M^+ A^-
M^+ A^-	¦	$[X]$
$[Y]$	¦	

where X, Y and Z represent the concentrations of the main components and the broken line the semipermeable membrane. The values of X, Y and Z can usually be determined by analytical methods after equilibrium has been attained. For a given total volume, assuming that the counter ions are completely dissociated from the colloid, and that at equilibrium the chemical potential of MA must be the same on both sides of the membrane, $Y(Y+Z)=X^2$, and the negative adsorption (repulsion) of the anion in the colloid compartment is given by $X-Y$. From this equation a number of relationships can be derived for ions of different valency, including the effects of dilution.

Perhaps the first point to clarify when dealing with soil colloids is that unlike the classical Donnan system, there is no need for a semipermeable membrane between the 'inner' (colloid) and 'outer' solutions because sedimentation can at least theoretically lead to separation of colloid and supernatant liquid and there would thus be a constraint on ionic movement between 'inner' (colloid) and 'outer' (supernatant) solutions which may therefore be assumed to behave as if separated by a semipermeable membrane. Each colloidal particle with its electrical double layer may also be regarded as a micro-Donnan system in which the attractive forces between the colloid and counterion act as a restraint, taking the place of the membrane; one is thus able in various ways to consider relationships between 'inner' and 'outer' solutions, the former being the solution containing the colloid and counterions together with some additional salt, and the latter a solution in equilibrium with the former.

Using the concept of 'inner' (″) and 'outer' (′) solution on a sodium-calcium clay equilibrated with solution containing sodium and calcium chlorides, the following activity product relationships hold:

$$(a''_{\mathrm{Na}})\cdot(a''_{\mathrm{Cl}}) = (a'_{\mathrm{Na}})\cdot(a'_{\mathrm{Cl}})$$

$$(a''_{\mathrm{Ca}})^{1/2}\cdot(a''_{\mathrm{Cl}}) = (a'_{\mathrm{Ca}})^{1/2}\cdot(a'_{\mathrm{Cl}})$$

giving

$$\frac{(a''_{\mathrm{Na}})}{(a''_{\mathrm{Ca}})^{1/2}} = \frac{(a'_{\mathrm{Na}})}{(a'_{\mathrm{Ca}})^{1/2}}$$

If it is assumed that the volume of the inner solution is constant and the volume of the outer solution V varies inversely with its total anion concentration (i.e. the total electrolyte content of the system remains constant), then $a''_{\mathrm{Na}}/(a''_{\mathrm{Ca}})^{1/2}$ is proportional to $1/V^{1/2}$, which means dilution of the outer liquid would involve a decrease in the ratio of sodium to calcium adsorbed, that is the clay adsorbs more Ca and less Na. This is a well known effect which is at least qualitatively expressed by all predominantly negatively charged soils. Although variations in the activity coefficients for cations in the outer solution are expected to be very small, especially in dilute solution, variation in the coefficients for adsorbed cations in the 'inner' solution may be large and hence lessen the predictive value of the simple relationship involving $V^{1/2}$.

5.7.2 Donnan Principles Applied to Cation Exchange

A standard approach to cation exchange equilibria in predominantly negatively charged soils is, of course, possible. In brief, an exchange equilibrium for a soil mono–divalent cation system of the type

$$2\text{Na soil} + \text{CaCl}_2 \rightleftharpoons \text{Ca soil} + 2\text{NaCl}$$

using the symbols X_{Na} and X_{Ca} to represent the outer solution concentrations and Y and Z, with appropriate subscripts, to represent the ions associated with the inner solution and the soil colloid respectively, can be formulated as follows

Firstly,

$$Y_{Cl}(Y_{Na}+Z_{Na}) = X_{Na}X_{Cl}$$

and, secondly

$$Y_{Cl}^{2}(Y_{Ca}+Z_{Ca}) = X_{Ca}\cdot X_{Cl}^{2}$$

so that

$$Y_{Cl}(Y_{Ca}+Z_{Ca})^{1/2} = X_{Ca}^{1/2}\cdot X_{Cl}$$

and dividing,

$$\frac{Y_{Na}+Z_{Na}}{(Y_{Ca}+Z_{Ca})^{1/2}} = \frac{X_{Na}}{(X_{Ca})^{1/2}}$$

which for low salt concentration, i.e. Y much smaller than Z, gives

$$\frac{Z_{Na}}{(Z_{Ca})^{1/2}} = \frac{X_{Na}}{(X_{Ca})^{1/2}}$$

which relates the partition of ions between soil and the equilibrium solution. The approximate minimum relative size of Z and Y can be obtained by considering a Ca-saturated illitic clay with a surface area of $\sim 100\ \text{m}^2\ \text{g}^{-1}$ in equilibrium $1\times10^{-1}\ \text{mol dm}^{-3}$ CaCl_2. According to diffuse double layer theory the approximate maximum thickness of the double layer would be 0.5 nm and hence the volume of the double layer per g would be

$$0.5\times10^{-9}\times100\times10^{6}\ \text{cm}^3 = 0.05\ \text{cm}^3\ \text{g}^{-1}.$$

If the clay possesses an exchangeable Ca content of $0.2\ \text{meq g}^{-1}$, i.e. $0.1\ \text{m mol Ca g}^{-1}$, then the concentration of Ca in the double layer would on average be $2.0\ \text{mol dm}^{-3}$, that is 20 times the concentration in the outer solution. For $1\times10^{-3}\ \text{mol dm}^{-3}$ Ca in the outer solution and approximately 5 nm for the maximum thickness of the double layer, comparable calculation shows the inner solution to be about 200 times more concentrated than the outer solution. When diffuse layers are not developed to their maximum, which is normally true of predominantly Ca-saturated soils and clays, and when outer solutions are dilute, then $Y \ll Z$ and Donnan theory is very likely to apply.

Provided the overall concentration of salt is low, then from an analysis of the equilibrium solution and calculation of activities, the ratio, $a'_{Na}/(a'_{Ca})^{1/2}$, is obtained which characterizes the relative activities of the two ions in the whole soil–soil solution system. Teräsvuori (1930) on the basis of Donnan theory concluded that there should be a constant value for the ratio $(C_{Ca})^{1/2}/C_H$, where C_{Ca} and C_H are the concentrations of calcium and hydrogen ions in the outer solution, for all solutions in equilibrium with a given soil sample containing calcium as the dominant cation. Schofield (1947), following these ideas, established how a Donnan approach could be applied to soils containing a range of mono-, di- and even trivalent cations.

5.7.3 The Ratio Law

When Donnan theory holds, Schofield stated soils would obey the Ratio Law (p. 356). In a classical Donnan system (with physical membrane) there is a unique potential drop at the membrane, but for individual soil colloid particles, typified by the layer aluminosilicates, this is not true. Pressing the analogy between a single clay particle with a built in constraint on cation apportionment and a traditional Donnan system with membrane is likely to lead to some difficulties. However, for negatively charged soil colloids, if a sufficiently large average potential drop exists at the colloid solution boundary, and solutions are dilute, anions have a vanishingly small concentration compared to that of the cations associated with the colloid surface. Schofield (1947), using appropriate double layer theory (see Section 5.4), put a minimum figure of 60 mV for this potential drop and stated that if this condition is fulfilled then the Donnan principle would hold; dilution of solution would increase all potential differences by the same amount and so would not disturb the equilibrium.

In order to meet Donnan theory requirements, Schofield stressed that soil particle surfaces had to be predominantly and highly negatively charged and that solution concentrations should not be too high, with the upper limit varying principally with the charge on the cation (Schofield and Taylor, 1955a) and with the nature of the anion. The higher the charge on the cation the lower the concentration at which the surface potential becomes too low and the Ratio Law fails to hold; for several common cations in the form of their chlorides the upper concentration limit often lies between about 0.02 and 0.1 mol dm^{-3}. The application of Donnan principles can be very successful provided certain conditions are met. If, however, the experimental data do not fit the theory there is no possibility of compromise because Donnan theory does not carry with it a model which can be elaborated or modified to meet special circumstances, in marked contrast with double layer theory.

5.7.4 The Relative Chemical Potentials of Adsorbed Ions

One of the most useful outcomes of the Ratio Law is that provided soils are predominantly negatively charged and the equilibrium electrolyte solutions are not too concentrated, then the values obtained for activity ratios of the types

$$\frac{a'_{M^+}}{(a'_{M^{2+}})^{1/2}} \quad \text{and} \quad \frac{a'_{M^+}}{(a'_{M^{3+}})^{1/3}}$$

are independent of the total electrolyte concentration in the equilibrium solution. At particular temperatures, unique values are obtained for the activity ratios which depend only on the amounts of cations adsorbed; for this to be so, anions must effectively be excluded from the inner parts of the double layer so that the environment in which the bulk of the adsorbed cations exists, is not materially affected by changes in the solution concentration. (This must not, of course, be confused with the effect of simple dilution, page 388, which causes $a'_{M^+}/(a'_{M^{2+}})^{1/2}$ to change and shifts the equilibrium between colloid and outer solution in favour of negatively charged colloid adsorbing more divalent and less monovalent cations.) The relative chemical potentials of adsorbed ions can be evaluated without taking arbitrary decisions about the precise methods of measurement. Hence, for example, where pCa, like pH, is the negative logarithm of the molar activity, $(\text{pH}-\frac{1}{2}\text{pCa})$ is a constant for a particular soil provided the method of measurement does not alter the composition of the labile cationic suite associated with the soil (Schofield and Taylor, 1955b). In using equilibrium activity ratios it must, however, be emphasized that one cannot go beyond measuring the potential of one ion relative to another (see Section 5.3). For soils in which calcium is the dominant exchangeable cation, it is convenient and reasonable to use it as a reference ion, although it is frequently more appropriate to use the sum of calcium and magnesium in solution as the reference 'ion'. This procedure is usually acceptable because Ca and Mg behave in similar ways when adsorbed on most soils. When the Ca or (Ca + Mg) potential is reasonably constant in different soils then a measure of the relative potentials of an ion such as potassium among the different soils can be obtained (Beckett, 1964b). This potential is then an approximation of the availability of the soil potassium to plants.

In many soils the natural content of anions in an equilibrium 'water extract' is enough to give analytically determinable amounts of the cations needed to calculate activity ratios, but the disadvantage is that water extracts are not always obtained sufficiently free from colloids for analysis, and the act of obtaining a water extract must to some extent deplete the soil of labile cations. Usually it is convenient to use Taylor's (1958) interpolation method for determining equilibrium cation activity ratios. When calcium is the dominant soil cation it is convenient to equilibrate soil with dilute calcium chloride solutions (usually within the range of 0.02–0.005 mol dm^{-3}) containing both slightly more and slightly less than the roughly estimated equilibrium concentration of the cation under consideration such as potassium; some of the particular cation will usually be either adsorbed or desorbed by the soil, but knowing the weight of soil and volume of solution, the composition of the solution which would not change on equilibration can be determined by interpolation. From the measured concentrations and appropriate activity coefficients the equilibrium activity ratio is obtained. In practice it is desirable to use sufficient total electrolyte to keep soil colloids flocculated, and hence facilitate analysis of the solution.

Ion activities a_i, are usually calculated from measured molar concentrations, C_i and activity coefficients γ_i are obtained using an appropriate form of Debye–Hückel expression. Thus

$$a_i = C_i \gamma_i$$

in which

$$-\log \gamma_i = \frac{A z_i^2 \sqrt{I}}{1 + \mathring{a}_i B_i \sqrt{I}}$$

the latter Debye–Hückel equation giving as precise an estimate of γ_i as is likely to be required for most purposes. In this equation z_i is the valency of the ion, A and B are constants for the solvent at specified temperature and pressure and $\mathring{a}_i$ has a value dependent on the effective diameter of the ion in question, and is almost invariably found by experiment to be larger than the ionic diameters existing in crystals. Robinson and Stokes (1970) have discussed this matter in some detail. The ionic strength I is given by

$$I = \tfrac{1}{2} \sum C_i z_i^2$$

in which C_i is the molarity (mol dm^{-3}) of the ion and z_i its valency. Thus for $CaCl_2$

$$I = \tfrac{1}{2} \{C_{Ca} \times 2^2 + C_{Cl} \times (-1)^2\}$$

When the ionic strength is very small the first approximation to the Debye–Hückel equation, $-\log \gamma_i = Az^2 \sqrt{I}$ can be used for calculating γ_i.

Those familiar with the use and limitations of the Gapon (1933) equation, which was originally derived to describe the adsorption of molecules on a liquid surface, will know that the validity of the Ratio Law is implicit in the equation when applied to cation exchange equilibria. In its simplest form, describing Na and Ca adsorption by predominantly negatively charged soil

$$\left(\frac{\text{Na}}{\text{Ca}}\right)_{\text{adsorbed}} = kG \left(\frac{\text{Na}}{(\text{Ca}/2)^{1/2}}\right)_{\text{solution}}$$

where the amounts of Na and Ca are expressed in equivalents or milliequivalents and kG is the Gapon constant.

5.8 THE MEASUREMENT OF pH AND THE SUSPENSION EFFECT

pH values of soil suspensions are measured using an electrode reversible to hydrogen ions (invariably glass) and a reference electrode. The reference

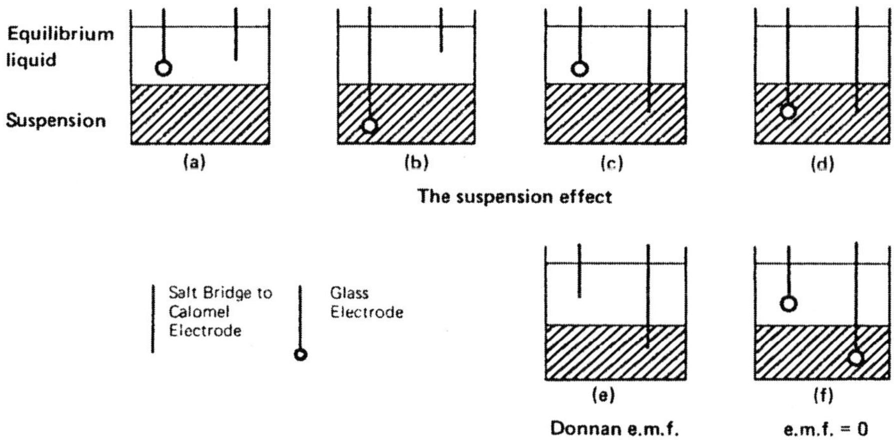

Figure 5.10 The suspension effect and Donnan electromotive force. Reproduced by permission of Interscience Publishers from *An Introduction to Clay Colloid Chemistry* by van Olphen, 1963

electrode is most often a saturated or 3.5 mol dm^{-3} KCl calomel electrode, although an Ag:AgCl electrode with KCl salt bridge can also be used. The reference electrode is assumed to be of invariant potential at a specified temperature, with insignificant variation in liquid junction potential when the standard or calibrating solution is replaced by a sample solution. There has at times been considerable uncertainty about what was actually being measured when pH readings were taken in a suspension of soil or clay.

There are four ways in which a reversible electrode, such as a glass electrode for measuring pH, and a reference electrode of (ideally) invariant potential can be arranged in a system consisting of a sediment and a clear supernatant liquid (Figure 5.10). Provided the sediment and clear liquid have attained equilibrium the position of the glass electrode should make no difference to the recorded potential. The readings in vessels (a) and (b) should be identical, otherwise we would have to conclude that work could be gained from a system in thermodynamic equilibrium which is contrary to the second law of thermodynamics. Clearly, the same argument must apply to vessels (c) and (d). The position of the salt bridge does, however, have an effect and potential measurements can be different between (a) and (b) on the one hand and (c) and (d) on the other.

When a soil is negatively charged the reference electrode connected to the salt bridge shows a lower potential when the bridge dips into the suspension than when in contact with the equilibrium solution (or less concentrated suspension). The reverse effect is obtained for positively charged colloids. Although some discussion will be needed on the so-called liquid junction potentials at the salt bridge test solution junctions, it is firstly necessary to consider the nature of the Donnan potential. (The following treatment stems from unpublished work of Drs T. D. Evans and J. R. Leal, who in turn have used the work of Overbeek (1953) and others.)

When a sediment of charged particles is in equilibrium with its supernatant liquid, the electrochemical potential of each ionic species must be the same in both cases. Thus

$$\bar{\mu}_i = \mu_i + z_i F\psi$$

where the symbols, and expression, are the same as set out on page 366. At equilibrium $\bar{\mu}_i'' = \bar{\mu}_i'$ where superscripts $''$ and $'$ represent the sediment and supernatant liquid, respectively. It follows that

$$\mu_i' + z_i F\psi' = \mu_i'' + z_i F\psi''$$

and therefore

$$\psi' - \psi'' = \frac{\mu_i'' - \mu_i'}{z_i F} \qquad (5.20)$$

The potential difference $\psi' - \psi''$ between the phases is known as the Donnan potential (ψ_{Donnan}). Regardless of what other effects are encountered in suspension-supernatant liquid systems, the ψ_{Donnan} is given by equation (5.20) which can be rewritten on the basis that

$$\mu_i = \mu_i^\circ + RT \ln a_i \qquad (5.21)$$

in which μ_i° represents the chemical potential of ion i in the standard state and a_i is the activity of the ion i. Equation (5.20) can be rewritten incorporating (5.21) to give

$$\psi_{\text{Donnan}} = \frac{1}{z_i F}(\mu_i^\circ + RT \ln a_i'') - \frac{1}{z_i F}(\mu_i^\circ + RT \ln a_i')$$

which simplifies to

$$\psi_{\text{Donnan}} = \frac{RT}{z_i F} \ln \frac{a_i''}{a_i'}$$

From what has been stated so far, the potential difference measured between two identical electrodes, Figure 5.10f, reversible to diffusible ionic species i (which can be H^+) is zero when equilibrium has been established, despite the difference in activity of i in the two phases (see page 366). The Nernst responses of the electrodes are

$$E_i'' = E_i^\ominus - \frac{RT}{z_i F} \ln a_i''$$

for the suspension and

$$E_i' = E_i^\ominus - \frac{RT}{z_i F} \ln a_i'$$

for the supernatant liquid where $E_i^\ominus$ is the standard potential. The potential difference between the electrodes in Figure 5.10f will be the difference in Nernst response, plus the Donnan potential which is able to interpose itself because the electrodes are electrical conductors measuring across the Donnan interface. Thus

$$\begin{aligned} E_{\text{Cell(f)}} &= (E_i'' - E_i') - \psi_{\text{Donnan}} \\ &= \frac{RT}{z_i F}\ln\frac{a_i'}{a_i''} + \frac{RT}{z_i F}\ln\frac{a_i''}{a_i'} \\ &= 0 \end{aligned}$$

and it has to be concluded that the net Nernst response of the electrodes in the equilibrium system is equal and opposite to the Donnan potential. The potential difference between two reference electrodes with salt bridges, Figure 5.10e has been termed the Donnan e.m.f. (Overbeek, 1953) to distinguish it from the Donnan potential. Thus

$$E_{\text{Donnan}} = E''_{\text{l.j.p.}} - E'_{\text{l.j.p.}} + \psi_{\text{Donnan}}$$

where $E_{\text{l.j.p.}}$ are the liquid junction potentials for the salt bridges in the different phases. At equilibrium the last equation simplifies to

$$E_{\text{Donnan}} = \Delta E_{\text{l.j.p.}} + \frac{RT}{z_i F}\, ln\, \frac{a_i''}{a_i'}$$

Referring back to Figure 5.10, it can be concluded that the difference between the cell potentials of the cells (a) and (d) normally thought of as giving the suspension effect is identical to that of cell (e). Furthermore the suspension effect is identical to the Donnan e.m.f. As already explained, the suspension effect is also measured between cells (b) and (c).

In the equilibrium system shown in Figure 5.11 (adapted from Overbeek, 1956) the distribution of small ions and the osmotic pressure are two thermodynamically well defined equilibrium properties. The Donnan e.m.f. is different in this respect as it can only be measured by introducing salt bridges into the two solutions and this adds irreversible processes to the system. In his original figure, Overbeek (1956) showed the difference in potential between E_1 and E_2 as equal to the Donnan potential. However, this was on the assumption that the liquid junction potentials at the salt bridges are zero or at least equal and opposite. With the experimental arrangement in Figure 5.11 it is the Donnan e.m.f. which is measured, i.e. the e.m.f. of the complete Donnan cell which includes the two liquid junction potentials, be they large or small. Stated in these terms, the true Donnan potential cannot be measured. The position can be summarized as follows; firstly, provided the two systems are at equilibrium there is no potential difference between the two reversible electrodes; The Donnan e.m.f. is measured by introducing two salt bridges into the two solutions; if there is a Donnan e.m.f. it is equal to $E_1 - E_2$ (Figure 5.11). In fact,

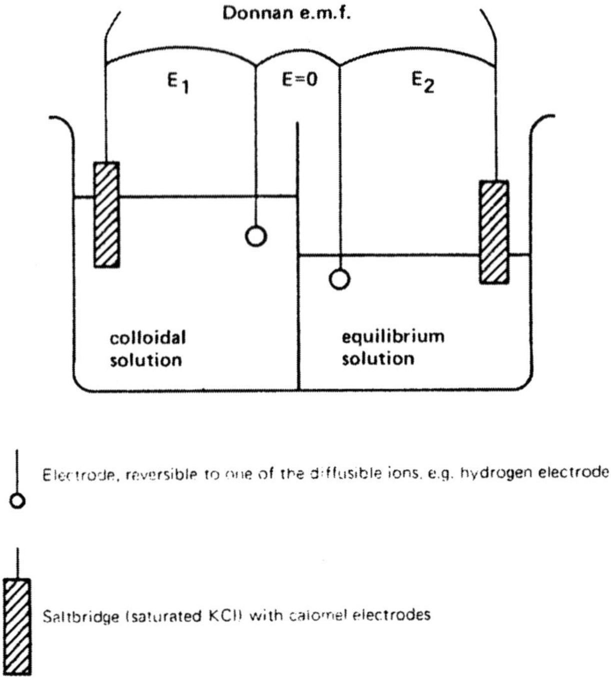

Figure 5.11 Direct and indirect measurement of the Donnan electromotive force (e.m.f.). Demonstration of equality of Donnan e.m.f. and suspension effect. In the original figure the Donnan e.m.f. was depicted as the Donnan potential on the basis that 'the liquid junction potentials are zero or at least of equal magnitude'. Reprinted with permission from J. Th. G. Overbeek in *Progress in Biophysics and Biophysical Chemistry* (Ed. J. A. V. Butler), Ch. 3 (1956), Pergamon Press Ltd

E_1 and E_2 are the e.m.f.s from which the pH values are calculated for the suspension and equilibrium solution, the difference in the values being given by

$$\mathrm{pH}_{\text{coll. susp.}} - \mathrm{pH}_{\text{equil. soln.}} = \frac{E_1 - E_2}{2.303RT/F}$$

This latter relationship shows that either the Donnan e.m.f. is measured directly $(E_1 - E_2)$, or indirectly through pH. The difference in pH values between the suspension and its equilibrium solution is referred to as the suspension effect (Pallman, 1930) and from what has been stated this effect is identical to the Donnan e.m.f.

It is inevitable that a diffusion potential is created at the liquid junction between a salt bridge and a solution of different composition. At liquid junctions, differences in the mobilities of the cation and anion create a diffusion potential which is also influenced by any differences in the ion activities which exist across the junction. Because factors involved in the liquid junction potential during pH calibration and during subsequent measurements are likely to be different, the liquid junction potential contribution to the total e.m.f. of a cell will be at least slightly different in calibration and during measurements. It is, of course, appropriate that an electrolyte in which the mobilities of the cation and anion are almost identical is used in the salt bridge, and, as is well known, KCl happens to fulfil the role rather well if the test solution is dilute and the pH between 2.5 and 11 (Bates, 1973). Because the current at the liquid junction is carried mainly by K and Cl ions, which happen to possess mobilities which are almost identical, the liquid junction potential should be small. So long as the ionic strengths of the standard and unknown solutions are small compared with the concentrated KCl bridge solution, any liquid junction potential should, as well as being small, tend to be rather constant in magnitude and hence can usually be ignored. When a salt bridge makes contact through a suspension instead of clear solution, the situation is complicated by a reduced mobility of counterions which arises from retarding effects created by the oppositely charged colloidal particles. This means that liquid junction potentials are likely to be much greater in suspensions than in supernatant liquids. It is for this reason that suspension effects can be reduced by introducing the liquid junction of the reference electrode into clear supernatant liquid and not into the suspension. Schofield and others suggested that when measuring soil pH the glass electrode should be in the suspension, arguing that it was better buffered than the supernatant liquid. However, the enhanced stability of readings obtained probably resulted from improved shielding of the glass electrode, and with shielded electrodes and modern pH meters this is usually unnecessary.

Despite many studies on the suspension effect, there is as yet no unequivocal method for apportioning Donnan e.m.f. measurements into liquid junction and Donnan potential components. The best that can be done is to reduce liquid junction potentials to a minimum.

5.9 APPARENT NON-STOICHIOMETRIC UPTAKE OF MONO- AND DIVALENT CATIONS

For predominantly negatively charged soils it has been customary to regard the cation exchange capacity, at a particular pH, to be invariant and independent of the index cation and almost independent of the electrolyte concentration used to determine the exchange capacity. In fact, for a soil of constant surface charge with an appreciable proportion of cations in a diffuse double layer, the surface excess of cations can change at least marginally as the deficit of anions changes. In practice, with the suppression of the diffuse part of the double layer, as a result of working at higher rather than lower electrolyte contents, the variations in the deficit of anions (negative adsorption) tends to be small. Thus, the value obtained for the cation exchange capacity (CEC) of most soils is largely independent of the nature of the cation used to saturate the soil. Despite this general rule it is well known that the CEC of some soils, notably some of the highly weathered soils from humid tropical regions of low organic matter content are apparently much higher when measured with a divalent cation than when measured with a monovalent cation (de Endredy and Quagraine, 1960). In other words, there is apparent non-stoichiometric uptake of mono- and divalent cations by soil surfaces, implying non-stoichiometric cation exchange or, as described by some, superequivalent adsorption of divalent cations. According to Carlson and Overstreet (1967) and Boehm and Herrmann (1967) the phenomenon is explained in terms of the adsorption of complexes of the type $CaOH^+$ on hydroxylic surfaces, illustrated by a reaction of the type

$$M{-}OH + Ca(OH)_2 \rightleftharpoons M{-}O^-Ca^{2+}OH^- + H_2O$$

For such soil types it is not unusual to find that cation exchange capacities measured with Ca or Ba are at least up to 50 per cent larger than those measured with Na or K. In extreme cases it would seem that equimolar cation exchange, monovalent for divalent, is possible.

Although some support for the above explanation has been forthcoming, others, for example, Kinniburgh *et al.* (1975) have been far from satisfied with the explanation and an alternative based on reasoning put forward by Lyklema (1971) is preferred. Imagine a reversible surface capable of specifically adsorbing a cation as shown in Figure 5.12. The cation occupying the inner (Stern) part of the electrical double layer promoted the adsorption of OH (or the desorption of H^+) by the solid surface. Because of the extra negative charge induced on the surface (the p.z.c. would move to lower pH) there must be extra counterions in the electrical double layer compared to the situation in which there is little or no specific adsorption. The outcome is that at a particular pH the surface accommodating specifically adsorbed cations e.g. Ca^{2+}, will possess a higher negative surface charge density and will therefore possess a higher CEC than the surface associated with cations showing much less or zero specific adsorption.

Figure 4.5 shows that particularly at lower electrolyte concentrations the overall adsorption of Ca^{2+} and Cs^+ is appreciably larger than Na^+ or Li^+ adsorption on Georgia kaolinite at any particular pH. It can be inferred that the particular kaolinite possesses a (mainly) constant potential surface exhibiting pronounced specific adsorption of Ca^{2+} and Cs^+. The phenomenon of so-called superequivalent adsorption of cations is peculiar to reversible surfaces and hence is most likely to be encountered in Oxisols and similar soils. It should, however, be remembered that soils which possess constant charge surfaces along with at least some reversible surfaces can be expected to exhibit the above phenomena to some extent, and this may occasionally explain trends obtained in experimental results.

It must be remembered that measurements of CEC on soils possessing reversible surfaces are bound to be concentration dependent, as illustrated in

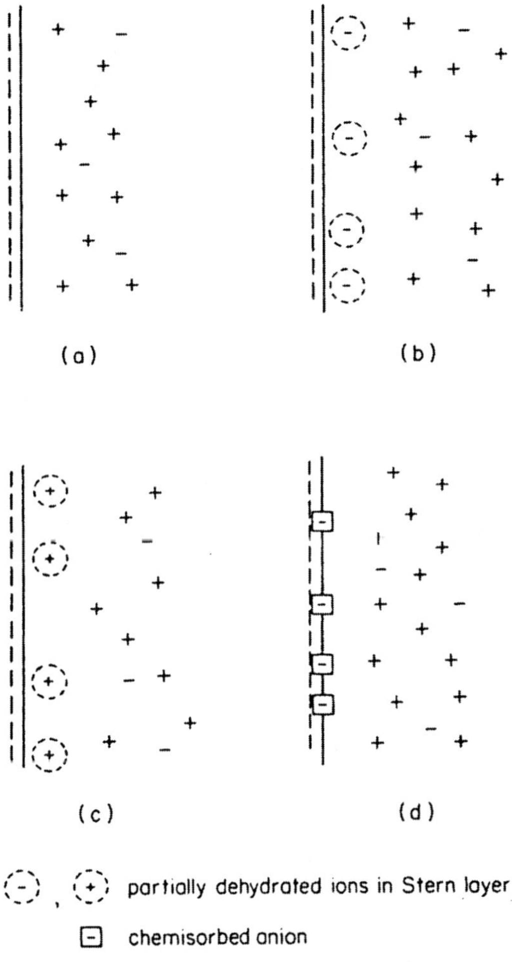

Figure 5.12 Some effects of ion adsorption into Stern layer and chemisorption on constant potential (reversible) surface at fixed pH: (a) Negative surface with indifferent counter and co-ions in diffuse double layer; (b) Specific adsorption of anion into Stern layer which induces decrease in negative surface charge (p.z.c.) moves to higher value compared to (a)); (c) Specific adsorption of cation which induces increase in negative surface charge (p.z.c. moves to lower value relative to (a)); (d) Chemisorption of anion (potential determining) which tends to increase negative charge on surface. Note: $CEC_{(c)} > CEC_{(a)}$, which is the most likely explanation for superequivalent adsorption of cations

Figure 5.4. The lower the electrolyte concentration the lower the surface charge, and hence the lower the indifferent cation retention capacity, or indifferent anion retention capacity when the soil is below its p.z.c. The implications of this dependence on concentration have not always been appreciated and have led to uncertainties about the significance of the washing step, particularly in CEC determinations when excess electrolyte is being removed prior to estimating the amount of index cation retained by the soil. The effects of 'alcohol washing' to remove excess electrolyte as commonly practised could have quite unpredictable results on ion adsorption by soils possessing constant potential surfaces.

5.10 ELECTROKINETIC PHENOMENA: USES AND LIMITATIONS IN SOIL–ELECTROLYTE STUDIES

In soil–electrolyte interaction studies, one of the main problems is the difficulty of making measurements which can be interpreted without ambiguity. No discussion of the electrical double layer would be complete without brief reference to electrokinetic principles because although useful measurable quantities can emerge from electrokinetic phenomena, considerable care has to be used in interpreting such data. All electrokinetic phenomena concern the relative movement of liquid with respect to charged surfaces or charged surfaces with respect to liquid, the latter being induced by applied potential gradients. There are four phenomena which come under the broad heading of electrokinetic processes, namely electrophoresis, electro-osmosis, streaming potentials and sedimentation potentials, the latter often being referred to as the Dorn effect. In the present context it will suffice to examine electrophoretic mobility because the general concept of a slipping or shear plane is clearly illustrated.

The picture of what happens to and around a charged particle in an aqueous medium in an electric field is probably as follows. The particle moves towards the electrode of opposite charge, carrying with it layers of adhering water. Counterions will move in the opposite direction except those located between the slipping (or shear) plane and the particle surface which will move with the particle. As well as the frictional force of the liquid on the particle, there will be two additional forces, namely the electrophoretic retardation force and the relaxation force. The electrophoretic retardation force arises from the water moved with the counterions in a direction opposite to that of the colloid. The relaxation force results from the distortion of the ionic atmosphere around a particle, which becomes slightly displaced with respect to the particle and hence slows the particle's movement. The potential at the slipping plane is known as the zeta potential, ζ, and many workers have suggested that the zeta potential may well be almost equal to the potential at the outer edge of the Stern layer; others surmise the shear plane lies outside the Stern layer (see Figure 5.6) but, as might be expected, it is impossible to resolve the problem with any precision. If such uncertainties did not exist, electrokinetic studies would be more useful tools for studying soil particle–ion interaction.

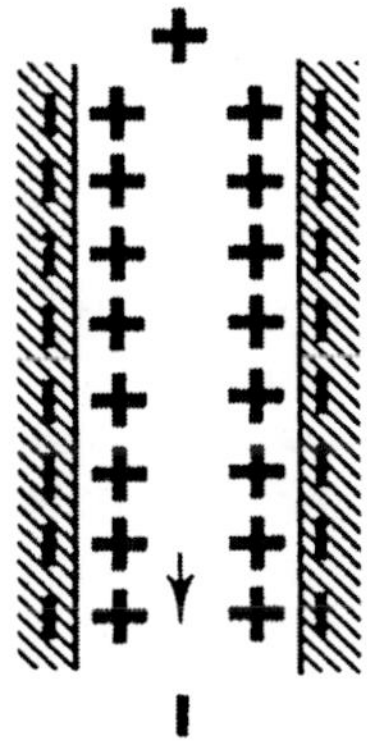

Figure 5.13 Electro-osmotic flow of water in glass capillary to negative electrode when the counterion is positive

Those familiar with micro-electrophoretic techniques will know that electro-osmosis usually takes place simultaneously with electrophoresis. Thus when electrodes are placed at either end of a capillary tube (glass) filled with water the liquid near the cell wall moves towards the negative electrode when the counterion is positively charged, Figure 5.13, or, when the counter charge is negative, the liquid moves towards the positive electrode. A colloidal particle showing no electrophoretic movement is said to be at its isoelectric point. In an electrophoretic cell electro-osmotic movement displaces liquid in the direction of one electrode and the resulting pressure difference causes a return flow of liquid, obeying the laws of laminar flow. Under dynamic equilibrium conditions the total transport of liquid is zero, with strong return flow far from the walls. If electrophoretic measurements are made at the points where liquid transport is zero, then no correction for liquid flow is required. For example in a cylindrical capillary tube the true electrophoretic velocity can be directly measured at a distance of $\frac{1}{2}(1-\frac{1}{2}\sqrt{2})$ ($= 0.146$) times the diameter from the wall of the tube. When Ka, the ratio of particle radius to the effective thickness of the electrical double layer is $\geqslant 300$, the Smoluchowski equation applies to within about 1 per cent. In this the electrophoretic mobility μ_E defined as electrophoretic mobility/potential gradient is given by

$$\mu_E = \frac{\varepsilon_r \varepsilon_0 \zeta}{1.5\,\eta}$$

where η is the viscosity of the medium and the other symbols have their usual meaning. If $Ka \leqslant 0.5$ the Hückel equation applies, namely

$$\mu_E = \frac{\varepsilon_r \varepsilon_0 \zeta}{\eta}$$

In practice, it is the Smoluchowski equation which finds most application, although it has to be admitted that many soil colloid problems involve particles which fall between $Ka \leqslant 0.5$ and $\geqslant 300$.

The main difficulty in exploiting electrokinetic measurements lies in the uncertainty of the position of the plane of shear. When particles are non-spherical, additional complications are bound to enter into the problem of interpreting electrophoretic data, and situations can be visualized where extra difficulties arise because single particles have two kinds of double layer, for example, the one associated with the plane faces of clay sheets which are negative and the other associated with their edges, and adsorbed hydrous oxides which have a charge dependent on the ionic constitution of the suspending medium, and may be positive.

REFERENCES

Aveyard, R. and Haydon, D. A. (1973), *An Introduction to the Principles of Surface Chemistry*. Cambridge University Press.

Bates, R. G. (1973), *Determination of pH, Theory and Practice*, 2nd edition. Wiley, New York.

Beckett, P. H. T. (1964a), Studies on soil potassium. I. Confirmation of the ratio law: measurements of potassium potential. *J. Soil Sci.*, **15**, 1–8.

Beckett, P. H. T. (1964b), Studies on soil potassium. II. The 'immediate' Q/I relations of labile potassium in the soil. *J. Soil Sci.*, **15**, 9–23.

Boehm, H. P. and Herrmann, M. (1967), Über die Chemie der Oberfläche des Titandioxids. I. Bestimmung des aktiven Wasserstoffs, thermische Entwasserung und Rehydroxylierung. *Z. anorg. allgem. Chem.*, **252**, 156–167.

Bolt, G. H. (1955), Ion adsorption by clays. *Soil Sci.*, **79**, 267–276.

Borggaard, O. K. (1974), Experimental conditions concerning potentiometric titration of humic acid. *J. Soil Sci.*, **25**, 189–195.

Breeuwsma, A. and Lyklema, J. (1971), Interfacial electrochemistry of haematite (α-Fe_2O_3). *Discuss. Faraday Soc.*, **52**, 324–333.

Chapman, D. L. (1913), A contribution to the theory of electrocapillarity. *Phil. Mag.*, **25**, (6), 475–481.

Carlson, R. M. and Overstreet, R. (1967), A study of the ion exchange behaviour of alkaline earth metals. *Soil Sci.*, **103**, 213–218.

de Endredy, A. S. and Quagraine, K. A. (1960), A comprehensive study of cation exchange in tropical soils. *Seventh Inter. Congr. Soil Science*, **2**, 312–320.

Derjaguin, B. V. and Landau, L. (1941), Theory of the stability of strongly charged lyophobic sols and of the adhesion of strongly charged particles in solutions of electrolytes. *Acta Physicochim. URSS*, **14**, 633–662.

Donnan, F. G. (1911), Theorie der Membrangleichgewichte und Membranpotentiale bei Vorhandensein von nicht dialysierenden Elektrolyten. *Z. Elektrochem.*, **17**, 572–581.

Epstein, E. (1972), *Mineral Nutrition of Plants: Principles and Perspectives*. Wiley, New York and Chichester.

Eriksson, E. (1952), Cation exchange equilibria on clay minerals. *Soil Sci.*, **74**, 103–113.

Everett, D. H. (1972), Manual of symbols and terminology for physicochemical quantities and units. Appendix II. Definitions, terminology and symbols in colloid and surface chemistry, Part I. *Pure Appl. Chem.*, **31**, 578–638.

Fried, M. and Broeshart, H. (1967), *Soil–Plant Systems, in Relation to Inorganic Nutrition*, Atomic Energy Comm. Monograph No. 67, Academic Press.

Gapon, E. N. (1933), Theory of exchange adsorption in soils. *J. Gen. Chim. URSS*, **3**, 144–163.

Gouy, G. (1910), Sur la constitution de la charge électrique à la surface d'un électrolyte. *J. Phys. (Paris)*, **9**, 457–468.

Grahame, D C. (1947), The electrical double layer and the theory of electrocapillarity. *Chem. Rev. (London)*, **41**, 441–501.

Kinniburgh, D. G., Syers, J. K. and Jackson, M. L. (1975), Specific adsorption of trace amounts of calcium and strontium by hydrous oxides of iron and aluminium. *Soil Sci. Soc. Amer. Proc.*, **39**, 464–470.

Langmuir, I. (1918), The adsorption of gas on plane surfaces of glass, mica and platinum. *J. Amer. Chem. Soc.*, **40**, 1361–1403.

Langmuir, I. (1938), The role of attractive and repulsive forces in the formation of tactoids, thixotrophic gels, protein crystals, and coacervates. *J. Chem. Phys.*, **6**, 873–896.

London, F. (1930), Zur Theorie und Systematik der Molekularkräfte. *Z. Phys.*, **63**, 245–279.

Lyklema, J. (1971), *Discuss. Faraday Soc.*, **52**, p. 318.

Marshall, C. E. (1956), Thermodynamic, quasithermodynamic and nonthermodynamic methods as applied to the electrochemistry of clays. Proc. fourth Natl. Conf. Clays and Clay Minerals, Natl. Res. Council Publ. No. 456, pp. 288–300.

McNeal, B. L. (1970), Prediction of interlayer swelling of clays in mixed salt solutions. *Soil Sci. Soc. Amer. Proc.*, **34**, 201–206.

Mekaru, T. and Uehara, G. (1972), Anion adsorption in ferruginous tropical soils. *Soil Sci. Soc. Amer. Proc.*, **36**, 296–300.

Norrish, K. (1954), The swelling of montmorillonite. *Discuss. Faraday Soc.*, **18**, 120–134.

Nye, P. H. and Tinker, P. B. (1977), *Solute Transport and Plant Growth in Soil.* Blackwell Scientific Publications, Oxford.

O'Connor, G. A. and Kemper, W. D. (1969), Quasi-crystals in Na-Ca systems. *Soil Sci. Soc. Amer. Proc.*, **33**, 464–469.

Overbeek, J. Th. G. (1952), The interaction between colloidal particles, in *Colloid Science*, **I**, pp. 245–277. Editor Kruyt, H. R.. Elsevier Publ. Co., Amsterdam.

Overbeek, J. Th. G. (1953), Donnan-EMF and suspension effect. *J. Colloid Sci.*, **8**, 593–605.

Overbeek, J. Th. G. (1956), The Donnan Equilibrium, in J. A. V. Butler (ed.), *Progress in Biophysics and Biochemical Chemistry*, **6**, pp. 57–84. Pergamon Press, Oxford.

Pallman, H. (1930), Die Wasserstokkaktivität in Dispersionen und Kolloiddispersen Systemen. *Kolloidchem. Beih.*, **30**, 334–405.

Posner, A. M. (1964), Titration curves of humic acid. *Eighth Inter. Congr. Soil Science*, Bucharest, **3**, 161–174.

Pratt, P. F. and Grover, B. L. (1964), Monovalent-divalent cation-exchange equilibria in soils in relation to organic matter and type of clay. *Soil Sci. Soc. Amer. Proc.*, **28**, 32–35.

Quirk, J. P. (1968), Particle interaction and soil swelling. *Israel J. Chem.*, **6**, 213–234.

Quirk, J. P. and Schofield, R. K. (1955), The effect of electrolyte concentration on soil permeability. *J. Soil Sci.*, **6**, 163–178.

Robinson, R. A. and Stokes, R. H. (1970), *Electrolyte Solutions*, 2nd edition, Butterworths, London.

Rowell, D. L., Payne, D. and Ahmad, N. (1969), The effects of the concentration and movement of solutions on the swelling, dispersion, and movement of clay in saline and alkali soils. *J. Soil Sci.*, **20**, 176–188.

Russell, E. W. (1934), The interaction of clay with water and organic liquids as measured by specific volume changes and its relation to the phenomena of crumb formation in soils. *Phil. Trans.*, **233A**, 361–389.

Schofield, R. K. (1947), A ratio law governing the equilibrium of cations in the soil solution. *Proc. 11th Intern. Congr. Pure and Applied Chem.* (London), **3**, 257–261.

Schofield, R. K. (1949), Effect of pH on electric charges carried by clay particles. *J. Soil Sci.*, **1**, 1–8.

Schofield, R. K. and Samson, H. R. (1954), Flocculation of kaolinite due to the attraction of oppositely charged crystal faces. *Discuss. Faraday Soc.*, **18**, 135–145.

Schofield, R. K. and Taylor, A. W. (1955a), Measurements of the activities of bases in soils. *J. Soil Sci.*, **6**, 137–146.

Schofield, R. K. and Taylor, A. W. (1955b), The measurement of soil pH. *Soil Sci. Soc. Amer. Proc.* **19**, 164–167.

Stern, O. (1924), Zur Theorie der Elektrolytischen Doppelschicht. *Z. Elektrochem.*, **30**, 508–516.

Taylor, A. W. (1958), Some equilibrium solution studies on Rothamsted soils. *Soil Sci. Soc. Amer. Proc.*, **22**, 511–513.

Teräsvuori, A. (1930), Über die Bodenazidität, mit besonderer Berücksichtigung des Elektrolytgehaltes der Bodenaufschlammungen. Akademische Abhandlung, Helsinki, Valtion Maalouskoet. Julkaisuja No. 29.

van Dyk, H. (1971), Colloidal Chemical Properties of Humic Matter, in A. D. McLaren and J. Skujins (eds.), *Soil Biochemistry* **Vol. II**, pp. 16–35, Marcel Dekker, New York.

van Olphen, H. (1963), *An Introduction to Clay Colloid Chemistry*, Interscience Publishers, New York.

van Raij, B. and Peech, M. (1972), Electrochemical properties of some oxisols and alfisols of the tropics. *Soil Sci. Soc. Amer. Proc.*, **36**, 587–593.

Verwey, E. J. W. (1935), The electrical double layer and the stability of lyophobic colloids. *Chem. Rev.* (London), **16**, 363–415.

Verwey, E. J. W. and Overbeek, J. Th. G. (1948), *Theory of the Stability Lyophobic Colloids.* Elsevier Publ. Co., Amsterdam.

Warkentin, B. P., Bolt, G. H. and Miller, R. D. (1957), Swelling pressure of montmorillonite. *Soil Sci. Soc. Amer. Proc.*, **21**, 495–497.

Wright, H. J. L. and Hunter, R. J. (1973a), Adsorption at solid-liquid interfaces. I. Thermodynamics and the adsorption potential. *Aust. J. Chem.*, **26**, 1183–1189.

Wright, H. J. L. and Hunter, R. J. (1973b), Adsorption at solid-liquid interfaces, II. Models of the electrical double layer at the oxide-solution interface. *Aust. J. Chem.*, **26**, 1191–1206.

CHAPTER 6

Water on particle surfaces

V. C. Farmer
Macaulay Institute for Soil Research, Aberdeen

6.1 INTRODUCTION

Water in soils is essential to plant growth, being the medium through which inorganic nutrients are translocated to the root surface and carried through the plant. Most of the water in soils at or near field saturation can be considered as equivalent to bulk water, and only bulk properties need be considered in discussing its movement through soil interstices and its retention in soil capillaries; this is the field with which soil physics deals. But, in the first molecular layer in contact with soil particles, and perhaps for several molecular layers beyond, the behaviour of water molecules is very different from that in the bulk. Strong adsorptive forces can reduce the volatility of these molecules and restrict their mobility. Such an adsorbed layer of water competes with other polar molecules for adsorption sites, and so has an important role in determining the release and retention of fertilizers and pesticides. Such strongly adsorbed layers confer wettability on soils, but hydrophobic surfaces can also be present and these lead to non-wettable soils with serious consequences in run off and inefficient use of precipitation. This region of distinctive molecular interactions between water and soil particles falls within the field of soil chemistry.

Soils can present a wide variety of types of particle surface, with hydration levels that vary from a maximum corresponding to immersion in water, to an almost anhydrous condition, such as that generated in surface layers by forest or scrub fires. We shall distinguish here inorganic and organic surfaces, while recognizing that many particles essentially inorganic in composition may have purely organic surfaces due to coatings. Within the inorganic field, it is necessary to distinguish surfaces that have a considerable cation exchange capacity, like those of layer silicates, from surfaces like those of many oxides and hydroxides whose exchange capacity depends on pH and electrolyte concentration (Chapter 4). Of all these surfaces, those of the expanding layer silicates have received most attention. Their internal surface areas are high, amounting to some 760 $m^2\ g^{-1}$, and their crystalline regularity imposes some regularity on their interlayer water structures, which have consequently been studied by a wide range of techniques, including X-ray crystallography, several spectroscopic methods, vapour pressure isotherms and conductivity measurements; so inevitably these surfaces will receive more space here. But it must not be forgotten that in surface soils and, of course, in peats, organic surfaces can play a more important role than inorganic surfaces.

In discussing the interaction of water with particle surfaces we shall try to understand the geometry and forces involved at the molecular level. We will begin by summarizing our knowledge of the behaviour of water in other environments, including bulk liquid, ice, and crystalline hydrates. These fields have been recently surveyed in the five-volume series edited by Franks (1972–75), a symposium proceedings (Luck, 1974) and other books and reviews (Eisenberg and Kauzmann, 1969; Safford and Leung, 1973).

The gaps revealed in our understanding by surveying these much more fully-investigated systems will save us from over-confidence in the more complex area of water on soil surfaces.

6.2 THE STRUCTURE OF LIQUID AND CRYSTALLINE WATER

6.2.1 Pure Water

The structures of ice in many of its various modifications have now been exactly analysed by X-ray diffraction technqiues. The form stable at atmospheric pressure has hexagonal symmetry and is termed ice Ih. In this structure (Figure 6.1) each water molecule has four nearest neighbours in a tetrahedral configuration, so that it can form strong hydrogen bonds with all four, acting as a proton donor to two of them, and as a proton acceptor to the other two. The tetrahedral configuration allows almost linear hydrogen bonds between adjacent water molecules, as the HOH angle, 104.5° for a gas molecule, is close to the tetrahedral angle of 109.5°. Moreover, negative charge on the oxygen is con-

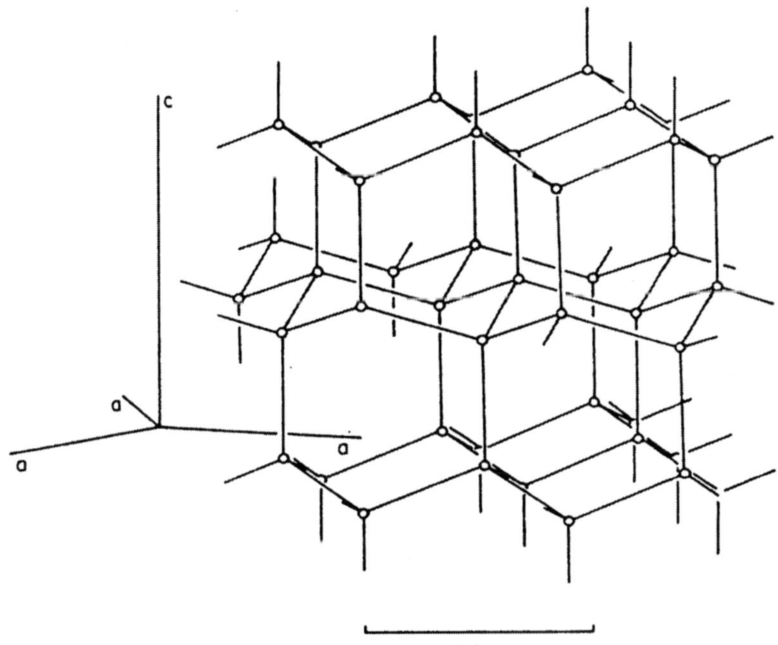

Figure 6.1 Arrangement of the oxygen atoms in Ice Ih in layers perpendicular to the hexagonal axis, showing the tetrahedral coordination. From Owston (1958), reproduced by permission of Taylor & Francis Ltd

centrated on lone-pair orbitals which make an angle of about 120° with each other and lie close to the ideal tetrahedral configuration. The resultant structure is a very open one, in which dense close-packing has been sacrificed to gain maximum strength in the intermolecular hydrogen bonds. The contrast between the crystal structure of water and neon, for example is striking, as each neon molecule, involved only in non-directive van der Waals interaction with its neighbours, makes contact with twelve nearest neighbours, compared with four for oxygen in ice.

Looking beyond the nearest-neighbour environment of the molecules in ice (Figure 6.1), one can recognize a common motif of six-membered rings in chair and boat configurations. The structure can be regarded as being built up from corrugated sheets of water molecules, lying perpendicular to the c-axis, which incorporate hexagonal rings in the chair configuration. Water molecules in the sheets alternately link upwards and downwards to adjacent sheets, and thereby build a new family of hexagons in the boat configuration. These form the walls of hexagonal channels that run parallel to the c-axis. One cell of the channel is bounded by six water molecules, and there is, in the structure as a whole, one cell for every two water molecules. In spite of this open structure, there appears to be little tendency for ice Ih to incorporate foreign molecules into these cells. The ice clathrate compounds, which do incorporate small molecules in open cells formed by hydrogen-bonded water molecules, have quite different structures that involve both five- and six-membered rings, and have larger cells that allow many more contacts between water molecules and the guest molecules.

The structure shown in Figure 6.1 defines the position of oxygens in ice but not that of protons, so that we cannot say, in any given O—O contact, to which oxygen the proton is attached. Over the structure as a whole, the orientation of the water molecules is completely random, although locally the orientation of any one water molecule places limitations on the orientation of its four immediate neighbours, two of which must present protons and the other two their lone-pair orbitals along the O—O contacts. A few orientational defects do occur, bringing a proton to face a proton or lone-pair orbitals to face each other, and although these play an important role in dielectric and proton diffusion mechanisms, on average they probably amount to only one defect per 10^7 molecules.

When ice melts, the resultant liquid undoubtedly retains some features of the open ice network, but the details of its structure continue to be the subject of lively dispute and conjecture. Any structure proposed must satisfy simultaneously a number of criteria derived from experimental observations. Some of these, such as radial distribution curves from X-ray diffraction, and hydrogen-bond strength distributions from Raman and infrared spectra, are closely related to the geometry of the structure, while others, such as thermodynamic properties, density, thermal expansion and dielectric constants are more remotely related. New techniques using neutron scattering and nuclear magnetic resonance yield information on molecular dynamics—i.e. diffusion rates, residence times of water molecules in particular configurations before they rotate or migrate into new environments, and frequencies associated with translational and rotational vibrations while the molecules lie in these short-lived potential wells. It is a considerable challenge to propose a model which can be checked against all these criteria, and many suggestions are aimed only at satisfying one or two of them.

The most informative evidence on local structure comes from oxygen—oxygen distances in the form of radial distribution curves deduced from X-ray scattering experiments (Figure 6.2). Throughout the 0–50 °C temperature range, the curves show a well defined maximum in the distribution curve of O—O separation near 0.285 nm, and two broader maxima near 0.45 and 0.70 nm. The curves are featureless beyond about 1.0 nm, indicating a continuous distribution of O—O separations beyond this distance. An estimate of the number of molecules involved in the 0.285 nm approach gives a coordination number of 4.4 suggesting the persistence of tetrahedral coordination like that in ice, and the second maximum near 0.45 nm is exactly what would be expected of second neighbours in a tetrahedral hydrogen-bonded complex. The hydrogen-bonding distances, however, are longer and less precisely defined than those in

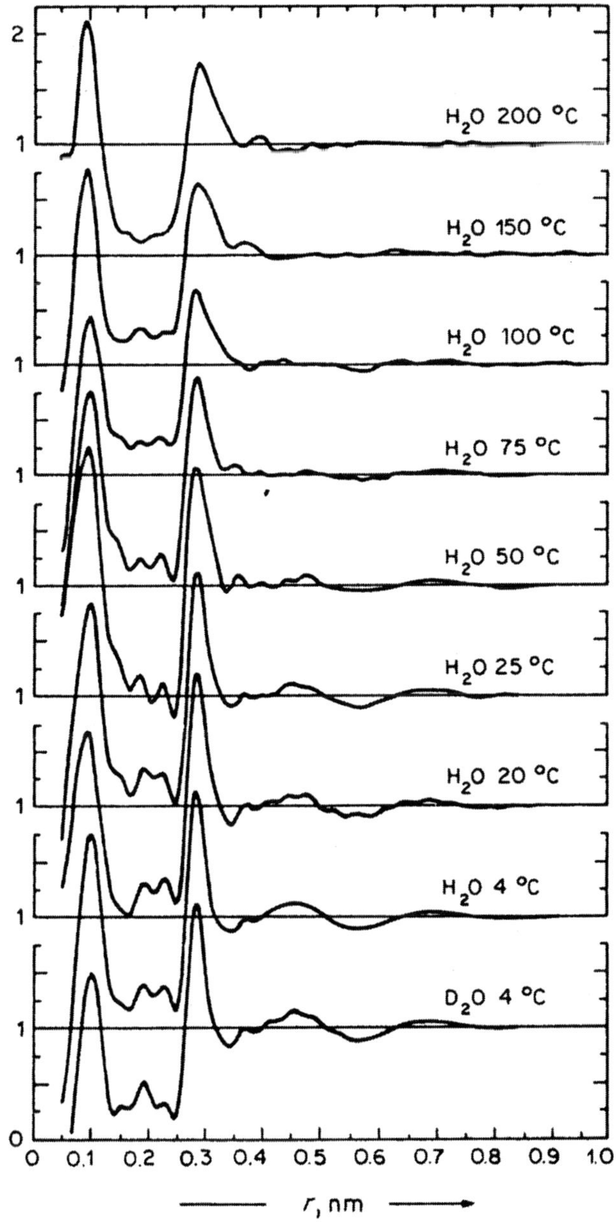

Figure 6.2 X-ray correlation functions for water. Curves represent a superposition of modified atom pair correlation functions $g(r)$ descriptive of O—O, O—H, and H···H interactions. From Narten and Levy (1972), reproduced by permission of Plenum Publishing Corp.

ice (0.276 nm), and there is also a considerable number of molecules with O—O separations between 0.285 and 0.45 nm in the liquid, whereas there are none in ice. Infrared and Raman spectra also indicate a weakening of the hydrogen bonds by a shift in the main OH stretching maximum of HDO in a D_2O matrix from 3277 cm^{-1} in ice to 3415 cm^{-1} in the liquid, and the greater bandwidth of the liquid points to a range in hydrogen-bond lengths and bond angles. A new feature of the liquid spectrum is a separate higher frequency component near 3625 cm^{-1} more obvious in Raman than infrared spectra. This separate absorption band is usually taken to indicate the presence of a separate category of broken bonds, distinct from the stronger hydrogen bonds that give rise to the main infrared absorption band, although Falk (1975) has argued that its presence is not inconsistent with a smooth continuous distribution of hydrogen-bond lengths. In any case, a frequency of 3625 cm^{-1} is still far below that characteristic of the gas (3707 cm^{-1}) and it can perhaps be best thought of as that of an OH group which does not find a lone pair orbital conveniently placed for formation of a strong hydrogen bond, but which still interacts with the electron cloud of a neighbouring water molecule.

One proposal to account for the radial distribution curves and the Raman spectra suggests that water has a distorted and expanded ice-I structure which incorporates a considerable number of molecules in interstitial positions. These additional molecules in the structure account for the 10 per cent increase in density when ice melts, for the O···O separations between 0.30 and 0.45 nm and for the presence of non-bonded OH groups, since the interstitial molecules are surrounded by water molecules all already involved in mutual hydrogen bonding, with no free orbitals to interact with the interstitial molecules. An unsatisfactory feature in such a picture is the very high proportion of the total water that must exist in weakly bonded interstitial forms (20%) to account for the observed density. Moreover, given such an initial structure, it is difficult to see why it would not rapidly break down to incorporate the interstitial molecules in a new pattern of hydrogen bonding involving four-, five- and seven-membered rings as well as the six-membered rings characteristic of ice I. It is clear, in fact, from the structures of ice clathrate compounds and the high-pressure forms of ice that there is no great loss of interaction energies on changing ring sizes, so long as each water can retain its fourfold coordination.

Estimates of the percentage of non-bonded OH groups from Raman spectroscopy, admittedly incorporating many assumptions, suggest that these amount to no more than 8% at 0°C, increasing to about 13% at 30 °C. In this picture it is proposed that there are no wholly unbonded molecules, but that the broken bonds are randomly distributed through the liquid, so that between 0 and 40 °C there are only four- or three-coordinated water molecules present, with two-coordinated molecules appearing at higher temperature. Thus the liquid at ambient temperatures is regarded as a hydrogen-bonded network that is locally tetrahedral on the average, but lacks long range order, and is composed of polygons of varying sizes and conformations. Within this random network about 20 per cent of the water molecules at 20 °C fail to find an acceptor for one of their protons, and so form a non-bonded OH···O interaction. Consequently, another 20 per cent fail to find a proton donor, and the structure as a whole therefore contains 40 per cent of 3-bonded water molecules. The spectroscopic measurements indicate that bond-breaking involves a ΔH value of about 10 kJ mol^{-1}, and the model as a whole gives satisfactory values for the dielectric constant and heat capacity.

Further support for this more random picture of the water structure comes from computer simulation studies, which use a computer to calculate the positions and dynamics of a set of water molecules which have been allowed to interact according to the laws of physics, assuming a simplified electrostatic interaction between spherical molecules that carry positive charges to represent the protons, and negative charges to represent the free orbitals; this simplified picture nevertheless strongly favours a local tetrahedral coordination.

The structure that results is difficult to describe, since it does not closely resemble any of the crystalline forms of water. Although neighbouring molecules are oriented into rough approximations to tetrahedral bonding, the bonds are often considerably bent, and the polygons formed are very variable in size and geometry. Dangling OH bonds exist that cannot be classed as hydrogen bonded. Nevertheless the structure does yield the main features of the radial distribution, and gives fair approximations to the self-diffusion coefficient, dipole relaxation rate, and certain vibration frequencies below 200 cm^{-1}. Clearly, the model is as yet only a first approximation, and probably requires rather stronger and more directional interatomic forces to make it more realistic. Its most important contribution is to show that it is not necessary to postulate interstitial molecules, or other mixtures of distinguishable classes of water, nor to assume distorted ice-like structures or other lattice-like models, in order to explain much of the statics and kinetics of water.

It has been possible here to discuss only two or three of the more strongly supported proposals that have been made, to give the reader some feeling for the present state of play in a tournament that has still some way to go.

6.2.2 Ion–Water Interactions

Just as it is reasonable to start from the crystalline forms of water before considering the liquid, so also is it reasonable to look first at crystalline salt hydrates before speculating on ionic solutions. One of the most striking features seen in such crystal structures is the occurrence of complete hydration shells around many cations. The coordination numbers vary from four for the small polar Be^{2+} cation up to eight for Sr^{2+}, with sixfold octahedral coordination widespread for di- and trivalent cations of the first row transition elements and Mg^{2+}. Generally, the water molecules are oriented so that the cation lies on the bisector of the lone-pair orbitals. Thus the cation, the oxygen and the two protons lie in the same plane, and no other electron acceptor can share the water

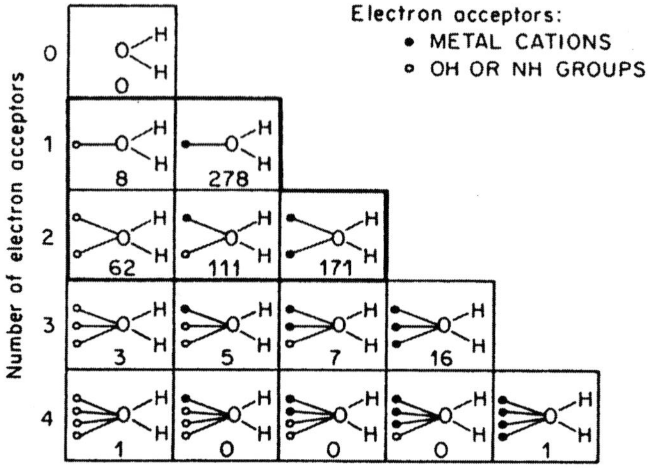

Figure 6.3 Classification of H_2O molecules in crystalline hydrates according to the number and type of electron-acceptor nearest neighbours. The number of cases observed is shown in each box. About 95 % of the reported cases correspond to coordinations shown inside the heavier frame. From Falk and Knop (1973), reproduced by permission of Plenum Publishing Corp.

molecule. This type of water coordination, although common, is by no means universal. The larger monovalent and divalent cations commonly incorporate both anions and water molecules in their first coordination sphere. In a survey of 282 salt hydrates, Falk and Knop (1973) confirmed that although coordination of water to a single cation was most common, there were very many examples of a water molecule coordinated to two cations (typically monovalent alkali metal cations) or to one cation and to a proton of another water molecule (Figure 6.3). Examples of three- or four-electron acceptors around a water molecule also occur. Another important conclusion of the survey was that each proton of water interacts at least weakly with an electron donor, which is either an anion or another water molecule. Very occasionally one proton interacts with two electron donors in a bifurcated hydrogen bond. Water molecules commonly link up in hydrogen-bonded chains two or three molecules long, and occasionally in longer, even infinite, chains, sheets and branching structures.

Given these many diverse structures, it is fair to ask what function the incorporation of water into an anhydrous crystal structure can play. One immediate conclusion that can be drawn is that for any stoichiometric hydrate that is stable above 0 °C, the hydration process must be exothermic, since otherwise the trapping of free water molecules into fixed positions, corresponding to a decrease in entropy, would not occur. This means that the overall process of separating the ions of the anhydrous salt, introducing water molecules, and bringing the hydrated ions back into contact must yield a structure of substantially lower energy. In this hypothetical sequence of events, attention is frequently concentrated on the energy of hydration of the cation and anion, but this cannot be of overriding importance, as can be seen by comparing the hydration tendencies of MgF_2 and $MgCl_2$. Both magnesium and fluoride ions have high hydration energies, but MgF_2 shows no tendency to hydrate, whereas the chloride ion, with a lower hydration energy, yields a salt which is difficult to dehydrate. This apparent anomaly is perhaps best understood in terms of the packing problems involved. In MgF_2, the eightfold coordination of F^- round Mg^{2+} appears to be just acceptable, while the fourfold tetrahedral coordination of Mg^{2+} round F^- is not unreasonable for the ion sizes involved, and is probably stabilized by a covalent component in the bonding. Neither coordination is at all acceptable when the larger Cl^- ion replaces F^-. However, by introducing six water molecules in a stable octahedral configuration round Mg^{2+}, the coordination number of the Cl^- also rises to six by forming $H \cdots Cl^-$ hydrogen bonds with the protons of the coordinated water. Most salt hydrates can be understood in these terms, and their widespread occurrence is simply a reflection of the fact that anions are usually larger than cations, so that they cannot form suitable coordination spheres for each other. The occurrence of extended chains of two or more water molecules can be thought of as a way of establishing indirect contact between a cation and a distant anion or a remote part of a large anion, and accordingly are found mostly commonly in salts of polyatomic anions such as sulphate and phosphate.

Before leaving the subject of salt hydrates, it is interesting to look at some of the hydrogen-bonding parameters revealed in Falk and Knop's review. In spite of the fact that the hydrogen bond often involves a negatively charged proton acceptor, the mean O—O distance observed (0.279 nm) is somewhat longer than in ice (0.276 nm), although the range extends from 0.255 up to 0.306 nm. The hydrogen-bond angle $OH \cdots Y$ (Y being any proton acceptor) ranges from 130° to 180°, the shorter bonds being less bent. There is a marked tendency for cation-coordinated water molecules to form increasingly stronger hydrogen bonds as the cation charge increases, O—O distances shortening from about 0.283 for mono- and divalent cations to 0.268 nm for tri- and tetravalent cations. Information on OH stretching frequencies in salt hydrates is as yet limited to relatively few compounds. Generally, results for isotopically dilute HDO are most suitable for intercomparison since these are not confused by coupling between the two OH(OD) groups in the same molecule, or coupling between adjacent molecules; such OH or OD frequencies correlate quite well with O—O distances and O—Cl distances in salt hydrates. Most of the salts examined incorporate group one or two metal cations, and sulphate, nitrate or halide anions, and generally the OH stretch lies in the range 3350–3500 cm^{-1}, substantially above the ice frequency, 3300 cm^{-1}. The highest frequency observed, 3605 cm^{-1}, appears to be associated with a very long (0.363 nm) $OH \cdots N$ interaction, which would not normally be considered to involve a hydrogen bond. Nevertheless the displacement of the OH frequency from that of the gas phase amounts to 100 cm^{-1}, and this certainly implies some form of interaction.

Turning our attention now to ionic solutions, it is reasonable to expect the ionic hydration patterns seen in crystals to persist in solutions. In the solid hydrates, the same water molecules usually serve to hydrate the cation and the anion, but this pattern is likely to appear only in concentrated solutions in the form of water-bridged cation–anion pairs. A direct link between cation and anion only seems likely with covalent cation–anion bonding, (e.g. $CuCl^+$ and $HgCl^+$), or with the larger mono- and divalent cations which have coordination numbers of eight and over. Such highly polarizing cations as Li^+, Be^{2+}, Mg^{2+} and Al^{3+} can be presumed to maintain a complete fourfold or sixfold hydration shell, according to their size.

In the older literature, the larger cations and anions have frequently been considered to be not hydrated: nevertheless, there can be no doubt from recent infrared and Raman evidence that water molecules in the immediate environment of large anions are oriented with some or all of their protons directed towards the anion and not available for hydrogen bonding with other solvent molecules. This conformation is revealed by the appearance of a high frequency component in the infrared and Raman spectra of HDO incorporated in solutions of BF_4^-, ClO_4^-, IO_4^-, PF_6^-, and other large monovalent anions (see, for example, Figure 6.4). For such anions, the high frequency component lies around 2650 cm^{-1} for OD stretching or 3620 cm^{-1} for OH stretching. With increasing anion size or higher charge these frequencies drop, for example, to 2580, 2545, 2525 and 2325 cm^{-1} for OD groups linked to I^-, Br^-, Cl^- and F^-. The frequency for water–chloride bonding is almost identical to that found for water–water hydrogen bonds in pure water, and there can be no disagreement with classifying these water–halide interactions as hydrogen bonding, although it is not universally agreed that the weak interaction between water and perchlorate should be so classified. Nevertheless, it is clear that even here there is at least an electrostatic ion–dipole interaction of sufficient strength to ensure the orientation of neighbouring water molecules with either one or both of their OH groups directed towards the anion.

Vibrational spectroscopy does not give such direct information on cation–water interactions, except in so far as the degree of polarization of coordinated water molecules by the cation affects their bonding with water in the second hydration shell. By this criterion, concentrated solutions of Al^{3+} cations show a proportion of hydrogen bonds much stronger than those in pure water, as would be expected for weakly acidic water molecules coordinated to Al^{3+}. Water coordinated to Mg^{2+} is slightly more acidic than pure water, whereas that coordinated to Na^+ is less acidic. Thus, with increasing concentration of $Mg(ClO_4)_2$ in solutions, water–water bonds strengthen very slightly as indicated by a shift of the OD stretch from 2510 cm^{-1} for pure water to 2500 cm^{-1}, but in $NaClO_4$ solutions the water–water bonds weaken more significantly, causing a 20 cm^{-1} rise in OD frequencies (Figure 6.4). This suggests that

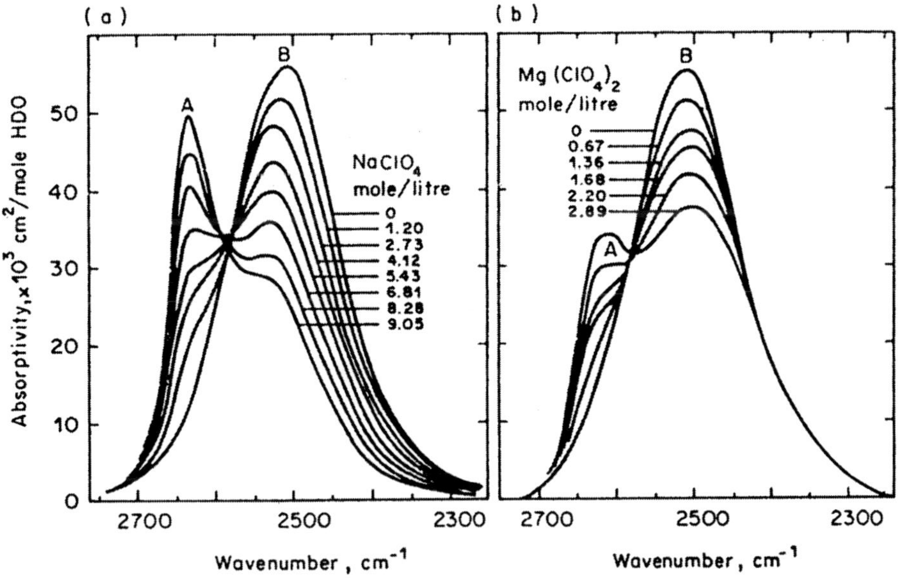

Figure 6.4 Molar absorptivities of the OD stretching bands of HDO in (a) aqueous $NaClO_4$ and (b) aqueous $Mg(ClO_4)_2$. From Brink and Falk (1970), reproduced by permission of the National Research Council of Canada from the *Canadian Journal of Chemistry*, **48**, 1970, pp. 3019–3025

Na—OH_2 bonding is weaker than OH_2—OH_2 bonding, and still weaker interaction would be expected for K^+, Rb^+ and Cs^+. These larger monovalent cations are unlikely to take up the full electron-donating capacity of water in the first sphere of coordination, which will therefore be in a position to interact both with the cation and with one or two adjacent protons, in configurations similar to those found in some crystals (Figure 6.3). Because of these additional interactions, water–water hydrogen bonds in salt solutions never become much weaker than those exhibited by $NaClO_4$ solutions.

The rather surprising conclusion from these infrared and Raman studies is that ion–water interactions are weaker than water–water interactions for anions less polarizing than Cl^-, and for cations equally or less polarizing than Na^+. Very similar conclusions have been drawn from proton magnetic relaxation studies, which suggest increased rotational and translational mobility of water in solutions containing the larger anions and cations and from neutron scattering measurements which show increased self-diffusion coefficients. These results tie in well with a classification of anions and cations into structure-making (smaller, highly polarizing species) and structure-breaking species based largely on thermodynamic entropy measurements, and on viscosity measurements. That ions should contribute to the structure of aqueous solutions by orienting and even trapping water molecules in their first and possibly even second hydration sphere is intuitively obvious, but the ability of large cations to weaken the structure of water has been a challenging puzzle, especially as neutral inert gas molecules in aqueous solution were found to enhance the surrounding water structure.

One strongly argued thesis has been that there is a mismatch between the structure of bulk water, distant from the ions, and the oriented molecules around the ions, and that it is in this transitional zone that water molecules may have additional translational and rotational degrees of freedom. However, the infrared, Raman, and proton relaxation studies indicate that the increased mobility and weaker bonding is already present in the first hydration shell of large ions. The more rigid surroundings of inert gas molecules in aqueous solution can be understood in terms of the structure of crystalline clathrate compounds, which show these gas molecules to be held in polyhedral cages built of water molecules interacting with each other so as to maintain a fourfold hydrogen-bonded configuration. The electrostatic fields associated with the larger ions ensure that some or all of the surrounding molecules redirect their lone-pair orbitals towards cations and their protons towards anions, in weak interactions that will not have the strongly directional and localized character of water–water bonds. Results subsequently reported for water–clay interactions support this interpretation of aqueous ionic solutions.

6.3 WATER ON SILOXANE SURFACES WITH ION-EXCHANGE PROPERTIES: SMECTITES AND VERMICULITES

6.3.1 Hydration Characteristics

These platy minerals are widely distributed in soils, and often form a major or even the sole constituent of the clay fraction. In the form of vermiculite, derived by weathering from micas, they can appear also in silts and sands. When present, their high internal surface area can be the major water-holding reserve in air-dry soils, although it is held too strongly for plants to extract. Their ability to undergo volume expansion at higher water contents can lead to major changes in soil volume, plasticity and porosity. Their internal surfaces can hold other polar molecules besides water, and they are efficient traps for certain inorganic and organic cations. These minerals consist of mica-like silicate unit layers about 0.95 nm thick which are interleaved with one or more sheets of water molecules that incorporate the exchangeable cations necessary to balance the negative charge on the silicate layers. These water molecules interact both with the exchangeable cations and with the oxygens on the silicate surface. Their role must therefore have close analogies with water in crystalline salts, and also with water in ionic solutions. The relationship to crystalline structures clearly appears in the fact that several well defined interlayer spacings are obtained by X-ray diffraction over a range of humidities or temperatures (Tables 2.6, 2.7 and 2.8). The minerals when anhydrous give a spacing of 0.95–1.0 nm, with one water sheet they give 1.15–1.20 nm, and with two water sheets 1.44–1.5 nm;

Table 6.1 Names and idealized formulae of typical expanding layer silicates (based on $O_{10}(OH)_2$; in Section 2.3 and Table 2.3 the unit used is $O_{20}(OH)_4$)

Charge and class	Dioctahedral	
	Tetrahedral substitution	Octahedral substitution
Smectite 0.25–0.6	Beidellite $M^{+}_{0.5}[Al_2(Si_{3.5}Al_{0.5})O_{10}(OH)_2]$	Montmorillonite $M^{+}_{0.4}[(Al_{1.6}Mg_{0.4})Si_4O_{10}(OH)_2]$
	Nontronite $M^{+}_{0.4}[Fe^{3+}_2(Si_{3.6}Al_{0.4})O_{10}(OH)_2]$	
Vermiculite 0.6–1.0	Altered muscovite $M^{+}[Al_2(Si_3Al)O_{10}(OH)_2]$	
Charge and class	**Trioctahedral**	
	Tetrahedral substitution	Octahedral substitution
Smectite 0.25–0.6	Saponite $M^{+}_{0.5}[Mg_3(Si_{3.5}Al_{0.5})O_{10}(OH)_2]$	Hectorite $M^{+}_{0.3}[(Mg_{2.7}Li_{0.3})Si_4O_{10}(OH)_2]$
Vermiculite 0.6–1.0	Vermiculite $M^{+}[Mg_3(Si_3Al)O_{10}(OH)_2]$	Altered Al-free phlogopite $M^{+}[Mg_{2.5}Si_4O_{10}(OH)_2]$

Compositions intermediate between beidellite, montmorillonite and nontronite are common, as are ferruginous forms of vermiculite and saponite.

spacings corresponding to water layers three or more molecules thick may also have regions of stability. In at least one of these phases (1.44 nm) the water molecules are sufficiently static to allow their positions to be identified by X-ray diffraction. At the other extreme there are grossly expanded phases with spacings of over 10 nm in which the interlayer water must be much more liquid-like away from the silicate surfaces and here analogies must be drawn with ionic solutions rather than with crystalline hydrates. However, whenever continuous layers of water molecules exist, a higher degree of mobility is possible than is usual for crystalline hydrates; on the other hand the extensive areas of regular crystalline anion surface will impose a greater degree of order than can occur in aqueous salt solutions.

The range of stability and the exact structure and properties of the interlayer water are a function both of the interlayer cation and of the composition of the silicate unit layers. Although the silicate layers are built to a common overall plan (Chapter 2) it is useful to distinguish two major families and some sub-groups, listed in Table 6.1. The main subdivision depends on the filling of the octahedral sites in the central sheet of cations and distinguishes the trioctahedral layer silicates, containing three divalent cations per structural unit, from the dioctahedral family containing only two predominantly trivalent cations and a vacant site per structural unit. Within each of these groups a further distinction can be made according to the site of the ionic substitution which gives rise to the layer charge. This substitution can be in the outer tetrahedral layer when Al^{3+} or Fe^{3+} replaces Si^{4+}, or in the central octahedral layer when Mg^{2+} or Fe^{2+} replaces Al^{3+} or Fe^{3+} in the dioctahedral series, and when Li^{+} or empty sites replace Mg^{2+} in the trioctahedral series. The site of substitution is of importance because of its effect on the localization of charge on the surface oxygens. With tetrahedral substitution of Al for Si, the aluminium is directly coordinated to three surface oxygens, and it is on these that the negative layer charge will principally reside: with octahedral substitution of Mg for Al, the Mg is not directly coordinated to any surface oxygens, but is linked to four silicon–oxygen tetrahedra which, in turn, expose ten of their oxygens on the layer surfaces, five on each side of the layer. This difference in charge localization is, as will be seen, confirmed by infrared spectroscopic studies and affects the hydration properties of the layers.

Vermiculite crystals are particularly attractive for scientific study. Their large size ensures that their hydration properties are little affected by edge effects, and makes it easy to distinguish experimentally the marked anisotropy of the crystal—to explore for example, differences in ion mobility or variations in hydrogen-bonding parallel and perpendicular to the water layers. Above all, they allow structure determinations by X-ray crystallography.

The first important conclusion of X-ray studies in this field was the existence of regions of stability of well defined interlayer spacings (Chapter 2 and Table 6.2). Associated with these changes in spacing there are of course changes in

Table 6.2 Mean values of basal spacings *d*(001) in nm of water solvation complexes of smectites and vermiculites. From Suquet *et al.* (1975), reproduced by permission of *Clays and Clay Minerals*

Clay mineral	Exchangeable cation					
	Li	Na	K	Ca	Mg	Ba
Montmorillonite	ind.	ind. [a]	{ind. 1.55	1.91	1.94	1.87
Beidellite						
from Rupsroth (Germany)[b]	ind.	1.52	1.27	{1.87 s 1.54 vw	1.85	
from Black Jack Mine, Idaho[c]	ind.	ind.	1.255	1.86	1.858	1.838
Saponite	ind.	1.52	1.26	{1.86 vw 1.54 s	1.89 s 1.57 vw	1.85 s 1.59 vw
Vermiculite	(1){ind. 1.5	1.49	(1){1.26 1.04	1.49	1.47	1.57

[a] According to the origin and the history of the sample.
[b] Unpublished data given by R. Glaeser.
[c] After Weir (1960).
ind. = indeterminate, highly swollen.

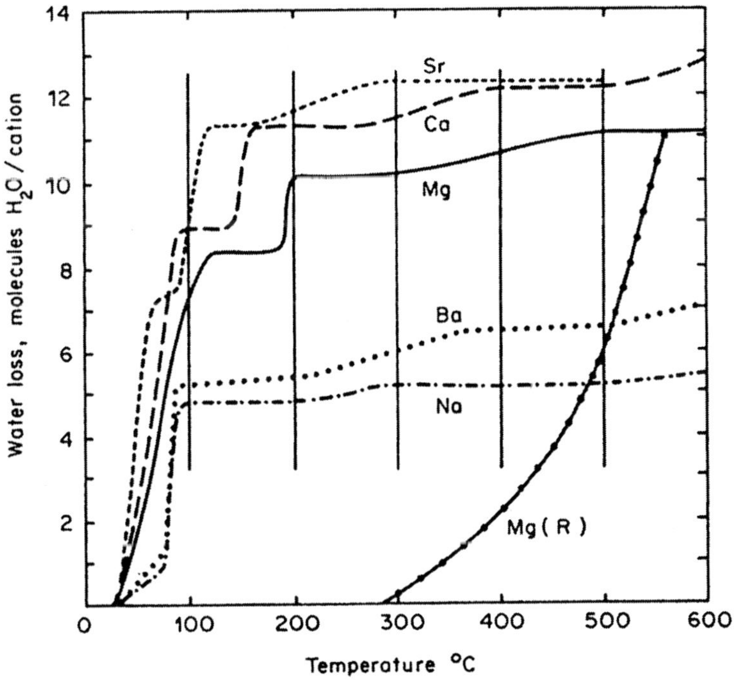

Figure 6.5 Dehydration curves of *World* vermiculite saturated with different cations, as indicated. The curve, Mg(R) indicates the weight loss exhibited by Mg-vermiculite after the sample has been allowed to rehydrate following heating to the temperatures indicated. From Keay and Wild (1961), reproduced by permission of the Mineralogical Society of Great Britain and Ireland

water content, and it will be seen in the equilibrium weight loss curve (Figure 6.5) that an initial water content of about $11H_2O$ per exchangeable Mg^{2+} cation drops continuously until a plateau is reached corresponding to $3H_2O$ per cation in the range of stability of the 1.16 nm phase, and then to one H_2O in the 1.05 nm phase before a final slow decomposition leading to an anhydrous condition (0.91 nm) above 500 °C. It will be noted that the transition from the 1.44 nm to the 1.16 nm phase, which occurs near $8H_2O$ per cation, is not distinguishable in the weight loss curve indicating that there is little difference in hydration energy between these two phases, whereas there are clearly major energetic differences between each of the 1.16, 1.05 nm and anhydrous 0.91 nm phases.

Many other hydrated inorganic cations in vermiculite give stable spacings near 1.5 and 1.2 nm as does Mg^{2+}, but the humidity (or temperature) at which the transition between these spacings occurs is a function of the cation. Broadly analogous behaviour is shown by all the expanding layer silicates; indeed, the transition from a 1.2 to 1.5 nm phase often occurs at an almost identical

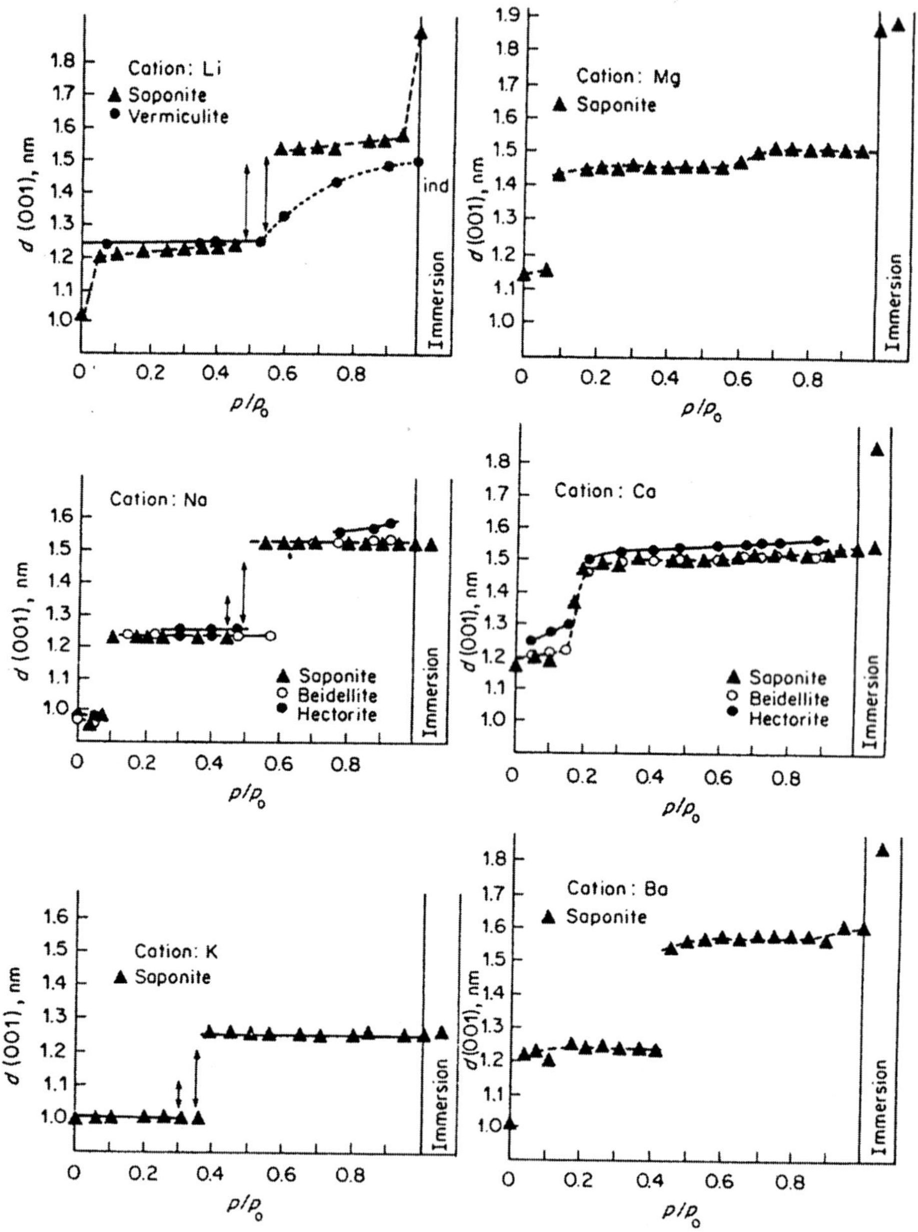

Figure 6.6 Variation of basal spacing with humidity exhibited by smectites and vermiculites saturated with the cations indicated. From Suquet *et al.* (1975), reproduced by permission of *Clays and Clay Minerals*

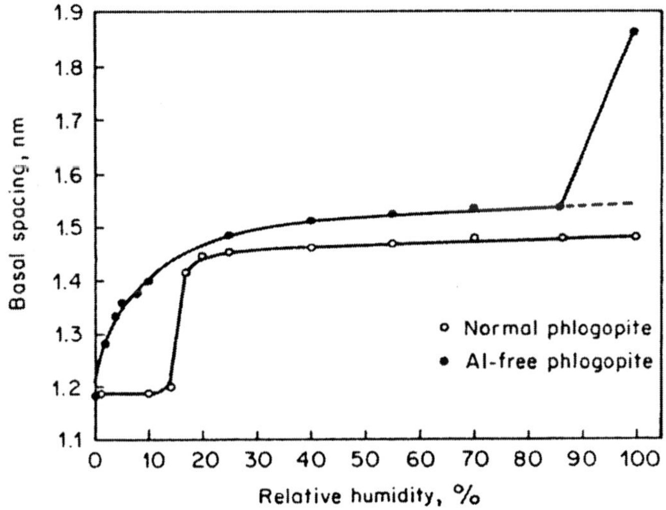

Figure 6.7 Variation of basal spacing with humidity exhibited by two synthetic phlogopites, following replacement of interlayer K^+ with Ca^{2+}. From Kodama *et al.* (1974), reproduced by permission of the Mineralogical Society of America

humidity for a particular saturating cation independent of the type of layer silicate involved. This is shown for Ca^{2+} in saponite, beidellite and hectorite in Figure 6.6; the same relative humidity ($p/p_0 = 0.2$) also induces this spacing change in tetrahedrally substituted vermiculite, but not in a vermiculite that derives its charge solely from vacancies in the octahedral layer (Figure 6.7). Other examples of approximate or partial analogies in behaviour can be seen for Na- and Li-saturated layer silicates in Figure 6.6. It is important to note that these well defined plateaux in interlayer spacing do not generally correspond to plateaux in water content. Both the 1.2 and 1.5 nm phases are stable over ranges of water contents which form a continuous sequence (Brown, 1961).

Taken together, these results point to an important role of both silicate structure and cation in determining the hydration properties of layer silicates. The behaviour of the cation can be seen to be related to its size and polarizing properties, so that the range of stability of the less hydrated 1.2 nm phases increases in the sequence Mg^{2+}, Ca^{2+} and Ba^{2+}. It will be noted, too, in Figure 6.6, that in zero humidity Ba-saponite collapses to a 1.0 nm phase at room temperature, whereas Mg-saponite, like Mg-vermiculite, retains a 1.15 nm spacing, collapsing to 1.0 nm only after heating to 300 °C. The roles of the different aspects of layer-silicate structure are more difficult to disentangle, partly because different structural features are often strongly correlated. Layer silicates of high charge are nearly all trioctahedral (vermiculite and saponite), derive their charge from tetrahedral substitution, and display some three-dimensional order: i.e. successive silicate layers are related to each other by specific displacements of $b/3$ or zero, and by rotations of $\pm 120°$ or zero (here b is a unit cell dimension in the plane of the layers). The low charge smectites (montmorillonite and hectorite) derive their charge principally from octahedral substitutions and have a turbostratic structure in which successive layers exhibit random displacements and rotations relative to each other. There is, however, very clear evidence for the importance of all of these structural features for certain hydration characteristics, particularly at high and low hydration levels, and for ion-trapping phenomena.

The potassium micas phlogopite and muscovite have no tendency to form interlayer hydrates, and vermiculites of high charge collapse to give broad 1.0–1.15 nm spacings when washed with solutions of potassium salts. The potassium is then exchangeable only with difficulty. Some water tends to get trapped between the layers when they collapse; although this can be reduced by boiling the solutions to effect more complete replacement of the hydrated cations originally present, and by heating the preparation above 100 °C to drive out pockets of trapped water, the final product is never quite identical with an anhydrous mica. Ammonium, rubidium and caesium ions are also trapped by trioctahedral vermiculites but sodium and lithium are not. Sodium could be trapped by a dioctahedral vermiculite of high charge as the corresponding anhydrous mica (paragonite) exists. When the layer charge falls to that of saponite or beidellite, potassium is readily exchangeable, and the clay readily hydrates to give a spacing near 1.2 nm. Natural vermiculites of moderate or low layer charge commonly have a proportion of expanding layers when saturated with potassium, and these contribute to the width of the 1.05–1.15 nm X-ray reflection that they exhibit in air, and sometimes give rise to a distinct 1.43 nm spacing (Harward *et al.*, 1968).

Layer charge is not the only factor contributing to trapping of potassium. The hydroxyl groups in the ideal phlogopite structure are oriented perpendicularly to the plane of the layers, bringing the hydroxyl proton close to the potassium ion. Mutual repulsion between their positive charges is thought to account for the greater ease of displacement of potassium from phlogopite than from muscovite where the hydroxyl is oriented nearly parallel to the layers. Accordingly re-orientation of the hydroxyl ion, or its replacement by fluoride or by oxide ions increases the difficulty in displacing potassium from biotites and vermiculites (Newman, 1969). Oxide ions are formed on oxidizing octahedral Fe^{2+} in the structure ($Fe^{2+}(OH)^- \rightarrow Fe^{3+}O^{2-}$); further oxidation of biotites can lead to ejection of Fe^{3+} (Farmer *et al.*, 1971), giving dioctahedral sites whose hydroxyl groups are oriented as in muscovite ($Fe_3^{2+}(OH)^- \rightarrow Fe_2^{3+}(OH)^- + Fe^{3+}$).

None of these features can induce trapping of potassium when the layer charge falls to the levels found in saponite and beidellite, i.e. around 0.5–0.6 electrons per unit cell. This hydration behaviour can easily be rationalized along the lines previously indicated in discussing the hydration of ions in crystals, where it was suggested that water of crystallization served to link

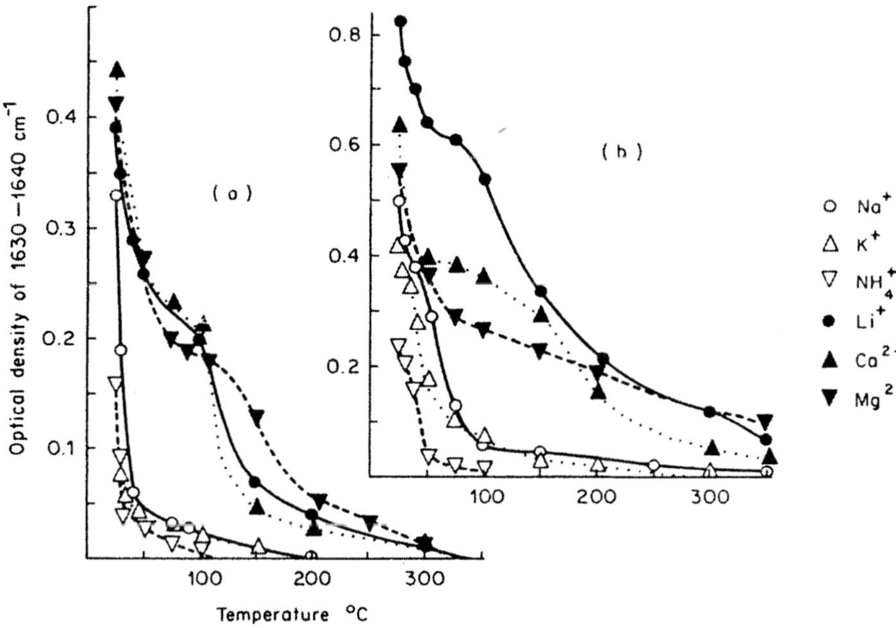

Figure 6.8 Variation in the optical density of the 163–164 nm water absorption band with temperature in (a) montmorillonite and (b) saponite saturated with the cations given in the key. From Russell and Farmer (1964), reproduced by permission of the Mineralogical Society of Great Britain and Ireland

cations to distant anions or to remote parts of large anions. In the micas and in the K-vermiculites of high layer charge each potassium is coordinated to twelve surface oxygens and each oxygen is coordinated to two potassium ions, so that hydration of the potassium ion cannot result in more effective links between the cation and surface oxygens. In saponite and beidellite only about half the hexagonal holes can be occupied by potassium in the anhydrous condition, and so some of the charge-bearing oxygens (i.e. those coordinated to aluminium) will have only one contact with a cation, and some will have none. Here interlayer water can reduce the overall energy of the structure by establishing links between potassium and surface oxygens.

6.3.2 Dehydration Characteristics

When hydrated layer silicates are subjected to forced drying by heating to progressively higher temperatures, there is a marked variation both in their retention of water and in their ability to rehydrate when re-exposed to moisture; this variation is dependent on both the saturating cation and the structure of the silicate. The process of dehydration is very readily followed in a qualitative manner by measuring the intensity of the HOH angle-bending vibration of water in the 1630–1640 cm^{-1} region (Figure 6.8) using self-supporting films of montmorillonite and saponite. In interpreting these curves it should be noted that the intensity of this band is not proportional to water content: for example, Li-montmorillonite loses 80 per cent of its water on heating to 100 °C, but the 1630–1640 cm^{-1} absorption band loses only half its intensity. At 100 °C essentially all the residual water is coordinated to interlayer lithium ions, each of which is coordinated to only one or two water molecules, causing an intense polarization which enhances the intensity of the bending absorption band of water. Similar effects operate with the divalent cations Mg^{2+} and Ca^{2+}, and serve to emphasize the marked inflections beginning around 80 °C that indicate stable coordination complexes for Li^+, Mg^{2+} and Ca^{2+} in both montmorillonite (Figure 6.8a) and saponite (Figure 6.8b). These complexes break down above 120 °C in montmorillonite, but decompose much more slowly in saponite where coordinated water is retained above 300 °C, a temperature at which montmorillonite is essentially anhydrous. X-ray spacings of saponite and equilibrium weight-loss curves do not show such well defined plateaux with rising temperature as do those of vermiculite but there is clear evidence for an analogous behaviour (Russell and Farmer, 1964). Dynamic thermogravimetric curves do show well-defined plateaux corresponding approximately to trihydrates of Mg^{2+} and Ca^{2+} in the ranges 130–190 °C and 110–150 °C respectively, and to monohydrates stable up to 350–400 °C. In Li-saponite the plateaux are less well defined but could again be interpreted as corresponding to a trihydrate and a monohydrate. In montmorillonite, weight-loss curves indicate a trihydrate of Mg^{2+} but only a monohydrate of Li^+ near 100 °C, and there is no evidence for a region of stability corresponding to a monohydrate of Mg^{2+}. The larger monovalent cations do not retain water so strongly as Li^+ and the divalent cations, but here again Na- and K-saponites dehydrate less readily than the corresponding montmorillonites.

This difference between saponite and montmorillonite is again evident in their rehydration behaviour following heating and exposure to moist air, as monitored by recovery of the water bending absorption band (Figure 6.9). In saponite, the monovalent ions, Li^+, Na^+ and K^+, rehydrate readily until the mineral begins to decompose at 700 °C, and the divalent cations only fail to rehydrate substantially when their monohydrate complexes decompose above 450 °C. In montmorillonite, none of the cations rehydrate after heating to 450 °C and the smaller cations Li^+ and Mg^{2+} fail to rehydrate significantly after heating to 250 °C. This behaviour of Li- and Mg-montmorillonite is due to the penetration of a proportion of these small cations, after dehydration, into the vacant sites in the octahedral sheet where they partially neutralize the layer charge. The reaction is used to distinguish montmorillonite with an octahedral deficiency in positive charge from beidellite with a tetrahedral deficiency, as the latter continues to re-expand readily when Li-saturated and heated to 300 °C. The larger cations Ca^{2+}, Na^+ and K^+ cannot penetrate the octahedral layer of montmorillonite, and their failure to rehydrate is probably associated with the fact that the negative charge of montmorillonites is widely but flexibly distri-

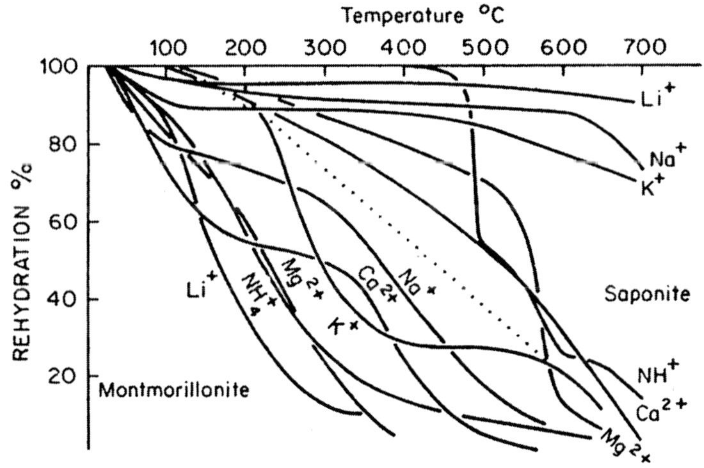

Figure 6.9 Percentage recovery of the $1640\,cm^{-1}$ water absorption band in montmorillonite and saponite on exposure to room humidity for 18 hours. From Russell and Farmer (1964), reproduced by permission of the Mineralogical Society of Great Britain and Ireland

buted over the oxygen surface, so that on dehydration this negative charge can be concentrated on the oxygens in direct contact with interlayer cations. In saponite, the negative charge is necessarily concentrated on the three surface oxygens coordinated to an aluminium, and a cation lying in the hexagonal holes in the surface can make direct contact with only two of these. Certainly, the trioctahedral character of saponite plays no role, since K-hectorite, also trioctahedral, fails to rehydrate in air after heating to 500 °C (Farmer and Russell, 1971). Vermiculite resembles saponite in its dehydration characteristics, except that its K^+ and NH_4^+ forms do not hydrate. The rehydration behaviour of NH_4^+-saturated montmorillonite and saponite (Figure 6.9) resembles that of Mg^{2+}; this is due to the decomposition of NH_4^+, liberating a proton, which enters the silicate sheet, and NH_3 which is lost.

The dehydration and re-expansion properties of layer silicates are commonly used to distinguish them from each other, and from the non-contracting chlorites. It is important to realize, however, that the collapse of montmorillonite to give 1.0 nm spacings can be reversed under fairly gentle conditions, e.g. treatment with concentrated aqueous ammonia (Farmer and Russell, 1966), provided the mineral has not lost its structural hydroxyl groups. The regenerated forms of Li- and Mg-montmorillonite have a reduced exchange capacity, because of the migration of some of these cations into the octahedral sheet during thermal treatment. After dehydroxylation at around 600 °C a 0.95–1.0 nm product is obtained which does not recover its expanding properties even after hydrothermal treatment, although it recovers structural hydroxyl groups to form a neutral pyrophyllite-like product. In contrast beidellite can be largely regenerated following dehydroxylation. After Mg-vermiculite is dehydrated at 600 °C it requires hydrothermal treatment to recover its expanding properties, and this may well be true for Mg-saponite.

6.3.3 Maximum Hydration States

On immersion in salt-free water the expanding layer silicates imbibe water to give a maximum separation of the layers which is again a function of saturating cation and layer silicate structure, as shown in Table 6.2. Layer charge is an important factor, as vermiculite containing divalent cations does not imbibe more than two sheets of water into the interlayer space to give spacings around 1.5 nm, whereas saponite, like all the other smectites, takes up three or four sheets of water to give spacings around 1.9 nm. Site of layer charge is also important in that an expanded phlogopite in which the layer charge arises from octahedral vacancies, not from tetrahedral substitution, gives a 1.86 nm spacing when Ca-saturated (Figure 6.7). Degree of order has an effect, as Ca-saponite only expands fully to 1.86 nm after grinding, which disturbs the ordered arrangement of successive silicate layers. Ion charge and size are of crucial importance as shown by the unlimited swelling exhibited by all the layer silicates (other than vermiculites of very high charge) when saturated with Li^+. Montmorillonite, and some beidellite and saponite samples show similar expansion when sodium saturated, but only montmorillonite appears to undergo expansion when potassium saturated, beidellite and saponite imbibing only one layer of water. With macroscopic Li-saturated vermiculite crystals, the swelling is strikingly visible as their thickness increases over tenfold (Garrett and Walker, 1962). Clay films show broad X-ray reflections corresponding to a range of spacings, the mean value of which is a function of salt concentration (Figure 6.10). Agitation of these expanded forms in dilute suspension results in a complete separation of the layers in random orientation to give highly disperse sols, which are immediately coagulated on adding a divalent cation. Thus sodium saturation is widely used to disperse clays and separate them from coarser mineral impurities, but this fails for smectites of higher layer charge, where Li-saturation is necessary for complete dispersion (Table 6.2).

The explanation of the expanding behaviour of clays has been frequently discussed, but never quantitatively explained. The degree of expansion achieved is often an equilibrium state controlled by finely balanced opposing forces, where quite minor changes in pretreatment of the sample can considerably modify the end result. The initial entry of water into nearly anhydrous clays is strongly exothermic, and the water is firmly held in the coordination sphere of the cation and in contact with the surface oxygens, but after one or two water layers have been introduced, further expansion is probably driven by osmotic forces, i.e. the entry of water dilutes the concentrated interlayer ionic solution. In terms of entropy changes, this expansion increases the entropy of the interlayer cation by allowing it greater freedom of movement. In fully dispersed

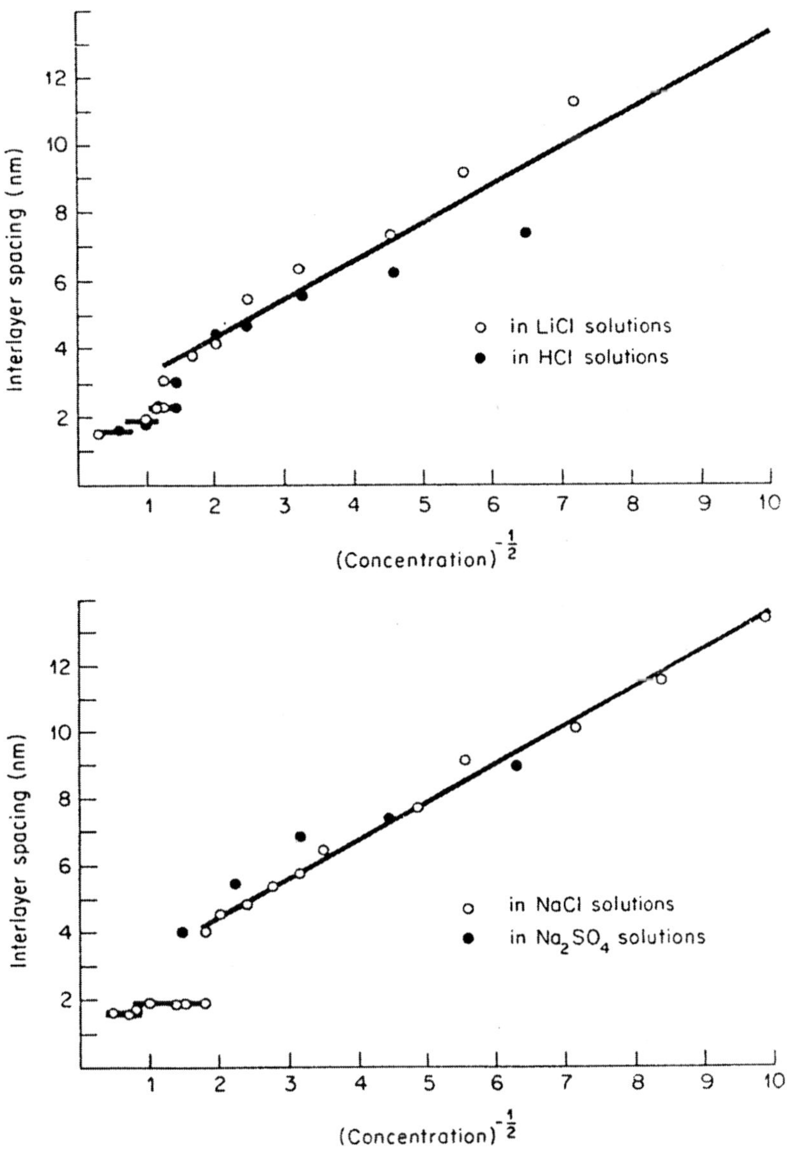

Figure 6.10 Change of approximate mean spacing between layers of montmorillonite with reciprocal of square root of concentration for Na^+, Li^+, and H^+ clays. From Norrish (1954), reproduced by permission of the Chemical Society

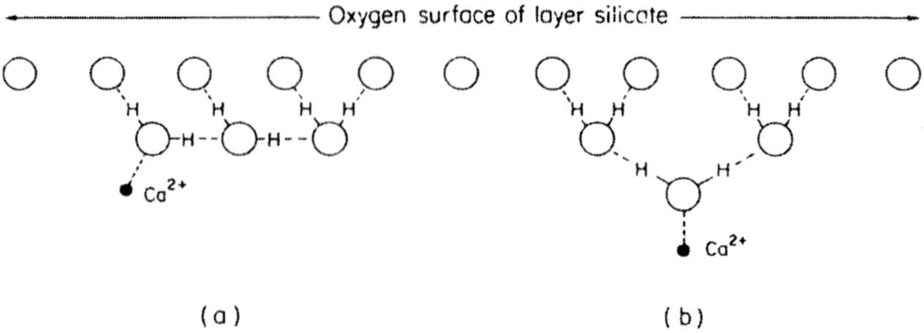

Figure 6.11 Water chains linking a divalent cation to surface oxygens: (a) linear chain in a water layer one or two molecules thick; (b) branching chain in a water layer three or more molecules thick

suspensions of Na-montmorillonite, the sodium ions constitute the diffuse part of a double layer in good agreement with Gouy–Chapman theory (Chapter 5.4). In this theory, repulsion between the sodium ions in the diffuse layers helps to keep the particles apart. The forces restricting expansion are probably also essentially electrostatic, i.e. a resistance to increased separation of the positive cation from the negative surface oxygens, although van der Waals attraction between adjacent layers has also been proposed.

Divalent and trivalent cations provide a sufficient attraction to the adjacent negatively charged surfaces to prevent highly dispersed sols possible with sodium- or lithium-saturated preparations being stable. When the surface charge is low and widely diffused over the surface, the chains of water molecules that link a divalent cation to distant charged oxygens become quite extended, and do not necessarily increase when an additional water sheet is introduced. Indeed Figure 6.11 shows that the chain length can be shortened from three water molecules (Figure 6.11a) to two (Figure 6.11b) and still reach four surface oxygens. This can account for the expansion of smectite containing divalent cations to around 1.9 nm. Vermiculite generally retains a 1.5 nm spacing in water, as the cations are more closely spaced, and most charged sites can be reached in one- or two-link water chains, so that further expansion necessarily increases their lengths. It is probably along these lines that a quantitative account of the expansion phenomena will finally be achieved. Certainly a molecular theory is essential to their understanding, since theories that treat water as a continuous dielectric medium and the oxygen surfaces as featureless planes can never account for the discrete spacings observed.

The above discussion has been restricted to inorganic cations, but macroscopic expansion of vermiculite crystals in water is also achieved by saturating them with propylammonium or butylammonium cations (Walker and Garrett, 1967). The former is most effective for vermiculites of very high charge (0.9–1.0 electrons/unit cell), and the latter for those of lower charge. Unlimited expansion of vermiculites can also be achieved in solutions of some amino acids in the zwitterion form.

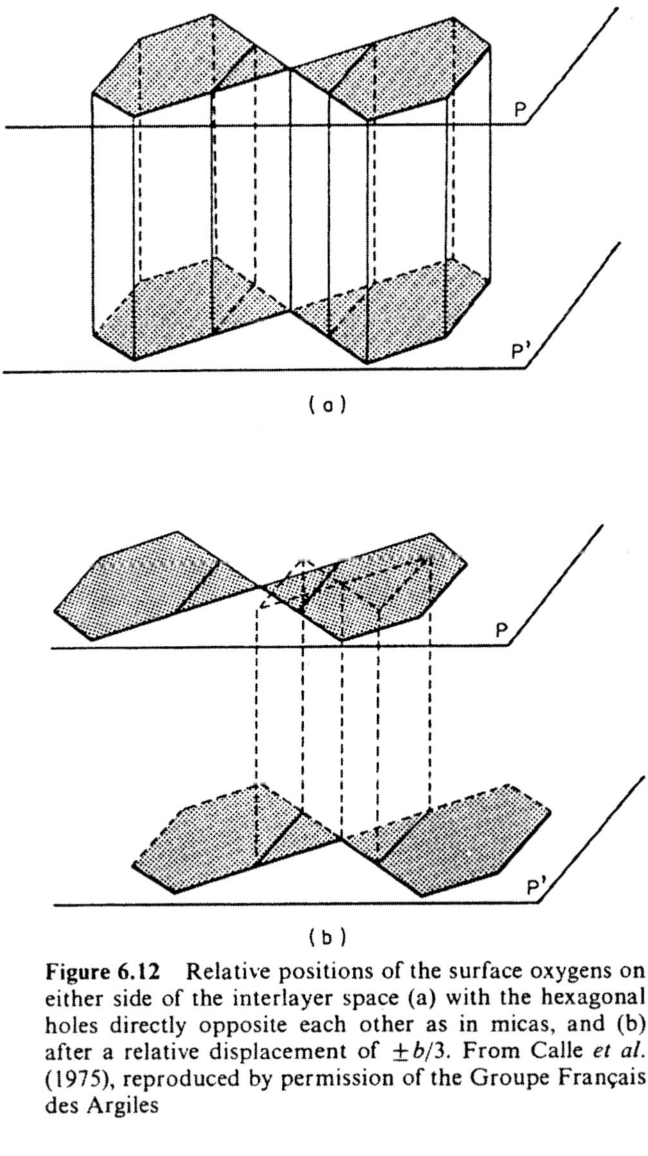

Figure 6.12 Relative positions of the surface oxygens on either side of the interlayer space (a) with the hexagonal holes directly opposite each other as in micas, and (b) after a relative displacement of $\pm b/3$. From Calle *et al.* (1975), reproduced by permission of the Groupe Français des Argiles

6.3.4 Nature of Interlayer Water

It is quite clear from the preceeding discussion of hydration and dehydration phenomena in expanding layer silicates that interlayer water interacts with both cations and surface oxygens. More detailed information on the types of interaction that occur, and on the mobility of cations and water in the interlayer space, can be obtained by the application of several physico-chemical techniques, including X-ray diffraction, electron paramagnetic resonance, infrared spectroscopy, neutron scattering and nuclear magnetic resonance. The results of each technique will be discussed separately.

X-ray diffraction

An identification of the location of cations and water molecules is potentially possible if these are regularly distributed and if single crystals exhibiting three-dimensional order are available. These conditions are only achieved in vermiculites that are derived, artificially or naturally, from well-ordered crystals of phlogopite and biotite. In the parent mica, the oxygen surfaces of successive layers are superimposed so that the interlayer potassium can settle into the Si_6O_6 'hexagons' in the opposing faces. When potassium is replaced by the hydrated cations Na^+, Ca^{2+} and Ba^{2+} the layers separate without shifting relative to each other (Figure 6.12a) to give a 1.5 nm spacing at full hydration, involving two sheets of water. On substituting Mg^{2+} or Ni^{2+} in the interlayer space opposing layers move relative to each other as shown in Figure 6.12b. This shift is reversible, and is proof for a directive influence of one layer on the next which in this case can only be carried by molecular interactions between the interlayer water-cation configuration and surface oxygens. This configuration is as yet known only for Mg-vermiculite, where a three-dimensional X-ray structure analysis has indicated the structure shown in Figure 6.13. The Mg^{2+} cation is octahedrally coordinated by six water molecules, three of which form hydrogen bonds to oxygens of the upper, and three to oxygens of the lower face. In Figure 6.13, only the upper set of water molecules and surface oxygens is shown. The magnesium lies on a centre of symmetry below the tetrahedron marked T_1 and the lower set of water molecules and surface oxygens can be found by inversion through this centre. This X-ray structural determination represents an averaged structure with a higher symmetry than the real structure, since there are only sufficient cations to occupy between half and one-third of the sites below T_1 tetrahedra. The water structure found must be assumed to represent the six molecules directly coordinated to Mg^{2+} and not those (from 2–6 per cation) that lie in a second sphere of coordination.

This particular articulation between cation, water and surface oxygens cannot apply to larger cations in vermiculite, although some other specific structural relationship is necessary to account for the ordering effect transmitted from one silicate layer to the next. It is unlikely to exist in montmorillonite where successive layers are randomly oriented, but it seems likely that all the 1.43–1.5 nm spacings observed in expanding layer silicates represent spacings determined by the requirement that octahedrally coordinated water molecules should make hydrogen bonding contact with surface oxygens. X-ray structural studies on the single-sheet 1.15–1.25 nm phases have not yet yielded exact locations for water molecules or cations, although here again a reversible shift in the silicate layers relative to each other has been shown to occur on formation of the 1.15 nm Mg-vermiculite complex (Walker, 1956).

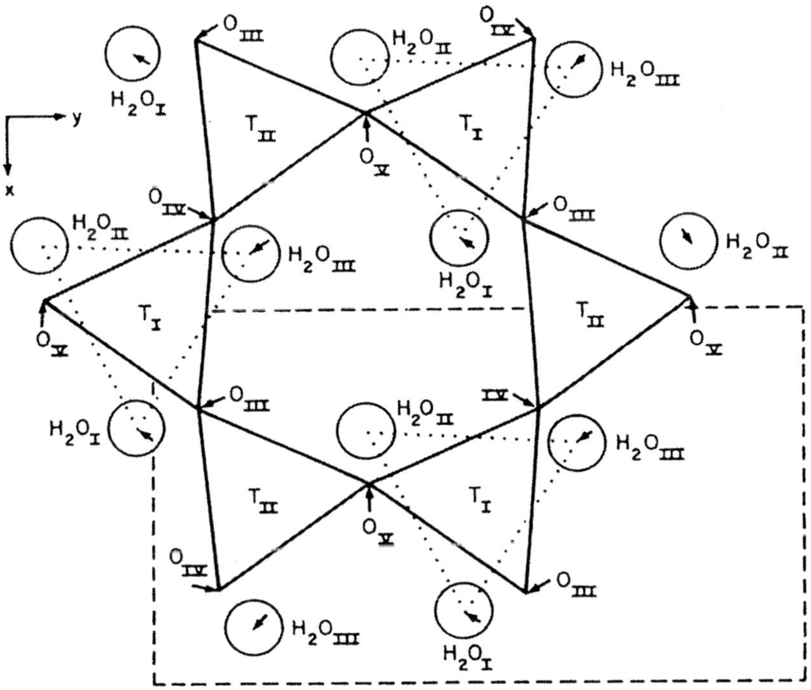

Figure 6.13 Disposition of water molecules in vermiculite relative to the plane of surface oxygens. The interlayer Mg^{2+} ions lie on centres of symmetry below the T_1 tetrahedral sites, where they are octahedrally coordinated by water molecules. From Shirozu and Bailey (1966), reproduced by permission of the Mineralogical Society of America

Electron paramagnetic resonance

Electron paramagnetic resonance (e.p.r.) spectra are obtained only from those transition-metal cations that contain unpaired electrons, but when applicable it gives a unique insight into the immediate environment of the cation. Particularly interesting results have been obtained with Cu^{2+}, which generally forms distorted octahedral coordination complexes exhibiting four strong bonds in a square coplanar conformation and two weak bonds perpendicular to this plane. The Cu^{2+} ion forms a single-layer water complex in montmorillonite with a 1.28 nm spacing, and e.p.r. spectra (Figure 6.14) show that the Cu^{2+} forms with four water molecules a square planar coordination complex, which lies parallel to the plane of silicate layers. The weak bonds are therefore directed perpendicular to the layers, and could be largely ionic in character, interacting with two or three surface oxygens rather than any particular one.

Cu-vermiculite forms a 1.42 nm complex in which the Cu^{2+} must be octahedrally complexed by six water molecules, and e.p.r. spectra show that the configuration of this complex must be essentially that of the $Mg(OH_2)_6$ octahedron in Figure 6.13 since neither the weak bonds nor the plane of strong

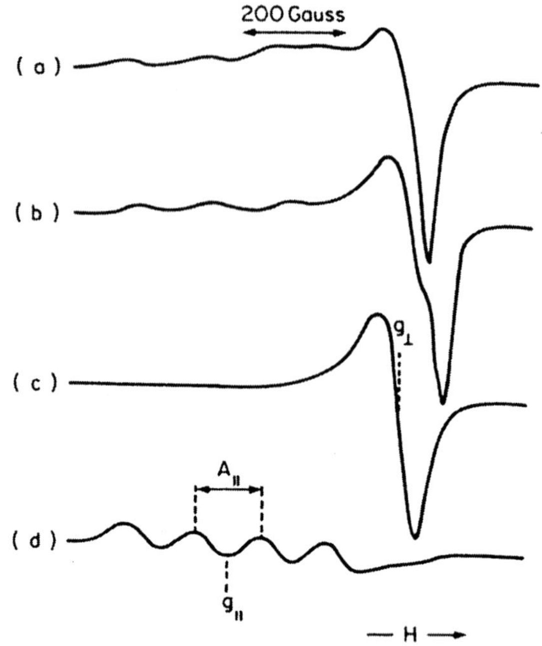

Figure 6.14 First derivative e.p.r. spectra for Cu(II)-hectorite. Spectra (a) and (b) are for randomly oriented powders at 300 and 77 K, respectively. Spectrum (c) is for an oriented film at 300 K with the silicate layers parallel to *H*. In spectrum (d) the layers lie perpendicular to *H*. Reprinted with permission from Clementz *et al.*, *J. Phys. Chem.*, 77, 196 (1973). Copyright by the American Chemical Society

bonds lie either parallel or perpendicular to the silicate layers (Clementz *et al.*, 1973).

Cu^{2+} does not form a 1.5 nm complex in montmorillonite but, when it is introduced in small amount into the 1.5 nm Mg-hectorite complex, e.p.r. spectra (McBride *et al.*, 1975) indicate an orientation identical to that of the 1.28 nm Cu-complex. Thus, if a $Cu(H_2O)_6{}^{2+}$ ion is present, it must carry axial water ligands above and below the plane of the four strongly coordinated water molecules, and McBride *et al* have suggested that the surrounding $Mg(H_2O)_6{}^{2+}$ ions will also have this orientation. It is possible, however, that the $Cu(H_2O)_4{}^{2+}$ plane is in contact with one of the clay layers, and that the local water structure is not representative of the whole sheet.

Infrared spectroscopy

Infrared spectra of interlayer water provide information on the orientation of water hydroxyl groups and the strengths of hydrogen bonds formed. They pro-

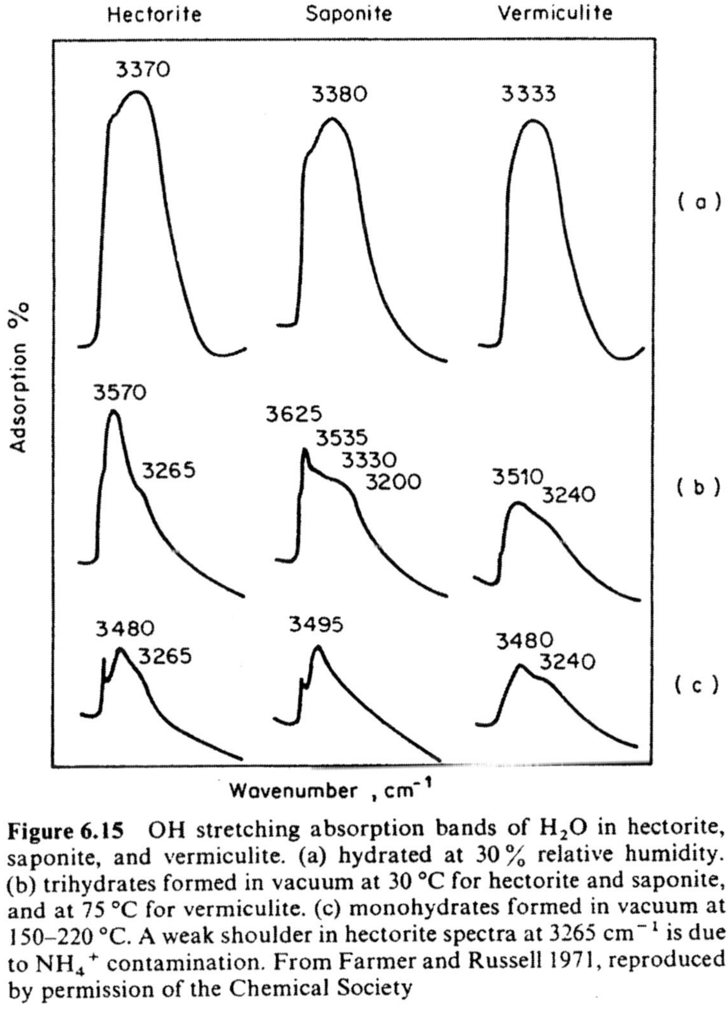

Figure 6.15 OH stretching absorption bands of H_2O in hectorite, saponite, and vermiculite. (a) hydrated at 30% relative humidity. (b) trihydrates formed in vacuum at 30 °C for hectorite and saponite, and at 75 °C for vermiculite. (c) monohydrates formed in vacuum at 150–220 °C. A weak shoulder in hectorite spectra at 3265 cm^{-1} is due to NH_4^+ contamination. From Farmer and Russell 1971, reproduced by permission of the Chemical Society

vide the most direct evidence for the differing distribution of layer charge on surface oxygens associated with tetrahedral and octahedral substitution. This is most clearly seen by comparing the spectra of Mg-saturated specimens in vacuum (Figure 6.15b). Under these conditions, each Mg^{2+} cation retains about three directly coordinated water molecules which can form hydrogen bonds only with surface oxygens, as water in outer spheres of coordination is absent. In hectorite, where the charge derives from octahedral substitution and is therefore quite uniformly distributed over surface oxygens, water gives a fairly narrow absorption band centred on 3570 cm^{-1}, representing a weak bond of about 0.295 nm length. In saponite, where the charge is expected to be largely localized on the oxygens coordinated to tetrahedral aluminium it is obvious

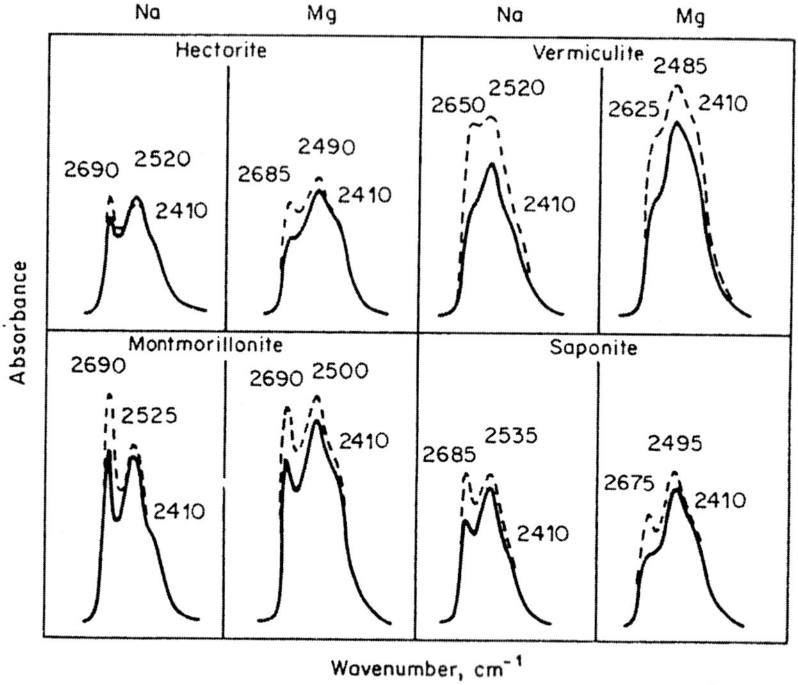

Figure 6.16 OD stretching absorption bands of D_2O in hectorite, montmorillonite, vermiculite and saponite, at 30% relative humidity. The marked increase in intensity of the higher frequency bands at 2625–2690 cm^{-1} when the clay films are turned at 45° to the beam (dashed curves) shows that the OH groups involved in weaker hydrogen bonds are directed towards the surface of the silicate layers. From Farmer and Russell 1971, reproduced by permission of the Chemical Society

that there is a much wider distribution of hydrogen bond strengths, extending from very weak (3625 cm^{-1}) to moderately strong (3330 cm^{-1}). In vermiculite, the weak hydrogen bonds associated with the 3625 cm^{-1} band are totally absent, indicating that all surface oxygens carry some negative charge. Since not all oxygens are coordinated to aluminium in vermiculite, it must be concluded that there is some transfer of charge to oxygens adjacent to but not directly coordinated to aluminium.

In clays examined at room temperature at 50% humidity, Mg^{2+} forms a hexahydrate and, depending on the space available, takes up from four to eighteen additional water molecules in outer spheres of coordination forming a double sheet, whereas Na^+ probably forms a trihydrate in the single sheet of water, with 2–4 additional molecules in outer spheres of coordination. Only in Na-vermiculite are the cations so closely spaced that one water molecule can be coordinated to two Na^+ cations. The spectra of such clays, especially in the

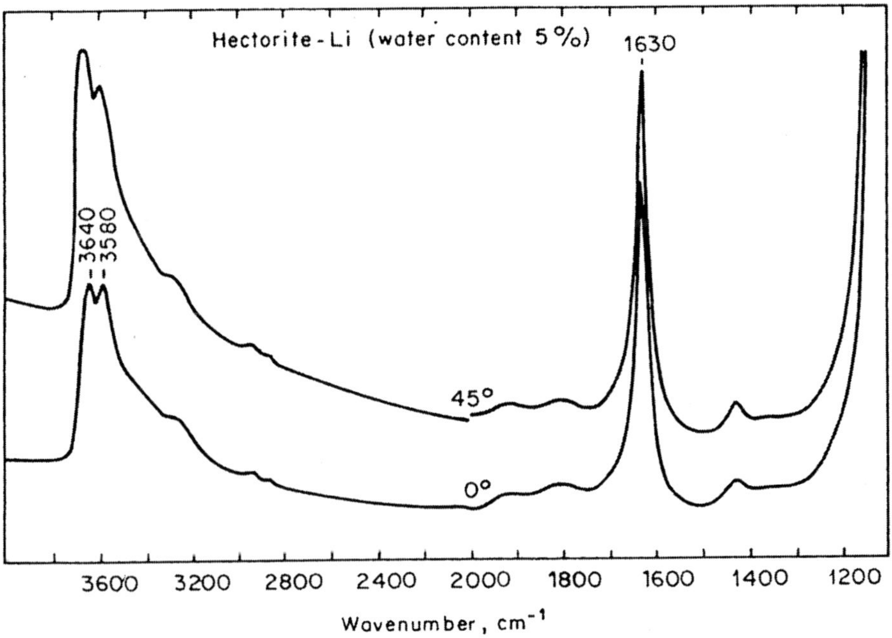

Figure 6.17 Infrared absorption bands of water molecules associated with Li^+ in the interlayer space of hectorite. The infrared beam makes an angle of 0° or 45° with the normal to the layers, as indicated in the figure. From Prost (1971). Reproduced by permission of the Institut de France, Académie des Sciences

D_2O form (Figure 6.16) show clear evidence of two families of hydrogen bonds, the weaker bonds being directed to the surface of the layers, the stronger being mostly parallel to the layers and so involved in water to water bonds. The overall picture for the smectites is consistent with the presence of chains of water molecules radiating from cations as shown in Figure 6.11a. The lengths of the water—water hydrogen bonds are comparable to those in aqueous salt solutions (0.283–0.288 nm) and tend to shorten with increasing polarization of coordinated water by the smaller polyvalent cations. Bonds to the surface range from about 0.282 nm for those directed towards the oxygens of Al—O—Si bridges in saponite and vermiculite, to over 0.3 nm for the interactions with uncharged oxygens in saponite, and the weakly charged oxygens in hectorite and montmorillonite. It has even been proposed that water molecules do not form hydrogen bonds to surface oxygens of hectorite and montmorillonite but are keyed into hexagonal holes in the oxygen surface with their protons directed towards the structural hydroxyl groups that lie within the silicate layers; this would require a proton-to-proton approach in hectorite and saponite, whose structural hydroxyls lie perpendicular to the layers (Prost, 1976).

The pattern of water-lattice bonding undoubtedly changes when the clay is partially dehydrated, and infrared spectroscopy has been able to show that residual water molecules coordinated to lithium in hectorite lie in an orientation such that a line joining the two protons in the same molecule is nearly perpendicular to the layers (Prost, 1971). Thus one OH group is directed towards the upper layer and one towards the lower at angles of almost 50° to the plane of the layers. The hydrogen bonds formed, if they can be so called, are very weak, and both correspond to O—O distances of 0.3 nm. The orientation is deduced from the marked dichroism (sensitivity to orientation) of the absorption band due to the antisymmetric stretching vibration (3640 cm^{-1}, Figure 6.17) which in this case is clearly resolved from that of the symmetric stretching vibration (3580 cm^{-1}). The dipole oscillation in the first is along the H—H direction, whilst that in the second bisects the H—O—H angle.

Proton and water mobility

X-ray structural analysis and infrared techniques give essentially static pictures of water and cation distributions. The first requires that cations and water spend the greater part of their time on discrete lattice sites; the second gives a snap-shot picture of the pattern of short-range order due to coordination and hydrogen bonding at one instant, since many internal vibrations of a molecule can occur before its environment changes. Undoubtedly, however, the spatial distributions are not static; water molecules are interchanging positions, interchanging protons, rotating, and redirecting their hydrogen bonds. Evidence on these fluctuations comes from nuclear magnetic resonance, neutron scattering, electrical conductivity, dielectric properties, and diffusion studies. Taken together, they all point to substantially less mobility of ions and water molecules in the interlayer space than for normal aqueous solutions, but nuclear magnetic resonance points to higher proton mobility.

For example, an n.m.r. study of the double-sheet hydrate of Na-vermiculite (Hougardy *et al.*, 1976) is consistent with an octahedral distribution of water round the cation, with the water rotating frequently—but not necessarily continuously—about the Na–OH_2 axes. The lifetime of a proton in the hydration shell is 10^{-5} s, 100 times shorter than for the same protons in bulk solution. Pulse n.m.r. spectroscopy gives information on dynamic aspects of the water structure, and can be interpreted in terms of a water diffusion coefficient of 10^{-6} cm^2 s^{-1} and a cation diffusion coefficient of 0.5×10^{-8} cm^2 s^{-1}, both of which are substantially less than for bulk solutions (2.3 and 1.3 $\times$ 10^{-5} cm^2 s^{-1} respectively).

Neutrons scattered from water in the interlayers can pick up or lose energy because of a Döppler effect arising from the diffusional motion of the water molecules. Experimentally, this is seen as a broadening of the peak corresponding to elastically scattered neutrons, and the broadening can be used to derive a water diffusion constant, D. Using Li-vermiculite, Li-montmorillonite, and Na-montmorillonite, all of which are capable of controlled expansion from one layer of water to over twenty, it was found that at one, two and three water sheets, the value of D increased rapidly from 0.05×10^{-5} to 0.2×10^{-5} to 0.8×10^{-5} and thereafter tended towards the value 2×10^{-5} $cm^2\,s^{-1}$ found in bulk water (Olejnik and White, 1972). The results indicated that the surface had no effect on water diffusion beyond about 1.5 nm and showed no detectable difference between montmorillonite and vermiculite, in spite of the wide divergence in their surface charge concentration and distribution.

The chemistry of interlayer water

Water in expanding layer silicates can be displaced, completely or partially, by other polar molecules, and can donate protons to bases in the interlayer space. Replacement reactions can be rationalized if we recall some of the unique characteristics of water which make it so effective a link between cations and the negatively charged lattice. These include its ability to donate electrons to two neighbours, and to accept electrons from two neighbours, its ability to form branching hydrogen-bonded chains, its great range of polarizability reflected in hydrogen bonds covering a great range of strengths that finally merge into complete proton transfer from water to the electron donor. No other polar molecule can replace it in all its different roles, but particular molecules are well adapted to perform some of its functions. Thus when the layer charge is low and widely diffused over surface oxygens, as in montmorillonite and hectorite, strong hydrogen bonds to the layer are not necessary, and when the cation is large and univalent a strong coordinate bond is not required. Under these conditions, the much less polar nitrobenzene molecule can completely replace interlayer water in potassium montmorillonite, with the slightly negative oxygens of the nitro group coordinated to the potassium, and the slightly positive hydrogens on the benzene ring in contact with the oxygens of the silicate surface. Nitrobenzene cannot easily replace the highly polarized directly coordinated water around the smaller divalent cations, such as Mg^{2+}, but can substitute for water in outer spheres of coordination to form a water-bridged complex, which can be formulated $Mg(OH_2 \cdot O_2N \cdot C_6H_5)_6$. This type of coordination complex is the most common form in which polar organic molecules are held in montmorillonite and hectorite containing divalent cations, and is formed even with bases as strong as pyridine. The ease of replacement of water by these large organic molecules is probably the result of an entropy effect, as up to four water molecules are liberated by one organic molecule.

Water on the primary hydration shell of divalent cations can only be easily replaced by molecules capable of equally strong coordination; ammonia and organic amines can often do so because of their affinity for transition metal cations, and alcohols can readily replace water round Mg^{2+} and Ca^{2+} in montmorillonite. When the surface charge is localized on oxygens linked to tetrahedral aluminium, most organic molecules cannot provide a localized positive charge to match the localized negative charge, and so organic complexes of saponite and vermiculite are less stable than those of montmorillonite.

Even alcohols cannot effectively substitute for water in vermiculite, as a chain of hydrogen-bonded alcohol molecules can only terminate in one hydrogen bond to the surface. In a corresponding water chain, each link can itself form a hydrogen bond to the surface, or originate a branch in the chain. Accordingly, alcohols do not completely replace water in vermiculite.

Proton transfer from water to bases often occurs to a greater extent within the interlayer space than would be expected in bulk solutions. The reaction is promoted by partial dehydration of the system, since then the residual water molecules around cations are more highly polarized and hence more acidic; for example, the water-bridged Mg-pyridine complex in montmorillonite forms magnesium hydroxide and pyridinium ions reversibly on gentle dehydration. Even in fully hydrated systems, however, formation of protonated bases is often favoured in the interlayer compared with the same reaction in bulk solution. This tendency has been ascribed to an enhanced acidity of interlayer water, and correlated with its greater degree of ionization as indicated by n.m.r. studies. However, strong adsorption of the protonated base (HB) in the interlayers can also push the reaction

$$M^+(OH_2)_n + B \rightarrow MOH + H^+B$$

to the right, and this is sufficient to account for the formation of more pyridinium ion in Na-montmorillonite than in Na-saponite when exposed to pyridine. It is not known however, whether this can account for more ammonium ion appearing in Ca-nontronite than in Ca-montmorillonite exposed to ammonia.

These and other aspects of interlayer complexes in layer silicates have been more fully reviewed by Farmer (1971), Mortland (1970) and Theng (1971).

6.4 WATER ON SURFACES OF IMPERFECT AND NON-EXPANDING LAYER SILICATES

The highly detailed information presented so far refers to selected layer silicate species that approach ideal compositions. In this ideal picture the internal surfaces constitute smooth planes of regularly packed oxygens, with water interlayers that contain only freely diffusible cations. In many soils, however, the interlayers of layer silicates incorporate islands of aluminium or iron hydroxides that profoundly modify the dehydration characteristics of the layer silicates, as they prevent the layers closing down on the interlayer cations. Water is then retained to much higher temperatures than in a perfect structure, partly by the exchangeable cations which can no longer coordinate directly to both oxygen surfaces, but also by iron and aluminium ions incorporated at the edges of the hydroxide interlayer islands.

A tendency to retain more water than would be expected from the ideal structure is also true of many non-expanding layer silicates of the kaolin and mica groups. The surfaces of such minerals, even of well crystallized forms, are difficult to characterize experimentally, and are in any case liable to variation because of surface adsorption (Chapter 4). Mica-like layers exposed on the surface of illites and chlorites carry a negative charge, and so are likely to be associated with exchangeable cations, or with positively charged aluminium or iron hydroxide sheets partially or wholly covering their surfaces. Kaolin minerals, if their surface layers are of ideal composition, should expose a gibbsite-like surface and a siloxane (Si—O—Si) surface with little or no exchange capacity, but most specimens have a pH-independent charge which probably arises from mica or smectite impurities (Mashali and Greenland, 1976; Ferris and Jepson, 1975). In all these non-expanding minerals, an important proportion of the total surface consists of edge faces on which are exposed hydroxyl groups and coordinated water molecules which may be arranged in regular or irregular (gel-like) structures. Such surfaces exhibit a pH dependent charge, arising from the adsorption and release of protons.

Most soil clays, even if characterized as kaolinite, illite or chlorite, are not well crystallized, and frequently contain expanding interlayers of the smectite or halloysite type. Thus their effective surface areas and exchange capacities are substantially greater than that expected from their external surface, and so also are their water retention capacities.

The principal experimental methods for exploring the nature of adsorbed water on imperfect and non-expanding layer silicates include adsorption isotherms, calorimetry, gravimetry, and infrared spectroscopy. The results obtained up to about 1965 have been reviewed by van Olphen (1975). Only sepiolite, palygorskite, and the kaolin minerals have been extensively studied, and considerable uncertainties remain in the interpretation of the results for kaolin minerals.

It is a common assumption in adsorption isotherm studies that samples evacuated at room temperature or slightly above will be free of adsorbed water, but infrared studies have clearly shown that this is not true for smectites carrying such strongly polarizing cations as Li^+, Ca^{2+}, and Mg^{2+} in their interlayer spaces, and it is unlikely to be true for illites or kaolinites carrying such cations on their surfaces. Infrared studies of water on kaolinite (Gribina and Tarasevich, 1971) clearly show persistence of the H_2O deformation vibration near 1640 cm^{-1} in spectra of Ca-kaolinite in vacuum at 100 °C, although it is lost from spectra of Na-kaolinite in vacuum at 20 °C. Blanc and Escubes (1975) found that attempts to remove water from Ca-kaolinite by heating to 120 °C caused a marked diminution in the heat of adsorption of water, indicating a deactivation of the surface.

Inevitably, therefore, there are uncertainties concerning the initial state of hydration of surfaces studied by classical adsorption isotherms, and consequent uncertainties in the thermodynamic conclusions drawn from them (van Olphen, 1975). Qualitatively, however, there is a clear-cut effect of exchangeable cations on the adsorption isotherms of kaolinite, and Blanc and Escubes (1975) have found a linear relationship between the energy of hydration of the cations and

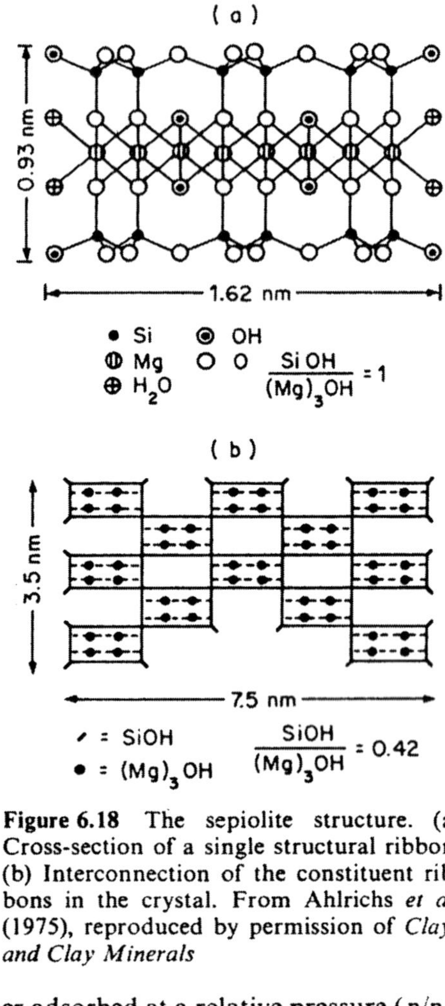

Figure 6.18 The sepiolite structure. (a) Cross-section of a single structural ribbon. (b) Interconnection of the constituent ribbons in the crystal. From Ahlrichs *et al.* (1975), reproduced by permission of *Clays and Clay Minerals*

the amounts of water adsorbed at a relative pressure (p/p_0) of 0.1. The infrared studies of Gribina and Tarasevich (1971) also show that at least a proportion of the adsorbed water must be associated with exchangeable cations, as they modify the position of the infrared bands of adsorbed water.

Other sites in addition to exchangeable cations must play a role in water adsorption on kaolinite. The siloxane surface has very little affinity for water, and the gibbsite-like surface, which adsorbs water by hydrogen bonding, should be readily dehydrated in vacuum (Russell *et al.*, 1974). Much stronger retention of water is to be expected on edge faces by coordination to exposed octahedral cations, although it is difficult to distinguish the role they play in the overall water adsorption characteristics of kaolinite. Their role has been more clearly defined for palygorskite, in which edge sites are numerous and regular. This mineral consists of narrow strips of layer silicate structure which are cross-linked at their corners (Figure 6.18) so that magnesium ions of the octahedral layer are exposed along each side of every strip, where they interact with water in the open channels between the strips (see also Section 2.3.7). Each ion is linked to four oxygens of the layer silicate structure, and completes its octahedral coordination with two water molecules which in turn are hydrogen bonded to zeolitic water filling the channels. The zeolitic water can be removed at room temperature under vacuum, but the first of the coordinated water molecules is lost only at 200 °C in vacuum, when the structure folds so that the coordination of magnesium ions remains sixfold by involving an oxygen of an Si—O—Si bridge on the surface of an adjacent strip. The remaining water molecule is lost only at 500 °C in vacuum (Serna *et al.*, 1975).

Of the more poorly ordered clay minerals, only halloysite and metahalloysite have been examined in any detail. Halloysite consists of kaolinite-like sheets which are separated by a monolayer of water, giving a 1.0 nm spacing (Section 2.3.5). This monolayer is irreversibly lost on exposure of the mineral to air, but the product, metahalloysite, has a wider spacing (0.73–0.75 nm) than kaolinite (0.715 nm). Temperatures up to 400 °C are necessary to reduce this spacing to 0.72 nm; this slow collapse is usually ascribed to the gradual release of water trapped in pockets between the layers, or as individual molecules in the hexagonal holes of the tetrahedral sheet, and infrared spectroscopy (Tarasevich and Gribina, 1972) confirms that water is strongly retained on heating to over 200 °C. Most poorly crystalline and irregularly interstratified clays retain water on heating much more strongly than do smectites, when both are potassium saturated to avoid retention of water by exchangeable cations. As suggested in the first paragraph of this section, this could be due to irregularities in the silicate layers and interstratified hydroxide sheets which cause gaps and wedges in which water is firmly retained. Study of the accessibility of water in such systems to exchange with D_2O should provide information on whether the water is indeed trapped in pockets or hexagonal holes, but this aspect has not yet been systematically explored.

6.5 WATER ON SURFACES OF HYDROUS OXIDES

Oxides and hydroxides of aluminium and iron occur as discrete clay-size minerals in soils, and also as coatings on the surfaces of other minerals, including clay-size layer silicates. Their clean surfaces consist principally of hydroxyl groups, but can include unprotonated oxide bridges and water coordinated to structural cations. It is important to emphasize the word *clean* as these oxides strongly adsorb polyvalent anions, and in soil are likely to be considerably modified by adsorbed organic matter, phosphate, and silica (Chapter 4). Similarly, clean silica surfaces have been fully characterized as consisting solely of silanol (SiOH) groups and Si—O—Si bridges, but this surface strongly adsorbs aluminium ions, and must be so modified in soils. If clean silica gel surfaces were present, the equilibrium solubility of nearly 100 p.p.m. of silica in the soil solution would be attained, whereas values of 10–15 p.p.m. are seldom exceeded.

The affinity of silica for alumina leads to their mutual interaction in soils to give amorphous allophanes and delicately ordered imogolite which form gel-like structures capable of high water retention. Their surfaces appear to resemble those of the individual oxides, exhibiting AlOH and SiOH groups, and perhaps also water coordinated to aluminium. Laboratory preparations of silica-alumina precipitates often exhibit considerable cation exchange capacities associated with tetrahedral aluminium; although natural allophane may exhibit some such cation exchange properties, imogolite certainly does not (Figure 4.4 and Wada and Harward, 1974).

The behaviour of the more strongly bonded water on such surfaces can be followed by adsorption isotherms, infrared spectroscopy, heats of wetting and dielectric constant measurements (Zettlemoyer *et al.*, 1975). So far, interest has been concentrated largely on nearly anhydrous oxide systems, since these are of commercial interest as catalysts, but a general picture of the behaviour of hydrous oxides is now possible. The hydrogen-bonding interactions with water are never very strong. Some isolated silanol groups on silica gels are unbonded at room temperature or slightly above; two closely spaced silanol groups that can form hydrogen bonds to both unshared orbitals on a water oxygen hold it more strongly, but the surface is rendered essentially anhydrous simply by evacuation. Similarly, all water is removed from the hydroxylated surfaces of goethite (α-FeOOH) and gibbsite (γ-$Al(OH)_3$) under vacuum, although they are fully hydrated in air (Russell *et al.*, 1974). The initial heat of adsorption of water on a highly hydroxylated iron oxide surface is 40–60 kJ mol^{-1}, and the likely structure is one in which the water donates electrons to two surface hydroxyls, as dielectric relaxation indicates that its rotational frequency is seven decades below that of bulk water. In contrast, the heat of adsorption of water on isolated SiOH groups is only 25 kJ mol^{-1}. Oxygens of Si—O—Si bridges interact so weakly with water that a dehydroxylated silica surface is hydrophobic. Three-coordinated oxide ions on the goethite surface are sufficiently strong electron donors to form carbonate with carbon dioxide, but even these do not retain water under vacuum. Ferric ions exposed at edge sites on goethite crystals do retain coordinated water under vacuum; this water can be readily replaced by coordinated pyridine which in turn cannot be pumped off at room temperature (Parfitt *et al.*, 1976).

The gel-like aluminosilicates allophane and imogolite have highly porous structures with surface areas of about 500 and 700 $m^2\ g^{-1}$ respectively (Wada and Harward, 1974). Imogolite is known to exist in the form of tubes of about 0.7–1.0 nm internal diameter and 1.7 to 2.1 nm external diameter. The outer surface is thought to be covered with hydroxyl groups linked to aluminium, and the inner surface with hydroxyls linked to silicon. These surfaces are therefore hygroscopic, and all are accessible to water which fills pores that constitute 55–60% of the total volume (Figure 6.19). All this water is lost on

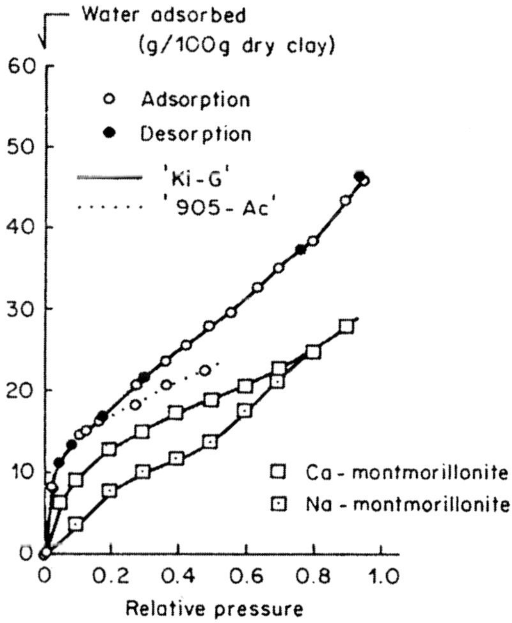

Figure 6.19 Adsorption of water vapour at 20 °C by imogolite from two sources (Ki-G and 905-Ac) and by Ca- and Na-montmorillonites. From Wada and Yoshinaga (1969), reproduced by permission of the Mineralogical Society of America

evacuation. Many soil allophanes consist of hollow spheres about 5.5 nm in diameter, and their hydration properties appear to be similar to those of imogolite.

The presence of these highly porous oxides confer unique properties on many volcanic ash soils.

6.6 WATER ON ORGANIC SURFACES

It is necessary to distinguish two forms in which organic matter appears in soil. One consists of discrete largely organic particles that result from the addition of peat, composts, crop residues and animal manures; the second is adsorbed in essentially molecular form on mineral surfaces. In the following discussion, we will take peats as representative of the first group. Adsorbed forms of organic matter and the modifications they cause in the surface properties of minerals are much less accessible for study, but are of no less importance.

Peats are used not only as soil additives, but can themselves constitute a soil body, suitable for agriculture or horticulture. Their water relations contrast in many ways with those of mineral soils. Volume for volume, peats can hold over

Table 6.3 Moisture holding capacities and wilting percentages of peat, soil, and 1 : 1 mixtures of these. Data selected by Allison (1973) from Feustel and Byers, *U.S. Dep. Agric. Tech. Bull.* **532**, 1–25, (1936). Reproduced by permission of Elsevier Scientific Publishing Co.

Material[a]	Weight of 100 cm^3 of air-dry material (g)	Moisture-holding capacity (%)	Wilting percentage (%)	Available moisture as percentage of moisture-holding capacity (%)	Moisture required to saturate 100 cm^3 of dry material (g)	Moisture retained by 100 cm^3 of material at wilting point (g)
Moss peat	11	1057	82.3	92.2	101	8.0
Sedge peat	27	374	60.8	83.7	91	15.0
Reed peat	39	289	70.7	75.5	99	24.0
Clay loam	109	44.3	7.1	84.0	48	7.7
Loamy fine sand	135	30.9	2.1	93.2	42	2.8
Clay soil–moss peat mixture	60	114	14.5	87.3	67	8.5
Clay soil–sedge peat mixture	68	95.7	19.2	79.9	63	13.0
Clay soil–reed peat mixture	74	94.1	21.2	77.5	67	15.0
Sandy soil–moss peat mixture	73	101	6.6	93.5	73	4.8
Sandy soil–sedge peat mixture	81	80.8	13.5	83.3	64	11.0
Sandy soil–reed peat mixture	87	79.3	16.2	79.6	67	14.0

[a] All mixtures were 50 % peat and 50 % soil on a volume basis.

twice as much water as do soils at saturation, and the denser forms of peat can retain considerable amounts of water at the wilting point (Table 6.3). Their moisture holding capacities exhibit striking irreversibility. After drying from the natural saturated condition they show substantial shrinkage, and become partially hydrophobic and slow to wet; even after prolonged soaking, they seldom recover their original volume and water content.

The large water-holding capacity of peats is associated with a wide range of pores varying in size and in environment. Some, for example those associated with persistent plant cell structures, are dimensionally stable, and can be emptied and refilled reversibly. An extreme example of such stable cell structures is found in sphagnum peats, and these cause the low density and high water-holding capacity of the moss peat in Table 6.3. Other much finer pores ranging down to molecular size occur in all peats, but especially in the more highly decomposed gel-like structures that characterize reed and sedge peats. The supporting structure of complex high molecular weight polymers probably forms a flexible cross-linked framework in which individual chains are themselves hydrated. Even partial dehydration of such a structure can lead to irreversible changes in conformation, and to the collapse of pores which will not refill on rehydration.

On the atomic scale, we can recognize a wide range of local structures in peats and soil organic matter, although we cannot yet define their overall organization. The principal polar sites where water adsorption is likely include carboxylic groups present both as free acids and in ionic form, phenolic and alcoholic hydroxyl groups, amides, amines, ketones, aldehydes and esters. Of these the carboxylic groups are numerically predominant, particularly in the more highly oxidized gel-like residues. Ionized carboxylic groups and their associated cations are likely to have the greatest affinity for water, but studies on biological polymers (Berendsen, 1975) have shown that free amide groups of peptides, and the hydroxyl of carbohydrates are hydrated at very low relative humidities. As drying proceeds, water held by single hydrogen bonds is likely to be lost first, while that held by two or more bonds will be less readily removed, leaving finally the most firmly held coordinated water. As polar sites are exposed by drying, the flexibility of the organic structure will allow at least a partial internal pairing of proton donors (OH and NH) and proton acceptors (C=O, COH and NH groups). Some of these new pairings will be strained, but many will prove perfectly stable, and not liable to rehydration. It is this process that leads to irreversibility in drying and wetting cycles.

In addition to polar sites, peats and soil organic matter incorporate an important proportion of hydrophobic regions. These include a range of aliphatic structures whose tendency to associate with each other and to exclude water (the hydrophobic effect) increases with increasing chain length. The aliphatic chains probably derive mostly from plant lipids and waxes, fungal metabolic products, and even animal (e.g. aphid) wax exudates. Aromatic structures derived from the lignin of plant residues also persist for some time, and their organophilic character is shown by their concentration in the alcohol–benzene soluble fraction of peat humic acids. In addition, polymers of overall hydrophilic character can include regions of hydrophobic character, as do proteins and peptides, for example. In the drying process, the mutual attraction of proton donors and proton acceptors will result in the exposure of an increasing proportion of hydrophobic sites on the surfaces of organic particles and this can account for their resistance to rewetting after drying.

Organic substances adsorbed on mineral particles can also confer marked hydrophobic properties on essentially mineral soils (Debanno and Letey, 1969). Such water-repellent soils suffer severe erosion because of run-off in heavy rain, and they present problems in seed germination for even very slight variations in ground level. Coarse-textured sands are most likely to be affected, and certain plant covers are more likely to contribute the necessary hydrophobic organic matter. The problem is most troublesome in warm low-rainfall areas, and is greatly exacerbated by forest and brush fires which generate from the litter organic vapours that penetrate two or three inches into the soil, rendering it water-repellent in depth. Waxes, fatty acids, fatty alcohols and ketones are most likely to be involved, as they contain polar groups for attachment to the hydroxylated surface of sands and other oxide minerals. In support of this thesis water repellency has been induced by treating sandy soils with decanol or catechol derivatives. Such compounds are not well adapted for firm attachment to the charge-bearing surfaces of ion-exchanging clay minerals, so that soils with more than 15 to 20 per cent clay are not liable to become water repellent naturally. Clay can be rendered hydrophobic by treatment with long-chain organic ammonium or pyridinium bases, and this process is used commercially to render clays organophilic. In nature, however, the most common bases are peptides incorporating basic amino acids, and it is these that are likely to cover and enter the interlayer space of layer silicates without decreasing their hydrophilic character.

In spite of their importance, the nature of mineral–organic complexes has been little investigated, nor have their consequences for mineral surface properties been deeply considered. This is perhaps not surprising: the complexity of soil minerals alone, or of soil organic matter alone, has led the soil mineralogist to simplify his problems by destroying adhering organic matter, while the soil organic chemist prides himself on his success in extracting ash-free organic matter. However, when these necessary preliminaries are completed we can expect a vigorous attack on the problem of natural mineral–organic complexes existing in soil, using our knowledge of the separate components as a springboard.

REFERENCES

Ahlrichs, J. L., Serna, C. and Serratosa, J. M. (1975), Structural hydroxyls in sepiolites, *Clays Clay Miner.*, **23**, 119.

Allison, F. E. (1973), *Soil Organic Matter and its Role in Crop Production*, Elsevier, Amsterdam.

Berendsen, H. J. C. (1975), Specific interactions of water with biopolymers, in Franks (1972–75), vol. 5, p. 293.

Blanc, R. and Escoubes, M. (1975), Adsorption of water on kaolinite. Effect of the nature of active sites, *Thermochim. Acta*, **11**, 115.

Brink, G. and Falk, M. (1970). Infrared spectrum of HDO in aqueous solutions of perchlorates and tetrafluoroborates, *Can. J. Chem.*, **48**, 3019.

Brown, G. (1961), *The X-Ray Identification and Crystal Structures of Clay Minerals*, Mineralogical Society, London.

Calle, C. de la, Suquet, H. and Pezerat, H. (1975), Displacement of the layers accompanying certain cationic exchanges in vermiculite monocrystals, *Bull. Groupe Franç. Argiles*, **27**, 31.

Clementz, D. M., Pinnavaia, T. J. and Mortland, M. M. (1973), Stereochemistry of hydrated copper(II) ions on the interlamellar surfaces of layer silicates. An electron spin resonance study, *J. Phys. Chem.*, **77**, 196.

Debanno, L. F. and Letey, J. (eds.) (1969), *Water-Repellent Soils. Proceedings of the Symposium on Water-Repellent Soils*, University of California, Riverside.

Eisenberg, D. and Kauzmann, W. (1969), *The Structure and Properties of Water*, Clarendon Press, Oxford.

Falk, M. (1975), Vibrational band profiles and the structure of water and of aqueous solutions, in W. A. Adams *et al.* (eds.). *Chem. Phys. Aqueous Gas Solutions, (Proc. Symp.)*, p. 19, Electrochem. Soc., Princetown.

Falk, M. and Knop, O. (1973), Water in stoichiometric hydrates, in Franks (1972–75), vol. 2, p. 55.

Farmer, V. C. (1971), The characterization of adsorption bonds in clays by infrared spectrometry, *Soil Sci.*, **112**, 62.

Farmer, V. C. and Russell, J. D. (1966), Infrared absorption spectrometry in clay studies, *Clays Clay Miner.*, **15**, 121.

Farmer, V. C. and Russell, J. D. (1971), Interlayer complexes in layer silicates, the structure of water in lamellar ionic solutions, *Trans. Faraday Soc.*, **67**, 2737.

Farmer, V. C., Russell, J. D., McHardy, W. J., Newman, A. C. D., Ahlrichs, J. L. and Rimsaite, J. Y. H. (1971), Evidence for loss of protons and octahedral iron from oxidized biotites and vermiculites, *Miner. Mag.*, **38**, 121.

Ferris, A. P. and Jepson, W. B. (1975), Exchange capacities of kaolinite and the preparation of homoionic clays, *J. Colloid Interface Sci.*, **51**, 245.

Franks, F. (1972–75), *Water, A Comprehensive Treatise*, Plenum, New York.

Garrett, W. G. and Walker, G. F. (1962), Swelling of some vermiculite–organic complexes in water, *Clays Clay Miner.*, **9**, 557.

Gribina, I. A. and Tarasevich, Yu. I. (1971), Infrared spectroscopic study of the state of water in kaolinite, *Geokhim.*, **7**, 878.

Harward, M. E., Castea, D. D. and Sayegh, A. H. (1968), Properties of vermiculites and smectites: expansion and collapse, *Clays Clay Miner.*, **16**, 437.

Hougardy, J., Stone, W. and Fripiat, J. J. (1976), Structure and motion of water molecules in the two layer hydrate of Na-vermiculite, *J. Chem. Phys.*, **64**, 3840.

Keay, J. and Wild, A. (1961), Hydration properties of vermiculite, *Clay Miner. Bull.*, **4**, 221.

Kodama, H., Ross, G. J., Iiyama, J. T. and Robert, J.-L. (1974), Effect of layer charge location on potassium exchange, *Amer. Mineral.*, **59**, 491.

Luck, W. A. P. (ed.) (1974), *Structure of Water and Aqueous Solutions. Proc. Int. Symp.* 1973, Verlag Chem. Marburg, Germany.

McBride, M. B., Pinnavaia, T. J. and Mortland, M. M. (1975), Electron spin resonance studies of cation orientation in restricted water layers on phyllosilicate (smectite) surfaces, *J. Phys. Chem.*, **79**, 2430.

Mashali, A. and Greenland, D. J. (1976), Dependence of charge characteristics of kaolinites on pH and electrolyte concentration, in S. W. Bailey (ed.), *Proc. Int. Clay Conf. 1975*, p. 240, Applied Publishing Ltd., Wilmette, Ill.

Mortland, M. M. (1970), Clay–organic complexes and interactions, *Advan. Agron.*, **22**, 75.

Narten, A. H. and Levy, H. A. (1972), Liquid water: scattering of X-rays, in Franks (1972–75), vol. 1, p. 311.

Newman, A. C. D. (1969), Cation exchange properties of micas. I: The relation between mica composition and potassium exchange in solutions of different pH, *J. Soil Sci.*, **20**, 357.

Norrish, K. (1954), The swelling of montmorillonite, *Discuss. Faraday Soc.*, **18**, 120.

Olejnik, S. and White, J. W. (1972), Thin layers of water in vermiculites and montmorillonites—neutron diffusion coefficient, *Nature Phys. Sci.*, **236**, 15.

van Olphen, H. (1975), Water in soils, in J. E. Gieseking (ed.), *Soil Components*, vol. 2, p. 497, Springer-Verlag, New York.

Owston, P. G. (1958), The structure of ice-I, as determined by X-ray and neutron diffraction analysis, *Advan. Phys.*, **7**, 171.

Parfitt, R. L., Russell, J. D. and Farmer, V. C. (1976), Confirmation of the surface structures of goethite (α-FeOOH) and phosphated goethite by infrared spectroscopy, *J. Chem. Soc. Faraday Trans. I*, **72**, 1082.

Prost, R. (1971), Infrared spectra of water adsorbed on lithium-saturated clays for $<5\%$ water, *C.R. Acad. Sci., Ser. D*, **273**, 1467.

Prost, R. (1976), Interactions between adsorbed water molecules and the structure of clay minerals: hydration mechanism of smectites, in S. W. Bailey (ed.), *Proc. Int. Clay Conf. 1975*, p. 351, Applied Publishing Ltd., Wilmette, Ill.

Russell, J. D. and Farmer, V. C. (1964), Infrared spectroscopic study of the dehydration of montmorillonite and saponite, *Clay Miner. Bull.*, **5**, 443.

Russell, J. D., Parfitt, R. L., Fraser, A. R. and Farmer, V. C. (1974), Surface structures of gibbsite goethite and phosphated goethite, *Nature* (London), **248**, 220.

Safford, G. F. and Leung, P. S. (1973), The structure and molecular dynamics of water, *Progr. Surf. Memb. Sci.*, **7**, 231.

Serna, C., Ahlrichs, J. L. and Serratosa, J. M. (1975), Folding in sepiolite crystals, *Clays Clay Miner.*, **23**, 452.

Shirozu, H. and Bailey, S. W. (1966), Crystal structure of a two-layer Mg-vermiculite, *Amer. Mineral.*, **51**, 1124.

Suquet, H., Calle, C. de la and Pezerat, H. (1975), Swelling and structural organization of saponite, *Clays Clay Miner.*, **23**, 1.

Tarasevich, Yu. I. and Gribina, I. A. (1972), Infrared-spectroscopic study of the state of water in halloysite, *Kolloid. Zh.*, **34**, 405.

Theng, B. K. G. (1971), *The Chemistry of Clay–Organic Reactions*, Hilger, London.

Wada, K. and Harward, M. E. (1974), Amorphous clay constituents of soils, *Advan. Agron.*, **26**, 211.

Wada, K. and Yoshinaga, N. (1969), The structure of 'imogolite', *Amer. Mineral.*, **54**, 50.

Walker, G. F. (1956), The mechanism of dehydration of Mg-vermiculite, *Clays Clay Miner.*, **4**, 101.

Walker, G. F. and Garrett, W. G. (1967), Chemical exfoliation of vermiculite and the production of colloidal dispersions, *Science*, **156**, 385.

Weir, A. H. (1960), Relationship between physical properties, structure and composition of smectites, Ph.D. Thesis, London University.

Zettlemoyer, A. C., Micale, F. J. and Klier, K. (1975), Adsorption of water on well-characterized solid surfaces, in Franks (1972–75), vol. 5, p. 249.

Author Index

Subject Index